Molecular Mechanism in Epithelial-Mesenchymal Transition (EMT) and Fibrosis

Molecular Mechanism in Epithelial-Mesenchymal Transition (EMT) and Fibrosis

Editors

Sabrina Lisi
Margherita Sisto

Basel • Beijing • Wuhan • Barcelona • Belgrade • Novi Sad • Cluj • Manchester

Editors

Sabrina Lisi
Department of Translational
Biomedicine and
Neuroscience (DiBraiN)
University of Bari 'Aldo Moro'
Bari
Italy

Margherita Sisto
Department of Translational
Biomedicine and
Neuroscience (DiBraiN)
University of Bari 'Aldo Moro'
Bari
Italy

Editorial Office
MDPI
St. Alban-Anlage 66
4052 Basel, Switzerland

This is a reprint of articles from the Special Issue published online in the open access journal *International Journal of Molecular Sciences* (ISSN 1422-0067) (available at: www.mdpi.com/journal/ijms/special_issues/WM5KDT7E6V).

For citation purposes, cite each article independently as indicated on the article page online and as indicated below:

Lastname, A.A.; Lastname, B.B. Article Title. *Journal Name* **Year**, *Volume Number*, Page Range.

ISBN 978-3-7258-0702-4 (Hbk)
ISBN 978-3-7258-0701-7 (PDF)
doi.org/10.3390/books978-3-7258-0701-7

© 2024 by the authors. Articles in this book are Open Access and distributed under the Creative Commons Attribution (CC BY) license. The book as a whole is distributed by MDPI under the terms and conditions of the Creative Commons Attribution-NonCommercial-NoDerivs (CC BY-NC-ND) license.

Contents

About the Editors vii

Preface ix

Margherita Sisto and Sabrina Lisi
Epigenetic Regulation of EMP/EMT-Dependent Fibrosis
Reprinted from: *Int. J. Mol. Sci.* **2024**, *25*, 2775, doi:10.3390/ijms25052775 1

Susumu Yoshie, Shigeyuki Murono and Akihiro Hazama
Approach for Elucidating the Molecular Mechanism of Epithelial to Mesenchymal Transition in Fibrosis of Asthmatic Airway Remodeling Focusing on Cl⁻ Channels
Reprinted from: *Int. J. Mol. Sci.* **2023**, *25*, 289, doi:10.3390/ijms25010289 27

Angélique Mottais, Luca Riberi, Andrea Falco, Simone Soccal, Sophie Gohy and Virginia De Rose
Epithelial–Mesenchymal Transition Mechanisms in Chronic Airway Diseases: A Common Process to Target?
Reprinted from: *Int. J. Mol. Sci.* **2023**, *24*, 12412, doi:10.3390/ijms241512412 40

Azine Datlibagi, Anna Zein-El-Din, Maxime Frohly, François Willermain, Christine Delporte and Elie Motulsky
Experimental Models to Study Epithelial-Mesenchymal Transition in Proliferative Vitreoretinopathy
Reprinted from: *Int. J. Mol. Sci.* **2023**, *24*, 4509, doi:10.3390/ijms24054509 76

Kirill V. Odarenko, Marina A. Zenkova and Andrey V. Markov
The Nexus of Inflammation-Induced Epithelial-Mesenchymal Transition and Lung Cancer Progression: A Roadmap to Pentacyclic Triterpenoid-Based Therapies
Reprinted from: *Int. J. Mol. Sci.* **2023**, *24*, 17325, doi:10.3390/ijms242417325 97

Jiwei Hou, Yanru Yang and Xin Han
Machine Learning and Single-Cell Analysis Identify Molecular Features of IPF-Associated Fibroblast Subtypes and Their Implications on IPF Prognosis
Reprinted from: *Int. J. Mol. Sci.* **2023**, *25*, 94, doi:10.3390/ijms25010094 132

Jean-Luc C. Mougeot, Thomas E. Thornburg, Braxton D. Noll, Michael T. Brennan and Farah Bahrani Mougeot
Regulation of STAT1 and STAT4 Expression by Growth Factor and Interferon Supplementation in Sjögren's Syndrome Cell Culture Models
Reprinted from: *Int. J. Mol. Sci.* **2024**, *25*, 3166, doi:10.3390/ijms25063166 148

Chin-Chuan Chen, Chi-Yuan Chen, Chau-Ting Yeh, Yi-Tsen Liu, Yann-Lii Leu and Wen-Yu Chuang et al.
Corylin Attenuates CCl₄-Induced Liver Fibrosis in Mice by Regulating the GAS6/AXL Signaling Pathway in Hepatic Stellate Cells
Reprinted from: *Int. J. Mol. Sci.* **2023**, *24*, 16936, doi:10.3390/ijms242316936 160

Marta Kinga Lemieszek, Marcin Golec, Jacek Zwoliński, Jacek Dutkiewicz and Janusz Milanowski
Cathelicidin Treatment Silences Epithelial–Mesenchymal Transition Involved in Pulmonary Fibrosis in a Murine Model of Hypersensitivity Pneumonitis
Reprinted from: *Int. J. Mol. Sci.* **2022**, *23*, 13039, doi:10.3390/ijms232113039 174

Dymph Klay, Karin M. Kazemier, Joanne J. van der Vis, Hidde M. Smits, Jan C. Grutters and Coline H. M. van Moorsel
New Insights via RNA Profiling of Formalin-Fixed Paraffin-Embedded Lung Tissue of Pulmonary Fibrosis Patients
Reprinted from: *Int. J. Mol. Sci.* 2023, 24, 16748, doi:10.3390/ijms242316748 192

Kosei Kunitatsu, Yuta Yamamoto, Shota Nasu, Akira Taniji, Shuji Kawashima and Naoko Yamagishi et al.
Novel Peritoneal Sclerosis Rat Model Developed by Administration of Bleomycin and Lansoprazole
Reprinted from: *Int. J. Mol. Sci.* 2023, 24, 16108, doi:10.3390/ijms242216108 204

Yanxia Li, Jing Zhao, Yuan Yin, Chenchen Zhang, Zhaoying Zhang and Yajuan Zheng
The Role of STAT3 Signaling Pathway Activation in Subconjunctival Scar Formation after Glaucoma Filtration Surgery
Reprinted from: *Int. J. Mol. Sci.* 2023, 24, 12210, doi:10.3390/ijms241512210 217

Huiyuan Pang, Di Lei, Tingting Chen, Yujie Liu and Cuifang Fan
The Enzyme 15-Hydroxyprostaglandin Dehydrogenase Inhibits a Shift to the Mesenchymal Pattern of Trophoblasts and Decidual Stromal Cells Accompanied by Prostaglandin Transporter in Preeclampsia
Reprinted from: *Int. J. Mol. Sci.* 2023, 24, 5111, doi:10.3390/ijms24065111 232

Regina Komsa-Penkova, Adelina Yordanova, Pencho Tonchev, Stanimir Kyurkchiev, Svetla Todinova and Velichka Strijkova et al.
Altered Mesenchymal Stem Cells Mechanotransduction from Oxidized Collagen: Morphological and Biophysical Observations
Reprinted from: *Int. J. Mol. Sci.* 2023, 24, 3635, doi:10.3390/ijms24043635 250

Pablo Sacristán-Gómez, Ana Serrano-Somavilla, Lía Castro-Espadas, Nuria Sánchez de la Blanca Carrero, Miguel Sampedro-Núñez and José Luis Muñoz-De-Nova et al.
Evaluation of Epithelial–Mesenchymal Transition Markers in Autoimmune Thyroid Diseases
Reprinted from: *Int. J. Mol. Sci.* 2023, 24, 3359, doi:10.3390/ijms24043359 264

About the Editors

Sabrina Lisi

Sabrina Lisi is a Professor of Human Anatomy at the Faculty of Medicine, University of Bari, Department of Translational Biomedicine and Neuroscience (DiBraiN), Italy. After a double PhD in Life Science: Aspects of Molecular and Cellular Biology (Cotutelle de These) awarded by the Louis Pasteur University, University of Strasburg, France (summa cum laude), and University of Bari, Italy, she deepened her research by continuing in the Laboratory of Genetics and Molecular Biology of Eucaryotes (LGME) of the CNRS in Strasbourg and in the Laboratory of Cellular and Molecular Biology, IGBMC, Illkirsh, France. Subsequently, she increased her expertise in the Developmental Biology Laboratory, Department of Teratology, Institute of Experimental Medicine, Academy of Sciences of the Czech Republic in Prague, Czech Republic, and the Laboratory of Developmental Biology and Tooth Morphogenesis, INSERM UMR 977, Faculty of Medicine, Strasbourg University, France. Sabrina Lisi's research interests focus on the molecular processes underlying the interaction between receptors of the immune response and inflammation and, in particular, the mechanisms of epithelial–mesenchymal interactions in the development of organ fibrosis. Her publication list includes papers in international peer reviewed journals.

Margherita Sisto

Margherita Sisto is a Professor of Human Anatomy at the Faculty of Medicine, University of Bari, Italy, with a Biological Degree cum laude and a Ph.D. in "Human and Experimental Morphology (macroscopic, microscopic, and ultrastructural)", also from the University of Bari, Italy. Professor Sisto's research interests are primarily focused on the area of pathophysiology and molecular immunology applied to immunological research lines, with emphasis on the elucidation of the molecular processes underlying the interaction between receptors of the immune response, inflammation, and the characterization of new anti-inflammatory molecules. Her particle research has involved the in vitro analysis of dysregulated immunological responses during the pathogenesis of chronic inflammatory diseases such as the autoimmune Sjogren's syndrome. Professor Sisto's innovative research has been highlighted in a number of prestigious, peer reviewed journals.

Preface

This Special Issue of the *International Journal of Molecular Sciences*, entitled "Molecular Mechanism in Epithelial–Mesenchymal Transition (EMT) and Fibrosis", collected 15 original research papers (5 reviews and 10 articles) written by a panel of experts from different countries who highlight recent advances in the EMT process.

Navigating the complex field of EMT, this Special Issue introduces the current understanding of the underlying mechanisms of EMT in the evolution and progression of fibrogenesis and discusses potential strategies for attenuating EMT to prevent and/or inhibit fibrosis.

Overall, the 15 scientific articles in this Special Issue of the *International Journal of Molecular Sciences* provide valuable insights into the complex mechanisms governing the EMT process linked to fibrosis and have highlighted the potential of novel therapeutic strategies. In the last few years, the field of EMT has shown considerable promise, and there is still much to be learned. As our understanding continues to grow, we hope that this Special Issue serves as a catalyst for further research and innovation in this developing field.

In addition, we would like to thank the authors who have contributed their innovative research and valuable insights to this Special Issue. Their intellectual contributions and dedication to advancing multidisciplinary knowledge have been vital in the success of this Special Issue. We would like to express our appreciation to Mr Jerry Wang for his helpful support with this Special Issue.

Sabrina Lisi and Margherita Sisto
Editors

Review

Epigenetic Regulation of EMP/EMT-Dependent Fibrosis

Margherita Sisto * and Sabrina Lisi

Department of Translational Biomedicine and Neuroscience (DiBraiN), Section of Human Anatomy and Histology, University of Bari, Piazza Giulio Cesare 1, I-70124 Bari, Italy; sabrina.lisi@uniba.it
* Correspondence: margherita.sisto@uniba.it; Tel.: +39-080-547-8315; Fax: +39-080-547-8327

Abstract: Fibrosis represents a process characterized by excessive deposition of extracellular matrix (ECM) proteins. It often represents the evolution of pathological conditions, causes organ failure, and can, in extreme cases, compromise the functionality of organs to the point of causing death. In recent years, considerable efforts have been made to understand the molecular mechanisms underlying fibrotic evolution and to identify possible therapeutic strategies. Great interest has been aroused by the discovery of a molecular association between epithelial to mesenchymal plasticity (EMP), in particular epithelial to mesenchymal transition (EMT), and fibrogenesis, which has led to the identification of complex molecular mechanisms closely interconnected with each other, which could explain EMT-dependent fibrosis. However, the result remains unsatisfactory from a therapeutic point of view. In recent years, advances in epigenetics, based on chromatin remodeling through various histone modifications or through the intervention of non-coding RNAs (ncRNAs), have provided more information on the fibrotic process, and this could represent a promising path forward for the identification of innovative therapeutic strategies for organ fibrosis. In this review, we summarize current research on epigenetic mechanisms involved in organ fibrosis, with a focus on epigenetic regulation of EMP/EMT-dependent fibrosis.

Keywords: epigenetic; inflammation; fibrosis; DNA methylation; histone modification; ncRNA

Citation: Sisto, M.; Lisi, S. Epigenetic Regulation of EMP/EMT-Dependent Fibrosis. *Int. J. Mol. Sci.* **2024**, *25*, 2775. https://doi.org/10.3390/ijms25052775

Academic Editor: Riccardo Alessandro

Received: 30 December 2023
Revised: 23 February 2024
Accepted: 24 February 2024
Published: 28 February 2024

Copyright: © 2024 by the authors. Licensee MDPI, Basel, Switzerland. This article is an open access article distributed under the terms and conditions of the Creative Commons Attribution (CC BY) license (https://creativecommons.org/licenses/by/4.0/).

1. Introduction

Fibrosis, characterized by the deposition of connective tissue in a tissue or organ, represents a reaction to an injury and has reparative or pathological significance. The fibrotic evolution of a tissue or organ can have very negative consequences, leading to the inability to perform normal physiological functions and resulting in a pathological condition with high mortality [1–4]. Fibrosis is often associated with pathologies characterized by a chronic inflammatory state, such as autoimmune diseases or tumors. In these circumstances, the prolonged release of growth factors and/or pro-inflammatory factors such as transforming growth factor-β (TGF-β) or various cytokines mediate the activation of a cellular transformation process called epithelial–mesenchymal plasticity (EMP) [5]. When EMP is activated, the epithelial cells, which have a phenotype of adherent cells closely connected to each other and are not invasive, become transformed, assuming a hybrid epithelial/mesenchymal phenotype and/or a completely mesenchymal phenotype. In this case, the process is defined as epithelial-to-mesenchymal transition (EMT) [5]. These cells acquire much higher migratory capabilities and are able to deposit extracellular matrix (ECM) proteins. The triggering of various cascades of molecular interconnected events leads to an exacerbation of the inflammatory state or to tumor proliferation and metastasis, with serious consequences [6–8]. Despite the fact that fibrosis appears to be a partly reversible process in various clinical studies [9], unfortunately, therapeutic options are still very limited. In recent years, very innovative studies have demonstrated how epigenetic modifications, by triggering or inhibiting gene transcription depending on the circumstances, can reprogram gene expression by adapting it to exposure to various risk factors [10]. This has been demonstrated, for example, in idiopathic pulmonary fibrosis

(IPF) or in patients with non-alcoholic fatty liver disease, in which biopsy samples show higher expression of DNA methyltransferase, suggesting that DNA methylation could represent a predisposing factor for the onset of these pathologies [11]. The application of sequencing technology has demonstrated that the activation of fibroblasts, involved in collagen deposition during fibrogenesis, depends on various epigenetic modifications affecting the DNA to be transcribed [12]. Furthermore, epigenetic modifications appear to be largely involved in the modifications of epithelial cells towards the mesenchymal phenotype, a process essentially mediated by EMT [13,14].

Some authors have described in detail the mechanisms through which the main epigenetic modifications would act, inducing a regulation of the fibrotic evolution of the inflammatory processes and determining the transcription of pro-fibrotic genes [12]. In addition, epigenetics could explain the reversibility of the fibrosis [10]. However, although recent discoveries tend toward the involvement of epigenetic modifications in EMP/EMT-dependent fibrosis, the elucidation of the mechanisms involved still seems far from clear. This review aims to collect the latest discoveries made by studying the involvement of epigenetic modifications in the activation of EMP/EMT-dependent fibrosis, with the aim of suggesting new therapeutic perspectives.

2. The Dynamic Balance between EMT and EMP

EMT is a dynamic complex process during which epithelial cells reduce their epithelial properties, gradually dissolving cell–cell junctions and rebuilding cell–matrix connections to acquire characteristics typical of mesenchymal cells [15–17]. When the mechanism of EMT was identified, it was discovered that EMT was responsible for multiple processes involved in embryonic development, such as gastrulation, neural crest formation, and heart development [15,18]. But researchers soon demonstrated that the activation of EMT also affected physiological processes represented by wound healing [19], with the fibrotic evolution of diseases characterized by chronic inflammation, and with the formation of metastases from primary tumors. EMT is classified into three functional types: type I, involved in embryonic morphogenesis; type II, responsible for normal wound healing, but this type can enhance myofibroblast activation leading to the deposition of high levels of ECM proteins and fibrosis in chronic diseases; and type III, characteristic of malignant epithelial cells that acquire a migratory phenotype capable of invading and metastasizing [20–22]. EMT is a reversible phenomenon, and the resulting cells shift back from motile, multipolar mesenchymal types to polarized epithelial types via the mesenchymal–epithelial transition (MET) process [23]. Therefore, until now, EMT was considered as an "all or nothing" program wherein the cells can exist with an epithelial morphology or in a mesenchymal state. Interestingly, novel insights have shown that the cells that undergo to EMT present multiple intermediate phenotypes. This new concept, recently named as EMP, defines the capacity of the cells to interconvert between several states along the epithelial–mesenchymal spectrum, thereby acquiring hybrid epithelial/mesenchymal phenotypic features [24,25]. Intriguingly, this cellular plasticity is very pliable, and epithelial cells often undergo partial reorganization and combine epithelial and mesenchymal features following the EMT process [26] (Figure 1). Indeed, such cellular shifts and the resultant heterogeneity provide the cells with the flexibility to face different physiological (embryonic development, wound healing) and pathological (organ fibrosis, cancer) conditions [27]. The dynamics of EMT/EMP and MET are controlled by a complex network of transcription factors (TFs) [28]. TFs, in epithelial cells, determine the transcription of a variety of genes involved in the activation of EMT programs [29]. These changes in transcription, sometimes seen as gene reprogramming, involve three TFs families: *Snail* (*Snail1*) and *Slug/Snail2*, *ZEB1* and *ZEB2*, and *Twist* [28,30]. All of these TFs share the ability to repress epithelial genes like the E-cadherin encoding gene *CDH1* via binding to E-Box motifs in their cognate promoter regions [15]. In parallel, the EMT-TFs, directly or indirectly, activate genes associated with a mesenchymal phenotype, including *VIM* (Vimentin), *FN1* (Fibronectin), and *CDH2* (N-cadherin) [15,28]. Upon induction of an epithelial plasticity response, they are considered as "master" drivers of the EMT program, conferring cellular shift among the epithelial–mesenchymal spectrum [26,29]. Interestingly, with increasing data relating to the mechanisms of activation of EMT pathways, other than the signaling molecules regulating EMT, it becomes clear that activation and execution of EMT occur as a result of genetic and epigenetic processes. The study of epigenetic regulation is an important aspect of modulation of EMT [25,31], and various epigenetic mechanisms appear to be involved in the modulation of EMT, although it is still difficult to correlate all of the scientific data collected together [25]. Currently, most of the studies carried out concern the epigenetic control of EMT during cancer progression and metastases formation [16–18]. Similarly, recent discoveries have also attributed a key role to epigenetic modifications in the activation of the EMP process. Numerous pieces of evidence have demonstrated an altered expression of the main epigenetic modifications underlying the delicate balance between EMP and EMT, including histone modification, DNA methylation, and non-coding RNA, which could facilitate cancer metastasis [31].

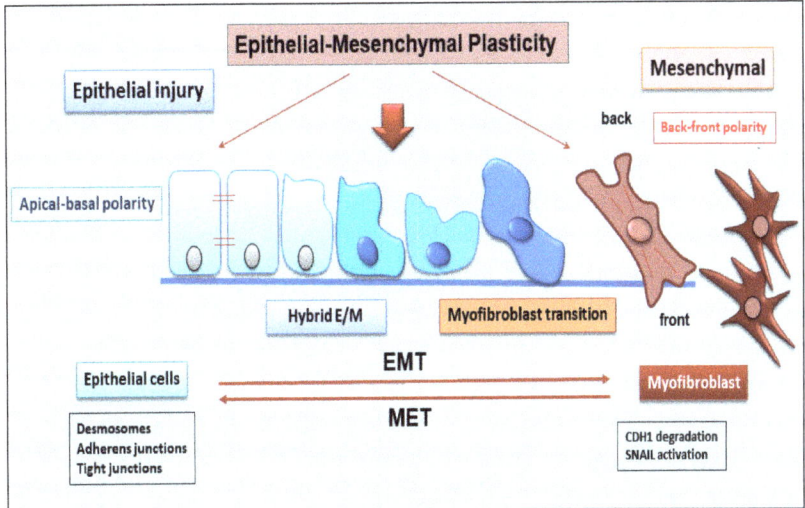

Figure 1. Schematic representation of epithelial to mesenchymal plasticity (EMP). EMP, in response to epithelial injury, allows cells to convert between multiple states across the epithelial to mesenchymal transformation, acquiring hybrid epithelial/mesenchymal phenotypic features.

3. Role of EMP/EMT in Organ Fibrosis

In recent years, our knowledge of the fibrotic process has been remarkably increased by the characterization of cellular mediators, key inflammatory and profibrogenic cytokines, molecular factors, and the evolution of new pathogenetic scenarios. A major determinant of fibrosis is the continuous spread of fibroblasts and myofibroblasts, which suggests the question of how this cellular system can be fed [32]. Experiments conducted in the last 2 years have shown that cellular plasticity, which also includes the phenomenon of EMP, is not limited exclusively to development; it also characterizes cells that undergo reprogramming that occurs during the repair of tissue damage, during fibrotic processes, and during carcinogenesis [26]. However, knowledge of the molecular mechanisms involved in the cell's ability to modify its phenotype by evolving into another cell type is still at the beginning. The mechanisms underlying EMT are much more explored and known, and numerous studies have been conducted to evaluate the key role of EMT in fibrosis. In the context of identifying cellular drivers of fibrosis, various in vitro and in vivo studies have reported that EMT is a key mechanism during fibrogenesis, substantially contributing to the increase in interstitial fibroblasts and myofibroblasts, and interrupting its progression can have a profound impact on the onset and progression of related diseases, particularly fibrosis [21,33,34]. In fact, a fibrotic evolution of chronic diseases can lead to pathological states affecting various organs, including the lungs, liver, kidneys, heart, and salivary glands [35].The following paragraph summarizes the most recent discoveries derived from evaluating the molecular mechanisms underlying the process of EMT-dependent fibrosis in various organs in pathological conditions in order to identify potential therapeutic targets.

Contribution of Epithelium to the Fibrotic Organ Process via EMT Activation

In the last few years, important findings have demonstrated that liver epithelial cells undergo the EMT process, contributing to their transformation into myofibroblasts. Indeed, hepatocytes in which the *Snail1* gene was deleted by using Cre-loxP technology showed a reduction in EMT factors and a decrease in the severity of the inflammatory response compared to controls [36]. Moreover, Rowe et al. have examined a panel of genes known to contribute to the progression of liver fibrosis, including interstitial collagen types I and III and fibroblast markers, demonstrating that *Snail1* determines an increased expression of profibrotic genes such as those encoding for type I or type II collagen, vimentin, and *FSP1* in the liver [36]. These data demonstrate that the hepatocyte Snail1 gene is a potent inducer in the progression of hepatic fibrosis. Indeed, explorations have been performed to determine the origin of hepatic myofibroblasts activated in response to the type of liver injury. In particular, hepatic stellate cells (HSCs) are capable of transforming into contractile myofibroblasts after liver injury. In a mouse model subjected to hepatotoxic CCL_4 liver injury, activated HSCs transformed almost totally into myofibroblasts, whereas cholestatic bile duct ligation treatment preferentially stimulated portal fibroblasts [37,38]. It is interesting to underline that HSCs, when not activated, predominantly express epithelial markers compared to mesenchymal ones, and, following damage, can become activated and undergo a change in phenotype driven by EMT [39].

Nowadays, it has been experimentally proven that the activation of an EMT program occurs in a variety of pulmonary fibrosis diseases [40]. A study highlighted the contribution of the bronchial epithelial cells that, when treated with TGF-β1, are able to acquire myofibroblast phenotypes, thereby leading to peribronchial fibrosis. This process drives airway epithelium remodeling, which is a feature of asthma [41]. In IPF, alveolar epithelial cells undergo EMT, inducing the formation of fibroblastic foci and thus triggering the fibrotic destruction of the lung architecture [42,43]. Interestingly, pleural mesothelial cells also undergo a special type of EMT, mesothelial-to-mesenchymal transition (MMT), during IPF pathogenesis. In the MMT process, the mesothelial cells, during serosal inflammation, acquire the mesenchymal phenotype and complete their transformation into myofibroblasts, thus contributing to the progression of parenchymal fibrosis that results in a progressive decline in lung function [44]. Indeed, an important report demonstrated the presence of pleural mesothelial cells exhibiting mesenchymal markers in the lung parenchyma of patients with IPF after fibrogenic stimulation in vivo and a correlation between disease severity and the degree of fibrosis [44].

It has now been widely demonstrated that tubular epithelial cells (TECs) are involved in a process of EMT-dependent fibrosis in chronic renal failure [45]. However, the percentage of TECs that transform into myofibroblasts during this process is not yet known. This has led to the proposal that renal epithelial cells would undergo a partial EMT (pEMT), resulting in renal fibrotic evolution [46,47]. During pEMT, TECs maintain some characteristics of epithelial cells and show, at the same time, typical markers of fibroblasts, acquiring an intermediate phenotype between the two cell types [46,47]. Therefore, TECs in this partial mesenchymal and epithelial phenotype remain attached to the basement membrane during the fibrotic process. Recently, it was demonstrated that Snail1 is able to trigger the pEMT process in TECs, relaying crucial signals for fibrogenic cytokine release and promoting differentiation into myofibroblasts, thus contributing to the exacerbation of the inflammatory response [46].

EMT is also activated in inflammatory bowel diseases (IBDs) such as ulcerative colitis (UC) and Crohn's disease (CD) [48]. In IBD patients, persistent intestinal inflammatory factors injure intestinal epithelial cells, determining reparative reactions that lead to the triggering of the EMT process and perpetuating a severe fibrotic condition [48,49]. Confirming this, the presence of high levels of tumor necrosis factor-like ligand 1A (TL1A) in the intestinal specimens of patients with UC and CD has been detected, which represents a potent inducer of EMT in intestinal fibrosis. As expected, the TGF-β1/Smad3 pathway may be involved in TL1A-induced EMT [49–51].

Recent discoveries have also shown that uncontrolled fibrosis is present in the heart and is triggered by EMT and its special type, endothelial to mesenchymal transition (EndMT) [52,53]. Epicardial EMT is activated after myocardial infarction, atherosclerosis, and valve dysfunction, and it determines angiogenesis and healing [54]. Under TGF-β stimuli, cardiac fibroblasts transdifferentiate into myofibroblasts, acquiring a phenotype similar to that of smooth muscle cells. Furthermore, fibroblasts can also originate from endothelial cells through EndMT, giving rise to the progression of cardiac fibrosis [54]. In addition, TGF-β-driven EMT responsible for cardiac fibroblast formation appears to be triggered by the Hippo pathway, an evolutionarily conserved kinase cascade [55], as seen in recent in vitro findings [56]. In addition, the process of cardiac fibrosis seems to be regulated by C-Ski protein, identified as an inhibitory regulator of TGF-β signaling [57].

A flourishing research field is focused on the evaluation of the pathways involved in the fibrotic process observed in the salivary glands (SGs) derived from Sjögren's Syndrome (SS) patients. Fibrogenesis observed in SGs can be considered the end result of chronic, intense inflammatory reactions induced by a variety of stimuli in this autoimmune disease [7,8,58,59]. Pioneering studies aimed at correlating SS with a fibrotic evolution of the salivary glands were conducted over 10 years ago, demonstrating a significant association between stimulated salivary flow, the focus score, and fibrosis in a high number of SS biopsy specimens. In comparison, unstimulated salivary flow appears to be weakly associated with the focus score and is not always correlated with fibrosis, which was considered an excellent measure of irreversible damage [58]. In all cases, SG fibrosis is linked with an evident impairment of organ function that leads to progressive atrophy and a decrease in quality of life for patients [60].

In this context, using technology to create transgenic mice that conditionally overexpress active TGF-β1, experimental data have confirmed that the overexpression of active TGF-β1 leads to an abnormal accumulation of ECM proteins and severe hyposalivation and acinar atrophy in the mutated mice [61]. More recently, studies have demonstrated an exuberant upregulation of TGF-β1 in SS SGs, which induces the loss of epithelial features and the acquisition of mesenchymal features in SG epithelial cells via triggering of the EMT program through the TGF-β1/Smad/Snail signaling pathway [62]. Indeed, TGF-β1 seems to be able to regulate EMT through both main pathways: the canonical Smad-dependent and non-canonical Smad-independent signaling pathways [7,8,62].

4. Main Epigenetic Mechanisms

Each phase of gene expression can undergo epigenetic modifications, thus leading to the synthesis or inhibition of certain downstream proteins. The epigenetic processes involve DNA methylation, histone modification, chromatin remodeling, and the effects of noncoding RNA.

The phenomenon of DNA methylation is an essential process for the physiological development of the individual and plays a key role in processes widely studied in recent years, such as genomic imprinting. DNA methylation and demethylation represent heritable epigenetic signatures that are evolutionarily conserved and do not involve an alteration of the DNA sequence but can, however, lead to widely modified gene expression [63,64].

Methylation of DNA is an epigenetic mechanism that consists of the transfer of a methyl group from *S*-adenyl methionine (SAM) to the C-5 position of a cytosine residue in a dinucleotide CG or polynucleotide CGGCGG context, also termed CpG islands, to form

5-methylcytosine catalyzed by DNA methyltransferases (DNMTs) [65]. The DNMT family comprises various elements: DNMT1, DNMT2, DNMT3A, DNMT3B, and DNMT3L [66]. Notably, the methylation of the promoter region or gene has different effects; excess promoter methylation silences the gene, while a reduction in promoter methylation causes increased gene expression [67]. On the contrary, at the gene level, an excess of methylation determines active transcription of the gene itself, while the methylation of the gene has a meaning that is not yet well known [68].

Similar to DNA methylation, post-translational histone modifications do not affect the DNA nucleotide sequence but alter its accessibility to the transcriptional machinery. Histones are small basic proteins assembled into nucleosomes and are essential to compact and stabilize DNA by making the DNA sites implicated in gene transcription accessible [69]. Each nucleosome is composed of approximately 150 base pairs of DNA and two copies of the four core histones: H2A, H2B, H3, and H4 [69]. In addition, H1 protein acts as a linker histone-compacting chromatin, and its role is to stabilize the internucleosomal DNA but does not form part of the nucleosome. Histone proteins, through post-translational modifications, control chromatin structure, triggering the transition from open chromatin, called euchromatin, which is actively transcribed, to a compacted chromatin structure called heterochromatin. In this compact form, DNA is not accessible to transcriptional machinery and thus cannot be transcribed, resulting in gene silencing [70].

Several of the best-known post-translational modifications of histones include acetylation, methylation, phosphorylation, and ubiquitylation. However, in recent years, other histone modifications have been identified, such as GlcNAcylation, citrullination, krotonylation, and isomerization, which still need to be further explored [71].

Acetylation modulates transcriptional activity through the neutralization of the positive charge present on the lysine residues of histone proteins. This action has the potential to weaken the interactions between histones and DNA, making them less stable, thus allowing gene transcription [72,73]. Acetylation consists of the addition of acetyl groups to lysine residues, neutralizing their positive charge. Thus, acetylation induces and enhances gene expression. Histone acetylation and deacetylation are catalyzed by histone acetyltransferases (HATs) and histone deacetylases (HDACs), respectively.

HDACs remove acetyl groups from acetylated proteins, consequently repressing gene expression. They are classified into four classes: class 1 (HDAC1,2,3,8), class 2 (2a: HDAC4,5,7,9; 2b: HDAC6,10), class 3 (SIRT), and class 4 (HDAC11). Therefore, sirtuin proteins, classified within class III HDACs, require nicotinamide adenine dinucleotide (NAD) as a cofactor for their catalytic activity. To date, 18 mammalian HDACs have been identified and classified into the above different classes [74].

Methylation occurs in both the lysine and arginine residues of histones H3 and H4 and, in particular, does not alter the charge of the histone protein. Arginine methylation, which requires arginine methyltransferase activity, induces gene transcription, while lysine methylation, which requires histone methyltransferase, can have either a positive or negative effect on transcription due to the site involved in the methylation [75–78]. Recently, it has been demonstrated that histone methylation is also a reversible event through the mechanism of histone demethylases [79].

Phosphorylation influences all core histones, with several effects on each. Phosphorylation of serine residues 10 and 28 of histone H3 and serine residue T120 of histone H2A is involved in chromatin condensation through the phases of cell replication during mitosis. Phosphorylation of the S139 residue in histone H2A evidences a landing point for the interaction with factors involved in the repair of DNA damage [80]. However, phosphorylation of histone H2B is not as well known but appears to induce chromatin compaction through several mechanisms such as apoptosis, DNA fragmentation, and cell necrosis [81].

All histone proteins can undergo a mechanism of ubiquitylation; however, in the last few years, studies have highlighted two well-characterized proteins, H2A and H2B, which are most frequently ubiquitinated in the nucleus [82]. Histone ubiquitination is linked with the activation of gene expression, but many studies have demonstrated that the presence of a single ubiquitin has different effects on H2A and H2B. Indeed, mono-ubiquitylated H2A is linked with gene silencing, while if the interaction concerns H2B, transcription activation is induced [82].

Epigenetic regulation also involves actively non-coding RNA (ncRNAs) and it has been widely discovered that ncRNAs are able to modulate gene expression at both transcriptional and post-transcriptional levels. ncRNA refers to a functional RNA molecule that is transcribed from DNA but is not translated into a protein [83]. NcRNA are divided into two broad categories based on their length: short ncRNAs, with a number of nucleotides less than 30, and long ncRNAs (lncRNAs), which include those RNAs with a number of nucleotides greater than 200 [83]. The three main classes of short noncoding RNAs include microRNAs (miRNAs), short interfering RNAs (siRNAs), and piwi-interacting RNAs (piRNAs).

MiRNAs and DNA methylation are the two epigenetic events that have emerged in recent years and correlate to the modulation of gene expression [84]. Notably, miRNAs act by linking to a specific target messenger RNA through a complementary sequence; this binding determines the fragmentation and degradation of the mRNA, consequently blocking the translation event. Interestingly, the presence of a mutual regulation between miRNAs and DNA methylation has been shown in human tumors. Indeed, miRNAs modulate DNA methylation by acting on the transcription of genes implicated in the synthesis of DNA methyltransferases [85].

SiRNAs represent small RNA molecules whose function is to repress the expression of a gene by binding to the mRNA, inducing its degradation, and thus preventing post-transcriptional gene and subsequent protein synthesis [85]. SiRNAs are gained from a long double-stranded RNA molecule that is cut into many small fragments by Dicer endoribonuclease. The siRNAs obtained are added to the so-called RISC complex (RNA-induced silencing complex) to form the inactive RISC-siRNA complex. Once activated, the siRNA loses one of the two strands and binds to the mRNA target messenger, i.e., the mRNA whose translation into protein is to be prevented [86,87]. The piRNAs are a complex class of sncRNAs that specifically interact with the PIWI protein subfamily of the Argonaute family [88]. Current research evidences that this interaction between PIWI proteins and piRNAs regulates novel epigenetic mechanisms such as DNA rearrangements; however, this has yet to be clarified. Recently, lncRNAs were discovered as important regulators of the epigenetic status of the human genome. LncRNAs are RNA fragments longer than 200 nucleotides that have various activities, such as chromatin remodeling and transcriptional and post-transcriptional regulation, and act as precursors of siRNAs with the function of gene silencing [89–91]. Many lncRNAs form complexes with proteins, leading to modifications in the conformation of chromatin [92,93]. In addition, a novel member of the lncRNA class, circular RNAs (circRNAs), are characterized by a covalently closed loop. They are recognized to have distinct biogenesis and to regulate gene expression and biological processes through different mechanisms, with some miRNA-sponging circRNAs identified [94]. A schematic overview of epigenetic modifications is reported in Figure 2.

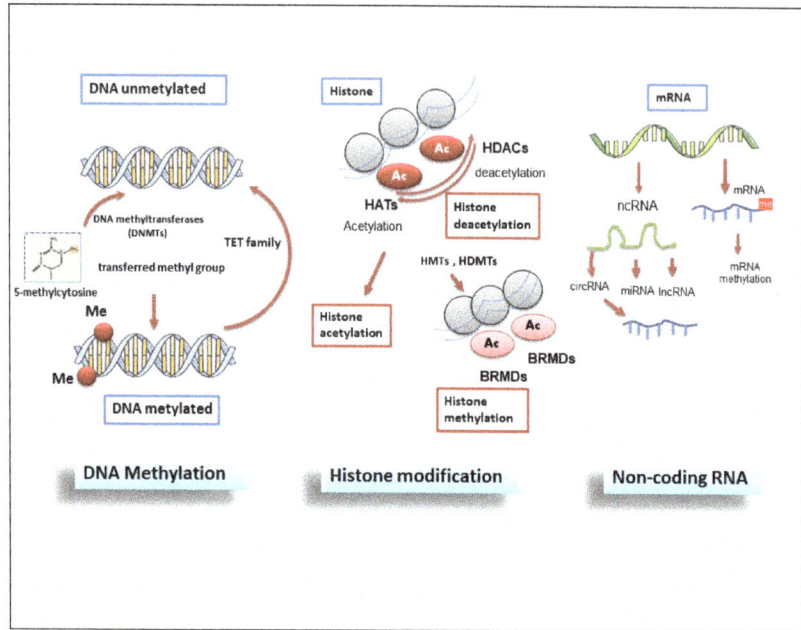

Figure 2. A schematic overview of epigenetic modifications. Epigenetic regulation involves DNA methylation, histone modification, and non-coding RNAs. DNMT family members mediate DNA methylation, which suppresses gene transcription by adding a methyl group to the cytosine position. Histone methylation is catalyzed by HMTs and HDMTs, and histone acetylation is regulated by HATs and HDACs. Non-coding RNAs include miRNAs, circRNAs, and lncRNAs. BET (Bromodomain and extraterminal); BRMDs (bromodomains); DNMT (DNA methyltransferase); HATs (Histone acetylases); HDACs (Histone deacetylases); HDMTs (Histone demethylases); HMTs (Histone methyltransferases); lncRNA (long non-coding RNA); TET (Ten-eleven translocation).

5. Epigenetics Regulation of EMP/EMT-Dependent Fibrosis

The studies conducted in order to evaluate a possible connection between epigenetics, EMP, and fibrosis are leading to the first discoveries, for example, in airway persistent inflammation [95], but require further investigation. On the contrary, the field of studies conducted on the role of epigenetic modifications in EMT-dependent fibrosis is much more flourishing. One of the hot topics in the last few years has been the association between epigenetic regulation and fibrotic processes triggered by EMT [13]. Epigenetic modulation of tissue–stroma interactions involves several types of alterations that have been shown in recent years to play a determining role in the activation of the EMT program and the consequent EMT-dependent fibrosis [13].The following paragraphs illustrate the current knowledge concerning the epigenetic aberrations involved in the fibrotic evolution induced by EMT during pathological processes.

5.1. DNA Methylation in EMT-Dependent Fibrosis

DNA methylation, catalyzed by DNMTs, represents one of the best-represented mechanisms of epigenetic control and modulation in EMT [96]. It is known that the downregulation of E-cadherin expression is fundamental to the evolution of the EMT process [97]. DNA methylation, regulated by DNMTs, seems to affect *CDH1* expression. The ten-eleven translocation (TETs) family, instead, is implicated in *CDH1* demethylation [98]. Transformations in *CDH1* promoter methylation lead to diminished E-cadherin protein expression in several fibrotic diseases [99–101]. Direct methylation of transcription factors by DNMT1 concurs with EMT program activation in renal epithelial cells [102] and, as expected, DNA methylation inhibition through specific inhibitors reversed EMT in arsenic-triggered renal fibrosis [103].

Therefore, interesting studies have evidenced the essential role of DNMTs in cardiac fibrosis, resulting in the activation of the EndMT process. The inhibition of the suppressor of cytokine signaling 3 (*SOCS3*) mediated by DNMT1 determines the activation of STAT3 that induces cardiac fibroblast activation and collagen deposition in cardiac fibrosis [104]. This phenomenon was also detected in the fibrotic skin of patients with systemic sclerosis, in which TGF-β induced the expression of DNMT3A and DNMT1 in fibroblasts in a SMAD-dependent manner, leading to a decreased expression of SOCS3 and facilitating activation of STAT3 to promote fibroblast-to-myofibroblast transition, collagen release, and fibrosis [104].

The transcriptional regulation driven by the DNA methylation pattern plays a pivotal role in liver fibrosis [10]. In particular, DNA methylation is involved in the differentiation of HSC during hepatic diseases characterized by a severe and progressive fibrotic process. These data were confirmed by downregulating *DNMT3a* and *DNMT3b* gene expression through the use of siRNA; in this case, DNA methylation was decreased, and HSC activation was subsequently suppressed [105]. Beyond DNMTs, another intriguing protein implicated in liver DNA methylation processes is glycine N-methyltransferase (GNMT). GNMT is the most abundant methyltransferase in the liver and hepatocytes. GNMT influences epigenetic regulatory determinants by competing with DNMT to regulate transmethylation flux [105]. The triggering of EMT in the liver appears to be related to activation by the hedgehog (Hh) pathway. Patched1 (PTCH1), a factor that negatively regulates Hh, is downregulated during the process of liver fibrosis and this appears to be related to its hypermethylation state. Recent studies have established the antifibrotic efficacy of salvianolic acid B (Sal B), attributing it to its ability to inhibit Hh-mediated EMT. An upregulation of *PTCH1* was noted due to a decrease in DNA methylation thanks to the inhibition of DNMT1. Interestingly, the increase in miR-152 in Sal B-treated cells was responsible for the hypomethylation of *PTCH1* by Sal B, and DNMT1 was found to be a direct target of miR-152 [106].

Several studies have mentioned altered DNA methylation linked to the evolution of chronic obstructive pulmonary disease and pulmonary fibrosis [104,107]. In addition, the effects of DNA methylation were linked to histone modifications and miRNA activity to induce or block gene expression in fibrotic progression [104,107] (Figure 3). An interesting study conducted on epithelial *cells* isolated from normal human bronchial epithelium exposed to nickel (NiCl2) demonstrated that, through the activation of the TLR4 signaling pathway and EMT, nickel is associated with the development of many chronic lung diseases, including pulmonary fibrosis [108]. It was experimentally demonstrated that NiCl2 exposure determines E-cadherin downregulation in normal bronchial epithelial cells associated with E-cadherin promoter DNA hypermethylation [109].

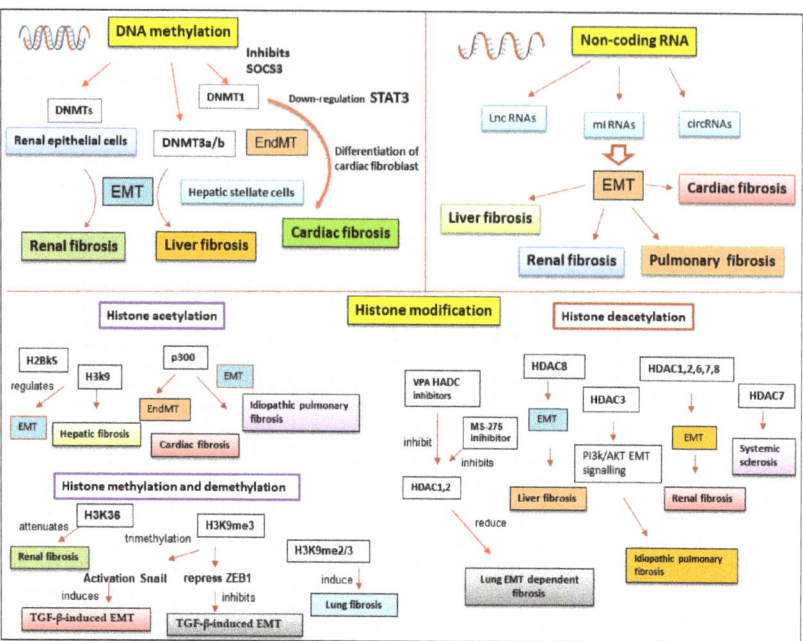

Figure 3. Scheme of the epigenetic modifications in EMP/EMT-dependent fibrotic diseases. DNMT (DNA methyltransferase); HAT (Histone acetylase); HDAC (Histone deacetylase); EndMT (Endothelial–mesenchymal transition); EMT (Epithelial–mesenchymal transition); SOCS3 (cytokine signaling 3); VAP (Valproic acid).

5.2. The Involvement of Histone Modifications in EMT-Dependent Fibrosis

5.2.1. Histone Acetylation and Deacetylation

Histone acetylation and deacetylation are widely analyzed histone modifications and have been recently linked to the activation of the EMT program in cancer and fibrosis [72]. In fact, tumors with stem cell features propagate and determine far away metastases by triggering the advancing EMT program. Recent studies have demonstrated that acetylation of histone H2BK5 is crucial in the control of EMT [14]. For example, in trophoblast stem cells, H2BK5 acetylation influences the expression of key genes implicated in the conservation of epithelial characteristics. These trophoblast stem cells share similar H2BK5 acetylation-regulated gene expression when compared with stem-like claudin-low breast cancer cells, thus linking EMT-dependent development and EMT observed in cancer cells [14]. In particular, HATs regulate gene silencing or transcription through the modulation of the acetylation of histones, thus orchestrating gene expression to induce liver fibrosis [110,111]. Additionally, histone acetylation has been implied in pulmonary and cardiac fibrosis [10], performed through the activity of p300 HAT [112,113]. An interesting discovery enriched the molecular scenario by demonstrating that p300 HAT induces the fibrotic process in IPF and cardiac fibrosis via EMT and EndMT, respectively [114].

HDACs, firstly identified in liver fibrosis, can enhance the cellular migration and ECM deposition by myofibroblasts [115,116]. HDACs belonging to each class appear to be implicated in EMT-dependent fibrosis activation, as reported in the following subparagraphs.

Class I Histone Deacetylase Involvement in EMT-Dependent Fibrosis

Histone deacetylation is largely studied in pulmonary fibrosis. Epigenetic histone modifications through deacetylation could explain the persistently activated state of IPF fibroblasts [117], which indicated a "cancer-like" upregulation. According to this hypothesis, almost all class I (and class II) HDAC enzymes result in overexpression and this could be responsible for the abnormal repression of pro-apoptotic genes [117,118]. Under conditions of hypoxia, HDAC3 combines with WD repeat domain 5 (WDR5) to recruit histone methyltransferase, leading to decreased acetylation of H3K4 and to an increased methylated form of H3K4 [119].

Interesting studies also concern the use of specific HDAC inhibitors. The short-chain fatty acid valproic acid (VPA), a class I-specific HDAC inhibitor, was able to reduce lung EMT-dependent fibrosis in bleomycin (BLM)-treated mice with Smad2/3 deactivation [120]. VPA seems to work without Akt cellular pathway deactivation, presumably due to the fact that VPA specifically inhibits the activities of HDAC1 and HDAC2 [121,122] but not of HDAC3 (directly involved in PI3K/Akt EMT signaling) [123]. Confirming this, the Class I HDAC inhibitor entinostat (MS-275), specific for HDAC1 and HDAC3, determines the inactivation of the PI3K/Akt pathway in TGF-β-stimulated lung fibroblasts [123]. Entinostat suppresses the TGF-β-induced expression of SPARC, a matricellular protein involved in the ECM turnover and apoptosis resistance of lung myofibroblasts restoring the expression of SPRC's negative regulator named ARHGEF3 (Rho guanine nucleotide exchange factor 3, also known as XPLN = exchange factor found in platelets and leukemic and neuronal tissues [124].

Recent data showed that HDACH3 inhibition results in the acetylation and degradation of a vector expressing the NOTCH1 intracellular domain (NICD1), thereby alleviating IPF [125]. This suggest that HDACs can deacetylate also non-histone targets with important consequences from the point of view of activated pathological mechanisms [126].

In addition, HDAC3 was upregulated in alveolar epithelial type 2 (AT2) cells from patients with IPF, and in AT2 cells from mice with BLM-induced pulmonary fibrosis. Moreover, HDAC3 deficiency in AT2 cells prevented mice from developing BLM-induced pulmonary fibrosis, characterized by a marked reduction of EMT in AT2 cells. In terms of mechanisms, we found that TGF-β1/SMAD3 can directly promote HDAC3 transcription and further inhibit GATA3 acetylation, thus promoting EMT in AT2 cells. GATA3 is a transcription factor and the most frequently mutated genes in breast cancer [127].

Recently, Chen et al. reported that inhibition of HDAC3 and Nuclear Factor Erythroid-Derived 2-Related Factor-2 (Nrf2) mitigates pulmonary fibrosis [128], and Zheng et al. suggested that HDAC3 accelerates pulmonary fibrosis by promoting EMT and inflammation through the Notch1 or STAT1 signaling pathway [72]. Actually, however, the main activator of EMT dependent fibrosis in lung is hypoxia that activate a signaling cascade that involves the activation of the transcription factor Snail [31]. HDAC3 significantly increased Snail expression under hypoxic conditions, and this effect was prevented by inhibition of Hypoxia-inducible factor 1-alpha (HIF1 α), a subunit of a heterodimeric transcription factor hypoxia-inducible factor 1 (*HIF-1*). HDAC3 increases the transcriptional activity of HIF-1α by promoting the binding of HIF-1α to the hypoxia-responsive element (HRE) sites of genes [129]. Since fibroblast migration is considered a critical contributor to lung fibrosis, it was recently demonstrated that HDAC3-miR224- Forkhead Box A1 (FOXA1) axis effectively regulated the migration and invasion of fibroblast cells under hypoxia [129]. The effective involvement of HDAC3 in EMT-dependent fibrosis in lung was demonstrated by the use of *HDAC3* siRNA that alleviated BLM-induced pulmonary fibrosis in mice. In addition, *FOXA1* gene was identified as the target gene of miR-224 in HDAC3-mediated alveolar EMT [129] and HDAC3 promotes hypoxia-induced alveolar EMT through stabilization of HIF-1α via the AKT pathway [129].

Interesting data were collected also for renal fibrosis. In human mesangial cells (HMC) treated with poly IgA1, HDAC1, HDAC2, and HDAC8 were upregulated, determining the subsequent activation of TGF-β/Smad2/3 and Jak2/Stat3 signalling pathways. These pathways activation leads to the proliferation of HMCs and facilitates ECM deposition and fibrosis progression [130]. Recent studies have used various HDAC inhibitors to evaluate the effects on EMT-related kidney fibrosis, demonstrating that class I HDAC inhibitors are more effective than class II HDAC inhibitors in regulating this process.

For example, using specific siRNAs against HDAC1, 2, 3 and 8, it was seen that the knockdown of HDAC1, HDAC2 or HDAC3 did not hinder the expression of ECMs and the initiation of TGF-β1-dependent EMT. This result seems to be due to a compensatory mechanism that intervenes when one HDAC does not function and which activates the others more [131]. Indeed, HDAC1 and HDAC2 show high amino acid homology and compensatory functions between them [132]. Using UUO mice as a model of renal fibrosis, the efficacy of another selective inhibitor of class I HDACs called FK228 was evaluated in EMT-dependent fibrosis [133]. Rat renal interstitial fibroblasts and renal tubular epithelial cells were treated in vitro with TGF-β1, in the presence or not of the inhibitor FK228. The results indicated that FK228 is able to reduce ECM protein deposition in both in vivo and in vitro. FK228 also blocked the activation and proliferation of renal fibroblasts and led to increased acetylation of histone H3 and seems to suppress renal interstitial fibrosis via canonical-Smad- and non-Smad-EMT pathways [133].

In the same experimental model UUO mice, Chen et al. [134] demonstrated a correlation between the high expression of HDAC3 in renal fibrotic tissues, and the low expression of the klotho protein, a membrane-bound protein that acts as a permissive co-receptor for Fibroblast Growth Facrot (FGF)-23. Silencing of the *HDAC3* gene has no effect on Klotho expression; on the contrary, inhibiting the TGF-beta receptor through the use of the specific inhibitor SB431542 led to a slowdown of the renal fibrotic process. By performing a more targeted inhibition using a specific inhibitor of SMAD3, SIS3, it was demonstrated that the TGF-β/Smad3 pathway led to an upregulation of HDAC3 [134]. In addition, it was been demonstrated that HDAC3 forms a complex with NCoR and NF-κB that acts on the klotho promoter, giving rise to a EMT-mediated pro-fibrotic renal signal transduction cascade that initiates from TGF-β and has HDAC3 as an intermediary [134].

A role of the complex formed by HDAC3 and Smad2/3 and NcoR in renal fibrosis has also been demonstrated in patients with focal segmental glomerulosclerosis who show high levels of elevated HDAC3. The complex appears to have an inhibitory effect on the miR-30d promoter, promoting renal fibrosis [135], although a concomitant activation of EMT in this fibrotic process has not yet been demonstrated.

HDAC3 also appears to regulate the expression of TIMAP protein (membrane-associated protein and inhibited by TGF-β). TIMAP expression is reduced in mice with renal fibrosis, probably due to the overexpression of TGF-β. The high concentration of TGF-β in fibrotic tissues would determine an overexpression of HDAC3 with the consequent activation of the TGF-β-HDAC3/Smad-TIMAP pathway which leads to a reduction in the expression of TIMAP [136]. This could be implicated in a regulatory loop of macrophage M2 phagocytosis. Given that hyperactive TGFβ often causes excessive macrophage phagocytic activities potentially leading to fibrotic evolution [136], the inhibition of the TGF-β-HDAC3/Smad-TIMAP pathway could represent a strategy to slow down the progression of renal fibrosis. Even in this case, it still remains to be discovered whether these mechanisms involve the activation of an EMT program.

Recent studies have also evaluated the role of HDAC8 on renal fibrosis, always belonging to class I of HDACs [137]. HDAC8 is overexpressed in UUO mice and the use of a selective inhibitor for this deacetylase, such as PCI34051, or gene silencing determines a slowdown of fibrotic progression by inhibiting the TGF-β-dependent EMT process. A very important fact is that the high expression of HDAC8 arrests renal tubular epithelial cells in the G2/M phase, determining the activation of Snail, a known pro-EMT and pro-fibrotic factor. The use of HDAC8 inhibitors leads to an inhibition of EMT and is capable of determining an increase in Klotho levels, thus acting as anti-fibrotic factors.

Class II Histone Deacetylases in EMT-Dependent Fibrosis

In the intertubular, extraglomerular, and extravascular spaces of the kidney, HDAC6 has been demonstrated to contribute to TGF-β-induced EMT. Although the mechanism of HDAC6 induction is not clear, this was enhanced by TGF-β, and subsequently, the activated HDAC6 deacetylated the renal tubular epithelial cytoskeletal protein (α-tubulin). α-tubulin deacetylation induces cytoskeletal rearrangement during the EMT process [138,139]. In addition, in unilateral ureteral obstruction (*UUO*) mice used as models of kidney fibrosis, HDAC1, 4, 5, 6, and 10 were overexpressed and HDAC8 was downregulated, and this seemed to determine pro-fibrotic events through a TGF-β1/Smad-independent pathway [140].

Supporting the role of histone acetylation in EMT-dependent fibrotic diseases, HDAC7 has also been demonstrated to regulate collagen deposition and other ECM protein accumulation in fibroblasts derived from patients affected by systemic sclerosis, and indeed, siRNA-mediated depletion of *HDAC7* reduced ECM production in these cells, resulting in an evident decrease in fibrogenesis [10]. However, the mechanism of action of class IIb HDACs in the fibrotic process will certainly require further investigation.

5.3. Histone Methylation and Demethylation Affects EMT-Dependent Fibrosis

Recently, it was demonstrated that the blocking of histone/lysine methyltransferases, such as the enhancer of zeste homolog 2 (EZH2), a histone H3 lysine 27 trimethylation methyltransferase, and SET and MYND domain-containing protein 2 (Smyd2), a histone H3 lysine 36 trimethylation methyltransferase, had a negative effect on EMT and consequently attenuated renal fibrosis [141]. However, knowledge of the role of demethylases is still limited and very recent. Lysine-specific demethylase 1 (LSD1), also known as KDM1A, is the first histone/lysine demethylase that specifically targets mono- and dimethylated H3K4 [142]. The upregulation of LSD1 was involved in renal tubular cell EMT, deposition of ECM proteins, and renal fibrosis progression in UUO model mice [143]. This was confirmed through the use of specific inhibitors or siRNA targeting LSD1, which reduced TGF-β1-induced EMT and blocked the activation of renal fibroblasts through TGF-β1/Smad3 and LSD1/PKC/α-Akt/STAT3 signaling pathways [143].

In addition, it was demonstrated that the different methylation status of specific lysines present in histones can lead to gene induction or suppression. In general, methylation of histone H3 lysine 4 (H3K4), H3K36, and H3K79 stimulates gene transcription, while H3K9, H3K27, and H4K20 methylation leads to gene suppression [144]. ZEB1 is a zinc-finger transcription factor implicated in the induction of EMT. It works on the promoter of the E-cadherin gene, suppressing its synthesis. The active constituent derived from Schisandra chinensis (SchB) appears to operate on TGF-β-mediated EMT through epigenetic inhibition of ZEB1 [145]. The ability of Sch B to enhance H3K9me3 levels of the ZEB1 promoter and to inhibit transcription of the *ZEB1* gene, which is crucial for TGF-β-induced EMT, was recently demonstrated [145]. SETDB1 (also known as ESET or KMT1E) is a specific histone methyltransferase able to block euchromatin genes by regulating the methylation of histone H3K9 [146]. SETDB1 seems to be able to trigger the trimethylation of histone H3K9 (H3K9me3) on the *Snai1* promoter region, leading to EMT reorganization, which is promoted by TGF-β-induced iron accumulation [147]. Confirming these observations, the knockdown of *SETDB1* reduced H3K9me3 and enhanced TGF-β/Snail-mediated EMT, which was accompanied by increased ferroptosis [147]. In radiation-induced lung fibrosis, Nagaraja et al. showed increased expression of H3K9me2/3 in irradiated cells [148]. This effect was reversed by the use of specific histone methyltransferase G9a inhibitor, which led to irradiated cells showing higher expression of epithelial markers and not of mesenchymal ones [148].

5.4. Epigenetic Involvement of ncRNAs in EMT-Related Fibrosis

NcRNAs were also recently recognized as potential key inductors of fibrogenesis [149]. Noncoding RNAs are micro-sequences of RNA transcribed but not translated into proteins, and they are grouped into short ncRNAs and long ncRNAs [150,151]. Increasing evidence has highlighted the involvement of noncoding RNAs in the EMT process.

5.4.1. MiRNAs in EMT-Dependent Fibrosis

Recently, important studies have reported that miRNAs are involved as regulators or activators of signaling pathways in the induction of fibrogenesis triggered by the EMT process. To investigate the correlation between miRNAs and EMT, Liang et al. have focused their investigations on the involvement of miRNAs in pulmonary fibrosis induced by EMT. In particular, the study investigated the impact and mechanism of miR-26a in mice affected by experimental pulmonary fibrosis. It was discovered that miR-26a is significantly downregulated and modulates high-mobility group protein A2 (HMGA2), a key regulator of the EMT process [152]. Therefore, inhibition of miR-26a in lung epithelial cells determines the induction of EMT, in which the epithelial cells acquire the features of the mesenchymal phenotype and thus transform into myofibroblasts. Interestingly, the overexpression of miR-26a reduced the EMT process triggered by TGF-β in adenocarcinoma A259 cells [153]. An interesting study has evidenced that, in mice exposed to silicone to induce pulmonary fibrosis, an overexpression of miRNA let-7d led to a reduction in HMGA2 expression and inhibition of EMT; while the suppression of miRNA let-7d augmented HMGA2 expression and triggered silica-induced EMT [154].

Additionally, Wang and colleagues have demonstrated that overexpression of miR-221 diminished the expression of HMGA2, reducing the EMT events in A549 and human bronchial epithelium (HBE) cell lines [155]. Therefore, using a mouse model of bleomycin (BLM)-induced pulmonary fibrosis confirmed the effect of miR-221 on the EMT process.

Recent experiments performed by Li and collaborators in radiation-induced pulmonary fibrosis (RIPF) have isolated extracellular vehicles (EVs) derived from mouse mesenchymal stem cells (mMSC-Exos) and examined their effects both in vitro and in vivo. The results obtained showed that mMSC-Exos protect cells and reverse the EMT process induced by radiation.

Interestingly and correlated with these data is the finding that miR-466f-3p in de-534 EVs separated from mMSC-Exos inhibited radiation-induced evolution by inhibiting AKT/GSK3β through c-MET [156].

In recent years, an interesting study conducted by Wang et al., using high-throughput sequencing, has shown that miR-155–5p is significantly downregulated in RIPF and acts as a key regulator of RIPF via the GSK-3β/NF-κB pathway. Blocking glycogen synthase kinase-3β (GSK-3β), a functional target of miR-155–5p, reversed radiation-induced EMT through the NF-κB pathway, preventing the progression of RIPF [157]. This suggests that ectopic expression of miR-155–5p reduces RIPF in mice through the GSK-3β/NF-κB pathway. These data were confirmed by recent findings in which it was demonstrated that ionizing radiation (IR) impeded the transcription of miR-486–3p, which determines the activation of EMT target genes such as *Snail*, leading to the induction of EMT in radiation-induced pulmonary fibrosis [158].

In addition, Liang et al. have demonstrated that miR-541-5p repression through Myeloid Zinc Finger 1 (MZF1) factor, an important transcriptional repressor, triggers the EMT process induced by irradiation, contributing to RIPF. In particular, miR-541-5p modulates the effect of the *Slug* gene on the EMT process. Irradiation activates MZF1 to downregulate miR-541-5p in alveolar epithelial cells, and hence induces EMT, contributing to RIPF by targeting *Slug* [158].

Therefore, using bioinformatics analysis, it was found that Fos-related antigen 1 (Fra-1) is a potential target of miR-34c-5p. The miR-34c-5p/Fra-1 axis represses the induction of the PTEN/PI3K/AKT signaling pathway and inhibits the EMT process [159]. This result can provide a promising therapeutic approach to alleviate the progression of pulmonary fibrosis. Recent studies have highlighted the role of miR-21 as a profibrotic factor since its upregulation was noted in patients with severe fibrosis kidney disease. MiR-21 influences the expression of different metalloproteinases (MMPs) through the downregulation of PTEN during TGF-β1-promoting EMT [160]. Furthermore, the inhibition of miR-21 reduces the induction of profibrotic genes in human podocytes and tubular cells in renal diseases [161,162]. Therefore, miR-21 modulates the TGF-β pathway, but TGF-β1 can also induce the expression of miR-21 through Smad signaling [163]. Microarray analysis has evidenced that upregulation of miR-21 in tubular epithelial cells depends on activation of TGF-β/Smad3 signaling. These data were confirmed in normal and Smad3 knockdown tubular epithelial cells wherein the suppression of Smad3, but not of Smad2, prevented cells from upregulating miR-21 in response to TGF-β [163,164]. Upregulation of miR-21 involves different mechanisms in promoting renal fibrosis, such as the decrease in dimethylarginine dimethylaminohydrolase 1 (DDAH1) expression. This leads to an elevated level of asymmetric dimethylarginine (ADMA), which is an endogenous repressor of nitric oxide synthase (NOS), thus diminishing the production of NO [165]. Emerging evidence has revealed an elevated expression of miR-34a in renal fibrosis. miR-34a directly links to the 3′ UTR of Klotho mRNA, leading to the downregulation of Klotho that preserves the kidneys against severe and chronic failure, thus contributing to renal fibrosis [165–167]. The overexpression of miR-34a-5p in the kidney tissues of patients suffering from type 2 diabetes mellitus and mice affected by diabetic nephropathy (DN) was associated with Ski-related novel gene (*SnoN*) downregulation and, thus, with the activation of EMT-dependent fibrosis [168]. MiR-130a-3p plays a pro-fibrotic role; indeed, it is implicated in the modulation of EMT and provokes severe TGF-β1-induced fibrosis in renal tubular epithelial cells [169]. Furthermore, the repression of miR-130a-3p was found to be linked with a marked reduction in α-SMA and vimentin, as well as an increase in levels of E-cadherin expression. Moreover, in vitro studies have evidenced the central role of miR-130a-3p in the modulation of Smads, which are implicated in the TGF-β signaling pathway in the course of renal fibrosis evolution [169].

Another study conducted by Bai et al. has evidenced that the overexpression of miR-27b-3p significantly reduced TGF-β1-dependent EMT through the decrease in mesenchymal factors in cell cultures [170]. These findings evidenced that augmented expression of miR-27b-3p avoids renal fibrosis through EMT suppression. Therefore, overexpression of miR-27b-3p diminished the expression of p-STAT1 and STAT1 [170], promoting renal fibrosis induced by TGF-β1. Moreover, both apoptosis and the EMT process were shown to be involved in miR-27b-3p-mediated control of renal fibrosis. In addition, the overexpression of miR-27b-3p repressed TGF-β1-induced and Fas-mediated apoptosis in human kidney cells through the downregulation of pro-apoptotic factors such as active caspase 3, Fas, and active caspase 8 [170].

MiR-30e expression is downregulated in diabetic nephropathy patients, and this event is linked with an augmented EMT process in renal tubule epithelial cells. In turn, the upregulation of this miRNA determines low levels of mesenchymal markers such as collagen I and vimentin and elevated levels of E-cadherin expression. These findings demonstrate that miR-30e plays an antifibrotic role and has protective activities [171].

The role of the miR-200 family in renal fibrosis is conflicting. It has been demonstrated that miR-200 may play an antifibrotic role [172], and in fact, findings confirmed that miR-200b suppresses TGF-β1-induced EMT through the reduction in fibronectin and the increase in E-cadherin [173]. The suppression of the TGF-β1-induced EMT pathway was demonstrated to be independent of TGF-β1-induced p-Smad2/3, phospho-p38, and p42/44 signaling. miR-200a is, furthermore, able to repress TGF-β2 expression, thus strongly reducing the evolution of renal fibrosis [173].

Emerging evidence has highlighted the profibrotic role of miR-32 in liver fibrosis. Indeed, hepatic fibrosis induced by the EMT process and influenced by MTA3 expression in hyperglycemic conditions was inhibited by anti-miR-32. These results demonstrated that miR-32 knockdown may ameliorate liver fibrosis progression, suggesting possible targets for the treatment of liver fibrosis [174].

An interesting study has evaluated the effect of miR-451 in diabetic cardiomyopathy influenced by endothelial to mesenchymal transition (EndMT). Growing evidence indicates that miR-451 silencing inhibits the EndMT process in diabetic mice hearts. This study also showed the effect of miR-451 knockdown on the reduction of hyperglycemia-induced EndMT in mouse cardiac endothelial cells, which led to an evident decrease in the progression of cardiac fibrosis [175].

5.4.2. CircRNA Regulation of Fibrosis Correlated to EMT

CircRNAs are RNA molecules with a loop structure primarily produced through back-splicing of pre-mRNA [176]. Most studies on non-coding RNAs have pointed to miRNAs and lncRNAs; however, with progress in bioinformatics analysis and the availability of advanced RNA-sequencing technologies, research has moved towards circRNAs. Several findings have shown that circRNAs participate in various physiological and pathological mechanisms through different processes [176]. Several studies have demonstrated that circRNAs play a regulatory role in EMT-dependent pulmonary fibrosis. In an experimental evaluation in which lung fibrosis was induced by silica, upregulation of the circRNA CDR1as served as a sponge and inhibitor for miR-7, leading to the release of TGFBR2 and promoting EMT [177]. Recently, Li et al. have demonstrated an elevated overexpression of the hsa_circ_0044226 in IPF tissue [178]. Data from Qi and collaborators supported the hypothesis that the downregulation of circRNAs has an antifibrotic effect, demonstrating that the reduced expression of hsa_circ_0044226 attenuates pulmonary fibrosis through the inhibition of *CDC27* expression, a parental gene for hsa_circ_0044226 [179]. New findings in this context have demonstrated that blocking circZC3H4 effectively inhibits the EMT process and blocks the progression of pulmonary fibrosis induced by SiO_2 [180].

Furthermore, circZC3H4 functions as a sponge for miR-212, modulating ZC3H4 expression and consequently inducing the EMT process [181]. These findings demonstrate the important role of circRNAs in regulating EMT-associated pulmonary fibrosis, even if further investigations and discoveries in this field are requested to fill these knowledge gaps (Figure 3).

5.4.3. LncRNA Regulation of EMT Related to Fibrosis

Long non-coding RNAs (lncRNAs) represent a class of non-coding transcribed RNA molecules of more than 200 nucleotides [182]. Accumulating evidence has highlighted the key role of lncRNAs in the control of EMT. Furthermore, several lncRNAs act as competitive endogenous RNAs (ceRNAs), preventing the link of miRNA to its target genes [183].

Interestingly, a recent study showed that, during the EMT process of lung epithelial cells, elevated levels of long non-coding RNA (lncRNA)-ATB were detected, lncRNA activated by TGF-β, which may lead to decreased levels of miR-200c. The inhibitory role of lncRNA-ATB on miR-200c was also demonstrated in silica-induced EMT-dependent pulmonary fibrosis [184]. Furthermore, it has been demonstrated that miR-200c is able to target ZEB1, to enhance E-cadherin levels, and subsequently inhibit the EMT process, thus potentially ameliorating silicosis [184].

Xu et al. (2021) have demonstrated that lncRNA-ATB has pro-fibrotic effects by downregulating miR-29b-2-5p and miR-34c-3p in vitro. This effect was supported by the observation that the overexpression of miR-29b-2-5p or miR-34c-3p inhibits the pro-fibrotic effects of lncRNA-ATB and thus attenuates the induction of EMT [185].

Sun et al. identified a large number of upregulated and downregulated long non-coding RNAs (lncRNAs) in lung tissue induced by paraquat (1,1-dimethyl-4,4-bipyridyl dichloride), a herbicide that can be fatal to humans and, in particular, can induce pulmonary fibrosis. The authors identified Zeb2 and Hoxa3, modulators of EMT, as target genes of two upregulated lncRNAs: uc.77 and 2700086A05Rik. These findings underline the crucial role of lncRNAs in the regulation of EMT during lung fibrosis [186].

Wang et al. (2022) discovered that the lncRNA miR99AHG functions as a competitive endogenous RNA (ceRNA) with miR-136–5p. As a consequence, ubiquitin-specific protease 4 (USP4) is less degraded by miR-136-5p. This determines the regulation of the expression of angiotensin-converting enzyme 2 (ACE2), a downstream target gene of USP4, and activation of the EMT process in the alveolar epithelial cells type II [187].

Zhan and collaborators have investigated the role of lncRNA MEG3 in a rat model of lung fibrosis induced by nickel oxide nanoparticles (NiO NPs) via activating TGF-β1, which was associated with the activation of the EMT process. The study showed that lncRNA MEG3 suppressed the TGF-β1 pathway, the EMT process, and collagen accumulation [188].

Intriguingly, findings have shown that the lncRNA ZEB1 antisense RNA 1 (ZEB1-AS1) was upregulated in a rat model of BLM-induced pulmonary fibrosis and in TGF-β1-induced alveolar type II epithelial (RLE-6TN) cells. Silencing of *ZEB1-AS1* inhibited the induction of EMT and reduced BLM-induced fibrogenesis [189].

Several other lncRNAs can be involved in liver fibrosis. In particular, GAS5 acts as a sponge platform for miR-23a through the PTEN/PI3K/Akt pathway, which could slow down the progression of hepatic fibrosis [190]. Chen and collaborators have discovered that Meg8 lncRNA is overexpressed in activated HSC, injured hepatocytes, and fibrotic livers. Furthermore, knockdown of Meg8 significantly inhibited the expression of epithelial factors and induced the expression of mesenchymal markers in hepatocytes [191]. (Figure 3). All of the data reported and related to epigenetic involvement of ncRNAs in EMT-related fibrosis are summarized in Table 1.

Table 1. microRNA regulation of EMT-dependent fibrosis.

ncRNAs	Target	Action Modes	Outcomes	References
miR-26a	*HMGA2*	Downregulation of miR-26a promotes EMT process	Pulmonary fibrosis	[152]
miRNA let-7d	reduces *HMGA2* expression	Overexpression of miRNA let-7d inhibits EMT process	Silicone-induced lung fibrosis	[154]
miR-221	reduces *HMGA2* expression	Overexpression of miR-221 inhibits EMT process	Bleomycin (BLM)-induced pulmonary fibrosis	[155]
miR-466f-3p	mMSCs-exo	Antifibrotic features of miR-466f-3p that prevent radiation-induced EMT	Radiation-induced pulmonary fibrosis	[156]
miR-155–5p	*GSK-3β*	Downregulation of miR-155–5p in radiation-induced pulmonary fibrosis	Radiation-induced pulmonary fibrosis	[157]
mir-486–3p	*Snail* gene	Downregulation of mir-486–3p promotes EMT process	Radiation-induced pulmonary fibrosis	[158]
miR-541-5p	*MZF1*, *Slug* gene	miR-541-5p repression induces EMT process	Radiation-induced pulmonary fibrosis	[158]
miR-34c-5p	*Fra-1*	miR-34c-5p/Fra-1 axis represses activation of EMT process	Pulmonary fibrosis	[159]
miR-21	MMPs, activation of TGF-β/Smad3 signaling	Upregulation of miR-21 induces TGF-β1, promoting EMT	Kidney fibrosis	[160,163,164]
miR-21	*DDAH1*	Upregulation of miR-21 decreases DDAH1 and increases ADMA, diminishing the production of NO	Kidney fibrosis	[165]
miR-34a	3' UTR of Klotho mRNA	Increased expression of miR-34a downregulates Kloto	Kidney fibrosis	[154]
miR-34a-5p	*SnoN*	Overexpression of miR-34a-5p determines the downregulation of SnoN and induction of EMT	Kidney fibrosis	[168]

Table 1. Cont.

ncRNAs	Target	Action Modes	Outcomes	References
miR-130a-3p	Smads	Downregulation of miR-130a-3p modulates Smads, induction of EMT	Kidney fibrosis	[169]
miR-27b-3p	α-SMA, fibronectin, collagen III, vimentin	Overexpression of miR-27b-3p reduces EMT process. Overexpression of miR-27b-3p reduces p-STAT1 and STAT1 induces EMT.	Kidney fibrosis	[170]
miR-30e		Downregulation of miR-30e expression in diabetic nephropathy, increases EMT	Kidney fibrosis	[171]
miR-200b		miR-200b inhibits TGF-β1-induced EMT, ameliorates renal fibrosis	Kidney fibrosis	[172]
miR200a		miR-200a represses TGF-β2 expression, inhibiting the development of renal fibrosis	Kidney fibrosis	[172,173]
miR-32	MTA3	Profibrotic role of miR-32	Liver fibrosis	[174]
miR-451		Induced by EndMT	Cardiac fibrosis	[175]
CDR1	sponge for miR-7	Upregulation of CDR1 promotes EMT process	Silica-induced pulmonary fibrosis	[177]
hsa_circ_0044226	CDC27	Upregulation of hsa_circ_0044226 promotes EMT	Idiopathic pulmonary fibrosis	[178]
circZC3H4	sponge for miR-212	Modulation of ZC3H4 expression induces EMT process	Idiopathic pulmonary fibrosis	[181]
lncRNA-ATB	sponge for miR-200c	lncRNA-ATB promotes EMT program	Silica-induced pulmonary fibrosis	[184]
uc.77 and 2700086A05Rik	Zeb2 and Hoxa3	Upregulation of uc.77 and 2700086A05Rik induces EMT process	Pulmonary fibrosis	[186]
lncRNA miR99AHG	miR-136–5p	miR99AHG regulates the expression of ACE2, a target gene of USP4, and induces EMT process	Pulmonary fibrosis	[187]
lncRNA MEG3		MEG3 induces NiO NPs via activating TGF-β1 and activation of EMT process	Pulmonary fibrosis	[188]
ZEB1-AS1		Upregulation of EB1-AS1 induces EMT process	Pulmonary fibrosis	[189]
GAS5	sponge for miR-23a	Activation of PTEN/PI3K/Akt pathway induces EMT process	Hepatic fibrosis	[190]
Meg8		Overepression of Meg8 promotes EMT process	Hepatic fibrosis	[191]

6. Conclusions and Future Perspective

This article represents an attempt to collect relevant data on the role of epigenetics in regulating EMP/EMT-dependent fibrosis. The purpose is to arouse growing interest in the scientific world and clarify the molecular mechanisms underlying this regulation of gene expression. This could identify factors predisposed to the onset of diseases related to the triggering of the EMP/EMT program, such as cancer or inflammatory diseases. After decades of studies, EMT is now considered a key physiological process active in embryo genesis and a pathological mechanism triggered in fibrotic diseases and cancer. Furthermore, recently, the meaning of EMT has also been revised by scientists, who have identified various degrees of cellular transformation toward a mesenchymal phenotype, so much so that they have coined the term EMP, underscoring cellular plasticity and the ability of an epithelial cell to assume intermediate phenotypes. Recent studies have unveiled the role of epigenetics in the control of EMP/EMT dynamics and, in general, in cellular plasticity. Several epigenetic modifications have been identified as capable of modifying gene transcription during the process of EMP/EMT related to fibrogenesis, adding even more complexity to the process of EMP/EMT-dependent fibrosis, which is finely regulated by various molecularly interconnected mechanisms. We hope that having summarized in this review the most recent discoveries in this very innovative field will provide new perspectives on molecular aspects and therapeutic approaches intended to regulate and reverse the EMP/EMT-dependent fibrotic process.

Author Contributions: M.S. and S.L. had full access to the data collected in the review, take responsibility for their integrity, and performed a critical reading. All authors have read and agreed to the published version of the manuscript.

Funding: This research received no external funding.

Institutional Review Board Statement: Not applicable.

Informed Consent Statement: Not applicable.

Data Availability Statement: Not applicable.

Conflicts of Interest: The authors declare no conflicts of interest.

References

1. Antar, S.A.; Ashour, N.A.; Marawan, M.E.; Al-Karmalawy, A.A. Fibrosis: Types, Effects, Markers, Mechanisms for Disease Progression, and Its Relation with Oxidative Stress, Immunity, and Inflammation. *Int. J. Mol. Sci.* **2023**, *24*, 4004. [CrossRef] [PubMed]
2. Mehal, W.Z.; Iredale, J.; Friedman, S.L. Scraping fibrosis: Expressway to the core of fibrosis. *Nat. Med.* **2011**, *17*, 552–553. [CrossRef] [PubMed]
3. Wick, G.; Grundtman, C.; Mayerl, C.; Wimpissinger, T.F.; Feichtinger, J.; Zelger, B.; Sgonc, R.; Wolfram, D. The immunology of fibrosis. *Annu. Rev. Immunol.* **2013**, *31*, 107–135. [CrossRef] [PubMed]
4. Hnderson, N.C.; Rieder, F.; Wynn, T.A. Fibrosis: From mechanisms to medicines. *Nature* **2020**, *587*, 555–566. [CrossRef] [PubMed]
5. Subhadarshini, S.; Markus, J.; Sahoo, S.; Jolly, M.K. Dynamics of Epithelial-Mesenchymal Plasticity: What Have Single-Cell Investigations Elucidated so Far? *ACS Omega* **2023**, *8*, 11665–11673. [CrossRef] [PubMed]
6. Miao, H.; Wu, X.Q.; Zhang, D.D.; Wang, Y.N.; Guo, Y.; Li, P.; Xiong, Q.; Zhao, Y.Y. Deciphering the cellular mechanisms underlying fibrosis-associated diseases and therapeutic avenues. *Pharmacol. Res.* **2021**, *163*, 105316. [CrossRef] [PubMed]
7. Sisto, M.; Ribatti, D.; Lisi, S. Organ Fibrosis and Autoimmunity: The Role of Inflammation in TGFβ-Dependent EMT. *Biomolecules* **2021**, *11*, 310. [CrossRef] [PubMed]
8. Sisto, M.; Lisi, S. Towards a Unified Approach in Autoimmune Fibrotic Signalling Pathways. *Int. J. Mol. Sci.* **2023**, *24*, 9060. [CrossRef]
9. Povero, D.; Busletta, C.; Novo, E.; di Bonzo, L.V.; Cannito, S.; Paternostro, C.; Parola, M. Liver fibrosis: A dynamic and potentially reversible process. *Histol. Histopathol.* **2010**, *25*, 1075–1091.
10. Liu, Y.; Wen, D.; Ho, C.; Yu, L.; Zheng, D.; O'Reilly, S.; Gao, Y.; Li, Q.; Zhang, Y. Epigenetics as a versatile regulator of fibrosis. *J. Transl. Med.* **2023**, *21*, 164. [CrossRef]
11. Distler, J.H.W.; Györfi, A.H.; Ramanujam, M.; Whitfield, M.L.; Königshoff, M.; Lafyatis, R. Shared and distinct mechanisms of fibrosis. *Nat. Rev. Rheumatol.* **2019**, *15*, 705–730. [CrossRef]
12. Xue, T.; Qiu, X.; Liu, H.; Gan, C.; Tan, Z.; Xie, Y.; Wang, Y.; Ye, T. Epigenetic regulation in fibrosis progress. *Pharmacol. Res.* **2021**, *173*, 105910. [CrossRef]
13. Wang, X.C.; Song, K.; Tu, B.; Sun, H.; Zhou, Y.; Xu, S.S.; Lu, D.; Sha, J.M.; Tao, H. New aspects of the epigenetic regulation of EMT related to pulmonary fibrosis. *Eur. J. Pharmacol.* **2023**, *956*, 175959. [CrossRef]
14. Mobley, R.J.; Abell, A.N. Controlling Epithelial to Mesenchymal Transition through Acetylation of Histone H2BK5. *J. Nat. Sci.* **2017**, *3*, e432.

15. Nieto, M.A.; Huang, R.Y.Y.J.; Jackson, R.A.A.; Thiery, J.P.P. EMT: 2016. *Cell* **2013**, *166*, 21–45. [CrossRef]
16. Bhatia, S.; Monkman, J.; Toh, A.K.L.; Nagaraj, S.H.; Thompson, E.W. Targeting epithelial-mesenchymal plasticity in cancer: Clinical and preclinical advances in therapy and monitoring. *Biochem. J.* **2017**, *474*, 3269–3306. [CrossRef] [PubMed]
17. Yuan, S.; Norgard, R.J.; Stanger, B.Z. Cellular plasticity in cancer. *Cancer Discov.* **2019**, *9*, 837–851. [CrossRef]
18. Thiery, J.P.; Acloque, H.; Huang, R.Y.; Nieto, M.A. Epithelial-mesenchymal transitions in development and disease. *Cell* **2009**, *139*, 871–890. [CrossRef] [PubMed]
19. Arnoux, V.; Nassour, M.; L'Helgoualc'h, A.; Hipskind, R.A.; Savagner, P. Erk5 controls Slug expression and keratinocyte activation during wound healing. *Mol. Biol. Cell* **2018**, *19*, 4738–4749. [CrossRef] [PubMed]
20. Acloque, H.; Adams, M.S.; Fishwick, K.; Bronner-Fraser, M.; Nieto, M.A. Epithelial-mesenchymal transitions: The importance of changing cell state in development and disease. *J. Clin. Investig.* **2009**, *119*, 1438–1449. [CrossRef] [PubMed]
21. Kalluri, R.; Weinberg, R.A. The basics of epithelial-mesenchymal transition. *J. Clin. Investig.* **2009**, *119*, 1420–1428. [CrossRef] [PubMed]
22. Willis, B.C.; duBois, R.M.; Borok, V. Epithelial origin of myofibroblasts during fibrosis in the lung. *Proc. Am. Thorac. Soc.* **2006**, *3*, 377–382. [CrossRef] [PubMed]
23. Pei, D.; Shu, X.; Gassama-Diagne, A.; Thiery, J.P. Mesenchymal-epithelial Transition in Development and Reprogramming. *Nat. Cell Biol.* **2019**, *21*, 44–53. [CrossRef] [PubMed]
24. Haerinck, J.; Goossens, S.; Berx, G. The epithelial–mesenchymal plasticity landscape: Principles of design and mechanisms of regulation. *Nat. Rev. Genet.* **2023**, *24*, 590–609. [CrossRef] [PubMed]
25. Dong, B.; Qiu, Z.; Wu, Y. Tackle Epithelial-Mesenchymal Transition with Epigenetic Drugs in Cancer. *Front. Pharmacol.* **2020**, *11*, 596239. [CrossRef] [PubMed]
26. Pastushenko, I.; Blanpain, C. EMT transition states during tumor progression and metastasis. *Trends Cell Biol.* **2019**, *29*, 212–226. [CrossRef]
27. Yang, J.; Antin, P.; Berx, G.; Blanpain, C.; Brabletz, T.; Bronner, M.; Campbell, K.; Cano, A.; Casanova, J.; Christofori, G.; et al. Guidelines and Definitions for Research on Epithelial-Mesenchymal Transition. *Nat. Rev. Mol. Cell Biol.* **2020**, *21*, 341–352. [CrossRef]
28. Dongre, A.; Weinberg, R.A. New insights into the mechanisms of epithelial–mesenchymal transition and implications for cancer. *Nat. Rev. Mol. Cell Biol.* **2019**, *20*, 69–84. [CrossRef]
29. Puisieux, A.; Brabletz, T.; Caramel, J. Oncogenic roles of EMT-inducing transcription factors. *Nat. Cell Biol.* **2014**, *16*, 488–494. [CrossRef]
30. Lamouille, S.; Xu, J.; Derynck, R. Molecular mechanisms of epithelial-mesenchymal transition. *Nat. Rev. Mol. Cell Biol.* **2014**, *15*, 178–196. [CrossRef]
31. Liu, Q.L.; Luo, M.; Huang, C.; Chen, H.N.; Zhou, Z.G. Epigenetic Regulation of Epithelial to Mesenchymal Transition in the Cancer Metastatic Cascade: Implications for Cancer Therapy. *Front. Oncol.* **2021**, *11*, 657546. [CrossRef] [PubMed]
32. Verstappe, J.; Berx, G. A role for partial epithelial-to-mesenchymal transition in enabling stemness in homeostasis and cancer. *Semin. Cancer Biol.* **2023**, *90*, 15–28. [CrossRef] [PubMed]
33. Lovisa, S.; LeBleu, V.S.; Tampe, B.; Sugimoto, H.; Vadnagara, K.; Carstens, J.L.; Wu, C.C.; Hagos, Y.; Burckhardt, B.C.; Pentcheva-Hoang, T.; et al. Epithelial-to-mesenchymal transition induces cell cycle arrest and parenchymal damage in renal fibrosis. *Nat. Med.* **2015**, *21*, 998–1009. [CrossRef] [PubMed]
34. Di Gregorio, J.; Robuffo, I.; Spalletta, S.; Giambuzzi, G.; De Iuliis, V.; Toniato, E.; Martinotti, S.; Conti, P.; Flati, V. The Epithelial-to-Mesenchymal Transition as a Possible Therapeutic Target in Fibrotic Disorders. *Front. Cell Dev. Biol.* **2020**, *8*, 607483. [CrossRef] [PubMed]
35. Marconi, G.D.; Fonticoli, L.; Rajan, T.S.; Pierdomenico, S.D.; Trubiani, O.; Pizzicannella, J.; Diomede, F. Epithelial-Mesenchymal Transition (EMT): The Type-2 EMT in Wound Healing, Tissue Regeneration and Organ Fibrosis. *Cells* **2021**, *10*, 1587. [CrossRef] [PubMed]
36. Rowe, R.G.; Lin, Y.; Shimizu-Hirota, R.; Hanada, S.; Neilson, E.G.; Greenson, J.K.; Weiss, S.J. Hepatocyte-derived Snail1 propagates liver fibrosis progression. *Mol. Cell. Biol.* **2011**, *31*, 2392–2403. [CrossRef] [PubMed]
37. Karin, D.; Koyama, Y.; Brenner, D.; Kisseleva, T. The characteristics of activated portal fibroblasts/myofibroblasts in liver fibrosis. *Differentiation* **2016**, *92*, 84–92. [CrossRef]
38. Iwaisako, K.; Jiang, C.; Zhang, M.; Cong, M.; Moore-Morris, T.J.; Park, T.J.; Liu, X.; Xu, J.; Wang, P.; Paik, Y.H.; et al. Origin of myofibroblasts in the fibrotic liver in mice. *Proc. Natl. Acad. Sci. USA* **2014**, *111*, E3297–E3305. [CrossRef]
39. Michelotti, G.A.; Xie, G.; Swiderska, M.; Choi, S.S.; Karaca, G.; Kruger, L.; Premont, R.; Yang, L.; Syn, W.K.; Metzger, D.; et al. Smoothened is a master regulator of adult liver repair. *J. Clin. Investig.* **2013**, *123*, 2380–2394. [CrossRef]
40. Salton, F.; Volpe, M.C.; Confalonieri, M. Epithelial-Mesenchymal Transition in the Pathogenesis of Idiopathic Pulmonary Fibrosis. *Medicina* **2019**, *55*, 83. [CrossRef]
41. Yang, Z.C.; Yi, M.J.; Ran, N.; Wang, C.; Fu, P.; Feng, X.Y.; Xu, L.; Qu, Z.H. Transforming growth factor-beta1 induces bronchial epithelial cells to mesenchymal transition by activating the snail pathway and promotes airway remodeling in asthma. *Mol. Med. Rep.* **2013**, *8*, 1663–1668. [CrossRef]

42. Yamaguchi, M.; Hirai, S.; Tanaka, Y.; Sumi, T.; Miyajima, M.; Mishina, T.; Yamada, G.; Otsuka, M.; Hasegawa, T.; Kojima, T.; et al. Fibroblastic foci, covered with alveolar epithelia exhibiting epithelial-mesenchymal transition, destroy alveolar septa by disrupting blood flow in idiopathic pulmonary fibrosis. *Lab. Investig.* **2017**, *97*, 232–242. [CrossRef]
43. Liu, L.; Sun, Q.; Davis, F.; Mao, J.; Zhao, H.; Ma, D. Epithelial-mesenchymal transition in organ fibrosis development: Current understanding and treatment strategies. *Burn. Trauma* **2022**, *10*, tkac011. [CrossRef]
44. Zolak, J.S.; Jagirdar, R.; Surolia, R.; Karki, S.; Oliva, O.; Hock, T.; Guroji, P.; Ding, Q.; Liu, R.M.; Bolisetty, S.; et al. Pleural mesothelial cell differentiation and invasion in fibrogenic lung injury. *Am. J. Pathol.* **2013**, *182*, 1239–1247. [CrossRef]
45. Luo, G.H.; Lu, Y.P.; Yang, L.; Song, J.; Shi, Y.J.; Li, Y.P. Epithelial to mesenchymal transformation in tubular epithelial cells undergoing anoxia. *Transpl. Proc.* **2008**, *40*, 2800–2803. [CrossRef]
46. Grande, M.T.; Sanchez-Laorden, B.; Lopez-Blau, C.; De Frutos, C.A.; Boutet, A.; Arevalo, M.; Rowe, R.G.; Weiss, S.J.; López-Novoa, J.M.; Nieto, M.A. Snail1-induced partial epithelial-to-mesenchymal transition drives renal fibrosis in mice and can be targeted to reverse established disease. *Nat. Med.* **2015**, *21*, 989–997. [CrossRef]
47. Sheng, L.; Zhuang, S. New Insights Into the Role and Mechanism of Partial Epithelial-Mesenchymal Transition in Kidney Fibrosis. *Front. Physiol.* **2020**, *11*, 569322. [CrossRef] [PubMed]
48. Park Kim, J.; Lee, Y.J.; Bae, S.U.; Lee, H.W. Inflammatory bowel disease-associated intestinal fibrosis. *J. Pathol. Transl. Med.* **2023**, *57*, 60–66. [CrossRef] [PubMed]
49. Wenxiu, J.; Mingyue, Y.; Fei, H.; Yuxin, L.; Mengyao, W.; Chenyang, L.; Jia, S.; Hong, Z.; Shih, D.Q.; Targan, S.R.; et al. Effect and mechanism of TL1A expression on epithelial-mesenchymal transition during chronic colitis-related intestinal fibrosis. *Mediat. Inflamm.* **2021**, *2021*, 1–21. [CrossRef]
50. Li, M.; Luan, F.; Zhao, Y.; Hao, H.; Zhou, Y.; Han, W.; Fu, X. Epithelial-mesenchymal transition: An emerging target in tissue fibrosis. *Exp. Biol. Med.* **2016**, *241*, 1–13. [CrossRef] [PubMed]
51. Ortiz-Masia, D.; Gisbert-Ferrandiz, L.; Bauset, C.; Coll, S.; Mamie, C.; Scharl, M.; Esplugues, J.V.; Alós, R.; Navarro, F.; Cosín-Roger, J.; et al. Succinate activates EMT in intestinal epithelial cells through SUCNR1: A novel protagonist in fistula development. *Cell* **2020**, *9*, 1104. [CrossRef]
52. Blom, J.N.; Feng, Q. Cardiac repair by epicardial EMT: Current targets and a potential role for the primary cilium. *Pharmacol. Ther.* **2018**, *186*, 114–129. [CrossRef]
53. Zeisberg, E.M.; Tarnavski, O.; Zeisberg, M.; Dorfman, A.L.; McMullen, J.R.; Gustafsson, E.; Chandraker, A.; Yuan, X.; Pu, W.T.; Roberts, A.B.; et al. Endothelial-to-mesenchymal transition contributes to cardiac fibrosis. *Nat. Med.* **2007**, *13*, 952–961. [CrossRef] [PubMed]
54. Travers, J.; Kamal, F.; Robbins, J.; Yutzey, K.; Blaxall, B. Cardiac fibrosis: The fibroblast awakens. *Circ. Res.* **2016**, *118*, 1021–1040. [CrossRef] [PubMed]
55. Mia, M.M.; Singh, M.K. New Insights into Hippo/YAP Signalling in Fibrotic Diseases. *Cells* **2022**, *11*, 2065. [CrossRef] [PubMed]
56. Aharonov, A.; Shakked, A.; Umanski, K.; Savidor, A.; Genzelinakh, A.; Kain, D.; Lendengolts, D.; Revach, O.Y.; Morikawa, Y.; Dong, J.; et al. ERBB2 drives YAP activation and EMT-like processes during cardiac regeneration. *Nat. Cell Biol.* **2020**, *22*, 1346–1356. [CrossRef]
57. Ling, Y.; Cai, Z.; Jin, W.; Zhuang, X.; Kan, L.; Wang, F.; Ye, X. Silencing of c-Ski aug-ments TGF-b1-induced epithelial-mesenchymal transition in cardiomyocyte H9C2 cells. *Cardiol. J.* **2019**, *26*, 66–76. [CrossRef]
58. Bookman, A.A.M.; Shen, H.; Cook, R.J.; Bailey, D.; McComb, R.J.; Rutka, J.A.; Slomovic, A.R.; Caffery, B. Whole stimulated salivary flow: Correlation with the pathology of inflammation and damage in minor salivary gland biopsy specimens from patients with primary Sjögren's syndrome but not patients with sicca. *Arthritis Rheumatol.* **2011**, *63*, 2014–2020. [CrossRef]
59. Llamas-Gutierrez, F.J.; Reyes, E.; Martínez, B.; Hernández-Molina, G. Histopathological environment besides the focus score in Sjögren's syndrome. *Int. J. Rheum. Dis.* **2014**, *17*, 898–903. [CrossRef]
60. Altrieth, A.L.; O'Keefe, K.J.; Gellatly, V.A.; Tavarez, J.R.; Feminella, S.M.; Moskwa, N.L.; Cordi, C.V.; Turrieta, J.C.; Nelson, D.A.; Larsen, M. Identifying fibrogenic cells following salivary gland obstructive injury. *Front. Cell Dev. Biol.* **2023**, *11*, 1190386. [CrossRef]
61. Hall, B.E.; Zheng, C.; Swaim, W.D.; Cho, A.; Nagineni, C.N.; Eckhaus, M.A.; Flanders, K.C.; Ambudkar, I.S.; Baum, B.J.; Kulkarni, A.B. Conditional overexpression of TGF-beta1 disrupts mouse salivary gland development and function. *Lab. Investig.* **2010**, *90*, 543–555. [CrossRef]
62. Sisto, M.; Lorusso, L.; Ingravallo, G.; Ribatti, D.; Lisi, S. TGFβ1-Smad canonical and -Erk noncanonical pathways participate in interleukin-17-induced epithelial-mesenchymal transition in Sjögren's syndrome. *Lab. Investig.* **2020**, *100*, 824–836. [CrossRef]
63. Ciechomska, M.; O'Reilly, S. Epigenetic Modulation as a Therapeutic Prospect for Treatment of Autoimmune Rheumatic Diseases. *Mediat. Inflamm.* **2016**, *2016*, 9607946. [CrossRef]
64. Robertson, K.D. DNA methylation and human disease. *Nat. Rev. Genet.* **2005**, *6*, 597–610. [CrossRef]
65. Gopalakrishnan, S.; Van Emburgh, B.O.; Robertson, K.D. DNA methylation in development and human disease. *Mutat. Res.* **2008**, *647*, 30–38. [CrossRef]
66. Del Castillo Falconi, V.M.; Torres-Arciga, K.; Matus-Ortega, G.; Díaz-Chávez, J.; Herre-ra, L.A. DNA Methyltransferases: From Evolution to Clinical Applications. *Int. J. Mol. Sci.* **2022**, *23*, 8994. [CrossRef]
67. Liang, Y.; He, L.; Yuan, H.; Jin, Y.; Yao, Y. Association between RUNX3 promoter methylation and non-small cell lung cancer: A meta-analysis. *J. Thorac. Dis.* **2014**, *6*, 694–705. [PubMed]

68. Zhang, X.; Hu, M.; Lyu, X.; Li, C.; Thannickal, V.J.; Sanders, Y.Y. DNA methylation regulated gene expression in organ fibrosis. *Biochim. Biophys. Acta Mol. Basis Dis.* **2017**, *1863*, 2389–2397. [CrossRef]
69. Onufriev, A.V.; Schiessel, H. The nucleosome: From structure to function through physics. *Curr. Opin. Struct. Biol.* **2019**, *56*, 119–130. [CrossRef] [PubMed]
70. Hergeth, S.P.; Schneider, R. The H1 linker histones: Multifunctional proteins beyond the nucleosomal core particle. *EMBO Rep.* **2015**, *16*, 1439–1453. [CrossRef] [PubMed]
71. Cavalieri, V. The Expanding Constellation of Histone Post-Translational Modifications in the Epigenetic Landscape. *Genes* **2021**, *12*, 1596. [CrossRef]
72. Roth, S.Y.; Denu, J.M.; Allis, C.D. Histone acetyltransferases. *Annu. Rev. Biochem.* **2001**, *70*, 81–120. [CrossRef]
73. Audia, J.E.; Campbell, R.M. Histone Modifications and Cancer. *Cold Spring Harb. Perspect. Biol.* **2016**, *8*, a019521. [CrossRef]
74. Park, S.Y.; Kim, J.S. A short guide to histone deacetylases including recent progress on class II enzymes. *Exp. Mol. Med.* **2020**, *52*, 204–212. [CrossRef]
75. Liu, M.; Jiang, J.; Han, Y.; Shi, M.; Li, X.; Wang, Y.; Dong, Z.; Yang, C. Functional Characterization of the Lysine-Specific Histone Demethylases Family in Soybean. *Plants* **2022**, *11*, 1398. [CrossRef] [PubMed]
76. Zhang, Q.; Ramlee, M.K.; Brunmeir, R.; Villanueva, C.J.; Halperin, D.; Xu, F. Dynamic and distinct histone modifications modulate the expression of key adipogenesis regulatory genes. *Cell Cycle* **2012**, *11*, 4310–4322. [CrossRef]
77. Rougeulle, C.; Chaumeil, J.; Sarma, K.; Allis, C.D.; Reinberg, D.; Avner, P.; Heard, E. Differential histone H3 Lys-9 and Lys-27 methylation profiles on the X chromosome. *Mol. Cell. Biol.* **2004**, *24*, 5475–5484. [CrossRef] [PubMed]
78. Huang, S.; Litt, M.; Felsenfeld, G. Methylation of histone H4 by arginine methyltransferase PRMT1 is essential in vivo for many subsequent histone modifications. *Genes Dev.* **2005**, *19*, 1885–1893. [CrossRef]
79. Greer, E.L.; Shi, Y. Histone methylation: A dynamic mark in health, disease and inheritance. *Nat. Rev. Genet.* **2012**, *13*, 343–357. [CrossRef] [PubMed]
80. Lowndes, N.F.; Toh, G.W. DNA repair: The importance of phosphorylating histone H2AX. *Curr. Biol.* **2005**, *15*, R99–R102. [CrossRef]
81. Lau, A.T.; Lee, S.Y.; Xu, Y.M.; Zheng, D.; Cho, Y.Y.; Zhu, F.; Kim, H.G.; Li, S.Q.; Zhang, Z.; Bode, A.M.; et al. Phosphorylation of histone H2B serine 32 is linked to cell transformation. *J. Biol. Chem.* **2011**, *286*, 26628–26637. [CrossRef]
82. Cao, J.; Yan, Q. Histone ubiquitination and deubiquitination in transcription, DNA damage response, and cancer. *Front. Oncol.* **2012**, *2*, 26. [CrossRef] [PubMed]
83. Kaikkonen, M.U.; Lam, M.T.; Glass, C.K. Non-coding RNAs as regulators of gene expression and epigenetics. *Cardiovasc. Res.* **2011**, *90*, 430–440. [CrossRef] [PubMed]
84. Bhaskaran, M.; Mohan, M. MicroRNAs: History, biogenesis, and their evolving role in animal development and disease. *Vet. Pathol.* **2014**, *54*, 759–774. [CrossRef] [PubMed]
85. Wang, S.; Wu, W.; Claret, F.X. Mutual regulation of microRNAs and DNA methylation in human cancers. *Epigenetics* **2017**, *12*, 187–197. [CrossRef] [PubMed]
86. Lam, J.K.; Chow, M.Y.; Zhang, Y.; Leung, S.W. siRNA Versus miRNA as Therapeutics for Gene Silencing. *Mol. Ther. Nucleic Acids* **2015**, *4*, e252. [CrossRef] [PubMed]
87. Agrawal, N.; Dasaradhi, P.V.; Mohmmed, A.; Malhotra, P.; Bhatnagar, R.K.; Mukherjee, S.K. RNA interference: Biology, mechanism, and applications. *Microbiol. Mol. Biol. Rev.* **2003**, *67*, 657–685. [CrossRef] [PubMed]
88. Han, Y.N.; Li, Y.; Xia, S.Q.; Zhang, Y.Y.; Zheng, J.H.; Li, W. PIWI Proteins and PIWI-Interacting RNA: Emerging Roles in Cancer. *Cell. Physiol. Biochem.* **2017**, *44*, 1–20. [CrossRef] [PubMed]
89. Quinn, J.J.; Chang, H.Y. Unique features of long non-coding RNA biogenesis and function. *Nat. Rev. Genet.* **2016**, *17*, 47–62. [CrossRef] [PubMed]
90. Kazimierczyk, M.; Wrzesinski, J. Long Non-Coding RNA Epigenetics. *Int. J. Mol. Sci.* **2021**, *22*, 6166. [CrossRef]
91. Geisler, S.; Coller, J. RNA in unexpected places: Long non-coding RNA functions in di-verse cellular contexts. *Nat. Rev. Mol. Cell Biol.* **2013**, *14*, 699–712. [CrossRef]
92. Kopp, F.; Mendell, J.T. Functional classification and experimental dissection of long noncoding RNAs. *Cell* **2018**, *172*, 393–407. [CrossRef]
93. Aliperti, V.; Skonieczna, J.; Cerase, A. Long Non-Coding RNA (lncRNA) Roles in Cell Biology, Neurodevelopment and Neurological Disorders. *Non-Coding RNA* **2021**, *7*, 36. [CrossRef]
94. Huang, S.; Yang, B.; Chen, B.J.; Bliim, N.; Ueberham, U.; Arendt, T.; Janitz, M. The emerging role of circular RNAs in transcriptome regulation. *Genomics* **2017**, *109*, 401–407. [CrossRef]
95. Brasier, A.R.; Qiao, D.; Zhao, Y. The Hexosamine Biosynthetic Pathway Links Innate Inflammation with Epithelial-Mesenchymal Plasticity in Airway Remodeling. *Front. Pharmacol.* **2021**, *12*, 808735. [CrossRef]
96. Galle, E.; Thienpont, B.; Cappuyns, S.; Venken, T.; Busschaert, P.; Van Haele, M.; Van Cutsem, E.; Roskams, T.; van Pelt, J.; Verslype, C.; et al. DNA methylation-driven EMT is a common mechanism of resistance to various therapeutic agents in cancer. *Clin. Epigenetics* **2020**, *12*, 27. [CrossRef]
97. Marrs, J.A.; Andersson-Fisone, C.; Jeong, M.C.; Cohen-Gould, L.; Zurzolo, C.; Nabi, I.R.; Rodriguez-Boulan, E.; Nelson, W.J. Plasticity in epithelial cell phenotype: Modulation by expression of different cadherin cell adhesion molecules. *J. Cell Biol.* **1995**, *129*, 507–519. [CrossRef]

98. Shenoy, S. CDH1 (E-Cadherin) mutation and gastric cancer: Genetics, molecu-larmechanisms and guidelines for management. *Cancer Manag. Res.* **2019**, *11*, 10477–10486. [CrossRef]
99. Bücker, L.; Lehmann, U. CDH1 (E-cadherin) Gene Methylation in Human Breast Cancer: Critical Appraisal of a Long and Twisted Story. *Cancers* **2022**, *14*, 4377. [CrossRef]
100. Kandimalla, R.; van Tilborg, A.A.; Zwarthoff, E.C. DNA methylation-based biomarkers in bladder cancer. *Nat. Rev. Urol.* **2013**, *10*, 327–335. [CrossRef]
101. Bechtel, W.; McGoohan, S.; Zeisberg, E.M.; Müller, G.A.; Kalbacher, H.; Salant, D.J.; Müller, C.A.; Kalluri, R.; Zeisberg, M. Methylation determines fibroblast activation and fibrogenesis in the kidney. *Nat. Med.* **2010**, *16*, 544–550. [CrossRef]
102. Chang, Y.W.; Singh, K.P. Arsenic induces fibrogenic changes in human kidney epithelial cells potentially through epigenetic alterations in DNA methylation. *J. Cell. Physiol.* **2019**, *234*, 4713–4725. [CrossRef]
103. Tao, H.; Shi, P.; Zhao, X.D.; Xuan, H.Y.; Gong, W.H.; Ding, X.S. DNMT1 deregulation of SOCS3 axis drives cardiac fibroblast activation in diabetic cardiac fibrosis. *J. Cell. Physiol.* **2021**, *236*, 3481–3494. [CrossRef]
104. Dees, C.; Pötter, S.; Zhang, Y.; Bergmann, C.; Zhou, X.; Luber, M.; Wohlfahrt, T.; Karouzakis, E.; Ramming, A.; Gelse, K.; et al. TGF-β-induced epigenetic deregulation of SOCS3 facilitates STAT3 signalling to promote fibrosis. *J. Clin. Investig.* **2020**, *130*, 2347–2363. [CrossRef]
105. Liu, R.; Li, Y.; Zheng, Q.; Ding, M.; Zhou, H.; Li, X. Epigenetic modification in liver fibrosis: Promising therapeutic direction with significant challenges ahead. *Acta Pharm. Sin. B* **2023**, in press. [CrossRef]
106. Yu, F.; Lu, Z.; Chen, B.; Wu, X.; Dong, P.; Zheng, J. Salvianolic acid B-induced microRNA-152 inhibits liver fibrosis by attenuating DNMT1-mediated Patched1 methylation. *J. Cell. Mol. Med.* **2015**, *19*, 2617–2632. [CrossRef]
107. Avci, E.; Sarvari, P.; Savai, R.; Seeger, W.; Pullamsetti, S.S. Epigenetic Mechanisms in Parenchymal Lung Diseases: Bystanders or Therapeutic Targets? *Int. J. Mol. Sci.* **2022**, *23*, 546. [CrossRef]
108. Mo, Y.; Zhang, Y.; Wan, R.; Jiang, M.; Xu, Y.; Zhang, Q. miR-21 mediates nickel nanoparticle-induced pulmonary injury and fibrosis. *Nanotoxicology* **2020**, *14*, 1175–1197. [CrossRef]
109. Wu, C.H.; Tang, S.C.; Wang, P.H.; Lee, H.; Ko, J.L. Nickel-induced epithelial mesenchymal transition by reactive oxygen species generation and E-cadherin promoter hypermethylation. *J. Biol. Chem.* **2012**, *287*, 25292–25302. [CrossRef]
110. Ning, L.; Rui, X.; Bo, W.; Qing, G. The critical roles of histone deacetylase 3 in the pathogenesis of solid organ injury. *Cell Death Dis.* **2021**, *12*, 734. [CrossRef]
111. Ghoneim, M.; Fuchs, H.; Musselman, C. Histone tail conformations: A fuzzy affair with DNA. *Trends Biochem. Sci.* **2021**, *46*, 564–578. [CrossRef]
112. Rubio, K.; Molina-Herrera, A.; Pérez-González, A.; Hernández-Galdámez, H.V.; Pi-ña-Vázquez, C.; Araujo-Ramos, T.; Singh, I. EP300 as a Molecular Integrator of Fibrotic Transcriptional Programs. *Int. J. Mol. Sci.* **2023**, *24*, 12302. [CrossRef]
113. Lim, Y.; Jeong, A.; Kwon, D.H.; Lee, Y.U.; Kim, Y.K.; Ahn, Y.; Kook, T.; Park, W.J.; Kook, H. P300/CBP-Associated Factor Activates Cardiac Fibroblasts by SMAD2 Acetylation. *Int. J. Mol. Sci.* **2021**, *22*, 9944. [CrossRef]
114. Chu, L.; Xie, D.; Xu, D. Epigenetic Regulation of Fibroblasts and Crosstalk between Cardiomyocytes and Non-Myocyte Cells in Cardiac Fibrosis. *Biomolecules* **2023**, *13*, 1382. [CrossRef]
115. de Ruijter, A.J.; van Gennip, A.H.; Caron, H.N.; Kemp, S.; van Kuilenburg, A.B. Histone deacetylases (HDACs): Characterization of the classical HDAC family. *Biochem. J.* **2003**, *370*, 737–749. [CrossRef]
116. Claveria-Cabello, A.; Colyn, L.; Arechederra, M.; Urman, J.M.; Berasain, C.; Avila, M.A.; Fernandez-Barrena, M.G. Epigenetics in Liver Fibrosis: Could HDACs be a Therapeutic Target? *Cells* **2020**, *9*, 2321. [CrossRef]
117. Huang, S.K.; Scruggs, A.M.; Donaghy, J.; Horowitz, J.C.; Zaslona, Z.; Przybranowski, S.; White, E.S.; Peters-Golden, M. Histone modifications are responsible for decreased Fas expression and apoptosis resistance in fibrotic lung fibroblasts. *Cell Death Dis.* **2013**, *4*, e621. [CrossRef]
118. Sanders, Y.Y.; Hagood, J.S.; Liu, H.; Zhang, W.; Ambalavanan, N.; Thannickal, V.J. Histone deacetylase inhibition promotes fibroblast apoptosis and ameliorates pulmonary fibro-sis in mice. *Eur. Respir. J.* **2014**, *43*, 1448–1458. [CrossRef]
119. Wu, M.Z.; Tsai, Y.P.; Yang, M.H.; Huang, C.H.; Chang, S.Y.; Chang, C.C.; Teng, S.C.; Wu, K.J. Interplay between HDAC3 and WDR5 is essential for hypoxia-induced epithelial-mesenchymal transition. *Mol. Cell* **2011**, *43*, 811–822. [CrossRef]
120. Chen, L.; Alam, A.; Pac-Soo, A.; Chen, Q.; Shang, Y.; Zhao, H.; Yao, S.; Ma, D. Pretreatment with valproic acid alleviates pulmonary fibrosis through epithelial-mesenchymal transition inhibition in vitro and in vivo. *Lab. Investig.* **2021**, *101*, 1166–1175. [CrossRef]
121. Korfei, M.; Skwarna, S.; Henneke, I.; MacKenzie, B.; Klymenko, O.; Saito, S.; Ruppert, C.; von der Beck, D.; Mahavadi, P.; Klepetko, W.; et al. Aberrant expression and activity of histone deacetylases in sporadic idiopathic pulmonary fibrosis. *Thorax* **2015**, *70*, 1022–1032. [CrossRef]
122. Eckschlager, T.; Plch, J.; Stiborova, M.; Hrabeta, J. Histone Deacetylase Inhibitors as Anticancer Drugs. *Int. J. Mol. Sci.* **2017**, *18*, 1414. [CrossRef]
123. Barter, M.J.; Pybus, L.; Litherland, G.J.; Rowan, A.D.; Clark, I.M.; Edwards, D.R.; Cawston, T.E.; Young, D.A. HDAC-mediated control of ERK- and PI3K-dependent TGF-beta-induced extracellular matrix-regulating genes. *Matrix Biol.* **2010**, *29*, 602–612. [CrossRef]
124. Kamio, K.; Azuma, A.; Usuki, J.; Matsuda, K.; Inomata, M.; Nishijima, N.; Itakura, S.; Hayashi, H.; Kashiwada, T.; Kokuho, N.; et al. XPLN is modulated by HDAC inhibitors and negatively regulates SPARC expression by targeting mTORC2 in human lung fibroblasts. *Pulm. Pharmacol. Ther.* **2017**, *44*, 61–69. [CrossRef]

125. Zheng, Q.; Lei, Y.; Hui, S.; Tong, M.; Liang, L. HDAC3 promotes pulmonary fibrosis by activating NOTCH1 and STAT1 signalling and up-regulating inflammasome components AIM2 and ASC. *Cytokine* **2022**, *153*, 155842. [CrossRef]
126. Sangshetti, J.N.; Sakle, N.S.; Dehghan, M.H.G.; Shinde, D.B. Histone deacetylases as targets for multiple diseases. *Mini Rev. Med. Chem.* **2013**, *13*, 1005–1026. [CrossRef]
127. Xiong, R.; Geng, B.; Jiang, W.; Hu, Y.; Hu, Z.; Hao, B.; Li, N.; Geng, Q. Histone deacetylase 3 deletion in alveolar type 2 ep-ithelial cells prevents bleomycin-induced pulmonary fibrosis. *Clin. Epigenetics* **2023**, *15*, 182. [CrossRef]
128. Chen, F.; Gao, Q.; Zhang, L.; Ding, Y.; Wang, H.; Cao, W. Inhibiting HDAC3 (Histone Deacetylase 3) Aberration and the Resultant Nrf2 (Nuclear Factor Erythroid-Derived 2-Related Factor-2) Repression Mitigates Pulmonary Fibrosis. *Hypertension* **2021**, *78*, e15–e25. [CrossRef]
129. Jeong, S.H.; Son, E.S.; Lee, Y.E.; Kyung, S.Y.; Park, J.W.; Kim, S.H. Histone deacetylase 3 promotes alveolar epitheli-al–mesenchymal transition and fibroblast migration under hypoxic conditions. *Exp. Mol. Med.* **2022**, *54*, 922–931. [CrossRef]
130. Dai, Q.; Liu, J.; Du, Y.; Hao, X.; Ying, J.; Tan, Y.; He, L.Q.; Wang, W.M.; Chen, N. Histone deacetylase inhibitors attenuate P-aIgA1-induced cell proliferation and extracellular matrix synthesis in human renal mesangial cells in vitro. *Acta Pharmacol. Sin.* **2016**, *37*, 228–234. [CrossRef]
131. Choi, S.Y.; Kee, H.J.; Kurz, T.; Hansen, F.K.; Ryu, Y.; Kim, G.R.; Lin, M.Q.; Jin, L.; Piao, Z.H.; Jeong, M.H. Class I HDACs specifically regulate E-cadherin expression in human renal epithelial cells. *J. Cell. Mol. Med.* **2016**, *20*, 2289–2298. [CrossRef]
132. Ma, P.; Pan, H.; Montgomery, R.L.; Olson, E.N.; Schultz, R.M. Compensatory functions of histone deacetylase 1 (HDAC1) and HDAC2 regulate transcription and apoptosis during mouse oocyte development. *Proc. Natl. Acad. Sci. USA* **2012**, *109*, E481–E489. [CrossRef]
133. Yang, M.; Chen, G.; Zhang, X.; Guo, Y.; Yu, Y.; Tian, L.; Chang, S.; Chen, Z.K. Inhibition of class I HDACs attenuates renal interstitial fibrosis in a murine model. *Pharmacol. Res.* **2019**, *142*, 192–204. [CrossRef]
134. Chen, F.; Gao, Q.; Wei, A.; Chen, X.; Shi, Y.; Wang, H.; Cao, W. Histone Deacetylase 3 Aberration Inhibits Klotho Transcription and Promotes Renal Fibrosis. *Cell Death Differ.* **2021**, *28*, 1001–1012. [CrossRef]
135. Liu, L.; Lin, W.; Zhang, Q.; Cao, W.; Liu, Z. TGF-β induces miR-30d down-regulation and podocyte injury through Smad2/3 and HDAC3-associated transcriptional repression. *J. Mol. Med.* **2016**, *94*, 291–300. [CrossRef] [PubMed]
136. Yang, J.; Yin, S.; Bi, F.; Liu, L.; Qin, T.; Wang, H.; Cao, W. TIMAP repression by TGFβ and HDAC3-associated Smad signalling regulates macrophage M2 phenotypic phagocytosis. *J. Mol. Med.* **2017**, *95*, 273–285. [CrossRef]
137. Zhang, Y.; Zou, J.; Tolbert, E.; Zhao, T.C.; Bayliss, G.; Zhuang, S. Identification of histone deacetylase 8 as a novel therapeutic target for renal fibrosis. *FASEB J.* **2020**, *34*, 7295–7310. [CrossRef]
138. Shan, B.; Yao, T.P.; Nguyen, H.T.; Zhuo, Y.; Levy, D.R.; Klingsberg, R.C.; Tao, H.; Palmer, M.L.; Holder, K.N.; Lasky, J.A. Requirement of HDAC6 for transforming growth factor-beta1-induced epithelial-mesenchymal transition. *J. Biol. Chem.* **2008**, *283*, 21065–21073. [CrossRef]
139. Gu, S.; Liu, Y.; Zhu, B.; Ding, K.; Yao, T.P.; Chen, F.; Zhan, L.; Xu, P.; Ehrlich, M.; Liang, T.; et al. Loss of α-tubulin acetylation is associated with TGF-β-induced epithelial-mesenchymal transition. *J. Biol. Chem.* **2016**, *291*, 5396–5405. [CrossRef]
140. Choi, S.Y.; Piao, Z.H.; Jin, L.; Kim, J.H.; Kim, G.R.; Ryu, Y.; Lin, M.Q.; Kim, H.S.; Kee, H.J.; Jeong, M.H. Piceatannol attenuates renal fibrosis induced by unilateral ureteral obstruction via downregulation of histone deacetylase 4/5 or p38-MAPK signalling. *PLoS ONE* **2016**, *11*, e0167340. [CrossRef]
141. Liu, L.; Liu, F.; Guan, Y.; Zou, J.; Zhang, C.; Xiong, C.; Zhao, T.C.; Bayliss, G.; Li, X.; Zhuang, S. Critical roles of SMYD2 lysine methyltransferase in mediating renal fibroblast activation and kidney fibrosis. *FASEB J.* **2021**, *35*, e21715. [CrossRef]
142. Shi, Y.; Lan, F.; Matson, C.; Mulligan, P.; Whetstine, J.R.; Cole, P.A.; Casero, R.A.; Shi, Y. Histone demethylation mediated by the nuclear amine oxidase homolog LSD1. *Cell* **2004**, *119*, 941–953. [CrossRef]
143. Zhang, X.; Li, L.X.; Yu, C.; Nath, K.A.; Zhuang, S.; Li, X. Targeting lysine-specific demethylase 1A inhibits renal epithelial-mesenchymal transition and attenuates renal fibrosis. *FASEB J.* **2022**, *36*, e22122. [CrossRef]
144. Barski, A.; Cuddapah, S.; Cui, K.; Roh, T.Y.; Schones, D.E.; Wang, Z.; Wei, G.; Chepelev, I.; Zhao, K. High-resolution profiling of histone methylations in the human genome. *Cell* **2007**, *129*, 823–837. [CrossRef]
145. Zhuang, W.; Li, Z.; Dong, X.; Zhao, N.; Liu, Y.; Wang, C.; Chen, J. Schisandrin B inhibits TGF-β1-induced epithelial-mesenchymal transition in human A549 cells through epigenetic silencing of ZEB1. *Exp. Lung Res.* **2019**, *45*, 157–166. [CrossRef]
146. Schultz, D.C.; Ayyanathan, K.; Negorev, D.; Maul, G.G.; Rauscher, F.J., 3rd. SETDB1: A novel KAP-1-associated histone H3, lysine 9-specific methyltransferase that contributes to HP1-mediated silencing of euchromatic genes by KRAB zinc-finger proteins. *Genes Dev.* **2002**, *16*, 919–932. [CrossRef]
147. Liu, T.; Xu, P.; Ke, S.; Dong, H.; Zhan, M.; Hu, Q.; Li, J. Histone methyltransferase SETDB1 inhibits TGF-β-induced epithelial-mesenchymal transition in pulmonary fibrosis by regulating SNAI1 expression and the ferroptosis signalling pathway. *Arch. Biochem. Biophys.* **2022**, *715*, 109087. [CrossRef]
148. Nagaraja, S.S.; Subramanian, U.; Nagarajan, D. Radiation-induced H3K9 methylation on E-cadherin promoter mediated by ROS/Snail axis: Role of G9a signalling during lung epithelial-mesenchymal transition. *Toxicol. Vitro* **2021**, *70*, 105037. [CrossRef] [PubMed]
149. Zhang, Y.; Luo, G.; Zhang, Y.; Zhang, M.; Zhou, J.; Gao, W.; Xuan, X.; Yang, X.; Yang, D.; Tian, Z.; et al. Critical effects of long non-coding RNA on fibrosis diseases. *Exp. Mol. Med.* **2018**, *50*, e428. [CrossRef]

150. Ling, H.; Fabbri, M.; Calin, G.A. MicroRNAs and other non-coding RNAs as targets for anticancer drug development. *Nat. Rev. Drug Discov.* **2013**, *12*, 847–865. [CrossRef]
151. Wu, C.; Bao, S.; Sun, H.; Chen, X.; Yang, L.; Li, R.; Peng, Y. Noncoding RNAs regulating ferroptosis in cardiovascular diseases: Novel roles and therapeutic strategies. *Mol. Cell. Biochem.* **2023**, in press. [CrossRef]
152. Li, H.; Zhao, X.; Shan, H.; Liang, H. MicroRNAs in idiopathic pulmonary fibrosis: Involvement in pathogenesis and potential use in diagnosis and therapeutics. *Acta Pharm. Sin. B* **2016**, *6*, 531–539. [CrossRef]
153. Liang, H.; Xu, C.; Pan, Z.; Zhang, Y.; Xu, Z.; Chen, Y.; Li, T.; Li, X.; Liu, Y.; Huangfu, L.; et al. The antifibrotic effects and mechanisms of microRNA-26a action in idiopathic pulmonary fibrosis. *Mol. Ther.* **2014**, *22*, 1122–1133. [CrossRef]
154. Yu, X.; Zhai, R.; Hua, B.; Bao, L.; Wang, D.; Li, Y.; Yao, W.; Fan, H.; Hao, C. miR-let-7d attenuates EMT by targeting HMGA2 in silica-induced pulmonary fibrosis. *RSC Adv.* **2019**, *9*, 19355–19364. [CrossRef]
155. Wang, Y.C.; Liu, J.S.; Tang, H.K.; Nie, J.; Zhu, J.X.; Wen, L.L.; Guo, Q.L. miR-221 targets HMGA2 to inhibit bleomycin-induced pulmonary fibrosis by regulating TGF-β1/Smad3-induced EMT. *Int. J. Mol. Med.* **2016**, *38*, 1208–1216. [CrossRef]
156. Li, Y.; Shen, Z.; Jiang, X.; Wang, Y.; Yang, Z.; Mao, Y.; Wu, Z.; Li, G.; Chen, H. Mouse mesenchymal stem cell-derived exosomal miR-466f-3p reverses EMT process through inhibiting AKT/GSK3β pathway via c-MET in radiation-induced lung injury. *J. Exp. Clin. Cancer Res.* **2022**, *41*, 128. [CrossRef]
157. Wang, D.; Liu, Z.; Yan, Z.; Liang, X.; Liu, X.; Liu, Y.; Wang, P.; Bai, C.; Gu, Y.; Zhou, P.K. MiRNA-155-5p inhibits epithelium-to-mesenchymal transition (EMT) by targeting GSK-3β during radiation-induced pulmonary fibrosis. *Arch. Biochem. Biophys.* **2021**, *697*, 108699. [CrossRef]
158. Liang, X.; Yan, Z.; Wang, P.; Liu, Y.; Ao, X.; Liu, Z.; Wang, D.; Liu, X.; Zhu, M.; Gao, S.; et al. Irradiation Activates MZF1 to Inhibit miR-541-5p Expression and Promote Epithelial-Mesenchymal Transition (EMT) in Radiation-Induced Pulmonary Fibrosis (RIPF) by Upregulating Slug. *Int. J. Mol. Sci.* **2021**, *22*, 11309. [CrossRef] [PubMed]
159. Pang, X.; Shi, H.; Chen, X.; Li, C.; Shi, B.; Yeo, A.J.; Lavin, M.F.; Jia, Q.; Shao, H.; Zhang, J.; et al. miRNA-34c-5p targets Fra-1 to inhibit pulmonary fibrosis induced by silica through p53 and PTEN/PI3K/Akt signalling pathway. *Environ. Toxicol.* **2022**, *37*, 2019–2032. [CrossRef] [PubMed]
160. Li, C.; Song, L.; Zhang, Z.; Bai, X.X.; Cui, M.F.; Ma, L.J. MicroRNA-21 promotes TGF-β1-induced epithelial-mesenchymal transition in gastric cancer through up-regulating PTEN expression. *Oncotarget* **2016**, *7*, 66989–67003. [CrossRef] [PubMed]
161. Chen, J.; Zmijewska, A.; Zhi, D.; Mannon, R.B. Cyclosporine-mediated allograft fibrosis is associated with micro-RNA-21 through AKT signalling. *Transpl. Int.* **2015**, *28*, 232–245. [CrossRef]
162. Bao, H.; Hu, S.; Zhang, C.; Shi, S.; Qin, W.; Zeng, C.; Zen, K.; Liu, Z. Inhibition of miRNA-21 prevents fibrogenic activation in podocytes and tubular cells in IgA nephropathy. *Biochem. Biophys. Res. Commun.* **2014**, *444*, 455–460. [CrossRef] [PubMed]
163. Zhong, X.; Chung, A.C.; Chen, H.-Y.; Meng, X.-M.; Lan, H.Y. Smad3-mediated upregulation of miR-21 promotes renal fibrosis. *J. Am. Soc. Nephrol.* **2011**, *22*, 1668–1681. [CrossRef]
164. Davis, B.N.; Hilyard, A.C.; Nguyen, P.H.; Lagna, G.; Hata, A. Smad proteins bind a conserved RNA sequence to promote microRNA maturation by Drosha. *Mol. Cell* **2010**, *39*, 373–384. [CrossRef] [PubMed]
165. Liu, X.-J.; Hong, Q.; Wang, Z.; Yu, Y.-y.; Zou, X.; Xu, L.-h. MicroRNA21 promotes interstitial fibrosis via targeting DDAH1: A potential role in renal fibrosis. *Mol. Cell. Biochem.* **2016**, *411*, 181–189. [CrossRef] [PubMed]
166. Satoh, M.; Nagasu, H.; Morita, Y.; Yamaguchi, T.P.; Kanwar, Y.S.; Kashihara, N. Klotho protects against mouse renal fibrosis by inhibiting Wnt signalling. *Am. J. Physiol. Ren. Physiol.* **2012**, *303*, F1641–F1651. [CrossRef]
167. Koh, N.; Fujimori, T.; Nishiguchi, S.; Tamori, A.; Shiomi, S.; Nakatani, T.; Sugimura, K.; Kishimoto, T.; Kinoshita, S.; Kuroki, T. Severely reduced production of klotho in human chronic renal failure kidney. *Biochem. Biophys. Res. Commun.* **2001**, *280*, 1015–1020. [CrossRef]
168. Gluba-Sagr, A.; Franczyk, B.; Rysz-Górzyńska, M.; Ławiński, J.; Rysz, J. The Role of miRNA in Renal Fibrosis Leading to Chronic Kidney Disease. *Biomedicines* **2023**, *11*, 2358. [CrossRef]
169. Ai, K.; Zhu, X.; Kang, Y.; Li, H.; Zhang, L. miR-130a-3p inhibition protects against renal fibrosis in vitro via the TGF-β1/Smad pathway by targeting SnoN. *Exp. Mol. Pathol.* **2020**, *112*, 104358. [CrossRef]
170. Bai, L.; Lin, Y.; Xie, J.; Zhang, Y.; Wang, H.; Zheng, D. MiR-27b-3p inhibits the progression of renal fibrosis via suppressing STAT1. *Hum. Cell* **2021**, *34*, 383–393. [CrossRef]
171. Zhao, D.; Jia, J.; Shao, H. miR-30e targets GLIPR-2 to modulate diabetic nephropathy: In vitro and in vivo experiments. *J. Mol. Endocrinol.* **2017**, *59*, 181–190. [CrossRef]
172. Patel, V.; Noureddine, L. MicroRNAs and fibrosis. *Curr. Opin. Nephrol. Hypertens.* **2012**, *21*, 410. [CrossRef] [PubMed]
173. Tang, O.; Chen, X.-M.; Shen, S.; Hahn, M.; Pollock, C.A. MiRNA-200b represses transforming growth factor-β1-induced EMT and fibronectin expression in kidney proximal tubular cells. *Am. J. Physiol. Ren. Physiol.* **2013**, *304*, F1266–F1273. [CrossRef]
174. Li, Q.; Li, Z.; Lin, Y.; Che, H.; Hu, Y.; Kang, X.; Zhang, Y.; Wang, L.; Zhang, Y. High glucose promotes hepatic fibrosis via miR-32/MTA3-mediated epithelial-to-mesenchymal transition. *Mol. Med. Rep.* **2019**, *19*, 3190–3200. [CrossRef] [PubMed]
175. Liang, C.; Gao, L.; Liu, Y.; Liu, Y.; Yao, R.; Li, Y.; Xiao, L.; Wu, L.; Du, B.; Huang, Z.; et al. MiR-451 antagonist protects against cardiac fibrosis in streptozotocin-induced diabetic mouse heart. *Life Sci.* **2019**, *224*, 12–22. [CrossRef] [PubMed]
176. Liu, C.X.; Chen, L.L. Circular RNAs: Characterization, cellular roles, and applications. *Cell* **2022**, *185*, 2016–2034. [CrossRef]
177. Yao, W.; Li, Y.; Han, L.; Ji, X.; Pan, H.; Liu, Y.; Yuan, J.; Yan, W.; Ni, C. The CDR1as/miR-7/TGFBR2 Axis Modulates EMT in Silica-Induced Pulmonary Fibrosis. *Toxicol. Sci.* **2018**, *166*, 465–478. [CrossRef]

178. Li, R.; Wang, Y.; Song, X.; Sun, W.; Zhang, J.; Liu, Y.; Li, H.; Meng, C.; Zhang, J.; Zheng, Q.; et al. Potential regulatory role of circular RNA in idiopathic pulmonary fibrosis. *Int. J. Mol. Med.* **2018**, *42*, 3256–3268. [CrossRef]
179. Qi, F.; Li, Y.; Yang, X.; Wu, Y.; Lin, L.; Liu, X. Hsa_circ_0044226 knockdown attenuates progression of pulmonary fibrosis by inhibiting CDC27. *Aging* **2020**, *12*, 14808–14818. [CrossRef]
180. Yang, X.; Wang, J.; Zhou, Z.; Jiang, R.; Huang, J.; Chen, L.; Cao, Z.; Chu, H.; Han, B.; Cheng, Y.; et al. Silica-induced initiation of circular ZC3H4 RNA/ZC3H4 pathway promotes the pulmonary macrophage activation. *FASEB J.* **2018**, *32*, 3264–3277. [CrossRef]
181. Jiang, R.; Zhou, Z.; Liao, Y.; Yang, F.; Cheng, Y.; Huang, J.; Wang, J.; Chen, H.; Zhu, T.; Chao, J. The emerging roles of a novel CCCH-type zinc finger protein, ZC3H4, in silica-induced epithelial to mesenchymal transition. *Toxicol. Lett.* **2019**, *307*, 26–40. [CrossRef] [PubMed]
182. Statello, L.; Guo, C.J.; Chen, L.L.; Huarte, M. Gene regulation by long non-coding RNAs and its biological functions. *Nat. Rev. Mol. Cell Biol.* **2021**, *22*, 96–118. [CrossRef]
183. Sebastian-delaCruz, M.; Gonzalez-Moro, I.; Olazagoitia-Garmendia, A.; Castellanos-Rubio, A.; Santin, I. The Role of lncRNAs in Gene Expression Regulation through mRNA Stabilization. *Non-Coding RNA* **2021**, *7*, 3. [CrossRef]
184. Liu, Y.; Li, Y.; Xu, Q.; Yao, W.; Wu, Q.; Yuan, J.; Yan, W.; Xu, T.; Ji, X.; Ni, C. Long non-coding RNA-ATB promotes EMT during silica-induced pulmonary fibrosis by competitively binding miR-200c. *Biochim. Biophys. Acta Mol. Basis Dis.* **2018**, *1864*, 420–431. [CrossRef]
185. Xu, Q.; Cheng, D.; Liu, Y.; Pan, H.; Li, G.; Li, P.; Li, Y.; Sun, W.; Ma, D.; Ni, C. LncRNA-ATB regulates epithelial-mesenchymal transition progression in pulmonary fibrosis via sponging miR-29b-2-5p and miR-34c-3p. *J. Cell. Mol. Med.* **2021**, *25*, 7294–7306. [CrossRef]
186. Sun, H.; Chen, J.; Qian, W.; Kang, J.; Wang, J.; Jiang, L.; Qiao, L.; Chen, W.; Zhang, J. Integrated long non-coding RNA analyses identify novel regulators of epithelial-mesenchymal transition in the mouse model of pulmonary fibrosis. *J. Cell. Mol. Med.* **2016**, *20*, 1234–1246. [CrossRef]
187. Wang, J.; Xiang, Y.; Yang, S.X.; Zhang, H.M.; Li, H.; Zong, Q.B.; Li, L.W.; Zhao, L.L.; Xia, R.H.; Li, C.; et al. MIR99AHG inhibits EMT in pulmonary fibrosis via the miR-136-5p/USP4/ACE2 axis. *J. Transl. Med.* **2022**, *20*, 426. [CrossRef]
188. Zhan, H.; Chang, X.; Wang, X.; Yang, M.; Gao, Q.; Liu, H.; Li, C.; Li, S.; Sun, Y. LncRNA MEG3 mediates nickel oxide nanoparticles-induced pulmonary fibrosis via suppressing TGF-β1 expression and epithelial-mesenchymal transition process. *Environ. Toxicol.* **2021**, *36*, 1099–1110. [CrossRef]
189. Qian, W.; Cai, X.; Qian, Q.; Peng, W.; Yu, J.; Zhang, X.; Tian, L.; Wang, C. lncRNA ZEB1-AS1 promotes pulmonary fibrosis through ZEB1-mediated epithelial–mesenchymal transition by competitively binding miR-141-3p. *Cell Death Dis.* **2019**, *10*, 129. [CrossRef]
190. Dong, Z.; Li, S.; Wang, X.; Si, L.; Ma, R.; Bao, L.; Bo, A. lncRNA GAS5 restrains CCl$_4$-induced hepatic fibrosis by targeting miR-23a through the PTEN/PI3K/Akt signalling pathway. *Am. J. Physiol. Gastrointest. Liver Physiol.* **2019**, *316*, G539–G550. [CrossRef]
191. Chen, T.; Lin, H.; Chen, X.; Li, G.; Zhao, Y.; Zheng, L.; Shi, Z.; Zhang, K.; Hong, W.; Han, T. LncRNA Meg8 suppresses activation of hepatic stellate cells and epithelial-mesenchymal transition of hepatocytes via the Notch pathway. *Biochem. Biophys. Res. Commun.* **2020**, *521*, 921–927. [CrossRef]

Disclaimer/Publisher's Note: The statements, opinions and data contained in all publications are solely those of the individual author(s) and contributor(s) and not of MDPI and/or the editor(s). MDPI and/or the editor(s) disclaim responsibility for any injury to people or property resulting from any ideas, methods, instructions or products referred to in the content.

Review

Approach for Elucidating the Molecular Mechanism of Epithelial to Mesenchymal Transition in Fibrosis of Asthmatic Airway Remodeling Focusing on Cl⁻ Channels

Susumu Yoshie [1], Shigeyuki Murono [2] and Akihiro Hazama [1,*]

[1] Department of Cellular and Integrative Physiology, Graduate School of Medicine, Fukushima Medical University, Fukushima 960-1295, Japan
[2] Department of Otolaryngology Head and Neck Surgery, Graduate School of Medicine, Fukushima Medical University, Fukushima 960-1295, Japan
* Correspondence: hazama@fmu.ac.jp

Abstract: Airway remodeling caused by asthma is characterized by structural changes of subepithelial fibrosis, goblet cell metaplasia, submucosal gland hyperplasia, smooth muscle cell hyperplasia, and angiogenesis, leading to symptoms such as dyspnea, which cause marked quality of life deterioration. In particular, fibrosis exacerbated by asthma progression is reportedly mediated by epithelial-mesenchymal transition (EMT). It is well known that the molecular mechanism of EMT in fibrosis of asthmatic airway remodeling is closely associated with several signaling pathways, including the TGF-β1/Smad, TGF-β1/non-Smad, and Wnt/β-catenin signaling pathways. However, the molecular mechanism of EMT in fibrosis of asthmatic airway remodeling has not yet been fully clarified. Given that Cl⁻ transport through Cl⁻ channels causes passive water flow and consequent changes in cell volume, these channels may be considered to play a key role in EMT, which is characterized by significant morphological changes. In the present article, we highlight how EMT, which causes fibrosis and carcinogenesis in various tissues, is strongly associated with activation or inactivation of Cl⁻ channels and discuss whether Cl⁻ channels can lead to elucidation of the molecular mechanism of EMT in fibrosis of asthmatic airway remodeling.

Keywords: asthma; airway; fibrosis; epithelial to mesenchymal transition; Cl⁻ channel; cell volume

1. Introduction

Asthma is one of the most common chronic diseases in the world; around 300 million people globally are asthmatic [1]. Patients with asthma experience respiratory symptoms such as wheezing, shortness of breath, chest tightness, and coughing. Asthma is a consequence of the complex interaction of genetic and environmental factors, and its attack and exacerbation are caused by various triggers, such as allergens, cold air, and tobacco [2,3]. The first choice for asthma treatment is the use of inhaled corticosteroids. Low to moderate doses of inhaled corticosteroids can be used to control symptoms in a large number of asthmatic patients. However, for approximately 5–10% of asthmatic patients, even if they inhale the maximum dose of corticosteroids, their symptoms cannot be relieved due to poor steroid responsiveness and/or persistent invasion of inflammatory cells into the airways [4]. The quality of life of these patients has been significantly reduced by the physical burdens of asthma, such as the frequent exacerbations of symptoms and the decrease in respiratory function. They have limited treatment options available, and severe asthma with uncontrolled symptoms leads to death [5]. Therefore, elucidation of the mechanism of asthma is urgently needed.

Asthma is a chronic inflammatory disease that causes airway remodeling, which is characterized by subepithelial fibrosis, goblet cell metaplasia, basement membrane thickening, smooth muscle cell hyperplasia, and angiogenesis [6–12]. Among them, subepithelial fibrosis worsens as the asthma disease progresses, and epithelial-to-mesenchymal transition (EMT) has been suggested as an important source of fibroblasts that contribute to subepithelial fibrosis [13,14]. This EMT process leads to the migration of an increased number of mesenchymal cells into the subepithelial fibroblast layers, leading to subepithelial fibrosis.

EMT is a phenomenon in which non-motile epithelial cells transdifferentiate into motile mesenchymal cells. Epithelial cells tightly adhere to neighboring cells by forming cell adhesion apparatus such as tight junctions, adherence junctions, and desmosome junctions [15]. On the basal side, epithelial cells are attached to the basement membrane by hemidesmosome junctions. These junctional complexes are critical for maintaining both apical-basal and cytoskeletal polarity within epithelial cells. On the other hand, mesenchymal cells lack apical-basal and cytoskeletal polarity. They exhibit a spindle-like morphology and extend actin-rich membrane projections that facilitate cellular motility. These projections contain sheet-like membrane protrusions called lamellipodia, on the edge of which are spike-like extensions called filopodia [16]. Actin-rich invadopodia cause the degradation of the extracellular matrix, thereby facilitating cell invasion [16,17]. Epithelial cells express cell adhesion molecules such as E-cadherin and ZO-1, while mesenchymal cells lack such expression and exhibit reduced intercellular adhesion. Therefore, during EMT, both polarity and adhesion to surrounding cells and basement membranes are greatly diminished. As a result, they gain enhanced migration and invasion capabilities, leading to their transformation into mesenchymal cells. EMT plays a critical role in diverse in vivo activities, including fibrosis, cancer metastasis, early embryonic development, and tissue repair [18]. A large number of studies have investigated the mechanism of EMT and identified TGF-β1 as an inducer of EMT. When epithelial cells are stimulated with TGF-β1, both the Smad and non-Smad signaling pathways are activated, and consequently, the expression of transcription factors such as SNAIL1, Slug, ZEB1, ZEB2, and TWIST is induced, leading to EMT. In addition, various signaling pathways including the Wnt signaling pathway are also reported to be involved in EMT [19]. However, since the molecular mechanism of EMT has not yet been fully clarified, the detailed elucidation of its molecular mechanism will help to suppress subepithelial fibrosis and ultimately lead to novel therapeutic agents for severe asthma.

The Cl^- channels have been reported to play important roles in various physiological phenomena that occur in vivo by transporting Cl^-. There are various types of Cl^- channels that open or close in response to cell membrane potential, intracellular Ca^{2+} concentration, cell volume changes, ligands, and cAMP. Cl^- channels are expressed in all types of cells and are widely involved in basic cell functions, such as cell volume regulation [20–22], cell migration [23], cell proliferation [22], cell death [22,24], and production [25]. In cell volume regulation, the transport of Cl^-, K^+, and Na^+ via channel, transporter, and/or pump induces passive water flow, leading to cell volume changes such as cell swelling or shrinkage [20,21]. In particular, the volume-sensitive outwardly rectifying Cl^- channels (VSOR) are activated after cell swelling caused by hypotonicity. When the cell is swollen by hypotonicity, the extracellular efflux of Cl^- through VSOR and that of K^+ cause the efflux of water molecules from the cell, returning it to its original cell volume [20,21]. It has been reported that Cl^- channels that function as VSOR include LRRC8, Ca^{2+}-dependent Cl^- channels such as some TMEM16 members and tweety homologs (TTYH1, TTYH2, and TTYH3), and voltage-dependent Cl^- channels such as ClC-2 and ClC-3. These Cl^- channels are deeply involved in the regulation of cell volume. On the other hand, it has also been reported that Cl^- channels such as ClC-2, ClC-3, and some TMEM16A members are not associated with a role as VSOR [20–22,26–28]. Since cell size in each organ and tissue is determined by developmental programs and exhibits a unique cell volume, there is a high possibility that Cl^- channels regulate the cell volume and are involved in cell fate decisions during cell differentiation, transdifferentiation, and embryogenesis. In recent years, it has been reported that Cl^- channels are involved in cell differentiation [29–31], transdifferentiation [32,33], and EMT that causes carcinogenesis and fibrosis [34–36]. Thus, EMT caused by the regulation of Cl^- channels is thought to be closely related to changes in cell volume. We previously reported that dysfunction of an unspecified number of Cl$^-$ channels changes cell volume and promotes EMT in oral squamous cell carcinoma through the activation of the Wnt/β-catenin signaling pathway [34]. Lamouille et al. also reported that cell volume is changed by TGF-β1 during EMT [37]. These reports raise the possibility that Cl$^-$ channels and TGF-β1 regulate cell volume and cause EMT. Some studies have reported that EMT, which causes fibrosis, is also closely associated with Cl$^-$ channels. Herein, we review recent EMT studies focused on Cl$^-$ channels and discuss whether Cl$^-$ channels provide clues for elucidating the molecular mechanisms of EMT in the fibrosis of asthmatic airway remodeling.

2. The Molecular Mechanism of EMT in Fibrosis of Asthmatic Airway Remodeling

2.1. TGF-β Signaling Pathway

TGF-β has been reported to be a key cytokine in the pathogenesis of fibroproliferative diseases of the lungs, kidneys, or livers [38–40]. There are three isoforms (TGF-β1, -β2, and -β3) in mammals [41,42], and most studies to date have focused on TGF-β1, which is the most prominent isoform. TGF-β1 is known to be a potent inducer of EMT, leading to fibrosis in tissues such as the airways [43], kidneys [44], and lungs [45]. In asthmatic patients, the expression levels of TGF-β1 are increased in both the airway epithelium and the airway submucosa [46,47]. It has also been reported that eosinophils are a source of TGF-β1 [47,48]. TGF-β1 binds to the constitutively active kinase type II TGF-β receptor, recruits type I TGF-β receptor, and causes the phosphorylation of Smad2/3 [40,43,49–51]. The phosphorylated Smad2/3 then translocates to the nucleus to regulate the transcription of target genes, leading to the EMT or airway remodeling. The expression levels of Integrin αvβ6 in epithelial cells have been reported to be increased in response to inflammation stimuli, and activation of TGF-β1 and/or its expression levels are increased [51,52], leading to the EMT. These findings suggest that EMT is caused by a complex interaction between eosinophils, Integrin αvβ6, and TGF-β1. During EMT, epithelial cells acquire the mesenchymal phenotype via downregulation of the expression of epithelial markers such as E-cadherin and up-regulation of the expression of mesenchymal markers such as SNAIL1, which is a well-known master regulator of EMT, as well as cytoskeletal markers such as fibronectin, αSMA, and vimentin, which are essential for enhanced motility [53,54]. On the other hand, it has also been reported that TGF-β1 activates not only the Smad signaling pathway but also the non-Smad signaling pathway to induce EMT. For example, TGF-β1 is known to play an important role in asthmatic airway remodeling by stimulating the PI3K/AKT/GSK-3β signaling pathway. TGF-β1 activates PI3K and AKT, and the activation of AKT phosphorylates GSK-3β, resulting in the inactivation of GSK-3β. Since GSK-3β negatively regulates SNAIL1, inactivation of GSK-3β leads to the activation and nuclear translocation of SNAIL1, as well as the subsequent down-regulation of E-cadherin, leading to the EMT. Yadav et al. reported that the inhibition of aldose reductase prevents TGF-β1-induced EMT in airway epithelial cells and airway remodeling in ovalbumin (OVA)-induced asthmatic model mice via inhibiting the TGF-β1/PI3K/AKT/GSK-3β signaling pathway [55]. Additionally, Liu et al. reported that Lok, which is a traditional folk medicine widely used in northwest China for asthma, inhibits EMT in OVA-induced asthmatic model mice and TGF-β1-induced EMT in airway epithelial cells through inhibiting the PI3K/AKT/HIF-1α signaling pathway [56]. These results indicate that TGF-β1 activates the PI3K/AKT signaling pathway in a Smad-independent manner during EMT, which causes fibrosis in asthmatic airway remodeling. Although TGF-β1 has not been shown to exert an epigenetic gene control mechanism in asthmatic airway remodeling, TGF-β1 causes EMT by inducing the expression of DNA methyltransferases (DNMTs) such as DNMT1, DNMT3A, and DNMT3B in upper airway remodeling caused by chronic rhinosinusitis, indicating that TGF-β1 exerts an epigenetic gene control mechanism. Conversely, the DNMT inhibitor 5-Aza suppresses TGF-β1-induced EMT [57].

2.2. Wnt Signaling Pathway

The Wnt/β-catenin signaling pathway has also been reported to contribute to EMT in the fibrosis of asthmatic airway remodeling. Wnt binds to Frizzled receptors, leading to the inhibition of the downstream component GSK-3β. Since GSK-3β negatively regulates β-catenin, inhibition of GSK-3β leads to the cytosolic accumulation of β-catenin and its translocation to the nucleus and subsequent up-regulation of transcriptional factors such as SNAIL1, leading to EMT [13]. It has been reported that high expression levels of Wnt family proteins and β-catenin have been detected in the airways of asthmatic model mice. These elevated expression levels are characterized by airway remodeling, such as subepithelial fibrosis and airway smooth muscle hyperplasia. Suppression of β-catenin expression in the airways of asthmatic model mice attenuated airway remodeling, including subepithelial fibrosis [58]. Furthermore, mesenchymal stem cell (MSC) injection or MSC-derived exosome reduced EMT in the airways of asthmatic model rats through the inhibition of the Wnt/β-catenin signaling pathway [59]. Taken together, these findings demonstrate that the Wnt/β-catenin signaling pathway is highly expressed in asthmatic airways and regulates the development of fibrosis. Furthermore, it has also been reported that the specific gene expression induced by β-catenin depends on the recruitment of the transcriptional co-activator CREB binding protein (CBP) [60]. Moheimani et al. have shown that inhibition of complex formation between β-catenin and CBP due to the use of the small molecule inhibitor ICG-001 results in suppression of EMT in airway epithelial cells [61]. This suggests that activation of β-catenin/CBP complexes contributes to EMT in asthmatic airway epithelial cells.

2.3. Other Signaling Pathways

Various signaling pathways other than the TGF-β1 signaling pathway and Wnt signaling pathway have been reported to be associated with EMT in fibrosis of asthmatic airway remodeling. Zou et al. reported that the combination exposure of TGF-β1 and house dust mites induces EMT in airway epithelial cells via activation of the SHH signaling pathway [62]. Feng et al. demonstrated that IL-24 contributes to EMT in asthmatic model mice via the activation of the ERK1/2 and STAT3 signaling pathways and further revealed that IL24-mediated EMT is significantly alleviated by the inhibition of the ERK1/2 and STAT3 signaling pathways [63]. Furthermore, the RhoA/ROCK signaling pathway has also contributed to EMT in OVA-induced asthmatic model mice [64]. Wang et al. reported that inhibition of the crosstalk between the TGF-β1/Smad3 and Jagged1/Notch1 signaling pathways attenuates EMT in OVA-induced asthmatic model mice [65]. These data mean that complex synergistic interactions between the TGF-β1/Smad3 and Jagged1/Notch1 signaling pathways facilitate EMT. Thus, the signaling pathways of EMT that cause fibrosis in asthmatic airway remodeling are diverse and interact with one another to form complex networks. The molecular mechanisms of EMT in fibrosis of asthmatic airway remodeling have not yet been fully elucidated, as new signaling pathways and molecules continue to be identified. There is a high possibility that previously unreported molecules and signaling pathways may contribute to EMT.

3. The Roles of Cl^- Channels on Morphological Changes Such as Cell Differentiation and Transdifferentiation

Cl^- channels have been reported to play important roles in cell volume regulation [20–22], cell differentiation [29–31], and transdifferentiation [32,33]. In the cell volume regulation mechanism, the transport of Cl^-, K^+, and Na^+ causes a passive flow of water, resulting in changes in cell volume such as cell swelling or shrinkage [20,21]. In particular, it has been reported that VSOR, which is a key player in vertebrate cell volume regulation, is activated by hypotonic stress in order to regulate cellular volume. The extracellular efflux of Cl^- through VSOR and that of K^+ cause the efflux of water molecules from the cell, returning it to its original cell volume [20,21]. Recently, members of the LRRC8 (leucine-rich repeat-containing 8) family have been identified as the central contributors to VSOR [20,21].

Additionally, it has been reported that TTYHs serve as LRRC8-independent VSOR [26–28]. Therefore, Cl⁻ channels are considered to be deeply involved in morphological changes, such as cell differentiation and transdifferentiation, that are related to cell volume changes. In previous reports, most studies of cell differentiation and transdifferentiation triggered by regulation of Cl⁻ channels have not been investigated with a focus on cell volume changes; however, those studies have suggested that Cl⁻ channels regulate the specific signaling pathways, the transcriptional factors, and the concentration of intracellular Cl⁻, and that they contribute to control cell differentiation and transdifferentiation in a variety of cells. For example, Hou et al. have reported that ClC-2, which is a voltage-dependent Cl⁻ channel, may function as an important positive regulator in oligodendrocyte precursor cell differentiation through the regulation of various transcriptional factors such as YY1, MRF, Sox10, and Sip1 [66]. Wang H et al. and Wang D et al. have suggested that ClC-3, which is also a voltage-dependent Cl⁻ channel, mediates osteogenic differentiation via the Runx2 pathway [29,67]. Furthermore, Yin et al. reported that ClC-3 plays a role in cell volume regulation as a VSOR and is associated with the fibroblast-to-myofibroblast transition [68]. Chen et al. have shown that the extracellular efflux of Cl⁻ caused by LRRC8 is activated during myogenic differentiation at an early stage, and a moderate amount of intracellular Cl⁻ is necessary for myoblast fusion [69]. It has also been reported that LRRC8 promotes myoblast differentiation by regulating hyperpolarization and intracellular Ca^{2+} signals [70]. These findings indicate that LRRC8 may control cell volume and be closely involved in cell differentiation and transdifferentiation. Additionally, it has been reported that cystic fibrosis transmembrane conductance regulator (CFTR) regulates mesendoderm differentiation from embryonic stem (ES) cells via the β-catenin signaling pathway [31]. CFTR has also been shown to control intestinal lineage differentiation from mouse ES cells [71]. In airway epithelial cells, defective TMEM16A, a Ca^{2+}-dependent Cl⁻ channel, promotes differentiation of secretory cells and goblet cells, resulting in goblet cell metaplasia [72,73]. On the other hand, Scudieri et al. suggested that the upregulation of TMEM16A is associated with the differentiation of goblet cells [74]. Furthermore, ClC-2 and the chloride intracellular channel, CLIC4, control the transdifferentiation from fibroblast to myofibroblast via the TGF-β1 signaling pathway [32,33]. These results indicate that Cl⁻ channels are deeply involved in morphological changes such as cell differentiation and transdifferentiation.

4. Relationship between Cl⁻ Channels and EMT That Causes Carcinogenesis, Migration, and Invasion on Various Tissues

As mentioned above, Cl⁻ channels have been reported to be closely related to morphological changes such as cell differentiation and transdifferentiation. On the other hand, there are many reports that Cl⁻ channels are also involved in EMT, which is one of the morphological changes and causes carcinogenesis, migration, and invasion on various tissues. For example, the expression levels of CLCA1, which is one of the Ca^{2+}-dependent Cl⁻ channels, are significantly lower in colorectal cancer tissues than in normal tissues. Increased expression levels of CLCA1 in colorectal cancer suppress growth and metastasis via inhibition of the Wnt/β-catenin signaling pathway in vitro and in vivo, whereas inhibition of CLCA1 causes the opposite results [75]. These results indicate that CLCA1 controls the EMT process via the Wnt/β-catenin signaling pathway. Xin et al. have reported that the expression levels of CLCA2, which is also a Ca^{2+}-dependent Cl⁻ channel, are significantly reduced in cervical cancer cells. Furthermore, the overexpression of CLCA2 inhibits EMT via the inactivation of the p38/JNK/ERK signaling pathway and also inhibits the proliferation, migration, and invasion of cervical cancer cells [35]. Additionally, the expression levels of CLCA2 are significantly lower in nasopharyngeal carcinoma tissues than in noncancerous nasopharyngeal tissues. Overexpression of CLCA2 significantly suppresses EMT through inactivation of the FAK/ERK1/2 signaling pathway. In contrast, knockdown of CLCA2 has the opposite effect [76]. Furthermore, CLCA4 has also been reported to suppress EMT in esophageal cancer, colorectal cancer, liver cancer, and breast cancer by regulating specific signaling pathways such as the PI3K/AKT pathway [77–80]. These results indicate that the Ca^{2+}-dependent Cl^- channels CLCA1, CLCA2, and CLCA4 function as tumor suppressors. Recently, TTYHs, which have been reported to act as VSOR, have contributed to EMT, including migration and invasion on cholangiocarcinoma through the Wnt/β-catenin signaling pathway [81]. TMEM16A, a Ca^{2+}-dependent Cl^- channel other than CLCA and TTYHs, has been associated with cell proliferation, migration, invasion, and tumor growth in various cancers such as glioblastoma [82], breast cancer [83], head and neck cancer [84], and gastric cancer [85]. CFTR, which is a cAMP-dependent Cl^- channel, is expressed in various epithelial cells, and CFTR mutations cause cystic fibrosis. CFTR has also been reported to be involved in the EMT of cancer cells [86,87]. Downregulation of CFTR in breast cancer cells enhances malignant phenotypes and is deeply involved in a poor prognosis for breast cancer [87]. Additionally, rather than focusing on a specific Cl^- channel, the inhibition of an unspecified number of Cl^- channels has been reported to promote EMT in oral squamous cell carcinoma by changing the cell volume and regulating the Wnt/β-catenin signaling pathway [34]. Taken together, the results of these reports indicate that Cl^- channels are closely involved in EMT, which causes carcinogenesis, migration, and invasion. Thus, Cl^- channels raise the possibility of contributing to EMT, which causes fibrosis in asthmatic airway remodeling.

5. Relationship between Cl^- Channels and EMT That Causes Fibrosis in the Airways and Other Tissues

In the airways and kidneys, previous studies have reported a tight relationship between Cl^- channels and EMT (Table 1). Quaresma et al. found that cystic fibrosis tissues or cells expressing mutant CFTR display several signs of EMT activation, including destructured epithelial proteins, defective cell junctions, increased levels of mesenchymal markers, and EMT-associated transcriptional factors. Furthermore, they suggested that mutant CFTR-triggered EMT is mediated by the transcription factor TWIST1 [88]. Thus, it is possible that temporary dysfunction or downregulation of CFTR may also cause EMT in the fibrosis of asthmatic airway remodeling. Additionally, it has been reported that CFTR expression decreases in unilateral ureteral obstruction (UUO)-induced kidney fibrosis in mice and kidney fibrosis in humans. The downregulation or dysfunction of CFTR in renal epithelial cells is a key event leading to EMT and kidney fibrosis via the aberrant activation of the β-catenin signaling pathway. Conversely, the overexpression of CFTR alleviates fibrotic phenotypes in the UUO model [36]. These results suggest that CFTR dysfunction is a trigger for EMT that causes fibrosis in the airways and kidneys. Furthermore, it has been reported that LRRC8, which functions as VSOR, is involved in EMT in renal tubular epithelial cells derived from fetal kidneys. The inhibition or defectiveness of LRRC8 attenuates TGF-β1-induced EMT phenotypes such as migration [89]. This finding suggests that cell volume changes may actually be linked to EMT, and LRRC8 may also cause EMT in fibrosis of the kidneys and asthmatic airways. Yang et al. have shown that overexpression of the voltage-dependent Cl^- channel ClC-5 in the UUO-induced kidney fibrosis mouse model and TGF-β1-treated human renal tubular epithelial cells restores E-cadherin expression, reduces vimentin expression, and inhibits EMT. Conversely, the downregulation of ClC-5 in TGF-β1-treated human renal tubular epithelial cells increases the acetylation of NF-κB and the expression of an invasion-related gene, MMP9, and further potentiates EMT [90]. This suggests that ClC-5 is strongly involved in EMT, which causes kidney fibrosis, through the NF-κB/MMP9 signaling pathway.

Table 1. Reports on fibrosis focusing on Cl^- channels and EMT.

Author (Year)	Channel	Reference Number
Zhang et al. (2017)	CFTR (cAMP-dependent Cl^- channel)	[36]
Quaresma et al. (2020)	CFTR (cAMP-dependent Cl^- channel)	[88]
Yang et al. (2019)	ClC-5 (voltage-dependent Cl^- channel)	[90]

Several EMT studies focused on TRP channels and K$^+$ channels have been reported (Table 2). Wang et al. and Xu et al. have revealed that TRP channels, which are non-selective cation channels that transmit not only Na$^+$ and K$^+$, but also Ca^{2+} and Mg^{2+}, are associated with EMT in asthma and chronic obstructive pulmonary disease [91–93]. TRP channels are widely recognized to respond to temperature, nociceptive stimuli, touch, osmotic pressure, pheromones, and other stimuli from within and outside the cell [94]. In particular, activation of TRPC1 among TRP channels increases intracellular Ca^{2+} concentration, subsequently downregulates the expression of cytokeratin 8 and E-cadherin, and upregulates the expression of αSMA, leading to EMT in airway epithelial cells [91]. Additionally, Pu et al. have shown that TRPC1 promotes EMT in house dust mite (HDM)-induced asthmatic model mice through the activation of the STAT3/NF-κB signaling pathway. It has also been suggested that airway remodeling is alleviated through the suppression of the STAT3/NF-κB signaling pathway in TRPC1$^{-/-}$ mice even after HDM challenge [93]. In addition to TRP channels, KCa3.1, a calcium-dependent K$^+$ channel, has been proposed as a new target for fibrosis of the airways and lungs. The expression levels of KCa3.1 are increased in the airway epithelium of asthmatic patients compared with those of healthy people, and the KCa3.1 current is larger in asthmatic airway epithelial cells compared with healthy airway epithelial cells. Several features of TGF-β1-induced EMT have been reported to be suppressed by selective blockers of KCa3.1 [95]. These findings indicate that various anion and cation channels control the flow of their respective ions and are involved in EMT through the activation or inactivation of specific signaling pathways.

Table 2. Reports on fibrosis focusing on ion channels other than Cl$^-$ channels and EMT.

Author (Year)	Channel	Reference Number
Pu et al. (2007)	TRPC1 (non-selective cation channel)	[93]
Arthur et al. (2015)	KCa3.1 (calcium-dependent K$^+$ channel)	[95]

These reports indicate that various Cl$^-$ channels, including LRRC8, CFTR, and ClC-5, may be closely involved in EMT that causes fibrosis in asthmatic airway remodeling.

6. Conclusions and Future Directions

Asthmatic airways are characterized by airway remodeling such as subepithelial fibrosis, goblet cell metaplasia, basement membrane thickening, angiogenesis, and smooth muscle cell hyperplasia. In particular, elucidation of the mechanism of EMT that causes fibrosis is urgently needed, since exacerbation of asthma is linked to fibrosis. The molecular mechanism of EMT in the fibrosis of asthmatic airway remodeling has been reported to be caused by diverse signaling pathways, including the TGF-β1 signaling pathway and the Wnt signaling pathway. However, the EMT mechanism is driven by complex interactions with various molecules and signaling pathways, and the molecular mechanism of EMT has not yet been fully clarified.

Cl$^-$ channels have been reported to play an important role in cell volume regulation [20–22]. In the cell volume regulation mechanism, the transport of Cl$^-$, K$^+$, and Na$^+$ causes a passive flow of water, resulting in changes in cell volume such as cell swelling or shrinkage [20,21]. In particular, it has been reported that VSOR plays a key role in vertebrate cell volume regulation. In short, Cl$^-$ transport mediated by those Cl$^-$ channels causes a passive flow of water, resulting in changes in cell volume. Since cell size in each organ and tissue is determined by developmental programs and exhibits a unique cell volume, there is a high possibility that Cl$^-$ channels regulate the cell volume and are deeply involved in cell fate decisions such as cell differentiation, transdifferentiation, and EMT. In fact, there have been many reports that Cl$^-$ channels are closely associated with cell differentiation, transdifferentiation, and EMT, which causes carcinogenesis in various tissues. Additionally, Cl$^-$ channels such as CFTR and ClC-5 have been shown to be strongly involved in EMT leading to kidney fibrosis; therefore, there is a high possibility that Cl$^-$ channels are involved in the fibrosis of asthmatic airway remodeling. In the near future, it is expected that Cl⁻ channels will provide a new clue for elucidating the mechanism of EMT that causes fibrosis in asthmatic airway remodeling and may become a new target for suppressing fibrosis in patients with severe asthma. However, although it has been suggested that Cl⁻ channels control EMT through the activation or inactivation of specific signal pathways such as the Wnt/β-catenin signaling pathway, there are only a limited number of EMT studies focused on cell volume changes. We [34] and Lamouille et al. [37]. have suggested that the change in cell volume is associated with EMT. Furthermore, LRRC8, which functions as VSOR, is involved in EMT in renal tubular epithelial cells. On the other hand, in hearts, myocardial necrosis is caused after ischemia/reperfusion-induced myocardial infarction. Uramoto et al. showed that the activation of endogenous CFTR channels in myocardial cells suppresses myocardial necrosis [96]. This finding suggests that chloride ions are released from myocardial cells via activated CFTR and that the cell swelling caused by ischemia/reperfusion-induced myocardial infarction is inhibited, thereby providing protection against necrotic myocardial injury. These reports indicate that Cl⁻ channels control cell volume and are involved in several phenomena. Thus, elucidating the EMT mechanism from the perspective of cell volume changes with a focus on Cl⁻ channels may also provide a new clue for elucidating fibrosis in the airways of patients with severe asthma (Figure 1). Investigating the direction of Cl⁻ transport via Cl⁻ channels using patch clamps or Cl⁻-sensitive fluorescent dyes before and after EMT stimulation will be the first step in clarifying the relationship between EMT and cell volume changes associated with Cl⁻ transport. Furthermore, during the EMT process, monitoring cell size and differentiation status in real time will help clarify the relationship between cell volume and EMT. In addition, intentional increases or decreases in cell volume caused by hypo- or hyper-osmolarity conditions may promote or suppress EMT. Strict control of cell volume in some way has the potential to control not only EMT but also various morphological changes, including cell differentiation and transdifferentiation. If a relationship between EMT that causes fibrosis in asthmatic airway remodeling and cell volume regulation via Cl⁻ channels is revealed, cell volume regulation via Cl⁻ channels will lead to a new treatment for fibrosis in asthmatic airways.

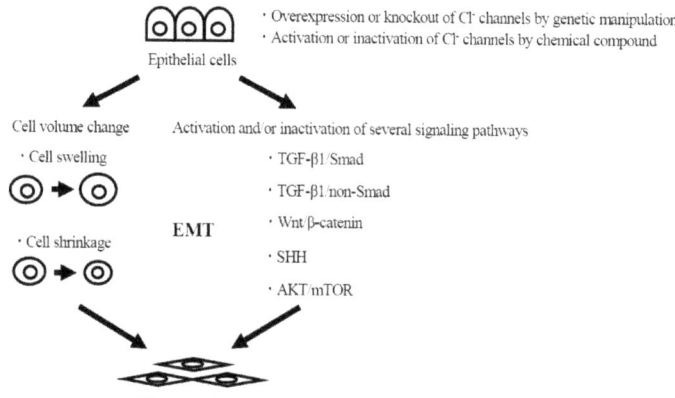

Figure 1. Elucidation of the EMT mechanism in fibrosis of asthmatic airway remodeling by Cl⁻ channel. Activation or inactivation of Cl⁻ channels by genetic manipulation or chemical compound changes the flow of Cl⁻ and water molecule, leading to cell swelling or cell shrinkage. Activation or inactivation of Cl⁻ channels by genetic manipulation or chemical compound also activates or inactivates specific signal pathways such as the TGF-β1/Smad signaling pathway, TGF-β1/non-Smad signaling pathway, Wnt/β-catenin signaling pathway, SHH signaling pathway, and AKT/mTOR signaling pathway. Consequently, Cl⁻ channels may contribute to EMT in fibrosis of asthmatic airway remodeling.

Author Contributions: S.Y., the first author, wrote and reviewed the manuscript. S.M., the co-author, conducted manuscript revision. A.H., the corresponding author, established the conception of the study, managed and supervised the entire process, and revised the manuscript. All authors have read and agreed to the published version of the manuscript.

Funding: This research received no external funding.

Institutional Review Board Statement: Not applicable.

Informed Consent Statement: Not applicable.

Data Availability Statement: Not applicable.

Conflicts of Interest: The authors declare no conflict of interest.

References

1. Porsbjerg, C.; Melén, E.; Lehtimäki, L.; Shaw, D. Asthma. *Lancet* **2023**, *401*, 858–873. [CrossRef] [PubMed]
2. Holgate, S.T.; Wenzel, S.; Postma, D.S.; Weiss, S.T.; Renz, H.; Sly, P.D. Asthma. *Nat. Rev. Dis. Primers* **2015**, *1*, 15025. [CrossRef] [PubMed]
3. Papi, A.; Brightling, C.; Pedersen, S.E.; Reddel, H.K. Asthma. *Lancet* **2018**, *391*, 783–800. [CrossRef] [PubMed]
4. Holgate, S.T.; Polosa, R. The mechanisms, diagnosis, and management of severe asthma in adults. *Lancet* **2006**, *368*, 780–793. [CrossRef] [PubMed]
5. Peters, S.P.; Ferguson, G.; Deniz, Y.; Reisner, C. Uncontrolled asthma: A review of the prevalence, disease burden and options for treatment. *Respir. Med.* **2006**, *100*, 1139–1151. [CrossRef]
6. Elias, J.A.; Zhu, Z.; Chupp, G.; Homer, R.J. Airway remodeling in asthma. *J. Clin. Invest.* **1999**, *104*, 1001–1006. [CrossRef]
7. Ordoñez, C.L.; Khashayar, R.; Wong, H.H.; Ferrando, R.; Wu, R.; Hyde, D.M.; Hotchkiss, J.A.; Zhang, Y.; Novikov, A.; Dolganov, G.; et al. Mild and moderate asthma is associated with airway goblet cell hyperplasia and abnormalities in mucin gene expression. *Am. J. Respir. Crit. Care Med.* **2001**, *163*, 517–523. [CrossRef]
8. Holgate, S.T. Epithelium dysfunction in asthma. *J. Allergy Clin. Immunol.* **2007**, *120*, 1233–1244. [CrossRef]
9. Liesker, J.J.; Ten Hacken, N.H.; Zeinstra-Smith, M.; Rutgers, S.R.; Postma, D.S.; Timens, W. Reticular basement membrane in asthma and COPD: Similar thickness, yet different composition. *Int. J. Chron. Obstruct. Pulmon. Dis.* **2009**, *4*, 127–135.
10. Bourdin, A.; Neveu, D.; Vachier, I.; Paganin, F.; Godard, P.; Chanez, P. Specificity of basement membrane thickening in severe asthma. *J. Allergy Clin. Immunol.* **2007**, *119*, 1367–1374. [CrossRef]
11. Joubert, P.; Hamid, Q. Role of airway smooth muscle in airway remodeling. *J. Allergy Clin. Immunol.* **2005**, *116*, 713–716. [CrossRef] [PubMed]
12. McDonald, D.M. Angiogenesis and remodeling of airway vasculature in chronic inflammation. *Am. J. Respir. Crit. Care Med.* **2001**, *164*, S39–S45. [CrossRef] [PubMed]
13. Pain, M.; Bermudez, O.; Lacoste, P.; Royer, P.J.; Botturi, K.; Tissot, A.; Brouard, S.; Eickelberg, O.; Magnan, A.; Moss, R.B. Tissue remodelling in chronic bronchial diseases: From the epithelial to mesenchymal phenotype. *Eur. Respir. Rev.* **2014**, *23*, 118–130. [CrossRef] [PubMed]
14. Hackett, T.L. Epithelial–mesenchymal transition in the pathophysiology of airway remodelling in asthma. *Curr. Opin. Allergy Clin. Immunol.* **2012**, *12*, 53–59. [CrossRef] [PubMed]
15. Huang, R.Y.; Guilford, P.; Thiery, J.P. Early events in cell adhesion and polarity during epithelial-mesenchymal transition. *J. Cell Sci.* **2012**, *125*, 4417–4422. [CrossRef] [PubMed]
16. Ridley, A.J. Life at the leading edge. *Cell* **2011**, *145*, 1012–1022. [CrossRef] [PubMed]
17. McNiven, M.A. Breaking away: Matrix remodeling from the leading edge. *Trends Cell Biol.* **2013**, *23*, 16–21. [CrossRef]
18. Yang, J.; Antin, P.; Berx, G.; Blanpain, C.; Brabletz, T.; Bronner, M.; Campbell, K.; Cano, A.; Casanova, J.; Christofori, G.; et al. Guidelines and definitions for research on epithelial-mesenchymal transition. *Nat. Rev. Mol. Cell Biol.* **2020**, *21*, 341–352. [CrossRef]
19. Lamouille, S.; Xu, J.; Derynck, R. Molecular mechanisms of epithelial-mesenchymal transition. *Nat. Rev. Mol. Cell Biol.* **2014**, *15*, 178–196. [CrossRef]
20. Jentsch, T.J. VRACs and other ion channels and transporters in the regulation of cell volume and beyond. *Nat. Rev. Mol. Cell Biol.* **2016**, *17*, 293–307. [CrossRef]
21. Okada, Y.; Sabirov, R.Z.; Sato-Numata, K.; Numata, T. Cell Death Induction and Protection by Activation of Ubiquitously Expressed Anion/Cation Channels. Part 1: Roles of VSOR/VRAC in Cell Volume Regulation, Release of Double-Edged Signals and Apoptotic/Necrotic Cell Death. *Front. Cell Dev. Biol.* **2020**, *8*, 614040. [CrossRef] [PubMed]
22. Guan, Y.Y.; Wang, G.L.; Zhou, J.G. The ClC-3 Cl$^-$ channel in cell volume regulation, proliferation and apoptosis in vascular smooth muscle cells. *Trends Pharmacol. Sci.* **2006**, *27*, 290–296. [CrossRef]
23. Schwab, A.; Fabian, A.; Hanley, P.J.; Stock, C. Role of ion channels and transporters in cell migration. *Physiol. Rev.* **2012**, *92*, 1865–1913. [CrossRef] [PubMed]

24. Maeno, E.; Ishizaki, Y.; Kanaseki, T.; Hazama, A.; Okada, Y. Normotonic cell shrinkage because of disordered volume regulation is an early prerequisite to apoptosis. *Proc. Natl. Acad. Sci. USA* **2000**, *97*, 9487–9492. [CrossRef] [PubMed]
25. Liu, C.L.; Shi, G.P. Calcium-activated chloride channel regulator 1 (CLCA1): More than a regulator of chloride transport and mucus production. *World Allergy Organ. J.* **2019**, *12*, 100077. [CrossRef] [PubMed]
26. Bae, Y.; Kim, A.; Cho, C.H.; Kim, D.; Jung, H.G.; Kim, S.S.; Yoo, J.; Park, J.Y.; Hwang, E.M. TTYH1 and TTYH2 serve as LRRC8A-independent volume-regulated anion channels in cancer cells. *Cells* **2019**, *8*, 562. [CrossRef]
27. Han, Y.E.; Kwon, J.; Won, J.; An, H.; Jang, M.W.; Woo, J.; Lee, J.S.; Park, M.G.; Yoon, B.E.; Lee, S.E.; et al. Tweety-homolog (*Ttyh*) Family Encodes the Pore-forming Subunits of the Swelling-dependent Volume-regulated Anion Channel (VRAC$_{swell}$) in the Brain. *Exp. Neurobiol.* **2019**, *28*, 183–215. [CrossRef]
28. Okada, Y.; Okada, T.; Sato-Numata, K.; Numata, T. Reexamination of the roles of LRRC8 and TTYH in the molecular identity of volume-sensitive outwardly rectifying anion channel VSOR. *J. Physiol. Sci.* **2020**, *70*, S150.
29. Wang, D.; Wang, H.; Gao, F.; Wang, K.; Dong, F. ClC-3 Promotes Osteogenic Differentiation in MC3T3-E1 Cell after Dynamic Compression. *J. Cell Biochem.* **2017**, *118*, 1606–1613. [CrossRef]
30. Lu, X.; Ding, Y.; Niu, Q.; Xuan, S.; Yang, Y.; Jin, Y.; Wang, H. ClC-3 chloride channel mediates the role of parathyroid hormone [1-34] on osteogenic differentiation of osteoblasts. *PLoS ONE* **2017**, *12*, e0176196. [CrossRef]
31. Liu, Z.; Guo, J.; Wang, Y.; Weng, Z.; Huang, B.; Yu, M.K.; Zhang, X.; Yuan, P.; Zhao, H.; Chan, W.Y.; et al. CFTR-β-catenin interaction regulates mouse embryonic stem cell differentiation and embryonic development. *Cell Death Differ.* **2017**, *24*, 98–110. [CrossRef] [PubMed]
32. Sun, L.; Dong, Y.; Zhao, J.; Yin, Y.; Zheng, Y. The CLC-2 Chloride Channel Modulates ECM Synthesis, Differentiation, and Migration of Human Conjunctival Fibroblasts via the PI3K/Akt Signaling Pathway. *Int. J. Mol. Sci.* **2016**, *17*, 910. [CrossRef] [PubMed]
33. Shukla, A.; Edwards, R.; Yang, Y.; Hahn, A.; Folkers, K.; Ding, J.; Padmakumar, V.C.; Cataisson, C.; Suh, K.S.; Yuspa, S.H. CLIC4 regulates TGF-β-dependent myofibroblast differentiation to produce a cancer stroma. *Oncogene* **2014**, *33*, 842–850. [CrossRef] [PubMed]
34. Kakinouchi, K.; Yoshie, S.; Tsuji, S.; Murono, S.; Hazama, A. Dysfunction of Cl⁻ channels promotes epithelial to mesenchymal transition in oral squamous cell carcinoma via activation of Wnt/β-catenin signaling pathway. *Biochem. Biophys. Res. Commun.* **2021**, *28*, 95–101. [CrossRef] [PubMed]
35. Xin, W.; Zhang, J.; Zhang, H.; Ma, X.; Zhang, Y.; Li, Y.; Wang, F. CLCA2 overexpression suppresses epithelial-to-mesenchymal transition in cervical cancer cells through inactivation of ERK/JNK/p38-MAPK signaling pathways. *BMC Mol. Cell Biol.* **2022**, *23*, 44. [CrossRef] [PubMed]
36. Zhang, J.T.; Wang, Y.; Chen, J.J.; Zhang, X.H.; Dong, J.D.; Tsang, L.L.; Huang, X.R.; Cai, Z.; Lan, H.Y.; Jiang, X.H.; et al. Defective CFTR leads to aberrant β-catenin activation and kidney fibrosis. *Sci. Rep.* **2017**, *7*, 5233. [CrossRef]
37. Lamouille, S.; Derynck, R. Cell size and invasion in TGF-beta-induced epithelial to mesenchymal transition is regulated by activation of the mTOR pathway. *J. Cell Biol.* **2007**, *178*, 437–451. [CrossRef]
38. Wynn, T.A. Common and unique mechanisms regulate fibrosis in various fibroproliferative diseases. *J. Clin. Investig.* **2007**, *117*, 524–529. [CrossRef]
39. Kim, K.K.; Kugler, M.C.; Wolters, P.J.; Robillard, L.; Galvez, M.G.; Brumwell, A.N.; Sheppard, D.; Chapman, H.A. Alveolar epithelial cell mesenchymal transition develops in vivo during pulmonary fibrosis and is regulated by the extracellular matrix. *Proc. Natl. Acad. Sci. USA* **2006**, *103*, 13180–13185. [CrossRef]
40. Guarino, M.; Tosoni, A.; Nebuloni, M. Direct contribution of epithelium to organ fibrosis: Epithelial-mesenchymal transition. *Hum. Pathol.* **2009**, *40*, 1365–1376. [CrossRef]
41. Kubiczkova, L.; Sedlarikova, L.; Hajek, R.; Sevcikova, S. TGF-β—An excellent servant but a bad master. *J. Transl. Med.* **2012**, *10*, 183. [CrossRef] [PubMed]
42. Yu, L.; Border, W.A.; Huang, Y.; Noble, N.A. TGF-beta isoforms in renal fibrogenesis. *Kidney Int.* **2003**, *64*, 844–856. [CrossRef] [PubMed]
43. Kamitani, S.; Yamauchi, Y.; Kawasaki, S.; Takami, K.; Takizawa, H.; Nagase, T.; Kohyama, T. Simultaneous stimulation with TGF-β1 and TNF-α induces epithelial mesenchymal transition in bronchial epithelial cells. *Int. Arch. Allergy Immunol.* **2011**, *155*, 119–128. [CrossRef]
44. Liu, Y. Epithelial to mesenchymal transition in renal fibrogenesis: Pathologic significance, molecular mechanism, and therapeutic intervention. *J. Am. Soc. Nephrol.* **2004**, *15*, 1–12. [CrossRef] [PubMed]
45. Kasai, H.; Allen, J.T.; Mason, R.M.; Kamimura, T.; Zhang, Z. TGF-beta1 induces human alveolar epithelial to mesenchymal cell transition (EMT). *Respir. Res.* **2005**, *6*, 56. [CrossRef] [PubMed]
46. Vignola, A.M.; Chanez, P.; Chiappara, G.; Merendino, A.; Pace, E.; Rizzo, A.; la Rocca, A.M.; Bellia, V.; Bonsignore, G.; Bousquet, J. Transforming growth factor-beta expression in mucosal biopsies in asthma and chronic bronchitis. *Am. J. Respir. Crit. Care Med.* **1997**, *156*, 591–599. [CrossRef] [PubMed]
47. Minshall, E.M.; Leung, D.Y.; Martin, R.J.; Song, Y.L.; Cameron, L.; Ernst, P.; Hamid, Q. Eosinophil-associated TGF-beta1 mRNA expression and airways fibrosis in bronchial asthma. *Am. J. Respir. Cell Mol. Biol.* **1997**, *17*, 326–333. [CrossRef] [PubMed]

48. Ijaz, T.; Pazdrak, K.; Kalita, M.; Konig, R.; Choudhary, S.; Tian, B.; Boldogh, I.; Brasier, A.R. Systems biology approaches to understanding Epithelial Mesenchymal Transition (EMT) in mucosal remodeling and signaling in asthma. *World Allergy Organ. J.* **2014**, *7*, 13. [CrossRef]
49. Sagara, H.; Okada, T.; Okumura, K.; Ogawa, H.; Ra, C.; Fukuda, T.; Nakao, A. Activation of TGF-beta/Smad2 signaling is associated with airway remodeling in asthma. *J. Allergy Clin. Immunol.* **2002**, *110*, 249–254. [CrossRef]
50. Wang, W.; Yang, Z.; Li, M.; Wang, Z.; Shan, Y.; Qu, Z. Six1 Promotes Epithelial-Mesenchymal Transition in Bronchial Epithelial Cells via the TGFβ1/Smad Signalling Pathway. *Int. Arch. Allergy Immunol.* **2021**, *182*, 479–488. [CrossRef]
51. Liu, W.; Sun, T.; Wang, Y. Integrin αvβ6 mediates epithelial-mesenchymal transition in human bronchial epithelial cells induced by lipopolysaccharides of Pseudomonas aeruginosa via TGF-β1-Smad2/3 signaling pathway. *Folia Microbiol.* **2020**, *65*, 329–338. [CrossRef] [PubMed]
52. Munger, J.S.; Huang, X.; Kawakatsu, H.; Griffiths, M.J.; Dalton, S.L.; Wu, J.; Pittet, J.F.; Kaminski, N.; Garat, C.; Matthay, M.A.; et al. The integrin alpha v beta 6 binds and activates latent TGF beta 1: A mechanism for regulating pulmonary inflammation and fibrosis. *Cell* **1999**, *96*, 319–328. [CrossRef]
53. Itoigawa, Y.; Harada, N.; Harada, S.; Katsura, Y.; Makino, F.; Ito, J.; Nurwidya, F.; Kato, M.; Takahashi, F.; Atsuta, R.; et al. TWEAK enhances TGF-β-induced epithelial-mesenchymal transition in human bronchial epithelial cells. *Respir. Res.* **2015**, *16*, 48. [CrossRef]
54. Doerner, A.M.; Zuraw, B.L. TGF-beta1 induced epithelial to mesenchymal transition (EMT) in human bronchial epithelial cells is enhanced by IL-1beta but not abrogated by corticosteroids. *Respir. Res.* **2009**, *10*, 100. [CrossRef] [PubMed]
55. Yadav, U.C.; Naura, A.S.; Aguilera-Aguirre, L.; Boldogh, I.; Boulares, H.A.; Calhoun, W.J.; Ramana, K.V.; Srivastava, S.K. Aldose reductase inhibition prevents allergic airway remodeling through PI3K/AKT/GSK3β pathway in mice. *PLoS ONE* **2013**, *8*, e57442. [CrossRef] [PubMed]
56. Liu, J.; Li, L.; Han, X.; Chen, Y.; Diao, J. Loke zupa decoction attenuates bronchial EMT-mediated airway remodelling in chronic asthma through the PI3K-Akt/HIF-1α signaling pathway. *Pharm. Biol.* **2023**, *1*, 1332–1342. [CrossRef] [PubMed]
57. Park, J.H.; Shin, J.M.; Yang, H.W.; Park, I.H. DNMTs Are Involved in TGF-β1-Induced Epithelial-Mesenchymal Transitions in Airway Epithelial Cells. *Int. J. Mol. Sci.* **2022**, *23*, 3003. [CrossRef] [PubMed]
58. Kwak, H.J.; Park, D.W.; Seo, J.Y.; Moon, J.Y.; Kim, T.H.; Sohn, J.W.; Shin, D.H.; Yoon, H.J.; Park, S.S.; Kim, S.H. The Wnt/β-catenin signaling pathway regulates the development of airway remodeling in patients with asthma. *Exp. Mol. Med.* **2015**, *47*, e198. [CrossRef]
59. Song, J.; Zhu, X.M.; Wei, Q.Y. MSCs reduce airway remodeling in the lungs of asthmatic rats through the Wnt/β-catenin signaling pathway. *Eur. Rev. Med. Pharmacol. Sci.* **2020**, *24*, 11199–11211.
60. Takemaru, K.I.; Moon, R.T. The transcriptional coactivator CBP interacts with beta-catenin to activate gene expression. *J. Cell Biol.* **2000**, *149*, 249–254. [CrossRef]
61. Moheimani, F.; Roth, H.M.; Cross, J.; Reid, A.T.; Shaheen, F.; Warner, S.M.; Hirota, J.A.; Kicic, A.; Hallstrand, T.S.; Kahn, M.; et al. Disruption of β-catenin/CBP signaling inhibits human airway epithelial-mesenchymal transition and repair. *Int. J. Biochem. Cell Biol.* **2015**, *68*, 59–69. [CrossRef] [PubMed]
62. Zou, Y.; Song, W.; Zhou, L.; Mao, Y.; Hong, W. House dust mite induces Sonic hedgehog signaling that mediates epithelial-mesenchymal transition in human bronchial epithelial cells. *Mol. Med. Rep.* **2019**, *20*, 4674–4682. [CrossRef] [PubMed]
63. Feng, K.N.; Meng, P.; Zou, X.L.; Zhang, M.; Li, H.K.; Yang, H.L.; Li, H.T.; Zhang, T.T. IL-37 protects against airway remodeling by reversing bronchial epithelial-mesenchymal transition via IL-24 signaling pathway in chronic asthma. *Respir. Res.* **2022**, *23*, 244. [CrossRef]
64. Huang, C.; Sun, Y.; Liu, N.; Zhang, Z.; Wang, X.; Lu, D.; Zhou, L.; Zhang, C. IL-27 attenuates airway inflammation and epithelial-mesenchymal transition in allergic asthmatic mice possibly via the RhoA/ROCK signalling pathway. *Eur. Cytokine Netw.* **2022**, *33*, 13–24. [PubMed]
65. Wang, Z.; Li, L.; Wang, C.; Piao, Y.; Jiang, J.; Li, L.; Yan, G.; Piao, H. Recombinant Pyrin Domain Protein Attenuates Airway Inflammation and Alleviates Epithelial-Mesenchymal Transition by Inhibiting Crosstalk Between TGFβ1 and Notch1 Signaling in Chronic Asthmatic Mice. *Front. Physiol.* **2020**, *11*, 559470. [CrossRef] [PubMed]
66. Hou, X.; Zhang, R.; Wang, J.; Li, Y.; Li, F.; Zhang, Y.; Zheng, X.; Shen, Y.; Wang, Y.; Zhou, L. CLC-2 is a positive modulator of oligodendrocyte precursor cell differentiation and myelination. *Mol. Med. Rep.* **2018**, *17*, 4515–4523. [CrossRef]
67. Wang, H.; Mao, Y.; Zhang, B.; Wang, T.; Li, F.; Fu, S.; Xue, Y.; Yang, T.; Wen, X.; Ding, Y.; et al. Chloride channel ClC-3 promotion of osteogenic differentiation through Runx2. *J. Cell Biochem.* **2010**, *111*, 49–58. [CrossRef]
68. Yin, Z.; Tong, Y.; Zhu, H.; Watsky, M.A. ClC-3 is required for LPA-activated Cl- current activity and fibroblast-to-myofibroblast differentiation. *Am. J. Physiol. Cell Physiol.* **2008**, *294*, C535–C542. [CrossRef]
69. Chen, L.; König, B.; Stauber, T. LRRC8 channel activation and reduction in cytosolic chloride concentration during early differentiation of C2C12 myoblasts. *Biochem. Biophys. Res. Commun.* **2020**, *532*, 482–488. [CrossRef]
70. Chen, L.; Becker, T.M.; Koch, U.; Stauber, T. The LRRC8/VRAC anion channel facilitates myogenic differentiation of murine myoblasts by promoting membrane hyperpolarization. *J. Biol. Chem.* **2019**, *294*, 14279–14288. [CrossRef]
71. Li, P.; Singh, J.; Sun, Y.; Ma, X.; Yuan, P. CFTR constrains the differentiation from mouse embryonic stem cells to intestine lineage cells. *Biochem. Biophys. Res. Commun.* **2019**, *510*, 322–328. [CrossRef] [PubMed]

72. He, M.; Wu, B.; Ye, W.; Le, D.D.; Sinclair, A.W.; Padovano, V.; Chen, Y.; Li, K.X.; Sit, R.; Tan, M.; et al. Chloride channels regulate differentiation and barrier functions of the mammalian airway. *eLife* **2020**, *14*, e53285. [CrossRef] [PubMed]
73. Centeio, R.; Cabrita, I.; Schreiber, R.; Kunzelmann, K. TMEM16A/F support exocytosis but do not inhibit Notch-mediated goblet cell metaplasia of BCi-NS1.1 human airway epithelium. *Front. Physiol.* **2023**, *14*, 1157704. [CrossRef] [PubMed]
74. Scudieri, P.; Caci, E.; Bruno, S.; Ferrera, L.; Schiavon, M.; Sondo, E.; Tomati, V.; Gianotti, A.; Zegarra-Moran, O.; Pedemonte, N.; et al. Association of TMEM16A chloride channel overexpression with airway goblet cell metaplasia. *J. Physiol.* **2012**, *590*, 6141–6155. [CrossRef]
75. Li, X.; Hu, W.; Zhou, J.; Huang, Y.; Peng, J.; Yuan, Y.; Yu, J.; Zheng, S. CLCA1 suppresses colorectal cancer aggressiveness via inhibition of the Wnt/beta-catenin signaling pathway. *Cell Commun. Signal.* **2017**, *15*, 38. [CrossRef]
76. Qiang, Y.Y.; Li, C.Z.; Sun, R.; Zheng, L.S.; Peng, L.X.; Yang, J.P.; Meng, D.F.; Lang, Y.H.; Mei, Y.; Xie, P.; et al. Along with its favorable prognostic role, CLCA2 inhibits growth and metastasis of nasopharyngeal carcinoma cells via inhibition of FAK/ERK signaling. *J. Exp. Clin. Cancer Res.* **2018**, *37*, 34. [CrossRef]
77. Song, X.; Zhang, S.; Li, S.; Wang, Y.; Zhang, X.; Xue, F. Expression of the CLCA4 Gene in Esophageal Carcinoma and Its Impact on the Biologic Function of Esophageal Carcinoma Cells. *J. Oncol.* **2021**, *2021*, 1649344. [CrossRef]
78. Chen, H.; Liu, Y.; Jiang, C.J.; Chen, Y.M.; Li, H.; Liu, Q.A. Calcium-Activated Chloride Channel A4 (CLCA4) Plays Inhibitory Roles in Invasion and Migration Through Suppressing Epithelial-Mesenchymal Transition via PI3K/AKT Signaling in Colorectal Cancer. *Med. Sci. Monit.* **2019**, *25*, 4176–4185. [CrossRef]
79. Liu, Z.; Chen, M.; Xie, L.K.; Liu, T.; Zou, Z.W.; Li, Y.; Chen, P.; Peng, X.; Ma, C.; Zhang, W.J.; et al. CLCA4 inhibits cell proliferation and invasion of hepatocellular carcinoma by suppressing epithelial-mesenchymal transition via PI3K/AKT signaling. *Aging (Albany NY)* **2018**, *10*, 2570–2584. [CrossRef]
80. Yu, Y.; Walia, V.; Elble, R.C. Loss of CLCA4 promotes epithelial-to-mesenchymal transition in breast cancer cells. *PLoS ONE* **2013**, *8*, e83943. [CrossRef]
81. Xue, W.; Dong, B.; Zhao, Y.; Wang, Y.; Yang, C.; Xie, Y.; Niu, Z.; Zhu, C. Upregulation of TTYH3 promotes epithelial-to-mesenchymal transition through Wnt/β-catenin signaling and inhibits apoptosis in cholangiocarcinoma. *Cell. Oncol.* **2021**, *44*, 1351–1361. [CrossRef] [PubMed]
82. Lee, Y.S.; Lee, J.K.; Bae, Y.; Lee, B.S.; Kim, E.; Cho, C.H.; Ryoo, K.; Yoo, J.; Kim, C.H.; Yi, G.S.; et al. Suppression of 14-3-3gamma-mediated surface expression of ANO1 inhibits cancer progression of glioblastoma cells. *Sci. Rep.* **2016**, *6*, 26413. [CrossRef] [PubMed]
83. Britschgi, A.; Bill, A.; Brinkhaus, H.; Rothwell, C.; Clay, I.; Duss, S.; Rebhan, M.; Raman, P.; Guy, C.T.; Wetzel, K.; et al. Calcium-activated chloride channel ANO1 promotes breast cancer progression by activating EGFR and CAMK signaling. *Proc. Natl. Acad. Sci. USA* **2013**, *110*, E1026–E1034. [CrossRef] [PubMed]
84. Shiwarski, D.J.; Shao, C.; Bill, A.; Kim, J.; Xiao, D.; Bertrand, C.A.; Seethala, R.S.; Sano, D.; Myers, J.N.; Ha, P.; et al. To "grow" or "go": TMEM16A expression as a switch between tumor growth and metastasis in SCCHN. *Clin. Cancer Res.* **2014**, *20*, 4673–4688. [CrossRef] [PubMed]
85. Liu, F.; Cao, Q.H.; Lu, D.J.; Luo, B.; Lu, X.F.; Luo, R.C.; Wang, X.G. TMEM16A overexpression contributes to tumor invasion and poor prognosis of human gastric cancer through TGF-beta signaling. *Oncotarget* **2015**, *6*, 11585–11599. [CrossRef]
86. Zhang, X.; Li, T.; Han, Y.N.; Ge, M.; Wang, P.; Sun, L.; Liu, H.; Cao, T.; Nie, Y.; Fan, D.; et al. miR-125b Promotes Colorectal Cancer Migration and Invasion by Dual-Targeting CFTR and CGN. *Cancers* **2021**, *13*, 5710. [CrossRef]
87. Zhang, J.T.; Jiang, X.H.; Xie, C.; Cheng, H.; Da, D.J.; Wang, Y.; Fok, K.L.; Zhang, X.H.; Sun, T.T.; Tsang, L.L.; et al. Downregulation of CFTR promotes epithelial-to-mesenchymal transition and is associated with poor prognosis of breast cancer. *Biochim. Biophys. Acta* **2013**, *1833*, 2961–2969. [CrossRef]
88. Quaresma, M.C.; Pankonien, I.; Clarke, L.A.; Sousa, L.S.; Silva, I.A.L.; Railean, V.; Doušová, T.; Fuxe, J.; Amaral, M.D. Mutant CFTR Drives TWIST1 mediated epithelial-mesenchymal transition. *Cell Death Dis.* **2020**, *11*, 920. [CrossRef]
89. Friard, J.; Corinus, A.; Cougnon, M.; Tauc, M.; Pisani, D.F.; Duranton, C.; Rubera, I. LRRC8/VRAC channels exhibit a noncanonical permeability to glutathione, which modulates epithelial-mesenchymal transition (EMT). *Cell Death Dis.* **2019**, *10*, 925. [CrossRef]
90. Yang, S.X.; Zhang, Z.C.; Bai, H.L. ClC-5 alleviates renal fibrosis in unilateral ureteral obstruction mice. *Hum. Cell.* **2019**, *32*, 297–305. [CrossRef]
91. Wang, J.; He, Y.; Yang, G.; Li, N.; Li, M.; Zhang, M. Transient receptor potential canonical 1 channel mediates the mechanical stress-induced epithelial-mesenchymal transition of human bronchial epithelial (16HBE) cells. *Int. J. Mol. Med.* **2020**, *46*, 320–330. [CrossRef] [PubMed]
92. Xu, F.; Liu, X.C.; Li, L.; Ma, C.N.; Zhang, Y.J. Effects of TRPC1 on epithelial mesenchymal transition in human airway in chronic obstructive pulmonary disease. *Medicine* **2017**, *96*, e8166. [CrossRef] [PubMed]
93. Pu, Q.; Zhao, Y.; Sun, Y.; Huang, T.; Lin, P.; Zhou, C.; Qin, S.; Singh, B.B.; Wu, M. TRPC1 intensifies house dust mite–induced airway remodeling by facilitating epithelial-to-mesenchymal transition and STAT3/NF-κB signaling. *FASEB J.* **2019**, *33*, 1074–1085. [CrossRef] [PubMed]
94. Venkatachalam, K.; Montell, C. TRP channels. *Annu. Rev. Biochem.* **2007**, *76*, 387–417. [CrossRef]

95. Arthur, G.K.; Duffy, S.M.; Roach, K.M.; Hirst, R.A.; Shikotra, A.; Gaillard, E.A.; Bradding, P. KCa3.1 K+ Channel Expression and Function in Human Bronchial Epithelial Cells. *PLoS ONE* **2015**, *10*, e0145259. [CrossRef]
96. Uramoto, H.; Okada, T.; Okada, Y. Protective role of cardiac CFTR activation upon early reperfusion against myocardial infarction. *Cell Physiol. Biochem.* **2012**, *30*, 1023–1038. [CrossRef]

Disclaimer/Publisher's Note: The statements, opinions and data contained in all publications are solely those of the individual author(s) and contributor(s) and not of MDPI and/or the editor(s). MDPI and/or the editor(s) disclaim responsibility for any injury to people or property resulting from any ideas, methods, instructions or products referred to in the content.

Review

Epithelial–Mesenchymal Transition Mechanisms in Chronic Airway Diseases: A Common Process to Target?

Angélique Mottais [1], Luca Riberi [2], Andrea Falco [2], Simone Soccal [2], Sophie Gohy [1,3,4,†] and Virginia De Rose [5,*,†]

1. Pole of Pneumology, ENT, and Dermatology, Institute of Experimental and Clinical Research, Université Catholique de Louvain, 1200 Brussels, Belgium; angelique.mottais@uclouvain.be (A.M.); sophie.gohy@uclouvain.be (S.G.)
2. Postgraduate School in Respiratory Medicine, University of Torino, 10124 Torino, Italy; luca.riberi@unito.it (L.R.); andrea.falco@unito.it (A.F.); simone.soccal@unito.it (S.S.)
3. Department of Pneumology, Cliniques Universitaires Saint-Luc, 1200 Brussels, Belgium
4. Cystic Fibrosis Reference Centre, Cliniques Universitaires Saint-Luc, 1200 Brussels, Belgium
5. Department of Molecular Biotechnology and Health Sciences, University of Torino, 10126 Torino, Italy
* Correspondence: virginia.derose@unito.it
† These authors contributed equally to this work.

Citation: Mottais, A.; Riberi, L.; Falco, A.; Soccal, S.; Gohy, S.; De Rose, V. Epithelial–Mesenchymal Transition Mechanisms in Chronic Airway Diseases: A Common Process to Target? *Int. J. Mol. Sci.* **2023**, *24*, 12412. https://doi.org/10.3390/ijms241512412

Academic Editors: Irmgard Tegeder, Margherita Sisto and Sabrina Lisi

Received: 15 May 2023
Revised: 30 July 2023
Accepted: 1 August 2023
Published: 3 August 2023

Copyright: © 2023 by the authors. Licensee MDPI, Basel, Switzerland. This article is an open access article distributed under the terms and conditions of the Creative Commons Attribution (CC BY) license (https://creativecommons.org/licenses/by/4.0/).

Abstract: Epithelial-to-mesenchymal transition (EMT) is a reversible process, in which epithelial cells lose their epithelial traits and acquire a mesenchymal phenotype. This transformation has been described in different lung diseases, such as lung cancer, interstitial lung diseases, asthma, chronic obstructive pulmonary disease and other muco-obstructive lung diseases, such as cystic fibrosis and non-cystic fibrosis bronchiectasis. The exaggerated chronic inflammation typical of these pulmonary diseases can induce molecular reprogramming with subsequent self-sustaining aberrant and excessive profibrotic tissue repair. Over time this process leads to structural changes with progressive organ dysfunction and lung function impairment. Although having common signalling pathways, specific triggers and regulation mechanisms might be present in each disease. This review aims to describe the various mechanisms associated with fibrotic changes and airway remodelling involved in chronic airway diseases. Having better knowledge of the mechanisms underlying the EMT process may help us to identify specific targets and thus lead to the development of novel therapeutic strategies to prevent or limit the onset of irreversible structural changes.

Keywords: epithelial-to-mesenchymal transition; fibrosis; chronic airway diseases; remodelling

1. Introduction

Epithelial-to-mesenchymal transition (EMT) is defined as a reversible cellular complex process, during which epithelial cells lose the classic epithelial markers, acquire a mesenchymal phenotype and are then able to produce extracellular matrix (ECM) [1]. This process, initially identified in embryonic development (type I), is also involved in tissue repair and fibrogenesis (type II) or tumour progression (type III) [2–5]. Several conditions, including chronic inflammation, can induce this molecular reprogramming, followed by the subsequent induction of self-perpetuating profibrotic, pro-remodelling and pro-neoplastic events [4]. EMT has been described in the pathophysiology of different lung diseases, such as lung cancer and its progression [6–8], interstitial lung diseases [9], asthma [10] and chronic obstructive pulmonary disease (COPD) [11]. More recently, it has also been described in other muco-obstructive lung diseases, such as cystic fibrosis (CF) [12] and non-cystic fibrosis bronchiectasis (NCFB).

Chronic obstructive airway diseases are characterised by different complex mechanisms underlying their airway changes and remodelling [13]. Although fibrosis does not play the key role observed in interstitial lung diseases [14], this process and its associated mechanisms, including EMT, appears to be crucial in pathological airway alterations leading to the remodelling, distortion and narrowing of the airways as well as the subsequent lung function impairment observed in these diseases.

The aim of this review is to focus on EMT molecular mechanisms and their associated airway remodelling involved in chronic airway diseases, including bronchial asthma and muco-obstructive lung diseases, such as CF, NCFB and COPD. We believe in fact that this is a quite neglected area of research and tried to highlight the fields on which future research should focus. The following MeSH terms were used for this review in the selected databases (PubMed and ResearchGate): epithelial mesenchymal transition; asthma; chronic obstructive pulmonary disease; cystic fibrosis; non-cystic fibrosis bronchiectasis; airway remodelling; airway fibrosis and airway diseases. We have excluded studies published in a language other than French, Italian or English. We have also excluded studies talking about cancer-associated EMT in lung disease.

2. Common Genetic and Biochemical Mechanisms of EMT

Three different types of EMT have been described depending on the biological or pathophysiological context [15,16]. Type I EMT is involved in embryogenesis and participates in organogenesis, such as the formation of the mesoderm during gastrulation and the delamination of the neural crest [17,18]. During embryogenesis, epithelial cells engage in several successive cycles of EMT and mesenchymal–epithelial transition. Type III EMT is associated with cancer progression [19]. Finally, type II EMT, which we will discuss in this review, is important for tissue regeneration and organ fibrosis. When exposed to pathogens or a toxic agent, the epithelium is damaged, and it will seek to rebuild itself. To repair the wound, basal cells migrate, proliferate and finally differentiate to reconstruct a pseudostratified respiratory epithelium. Chronic aggression and/or excessive inflammation leads to fibrogenesis. Fibrosis is characterised by the proliferation and activation of fibroblasts and myofibroblasts as well as the production of an excessive and abnormal extracellular matrix (ECM) [20]. Myofibroblasts originating from specialised epithelial cell populations during EMT have profibrotic and pro-inflammatory activity; these cells are involved in production of ECM components, as well as in matrix remodelling through the production of proteins, such as MMPs and the tissue inhibitors of metalloproteinases [21]. Myofibroblasts can originate from multiple other sources besides epithelial–mesenchymal transition, including differentiation from local fibroblasts, the recruitment of fibrocytes from bone marrow, endothelial–mesenchymal transition or from macrophages (i.e., macrophage–mesenchymal transition (MMT)) as well as from pericytes that are also able to differentiate into myofibroblasts [22].

Genetic and biochemical elements are common in the generation of different types of EMT, a process that is characterised by molecular reprogramming involving the loss of expression of epithelial proteins (e.g., E-cadherin, claudin and occludin) and the activation of mesenchymal genes, such as α-smooth muscle actin (α-SMA), N-cadherin and vimentin, which further lead to the loss of intercellular junctions (e.g., adherence junctions, tight junctions and desmosomes) and thus of epithelial cell polarity [23]. This molecular reprogramming is regulated by the activation of different transcription factors (TFs), such as SNAIL factors, Zinc-finger E-box-binding (ZEB) factors and TWIST factors [23,24].

The factors inducing EMT are multiple, and they are type- and organ-dependent. The pulmonary type II EMT results primarily from chronic lung inflammation, and it involves many pro-inflammatory actors (such as inflammatory cells, cytokines, chemokines and growth factors (GFs)) and different signal pathways, as summarised in Figure 1 [7,11].

The most widely described EMT-inducing factor is Transforming Growth Factor-β (TGF-β). The TGF-β family contains 33 TGF-β-related proteins, including 3 TGF-β isoforms (TGF-β1, 2 and 3). The proteins of the TGF-β family have multiple functions at the cellular and developmental level [25–27] as well as in the pathogenesis of diseases [28,29]. In the lung, TGF-β isoforms are involved in embryogenesis, including lung tissue development as well as in inflammation and tissue repair [30,31]. TGF-β1 is the first isoform to have been identified [25], and it has been found to play a crucial role in fibrotic lung diseases [26,32–37], particularly in airway remodelling [38,39]. Moreover, TGF-β1 is involved in the regulation of ECM components, as well as in fibroblast activation and in myofibroblast differentiation [40,41]. After synthesis, TGF-β dimerises via a disulphide bond and associates with the latency-associated peptide, which can attach to the latent TGF-β-binding protein [42]. These latent complexes are in a biologically inactive form that can be cleaved by various proteases to release active TGF-β [43]. The binding of TGF-β to its receptor (TGFBR1/2) can trigger two signalling pathways, the canonical SMAD-dependent pathway and the non-canonical pathways, which include phosphoinositide 3-kinase (PI3K/AKT) and mitogen-activated protein kinase (RAS/MAPK) (Figure 1). These two signal cascades lead to the repression of genes involved in the expression of epithelial phenotype and the activation of the expression of a mesenchymal phenotype.

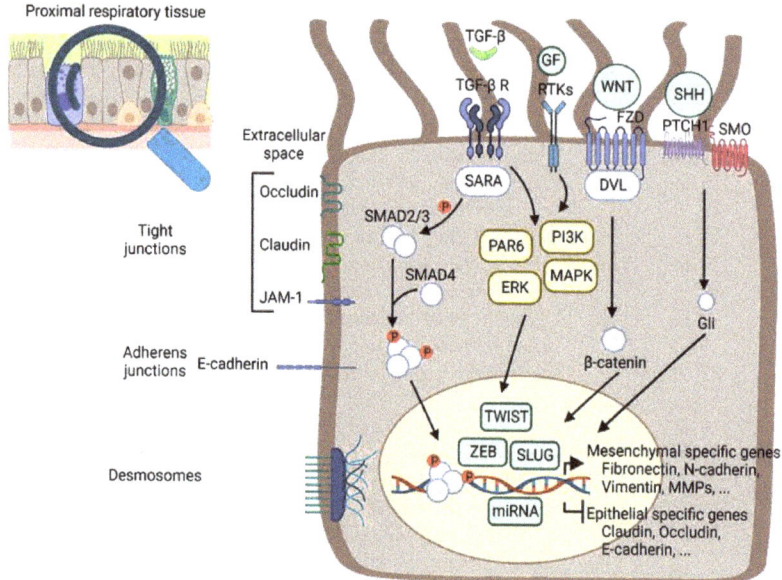

Figure 1. Signal pathways involved in type II EMT. TGF-β, other GF, WNT and SHH are particularly known as transcriptional-factor EMT inducers (e.g., SNAIL or ZEB, and miRNAs) that repress genes encoding epithelium-specific proteins, such as components of tight junctions (occludin, claudin and JAM-1), adhesion junctions and desmosomes, and activate mesenchyme-specific genes, such as N-cadherin and vimentin. These transcriptional changes lead to the reorganisation of the cytoskeleton and the production of ECM (created with BioRender.com and inspired by [7,11]). ERK, extracellular signal-regulated kinases; FZD, fizzled; GF, growth factor; JAM, junctional adhesion molecule; MAPK, mitogen-activated protein kinase; miRNA, microRNA; MMP, matrix metalloproteinase; PI3K, phosphoinositide 3-kinase; PTCH, Patched protein; RTK, receptor tyrosine kinase; SARA, SMAD anchor for receptor activation; SHH, Sonic Hedgehog; SMO, Smoothened protein; TGF, Transforming Growth Factor-β; WNT, wingless/integrated; ZEB, Zinc-finger E-box-binding.

Other signalling pathways activating TFs, such as the WNT/β-catenin [44,45] or Sonic Hedgehog (SHH) pathways may also be involved in the induction of EMT [46–49]. The binding of WNT to its receptor induces the release of β-catenin, which, once stabilised, migrates into the nucleus to regulate target genes. The binding of the SHH ligand to the Patched protein releases the Smoothened protein from the repressive action of the latter, resulting in the dissociation of the cytoplasmic complex, thus preventing the degradation of TFs that activate or repress the target genes.

Other GFs, such as the Platelet-derived Growth Factor (PDGF) [50], TGF-α [51–53], Fibroblasts Growth Factor-β (FGF-β) [54], Epidermal Growth Factor (EGF) [55–57] and Connective Tissue Growth Factor; and mediators, such as Transglutaminase-2 (TG2) [58–60] or reactive oxygen species (ROS) [61,62], participate in the EMT process and activate different pathways, such as JAK/STAT and PI3K/AKT, by binding to receptor tyrosine kinases. GFs also participate in the recruitment of neutrophils [63] and in the transformation of monocytes into macrophages. These cells produce cytokines (e.g., Tumour Necrosis Factor (TNF)-α, interleukin (IL)-1, IL-6 and GFs (e.g., TGF-β and PDGF)) that participate in the proliferation of fibroblasts and their differentiation into myofibroblasts [64].

3. Asthma

Asthma is a heterogeneous disease with a worldwide prevalence characterised by variable airflow obstruction, chronic airway inflammation and remodelling, and airway hyper-responsiveness [65].

Despite an increasing research interest in mechanisms of this disease and improved knowledge in this field, many questions still remain open. In particular, the processes leading to structural changes and airway remodelling are still not fully defined. The term airway remodelling summarises a wide variety of structural changes in both large and small airways of asthmatic patients; these include epithelium disruption [66], thickening of the reticular basement membrane (RBM) associated with subepithelial fibrosis, goblet metaplasia, angiogenesis, increased airway smooth muscle (ASM) mass [67–69] and ECM deposition [70]. Airway remodelling is progressively becoming a relevant research topic as it is present in all asthma phenotypes independently of disease severity, and it is also linked to progressive loss of lung function, airway hyper-responsiveness and the greater need for medications [71–73]. Furthermore, although airway remodelling has been considered a progressive consequence of chronic inflammation [74–76], more recent data have shown that features of remodelling are already observed in asthmatic children who are 2–4 years old, even in the absence of atopic inflammation or before clinical manifestations of the disease [77–79]. Broekema et al. studied a cohort of adult asthmatic patients followed up for 3 years and documented that the extent of structural changes due to airway remodelling remained unchanged independently of symptoms control or medication use [80].

EMT is believed to play an important contributory role to airway remodelling in asthma [10,81,82]; this process leads to the migration of increased numbers of mesenchymal cells into the subepithelial tissue and to enhanced production of ECM, contributing to airway wall fibrosis [83]. However, the potential drivers of EMT as well as its causal role in airway remodelling in asthma are a controversial issue, and some discrepancy exists in the available data.

3.1. Chronic Inflammation, EMT and Airway Remodelling

The chronic inflammatory process in response to persistent environmental triggers is a crucial driver of epithelial barrier injury and subsequent altered repair in asthma, leading to the loss of barrier function. EMT is one of the processes that have been suggested to contribute to the disruption of the epithelial barrier, however the intrinsic mechanisms of this process are not yet clearly defined. Furthermore, the evidence suggests that additional factors besides inflammatory insult may contribute to barrier dysfunction in asthma and trigger the EMT process (in vitro and in vivo evidence is summarised in Table 1). Lofredo et al. [84] used gene expression microarray datasets by several independent groups to gain insights into the processes involved in epithelial barrier dysfunction in asthma. Using this approach, they reported little evidence of classical EMT markers expression but identified a novel suite of potential biomarkers involved in epithelial–mesenchymal signal dysregulation, including Ephrin B2, FGF receptor 1, FGF receptor 2, Insulin Receptor, Insulin Receptor Substrate 2, NOTCH2, TLE family member 1 and neurotrophic receptor tyrosine kinase 2; this signature of asthma was present in mild to severe disease and seems to progress with disease severity. These findings thus suggest that factors others than

EMT driven by chronic inflammation might be involved in epithelial barrier disruption in asthma. Following the chronic aggression by noxious external agents, the airway epithelium is able to modulate the inflammatory/immune responses by interacting with inflammatory/immune cells and by releasing several cytokines and GFs, such as TGF-β, EGF, FGF and PDGF [85,86], thus contributing to structural alterations of the airway walls and the remodelling process. Subepithelial fibrosis, with an increase in fibroblasts in the airway epithelium, myofibroblasts hyperplasia and ECM deposition are major pathological features of the remodelling process in bronchial asthma. As already discussed, EMT is a relevant mechanism in fibroblast and myofibroblast proliferation and activation.

TGF-β1, a key molecule driving this process, induces the typical changes in cells from asthmatic patients [87,88]. Increased TGF-β1 levels have been observed in bronchoalveolar lavage fluid (BALF) and bronchial biopsies from asthmatic patients [89]. Interestingly, the Th2 cytokine IL-4 and the Th17-derived IL-17A, which are crucial mediators of the inflammatory process in severe asthma, also represent the two major cytokines inducing EMT and airway remodelling through TGF-β1 expression [90,91]. TGF-β1 synergises with IL-4 and IL-17 in suppressing E-cadherin expression and inducing α-SMA, vimentin and fibronectin expression in epithelial cells [92]. This mediator has also been shown to synergise with house dust mite (HDM) extracts in enhancing the expression of EMT markers [93,94].

The role of the EMT process in airway remodelling and subepithelial fibrosis in asthma has been further confirmed by the study of Johnson et al. [95], that in a transgenic murine model sensitised to HDM showed that large airway epithelial cells progressively lost their typical features, acquiring the expression of mesenchymal markers, such as vimentin, α-SMA and type I pro-collagen. An increased expression and nuclear translocation of SNAIL1, a transcriptional factor that is a potent inducer of EMT, was also observed in the airway epithelial cells of HDM-exposed mice. Interestingly, fate-mapping studies documented that epithelial cells migrated into the subepithelial compartment of the airway wall [95].

As already highlighted, GFs, such as TGF-β1, are also involved in EMT induced by inflammatory cells, and it has been reported in several studies that TGF-β1 expression correlates with the number of eosinophils and the degree of airway remodelling in bronchial asthma [96,97]. To evaluate the effect of eosinophils on EMT, Yasukawa et al. [98] carried out in vitro and in vivo studies. In particular, they assessed EMT in mice instilled with bone-marrow-derived eosinophils and inhuman bronchial epithelial cells (BEC) co-cultured with eosinophils. The intratracheal instillation of eosinophils induced enhanced bronchial inflammation, EMT features and fibrosis associated with increased concentration of GFs; interestingly, the instillation of eosinophils pre-treated with TGF-β1 siRNA was associated with reduced airway wall fibrosis. When co-cultured with BECs, eosinophils induced EMT in cells that were associated with enhanced TGF-β1 expression and SMAD3 phosphorylation. These findings suggest that eosinophils are capable of inducing fibrotic changes and EMT in airway epithelial cells, thus contributing to airway remodelling in asthma.

In an in vitro study using BECs, IL-1β was able to induce a decreased expression of E-cadherin, associated with an increased expression of some ECM component such as tenascin C; these effects were enhanced when the cells were co-stimulated with both TGF-β and IL-1β, further supporting the concept that the inflammatory context is crucially involved in the EMT process in asthma [99]. However, it is worth highlighting that in the same study, glucocorticoids were not able to induce any effect on EMT. Thus, further studies are needed to define the role of inflammatory pathways resistant to steroids in EMT.

IL-24 is a pleiotropic cytokine, member of the IL-10 family, that has been implicated in the induction of tissue fibrosis and remodelling [100], and increased levels of this cytokine have been reported in nasal secretions and in the induced sputum of asthmatic patients [101]. Very recently, Feng and co-workers [102] showed that IL-24 was able to promote the expression of EMT mesenchymal markers in BEAS-2B cells via the STAT3 and extracellular signal-regulated kinases (ERK)1/2 pathways. Furthermore, in vivo, IL-24 was highly expressed in mouse airway epithelium in a HDM-induced model of asthma, and this was associated with an upregulation of EMT markers. Interestingly, IL-37 reduced the airway remodelling by inhibiting IL-24-mediated EMT. These findings thus suggest that IL-24 contributes to airway remodelling and may represent a novel therapeutic target for preventing and treating this process in asthma, suggesting, on the other hand, a potential therapeutic effect of IL-37 on airway remodelling [102].

ECM, a key component of airway remodelling, is involved in a crosstalk with airway epithelial cells. Changes in its composition in fatal asthma have been demonstrated in various studies [103–109]. As already emphasised, fibroblasts and myofibroblasts are the most important cells involved in the production of ECM and both play a crucial role in the process of airway remodelling. An increase in myofibroblasts has been documented in the conducting airways of asthmatic patients [110,111] and is mostly responsible for the increase in type I collagen, type III collagen and fibronectin observed at this level; these cells have been found close to the RBM and ASM cells [112,113].

In vivo and in vitro observations in different asthma models have also documented the recruitment of circulating fibrocytes into the airways of asthmatic patients. Nihlberg and co-workers [114] have demonstrated that in bronchial biopsies of patients with mild asthma, fibrocytes were localised close to the RBM, and their numbers correlated with RBM thickness. These cells were also present in the BALF of these patients, supporting a potential role of fibroblasts progenitors in the early stage of airway remodelling in asthma.

Changes in ECM in asthma are not limited to the increase in fibrillar collagen but are also characterised by an altered deposition [103]; a study using nonlinear optical microscopy [72] described a lack of production of decorin by fibroblasts. This molecule is involved in collagen formation, and its deficiency causes a more fragmented and disorganised collagen deposition in the *lamina propria* of asthmatic airways. This altered deposition, in turn [72], stimulates myofibroblasts formation and the subsequent ECM synthesis. This aberrant mechanism thus induces a vicious cycle of persistent deposition of disorganised collagen and airway remodelling leading to increased basal membrane thickening [72,115].

The alterations of the ECM and the thickening of the RBM seem to be correlated with airway hyper-responsiveness [116,117]. Interestingly, a great variance in RBM thickness has been observed among patients as well as in different areas of the airway walls in the same patient, suggesting that ECM alteration is a dynamic process that may proceed at different rates in different areas of the airway wall. No differences were observed between fatal and non-fatal asthma or according to patients' age or gender [118]. Respiratory viruses, in particular *Rhinovirus*, contribute to the ECM remodelling in asthma, by increasing fibronectin, perlecan and type IV collagen deposition, as well as promoting ASM cell migration [119].

Another key component of airway remodelling—that is however less relevant in the context of this review—is the ASM [67–69]. In several studies carried out on bronchial biopsies, ASM mass correlates with asthma severity and airway hyper-responsiveness [120] and may increase from 5% up to 12% in patients with fatal asthma. This increase is mediated by both the hyperplasia and hypertrophy of ASM cells under the stimulus of various cytokines and GFs, including TGF-β1, EGF and PDGF [121–124]. ASM cells contribute to airway remodelling in asthma also through the modulation of the inflammatory process; in fact, these cells are able to release inflammatory cytokines such as IL-1β, TNF-α, IL-5, IL-13 and TGF-β [125–127]. Furthermore, it has been shown that repeated airway bronchoconstriction leads to a higher expression of TGF-β by epithelial cells, further enhancing subepithelial fibrosis and basal membrane thickening [128]. Bronchoconstriction may thus represent a trigger for airway remodelling independent of the inflammatory process [129].

Mucus hypersecretion is observed throughout the conducting airways and occurs at any stage of the disease from mild to fatal asthma [130–134]. It may be mediated by goblet cells hyperplasia and metaplasia, which is another feature of airway remodelling in asthma [68,135].

Table 1. Animal and cells culture models suggesting EMT implication in asthma.

Models	Techniques	EMT Program	AR	EMT Signal	EMT TFs	EMT Mark	Refs.	Year
HBECs of asthma and control subjects monolayer and ALI culture	RT-qPCR WB SDS-PAGE IHC and IF	+TGF-β1 10–50 ng/mL 72 h: spindle-shaped appearance ↓ E-cadherin and ZO-1 (protein) ↑ fibronectin, vimentin and α-SMA (mRNA and protein)	x	x		x	[87]	2009
16HBE 14o- cell line	RT-qPCR Cell proliferation WB ELISA	↑ collagen-1α1, fibrinogen, connective tissue growth factor and TGF-β1 (mRNA) Smad3-dependent process TGF-β1, IL-4 and IL-17A stimulation (72 h): proportion with a spindle-shape, fibroblast-like morphology with reduced cell–cell contact ↓ E-cadherin (mRNA and protein) ↑ α-SMA (mRNA and protein) ↑ pERK1/2	x	x		x	[90]	2013
16HBE 14o- cell line	IF WB SDS-PAGE	TGF-β1 5 ng/mL: ↓ E-cadherin (protein) ↑ vimentin and fibronectin (protein) ↑ glycogen synthase kinase-3β (protein)	x	x		x	[93]	2010
ALI normal HBECs	WB	+TGF-β1 + HDM 50 µg/mL: ↑ vimentin (protein) ↓ cytokeratin	x	x		x	[93]	2010
Male and female transgenic mice stably expressing LacZ in lung epithelial cells (SPC-Cre; R26stop-LacZ) 8–12 weeks	IF	↑ fibroblast-specific protein-1 (protein) delocalisation of E-cadherin +25 µg/day HDM intranasally (10–15 weeks): inflammation epithelial damage and thickening of the sub-epithelial contractile smooth muscle layer tissue ↓ occludin and E-cadherin (protein) ↑ vimentin, α-SMA and pro-collagen I	x			x	[95]	2011
Male C57/BL6 mice (9 weeks)	RT-qPCR WB ELISA	↑ TGF-β1 in BALF Activation of Smad-dependent TGF-β signalling pathways (p-Smad3 and SNAIL1 proteins in the nuclei) +bone-marrow-derived eosinophils (intratracheal instillation): inflammation in BALF ↑ TGF-β1 in BALF type I collagen deposition ↓ E-cadherin ↑ α-SMA and vimentin +TGF-β1 siRNA less airway remodelling	x	x		x	[98]	2013

Table 1. *Cont.*

Models	Techniques	EMT Program	AR	EMT Signal	EMT TFs	EMT Mark	Refs.	Year
BEAS-2B cell line	RT-qPCR WB ELISA	+primary human eosinophils or TGF-β1 or EoL-1 cells: fibroblast-like morphology and filamentous actin forming long stress fibres ↓ E-cadherin (mRNA and protein) ↑ vimentin (mRNA and protein) Need for cell-to-cell contact for induction of EMT	x	x		x	[98]	2013
BEAS-2B cell line	RT-qPCR WB IF Wound-healing assay	+PI3K and JNK inhibitors: EMT blocked rhTGF-β1 5 ng/mL: spindle-fibroblast-like morphology with reduced cell–cell contact ↑ α-SMA (mRNA and protein) ↓ E-cadherin (mRNA and protein) ↑ collagen type I, fibronectin-EDA and tenascin C (mRNA) ↑ motility ↑ MMP-2/9 (protein) +IL-1β: ↓ E-cadherin (mRNA ↑ tenascin C expression (mRNA)	x	x		x	[99]	2009
Primary normal HBEC monolayer	RT-qPCR WB	Corticosteroid pretreatment does not abrogate TGFβ1-induced EMT rhTGF-β1 2 ng/mL: ↑ α-SMA (mRNA and protein) ↑ vimentin (mRNA) ↓ E-cadherin (mRNA and protein) ↑ collagen type I, fibronectin-EDA and tenascin C (mRNA) ↑ MMP-2/9 release (protein)		x		x	[99]	2009
BEAS-2B cell line	RT-qPCR IF WB Wound-healing cell migration assays	+IL-24 (dose–response): ↑ migratory capability Acquisition of a larger and more spindle-shaped morphology ↑ vimentin and α-SMA (mRNA and protein) ↓ E-cadherin (mRNA and protein) Activation of STAT3 and ERK1/2 phosphorylation	x	x	x	x	[102]	2022

Table 1. Cont.

Models	Techniques	EMT Program	AR	EMT Signal	EMT TFs	EMT Mark	Refs.	Year
Female wild-type SPF BALB/c mice 6–8 weeks	IHC BALF Assessment of airway hyper-responsiveness	+IL-24 + inhibitors of JAK or ERK1/2: WT phenotype restored +25 µg/day HDM-induced asthma group 5 weeks: ↑ Airway resistance and inflammation ↑ IL-24 protein Collagen deposition ↑ TGF-β1 (protein in BALF) ↓ E-cadherin (protein) ↑ vimentin and α-SMA (protein) ↑ p-STAT3 and p-ERK1/2 (protein)	x	x	x	x	[102]	2022
Female C57BL/J mice 6–8 weeks CD146-KO C57BL/J mice IL-33 KO C57BL/J mice MLE-12 (mouse pulmonary epithelial cell line) and A549	WB IHC and IF ELISA WB IF RT-qPCR	+HDM + si-IL-24 or rhIL-37: ↓ HDM-induced dysregulations +25 µg intranasally administered HDM 5 weeks: ↓ inflammation in both KO mice ↓ collagen I (protein) in both KO mice +HDM 10–100 µg/mL: ↑ CD146 (mRNA and protein) ↑ MyD88, phosphorylation of NF-κB p65 and p38 (protein) ↑ TGF-β and pSMAD3 (protein)		x			[136]	2020
Primary alveolar epithelial cells from mice		+IL-33 0.1–100 ng/mL: ↑ CD146 (protein in cells and secreted)		x			[136]	2020
Normal primary HBEC monolayer	RT-qPCR IF ELISA	+siRNA CD146: ↑ E-cadherin; expression inversely correlated with CD146 +TGF-β1 10 ng/mL or neutrophils of asthma patients 48 h: ↓ E-cadherin (mRNA) ↑ N-cadherin, α-SMA and vimentin (mRNA) morphological changes	x			x	[137]	2019
BALB/c mice 6–8 weeks	ELISA IHC RT-qPCR WB	+25 µg HDM intranasal instillation 4 weeks: ↑ TF (mRNA and protein) +HDM + shRNA TF: ↓ inflammation Improved hyperplasia collagen biomarkers of EMT reversed	x	x		x	[138]	2021
16HBE14o- cells	RT-qPCR Wound-healing and invasion assay	+TGF-β1 and HDM: ↑ fibronectin 1, TF and TGF-β1 (mRNA and protein) shTF reverse the changes		x		x	[138]	2021

Table 1. Cont.

Models	Techniques	EMT Program	AR	EMT Signal	EMT TFs	EMT Mark	Refs.	Year
16HBE14o- cells	RT-qPCR WB	+TGF-β1: ↓ miR-448-5p (mRNA) ↑ Six1 (mRNA and protein) +TGF-β1 + miR-448-5p: ↓ TGF-β1, pSMAD3 ↑ E-cadherin (mRNA) ↓ vimentin (mRNA) +Six1 silencing: similar phenotype to miR-448-5p overexpression		x		x	[139]	2019
BEAS-2B cell line	WB IHC	LPS stimulation: ↑ TGF-β1, TGF-β RI and TGF-β RII (protein)	x	x		x	[140]	2014
Male BALB/c mice 6 weeks	IHC	+10 ng/mL TGF-β: change of morphology PAR-1 induction ↑ α-SMA (protein) +ovalbumin challenge: ↑ TGF-β1, TGF-β RI Epithelial thickening, collagen IV deposition ↓ E-cadherin (protein) ↑ α-SMA (protein) PAR-1 induction	x	x		x	[140]	2014

α-SMA, Alpha Smooth Muscle Actin; ALI, air–liquid interface culture; AR, airway remodelling; BALF, bronchoalveolar lavage fluid; ELISA, enzyme-linked immunosorbent assay; EMT, epithelial–mesenchymal transition; HBECs, human bronchial epithelial cells; HDM, house dust mite extract; IF, immunofluorescence staining; IHC, immunohistochemistry; IL, interleukin; JNK, c-Jun N-terminal kinases; KO, knock-out; Mark, EMT-markers; MMP, matrix metalloproteinase; pERK1/2, phospho extracellular signal-regulated kinase; PI3, phosphatidylinositol3; Rh, recombinant human; RT-qPCR, quantitative real-time-PCR; SDS-PAGE, sodium dodecyl sulphate polyacrylamide gel electrophores; Signals, EMT-inducing signals; STAT3, signal transducer and activator of transcription; TF, tissue factor; TFs, EMT-transcription factors; TGF, transforming growth factor; WB, Western blot; WT, wild-type; ZO, zonula occludens.

3.2. Potential Effect of Pharmacological Treatment on EMT and Airway Remodelling

Relevant progresses have been obtained in asthma treatment during the last decades; at present, this treatment is focused on chronic inflammation and bronchial hyperresponsiveness and aims to achieve the best possible disease control and to minimise exacerbation risk and the development of persistent airflow limitation, whereas unfortunately, no drugs are yet available that primarily target airway remodelling.

The cornerstone of pharmacological treatment of asthma is represented by inhaled corticosteroids (ICS) that decrease airway inflammation inducing disease control. Several studies evaluated the potential effects of ICS on airway remodelling, showing a decrease in RBM thickness [116] and type III collagen deposition [141] in some studies that, however, were not confirmed in others [107,142,143]. ICS treatment has also been shown to partially restore the epithelial damage through the inhibition of inflammatory responses [144]. Despite these potential effects, longitudinal studies have demonstrated that lung function impairment persists over time if present in childhood despite treatment with ICS and bronchodilators [145,146]. Therefore, it seems that ICS do not have a definite impact on airway remodelling.

Similarly, data on the effects on airway remodelling of other drugs used in asthma treatment, such as antileukotrienes and macrolides, are scarce and have been mostly obtained in murine models. Antileukotrienes have been documented to decrease ASM mass in the large airways and to induce a reduction in RBM thickness in animal models of asthma [147]. A decrease in goblet cells hyperplasia and collagen deposition mediated by TGF-β inhibition was also observed with these drugs [148]. Azithromycin seems to reduce ASM cells viability and proliferation as well as ASM thickness in both proximal and distal airways in murine models of asthma [149,150]. A deficit in vitamin D seems to correlate with worsening of severe asthma, and in vitro studies have shown that this compound slowed down airway remodelling by inhibiting Nuclear Factor-κB activation and decreasing ASM cells proliferation as well as the secretion of some inflammatory mediators; thus it has been suggested that it could represent a potential treatment for airway remodelling in asthma [151–153]; however, further studies are needed to confirm these findings.

In the last 20 years, biological therapies have become available for specific endotypes of severe asthma, with relevant impact on clinical outcome. However, only a few studies have evaluated their effects on structural changes and airway remodelling. The first drug available in clinical practice was omalizumab, an anti-IgE antibody. In vitro studies have demonstrated that this drug may prevent ASM cells proliferation and the deposition of type I collagen and fibronectin in response to IgE [154]. Other studies, using high-resolution computed tomography (HRCT), showed a reduction in airway wall thickness and an increase in airway luminal area in patients treated with this drug [155–157]. Haldar et al. evaluated the effects of the anti-IL-5 antibody mepolizumab by using HRCT and detected an improvement in total airway surface in treated patients [158]. Using a bronchial biopsy, Flood-Page et al. documented a significant reduction in tenascin, lumican and type III pro-collagen expression in mild asthmatic patients as well as a decrease in TGF-β1 mRNA in eosinophils and in TGF-β1 levels in BALF [159]. Chachi et al. studied the effects of benralizumab, an anti-IL-5R antibody, on bronchial biopsies, showing an increase in ASM cells apoptosis with a consequent ASM mass reduction and suggested that this effect was likely the consequence of an indirect effect on eosinophilic inflammation [160].

Thus, the current knowledge is limited, and further studies are needed to clearly define the effects of current asthma treatment on airway remodelling. The development of new drugs primarily targeting the structural changes in the airways is a very important goal of future asthma pharmacological therapies and may represent a revolutionary approach to the treatment of this disease.

4. COPD

COPD is a leading cause of morbidity and mortality worldwide, and it is associated with an increasing social and economic burden [161]. The most important risk factor for COPD development is cigarette smoke, although other factors are also important, such as indoor and outdoor air pollution, and occupational exposure [161,162]. The persistent exposure to potential noxious agents in patients with COPD, associated with an individual susceptibility, induces tissue injury and an aberrant repair that cause both airway and lung parenchymal damage in different combinations and lead to chronic respiratory symptoms and the progressive impairment of lung function. Although pathological alterations occur throughout the respiratory tract in COPD, it is well known that the major site of injury and obstruction are the small airways, with peribronchiolar fibrosis and airway narrowing and obliteration [163,164]. Large airways are also affected in COPD, and characterized by squamous metaplasia, mucus hypersecretion and ASM hyperplasia [163,164].

4.1. Chronic Inflammation in COPD

A chronic inflammatory process occurs both in the airways and the lung parenchyma in COPD as a consequence of persistent exposure to cigarette smoke and other noxious agents. A complex interplay of different inflammatory cells and mediators leads to small airway disease and emphysema, contributing to the progressive deterioration of lung function. Neutrophils are the main inflammatory cells present in the airway lumen and infiltrating the bronchial epithelium, glands and ASM; these cells largely mediate the structural damage through the release of several mediators, particularly proteolytic enzymes and ROS. Macrophages and T and B lymphocytes are also increased in small airways of patients with COPD and contribute to the enhancement of the inflammatory process as well as to the structural changes [163,164]. Macrophages are important effector cells in COPD; they produce several chemokines and cytokines, express high levels of mRNA for MMPs and release proteolytic enzymes and ROS, thus increasing tissue damage. These cells also release FGFs and other mediators involved in the proliferation of fibroblasts and their conversion to myofibroblasts and are thus major drivers of fibrosis in COPD. Moreover, macrophages contribute to the recruitment of other immune cells to the site of injury, thus enhancing the inflammatory process [22,165]. Lymphocytes and dendritic cells are organised into lymphoid follicles, which increase in severe disease, suggesting the relevance of the adaptive immune response in COPD. Furthermore, B cell infiltration in the walls of terminal bronchioles and alveoli is observed, which correlates with a loss of alveolar attachment to the airway walls. Increased numbers of type 1 and type 17 helper T cells have also been reported in patients with more severe disease [166]. Another cell type that may play a role in tissue alterations in COPD are mast cells, which have been implicated in airway vascular remodelling in this disease [163]. Mast cells secrete several bioactive molecules, including VEGF, FGF-β, TGF-β and also anti-angiogenetic factors [165,167]. Soltani and colleagues have documented that mast cell density in the *lamina propria*, and the RBM was greater in COPD patients compared to that of the controls and was related to the hypervascularity observed in the RBM, suggesting that perivascular mast cells may be involved in increased angiogenesis [168].

4.2. EMT and Fibrosis in COPD

As previously outlined, the pathogenesis of COPD is the result of complex and heterogeneous mechanisms that can interact each other; varying degrees of inflammation, tissue injury and abnormal repair with remodelling and fibrosis occur that involve the airways, lung parenchyma and lung vasculature. EMT has been suggested as an important mechanism contributing to the airway remodelling and fibrotic changes in COPD (recapitulated in Tables 2 and 3). Several noxious agents, in particular cigarette smoking, can induce typical features, such as altered epithelial barrier function, the acquisition of mesenchymal phenotype by epithelial cells and an increased production of ECM, and there is evidence to support that EMT is activated in the airway tissue of COPD patients (recently reviewed in [11]).

As we already mentioned, small airways are the site of major involvement in COPD; remodelling with peribronchiolar fibrosis at this level is related to bronchiolar distortion and airway obstruction. The mechanisms of small airways fibrosis are not completely defined and several factors may play a role; recent studies, however, suggest that EMT is involved in these structural changes. The expression of EMT-related TF SNAIL1 and TWIST is upregulated in smokers with and without COPD and has been found to be associated with EMT activity and the levels of airflow obstruction [169]. Interestingly, the expression of SNAIL1 was higher in COPD patients with α1-antitrypsin deficiency than in patients without the deficiency [170]. E-cadherin, a prominent hallmark of EMT, was found to be decreased in COPD airways in several studies [171,172]. Shirahata et al. showed that also plasmatic sE-cadherin levels were lower in COPD patients and were related to the severity of airflow limitation [173].

BECs from COPD patients showed upregulated mesenchymal markers, such as α-SMA, type I collagen, vimentin, and NADPH Oxidase 4 as well as a lower expression of epithelial markers, confirming an active EMT in these patients [174]. Mesenchymal markers, such as S100A4, vimentin and α-SMA proteins as well as ECM proteins were also found increased in vivo in smokers with normal lung function and COPD patients (Table 3).

A major feature of EMT activity in vivo is represented by RBM fragmentation, associated with mesenchymal markers expression [16]. Sohal et al. showed that these pathological features are present in endobronchial biopsies from smokers with normal lung function and were even more evident in smokers with COPD [175]. Interestingly, membrane fragmentation in COPD has been shown to correlate with smoking history [176]. In a subsequent study, these authors demonstrated that in the RBM of smokers with COPD, there are cells that double stain for both mesenchymal and epithelial markers, further supporting the evidence of an active EMT in smoking-related COPD [177]. Cigarette smoking is not only capable of activating several signalling pathways involved in EMT per se, but it is also associated with a high oxidative burden and a chronic inflammatory process into the COPD airways that significantly contribute to the EMT process and the structural changes in the airways and lung parenchyma. One of the most important pathways affected by cigarette smoking is represented by the TGF-β/SMAD signalling pathway. As we already emphasised, TGF-β, and in particular TGF-β1, plays a crucial role in fibrotic diseases, and it is considered a key regulator of airway remodelling in COPD [171]. TGF-β is involved in the regulation of ECM components, as well as in the fibroblasts activation and in myofibroblasts differentiation [178,179]. It has been shown that TGF-β1 is capable of inducing EMT in cultures of human bronchial and lung epithelial cells in vitro [11]. Interestingly, cigarette smoking similarly induces EMT in lung and BEC through the TGF-β1/SMAD signalling pathway both in vitro and in vivo [11]. Furthermore, Takizawa et al. documented that mRNA levels of TGF-β1 were significantly higher in small airway epithelium from smokers with and without COPD than in non-smokers and were related to the degree of small airway obstruction, suggesting a tight relationship between smoking and TGF-β1 expression in small airways [180]. As previously mentioned, TGF-β1 also induces the differentiation of fibroblasts into myofibroblasts, motile and contractile cells that, once activated, secrete excessive and altered ECM (reviewed in [31]). As already discussed, these cells, that are crucial actors in the fibrotic process into the airways, can also originate from resident macrophages; during lung injury, in fact, macrophages may acquire a pro-inflammatory phenotype that promotes the differentiation and activation of these cells [181]. The process of MMT, which involves the transformation of macrophages into myofibroblasts, has recently been described in the lung and may contribute to development of fibrotic changes [182,183]. Although the molecular mechanisms involved in MMT are still unknown, this new evidence suggests novel potential targets for the development of antifibrotic therapies. Several studies suggest that the contribution of myofibroblasts and the EMT process may be important in both small and large airways fibrotic changes in COPD [175,177,184,185].

Recently, Eapen and colleagues [186] documented an increase in α-SMA+ myofibroblasts in the small airways of patients with COPD that was associated with the increased deposition of ECM proteins, EMT activity in epithelial cells and thickening of the *lamina propria*. These changes were also related to lung function impairment, further supporting an important role of EMT in small airways remodelling and narrowing.

Fibroblasts play also a key role in the fibrotic process; interestingly, Togo et al. demonstrated that lung fibroblasts of COPD patients have impaired repair mechanisms and suggested that this defect could contribute to the development of emphysema [187]. However, it is not known whether small airways fibroblasts are also defective in repair mechanisms, and it is not yet known whether these cells differ from interstitial lung fibroblasts. In this context, it is worth mentioning the observation of Lynch and colleagues who emphasised the possibility of fibroblast heterogeneity even within the same tissue [188,189].

As we already mentioned, cigarette smoke induces a high oxidative burden into the airways of COPD patients that is further enhanced by the persistent recruitment of inflammatory cells [190]. The oxidative stress drives additional inflammatory mechanisms in COPD; induces the expression of senescence markers in small airway fibroblasts; promotes profibrotic markers, including TGF-β and COL3A1, and is also associated with an impairment of antioxidant defences superoxide dismutase 2 and 3. These findings suggest that oxidative stress may contribute to small airway fibrosis in COPD [188] and contributes to the promotion of EMT [179]. Furthermore, the interaction between the oxidative stress and TGF-β is crucial in promoting fibrosis, inducing a self-perpetuating process by which TGF-β favours the production of ROS with increased oxidative stress that, in turn, activates latent TGF-β (reviewed in [31,191]). TGF-β can also induce EMT via non canonical pathways, such as the ERK, p38 MAPK, PI3K, Notch and WNT signalling pathways [192]. Recently, several studies suggested that the WNT/β-catenin pathway is activated in smokers and COPD patients and appears to be related to the EMT activity and the airway obstruction. The expression of the genes and proteins involved in this pathway is increased in the airway epithelium of smokers with COPD [193]. Interestingly, in vitro studies demonstrated that cigarette smoking and nicotine were able to induce EMT in human BEC by activating the WNT-3A/β-catenin pathway [194]. On the other hand, in peripheral tissue, a decrease in WNT/β-catenin signalling was observed that was associated with parenchymal alterations and the disruption of repair mechanisms, leading to an increase of emphysema in COPD patients [195].

PI3K/AKT is a signalling pathway involved in the regulation of several biological functions, including EMT. Also, in this case, some evidence suggests that cigarette smoke induces EMT through this pathway in COPD (Tables 2 and 3). Milara and colleagues assessed in vitro the CSE-induced EMT in primary human BECs from small bronchi and suggested that the CSE effect is partially mediated by the activation of the PI3K/AKT/β-catenin pathway and the generation of ROS [196]. Recent studies added novel evidence on the potential mediators involved in EMT induction and airway remodelling in COPD: Jiang and co-workers [197] investigated the potential role of cathelicidin in inducing EMT in COPD. This protein is involved in various biological functions, including the regulation of inflammation and immunity and the promotion of tissue repair, and its overexpression in the airway epithelium has been implicated in mucus hypersecretion and fibroblast collagen production in smoking-related COPD [198,199]. Jiang and colleagues evaluated the expression of cathelicidin and EMT markers in human lung tissues from smokers with and without COPD, and in a COPD mouse model. They showed an upregulation of cathelicidin expression associated with EMT markers in the small airways of smokers with and without COPD. Significant smoking-induced EMT was also observed in the airways of mice. Interestingly, EMT was inhibited by the downregulation of CRAMP (the murine homologue of cathelicidin) in COPD mice. Finally, the authors demonstrated that cathelicidin promoted EMT by activating Tumour necrosis factor alpha (TNF-α) converting enzyme (TACE), Transforming growth factor alpha (TGF-α), and Epidermal growth factor receptor (EGFR) signalling pathways. Chu and colleagues evaluated the effect of CSE and IL-17A on bronchial EMT in a mice model of COPD and showed an increased expression of IL-17A in lung tissues and a synergistic effect of this cytokine and CSE on the induction of bronchial EMT [200]. Another pathway involved in EMT in COPD is the urokinase-type plasminogen activator (uPA)/urokinase-type plasminogen activator receptor (uPAR)-dependent cell signalling pathway. Wang Q et al. have documented an increased uPAR expression, associated with an increased EMT activity, in the small airway epithelium of COPD patients as compared with non-smokers and smokers with normal lung function [201]. Moreover, in a subsequent study, the same authors showed that uPA is also upregulated in human small airway epithelial cell lines (HSAEpiCs), as well as in the small airways epithelium of COPD patients and is correlated with vimentin expression at this level. uPA and uPAR inhibition was able to inhibit CSE-induced EMT in HSAEpiCs [202]. These findings thus suggest that the activation of the uPA/uPAR pathway might represent a novel mechanism involved in EMT development and in airway remodelling in COPD. Interestingly, in a retrospective study, uPAR expression was also found increased in pulmonary macrophages and alveolar cells from COPD patients compared to controls, and it was also positively correlated with the levels of collagen [203]. It is important to highlight that other pathways may be involved in the induction of EMT by cigarette smoking in COPD; moreover, as we already emphasised, in addition to cigarette smoke, other environmental stresses, including the high oxidative burden and the reduced antioxidant defences, as well as the signals induced by the excessive inflammatory response or the mechanical stress can also trigger mechanisms and processes involved in EMT. Therefore, the development of EMT and the progression of fibrotic changes in the airways of COPD patients are the result of a complex network involving different triggers and multiple signalling pathways. Finally, we would like to emphasise that EMT is also active in large airways of smokers with COPD. Interestingly, Malik and colleagues documented that type III EMT is characteristic of the large airways, where RBM hypervascularisation is also observed, while type II EMT is active in the small airways, where it is involved in the fibrotic changes, contributing to the remodelling and obliteration of these airways [202]. In large airways, the type III EMT process could induce, in the context of some microenvironment alterations, the development of lung cancer that is known to be associated with COPD. Recently, a genomic link was demonstrated among COPD, lung cancer and Hedgehog signalling, which is also involved in EMT induced by tobacco-smoke [204].

Table 2. Animal and cells culture models suggesting EMT implication in COPD.

Models	Techniques	EMT Program	AR	EMT Signal	EMT TFs	EMT Mark	Refs.	Year
ALI HBECs (2–5 weeks) from non-smokers, smoker controls, mild, moderate and severe to very severe COPD	IHC and IF RT-qPCR WB ELISA	↑ vimentin (mRNA and protein) for severe COPD and fibronectin (protein) ↓ E-cadherin and ZO-1 (protein) +TGF-β1 ↑ vimentin and fibronectin (protein) +blocking TGF-β1 Restoring a cobblestone shape compared with the spindle shape ↓ vimentin expression	x	x		x	[171]	2015
ALI HBECs from non-smokers, smokers and patients with COPD	IHC and IF RT-qPCR WB ELISA	↑ α-SMA, vimentin and collagen type I (mRNA and protein) ↓ E-cadherin and ZO-1 (mRNA and protein) ↓ KRT5 and KRT18 (mRNA)				x	[174]	2013
ALI HBECs (14–21 days) from non-smokers, smoker controls, mild, moderate and severe to very severe COPD	RNA-seq WB ELISA IHC and IF	Activation of WNT/β-catenin pathway with ↑ nuclear expression of β-catenin in the COPD airway epithelium ↑ vimentin (mRNA and protein) ↑ fibronectin release following WNT activation	x	x		x	[193]	2020
HBEC monolayer	RT-qPCR WB IHC and IF ELISA	+nicotine: ↑ Wnt3a (protein and mRNA) ↑ total β-catenin (protein) ↓ E-cadherin (protein)		x	x	x	[194]	2013
ALI HBECs from controls lung tissue	IF RT-qPCR WB ROS	↑ α-SMA, MMP-9 and collagen type I (protein) +2.5% CSE for up to 7 days: ↓ E-cadherin, ZO-1 (mRNA and protein) ↑ vimentin, collagen type I and α-SMA (mRNA and protein) ↑ GTP-Rac1 and pAKT (protein)	x	x		x	[196]	2015
Male BALB/c mice 6 weeks	IHC WB	+CSE: collagen deposition ↑ CRAMP and vimentin ↓ E-cadherin	x	x		x	[197]	2021
NCI-H292 cell line	IF	↑ TACE, TGF-α and EGFR (protein) +CSE 5%: ↓ E-cadherin ↑ vimentin				x	[197]	2021

Table 2. Cont.

Models	Techniques	EMT Program	AR	EMT Signal	EMT TFs	EMT Mark	Refs.	Year
male C57BL/6 mice 8 weeks	IHC and IF RT-qPCR WB	20 CS/day 12 or 24 weeks: inflammation ↑ECM, smooth muscle thickening, goblet cell hyperplasia and mucus secretion ↑IL-17A and C-EBPβ ↓E-cadherin (mRNA and protein) ↑vimentin (mRNA and protein)	x			x	[200]	2021
Murine bronchial epithelial cells	IF RT-qPCR WB	20% CSE 72 h: ↑IL-17R ↓E-cadherin (mRNA and protein) ↑vimentin (mRNA and protein)				x	[200]	2021
HSAEpiCs	RT-qPCR WB	5% CSE 24–96 h: ↑α-SMA, N-cadherin and uPAR (protein) ↓E-cadherin and α-catenin (protein)				x	[201]	2013
BEAS-2B cell line	WB RT-qPCR	+CSE 1% 24 h: ↓ZO-1 (protein and mRNA) ↑vimentin (mRNA) ↑TGF-β1 (mRNA) +TGF-β1 230 pg/mL ↓ZO-1 (mRNA)				x	[205]	2020
HSAEpiC	WB RT-qPCR	+CSE 1% 24 h: ↑vimentin (mRNA) ↓ZO-1 (mRNA) ↑TGF-β1 (mRNA) and p-Smad2/3 (protein) +TGF-β1 230 pg/mL, 2 h ↓ZO-1 (protein and mRNA)		x		x	[205]	2020
Male and female Sprague Dawley rats (8 weeks)	ELISA IHC WB	+48 CS/day inhalation 12 weeks: ↑IL-8, IL-6, TNF-α, sICAM-1 and ROS in BALF Airway fibrosis, airway epithelial thickness and ASM thickness ↑α-SMA (protein) ↓lung function ↑TGF-β1 and p-Smad2/3 (protein) ↓PPAR-γ (protein)	x	x		x	[206]	2021
Male C57BL/6 mice 26–28 weeks	ELISA RT-qPCR WB	+CSE intraperitoneally injected: airway epithelium thickening, enlargement of alveolus and inflammatory cell infiltration ↑IL-6 and TNF-α in BALF ↑TGF-β1, Smad2 and Smad3 (protein and mRNA)	x	x			[207]	2019

α-SMA, Alpha Smooth Muscle Actin; ALI, air–liquid interface culture; AR, airway remodelling; ASM, airway smooth muscle; BALF, bronchoalveolar lavage fluid; COPD, chronic obstructive pulmonary disease; CRAMP, cathelin-related antimicrobial peptide; CS, cigarette smoke; CSE, cigarette smoke extract; EGFR, epithelial growth factor receptor; ELISA, enzyme-linked immunosorbent assay; HBECs, human bronchial epithelial cells; HSAEpiC, human small airway epithelial cells; IF, immunofluorescence staining; IHC, immunohistochemistry; IL, interleukin; KRT, keratin; Mark, EMT-markers; MMP, matrix metalloproteinase; pERK, phospho extracellular signal-regulated kinase; PPAR, Peroxisome-proliferator-activated Receptor; RT-qPCR, quantitative real-time PCR; Signals, EMT-inducing signals; TNF, tumour necrosis factor; TFs, EMT transcription factors; TGF, Transforming Growth Factor; WB, Western blot; ZO, zonula occludens. Lung tissue was obtained from patients undergoing lung resection for appropriate clinical indications.

Table 3. In situ evidence for EMT implication in COPD.

Models	Techniques	EMT Program	AR	EMT Signal	EMT TFs	EMT Mark	Refs.	Year
Bronchial biopsy from non-smokers, smoker controls and COPD subjects	IHC	Correlation between β-catenin and SNAIL1 expression with both S100A4 and also airflow obstruction		x	x	x	[169]	2017
Lung tissue of α1-antitrypsin-deficiency-related COPD and non-α-1 antitrypsin deficiency COPD subjects	RT-PCR	↑ SNAIL homolog 1 in α1-antitrypsin-deficiency-related COPD group					[170]	2012
Lung tissue of non-smokers, smoker controls, mild, moderate and severe-to-very-severe COPD	IHC	↓ E-cadherin expression ↑ vimentin in large and small airways Negative correlation between vimentin and airway obstruction				x	[171]	2015
Lung tissue of COPD subjects	WB IHC	↓ E-cadherin ↑ α-SMA, N-cadherin and vimentin Fragmentation and clefts in RBM	x			x	[172]	2021
Bronchial biopsy from COPD patients	IHC	Cytokeratin-(s) and S100A4 double staining				x	[177]	2011
Brushing from non-smokers, smoker controls, COPD subjects	RT-PCR IHC	↑ TGF-β1 (protein and mRNA) Positive correlation between TGF-β1 mRNA levels and the extent of smoking history		x			[180]	2001
Bronchial biopsy	IHC	RBM fragmentation ↑ S100A4 and MMP-9 in RBM and/or basal epithelium +inhaled corticosteroids treatment: ↓ EGFR ↓ %RBM fragmentation ↓ S100A4 and MMP-9 in RBM and/or basal epithelium	x			x	[185]	2014
Lung tissue of non-smokers, smokers and patients with COPD (smoker or ex-smoker)	IHC	↑ lamina propria and adventitia thickness in small airways of COPD subjects ↑ α-SMA-positive cells (myofibroblasts) in SA ↑ collagen-1 and fibronectin deposition Negative correlation between increased SA wall thickening and decrease in airflow in the COPD groups Correlation between collagen-1 deposition in the SA *lamina propria* and lung function in the COPD-smokers group	x			x	[186]	2021
Lung tissue of non-smokers, smoker controls, mild, moderate and severe-to-very-severe COPD	IHC and IF RT-PCR	↑ β-catenin expression in the COPD airway epithelium β-catenin upregulation in COPD airway epithelium correlates with altered differentiation	x	x			[193]	2020

Table 3. Cont.

Models	Techniques	EMT Program	AR	EMT Signal	EMT TFs	EMT Mark	Refs.	Year
Lung tissue of non-smokers, smoker controls and COPD subjects	IHC	↗ vimentin ↗ uPAR Negative correlation between FEV1% and uPAR expression Positive correlation between uPAR and the number of vimentin-positive cells				x	[201]	2013
Lung tissue of controls or subjects with chronic airflow limitation	IHC	↗ vimentin and S100A4 in SA of COPD S100A4 expression associated with airflow obstruction in small airway				x	[202]	2015
Bronchial biopsy from non-smokers, smoker controls and COPD subjects	IHC	↗ TGF-β1 in large airway Correlations between pSmad 2/3 and pSmad 7 expression and both S100A4 and airflow obstruction		x		x	[208]	2017
Lung tissue of non-COPD and patients with COPD	IHC	↗ TGF-β1 (protein and mRNA)		x			[209]	1998

α-SMA, Alpha Smooth Muscle Actin; AR, airway remodelling; COPD, chronic obstructive pulmonary disease; EGFR: epithelial growth factor receptor; FEV, forced expiratory volume; IF, immunofluorescence staining; IHC, immunohistochemistry; Mark, EMT-markers; MMP, matrix metalloproteinase; NOX, nicotinamide adenine dinucleotide phosphate oxidase; uPAR, urokinase-type plasminogen activator receptor; RBM, reticular basement membrane; RT-qPCR, quantitative real-time PCR; SA, small airway; Signals, EMT-inducing signals; TFs, EMT transcription factors; TGF, Transforming Growth Factor; ZO, zonula occludens. Lung tissues were obtained from patients undergoing lung resection for appropriate clinical indications.

4.3. Potential Effect of Treatment on EMT and Small Airway Fibrosis in COPD

Pharmacological strategies for COPD include the use of bronchodilators and, when indicated, also ICS. Other drugs used in these patients are antibiotics when indicated, methylxanthines, mucolytics and antioxidant agents, and phosphodiesterase-4 (PDE4) inhibitors.

Despite these therapeutic approaches, it is known that COPD continue to worsen over time with the progression of lung damage and the impairment of lung function. Thus, it would be very important to develop novel therapeutic strategies addressing targets and mechanisms involved in the development of the EMT process and the fibrotic alterations in COPD airways that play a crucial role in airway remodelling and obstruction. However, few studies have investigated the effects of potential therapeutic strategies so far, either alone or in combination with other therapies, on these targets, and additional data as well as preclinical and clinical trials addressing this topic are needed.

Sohal et al. performed a proof-of-concept randomised controlled study with ICS administered for ≥6 months and showed a regression of typical EMT alterations, such as epithelial activation, RBM fragmentation and EMT biomarkers in treated patients compared to the placebo group [185].

As already discussed, it has been suggested that type III EMT may play a role in the development of cancer in COPD patients. From an analysis of nine prospective cohorts, Fan Ge et al. suggested a protective effect of ICS against lung cancer in COPD patients [210]. However, further studies on larger numbers of patients are needed to confirm these data.

Recently, Zhu et al. have shown that N-acetylcysteine, an antioxidant and mucolytic agent, was able to reduce α-SMA levels, collagen volume, wall thickness and bronchioles diameter in a COPD rat model. Furthermore, the drug was shown to inhibit the EMT process and promote immune response by acting on the VWF/p38 MAPK axis; the authors thus suggest that this drug might improve fibrotic changes in COPD [211].

Roflumilast is a PDE4 inhibitor that may be used in severe COPD associated with chronic bronchitis and frequent exacerbations. Martorana et al. demonstrated that this molecule may decrease lung damage and emphysematous changes induced by cigarette smoking in mice. They also observed a decrease in macrophage density in treated mice, suggesting that the anti-inflammatory effect of this drug might contribute to the hindering of EMT in COPD [212]. Milara et al. also showed that PDE4 inhibitors, in particular Roflumilast N-oxide (RNO), mediated a protective effect against EMT induced by cigarette smoking in bronchial epithelial cells; interestingly, this effect was also observed in primary human BECs isolated from the smokers and COPD patients' small bronchi. Moreover, RNO induced a reduction in ROS, in NADPH Oxidase 4 expression, in TGF-β1 release, as well as in SMAD3/ERK1/2 phosphorylation induced by cigarette smoke [213]. It has also been shown that the addition of statins (simvastatin) seems to enhance in vitro the inhibitory effect of RNO against cigarette-smoking-induced EMT in human BECs [213]. Several other molecules and novel therapeutic approaches with potential inhibitory effects on the EMT process have been recently studied, such as celecoxib, a selective COX2 inhibitor [214,215]; galunisertib, a TGF-β receptor 1 inhibitor [216]; all-trans retinoic acid [217] and N-cadherin antagonist ADH-1 [218]. Most of these molecules have emerged in the oncological field over the last years and are in the preclinical or clinical phase for various solid tumours; however, in the context of their effects on inhibiting or preventing EMT induction, some of these compounds might also have a potential role in inhibiting EMT development and fibrotic changes in COPD [11].

Moreover, antifibrotic drugs approved for the treatment of idiopathic pulmonary fibrosis (pirfenidone and nintedanib) have been reported to have an inhibitory effect on EMT mainly through the TGF-β pathway [219,220].

Thus, several EMT-targeted therapies have the potential to be effective in preventing or modulating the pathological changes induced by EMT and the other processes involved in fibrotic changes in COPD. However, further preclinical and clinical studies are needed to evaluate and validate the potential benefits of these therapies in COPD.

5. Cystic Fibrosis and NCFB

CF is a rare genetic disease affecting over 100,000 people worldwide [221]. Mutations in the *Cystic Fibrosis Transmembrane Conductance Regulator* (CFTR) gene lead to a defect in the expression or function of the CFTR protein involved in the regulation of transepithelial ions and fluid transports. Although it is a multi-systemic disease, the main cause of morbidity and mortality in patients with CF (pwCF) is lung disease. Airway involvement with thick mucus obstructing the bronchi and an impairment of mucociliary clearance promote cycles of inflammation and infection by pathogens, such as *Staphylococcus aureus* and/or *Pseudomonas aeruginosa* [222]. The chronicity of this infectious/inflammatory process results in progressive lung damage with the development of alterations of the airway walls and a structural remodelling observed on autopsy studies or CF bronchial biopsies or lung explants and involving squamous metaplasia [223], subepithelial fibrosis [224], submucosa mucus gland enlargement [225], hyperplasia of ASM [226] and RBM thickening [227]. The order of events leading to the structural airway changes in CF and their relationship to infection and inflammation has long been debated. The observation of an early inflammatory response [228] suggests that inflammation develops prior to infection and that CFTR dysfunction is associated with a dysregulation of inflammation and is involved in the alterations of airway structure.

5.1. CF Inflammation

CF is characterised by the dehydration and acidification of the airway surface liquid and the hyperconcentration of mucus in the lungs. CFTR dysfunction or defect severely compromises the airway microenvironment and ultimately leads to structural lung damage and obstruction of the airways which favours persistent infection and a chronic inflammatory process [229]. The formation of mucus plugs in CF airways can trigger airway inflammation per se, even in the absence of bacterial infection and very early in the natural history of the CF lung disease. In muco-obstructive diseases [228], such as CF, a "vicious" circle occurs: mucus plugs activate lung-resident macrophages, inducing the release of IL-1β and producing a hypoxic microenvironment with necrosis of airway epithelial cells that will release IL-1α. Both IL-1α and β will activate epithelial IL-1 receptors, inducing mucin biosynthesis and the expression of pro-inflammatory cytokines and chemokines, such as IL-8, thus further amplifying the inflammatory response.

CF airway inflammation is mainly characterised by a marked and persistent recruitment of neutrophils into the airways; these cells play a crucial role in lung damage in CF. In fact, they release several noxious mediators, including proteases, such as neutrophil elastase (NE), ROS, DNA and inflammatory mediators, thus enhancing and perpetuating the chronic inflammatory process and tissue damage. NE is a serine protease actively participating in the degradation of lung tissue, and its levels have been associated with a decline of lung function [230]. It is also a predictive biomarker for the development of bronchiectasis in children [231]. The addition of NE to cultures of BECs from pwCF results in a concentration-dependent delayed/inhibited repair process, and it has been shown that alpha-1-antitrypsin, by inhibiting NE, leads to faster tissue repair [232]. Furthermore, the high levels of NE released by neutrophils induce a bacteria-killing defect by cleaving proteins (i.e., elafin and SLPI) that are involved in antimicrobial and anti-inflammatory responses [233]. A protease/antiprotease imbalance [234] occurs in CF as a consequence of the high protease burden, mainly due to the release of NE and MMPs. Furthermore, antiproteases are degraded by the proteases released by inflammatory cells and pathogens, thus further reinforcing the imbalance and the subsequent structural damage. Neutrophils also produce defensins, which activate MMPs, which normally participate in ECM degradation (collagen, elastin and gelatin) and in cytokines expression [235]. In CF, the overexpression of MMPs (MMP-8, MMP-9 and MMP-12) correlates with the impairment of lung function; in particular, MMP-9 expression has been shown to correlate with RBM degradation, the onset of bronchiectasis and a decline of lung function in pwCF [236–239]. The proteolytic activity of MMPs is also directed against CFTR [240], affecting its function and which could further aggravate the lung damage.

Some evidence suggests that the excessive and abnormal inflammatory response in CF may lead to airway remodelling [241]. In CFTR knock-out mice, chronic *P. aeruginosa* LPS exposure induced an enhanced inflammatory response that was associated with an increased susceptibility to the development of lung remodelling (increase in goblet cells and fibrosis) [242].

5.2. Future Directions on the Role of EMT in CF

Few papers have investigated the involvement of EMT in CF airway remodelling, while in other pathologies, a link between CFTR and signalling pathways also implicated in EMT has been demonstrated (reviewed in [12,243] and described in Table 4). In breast cancer, the downregulation of CFTR expression seems to favour cancer development and EMT induction [244–247]. Furthermore, a decreased CFTR expression leads to increased WNT signalling in mouse lung development [248]. On the contrary, CFTR knockdown in HEK293 cells was associated with a significant reduction in WNT signalling in the context of haematopoiesis [249]. So far, although the CFTR protein interacts with several actors involved in type II EMT, there is no clear evidence of its direct involvement in this mechanism [250].

Table 4. Potential indicators of EMT implication in CF models and CF lung explants.

Models	Techniques	EMT Program	AR	EMT Signal	EMT TFs	EMT Mark	Refs.	Year
Primary HBECs (from explanted CF lung or from control subjects who underwent lung surgery)	TEER IF RT-qPCR WB Wound healing	↑ N-cadherin and vimentin (protein) ↓ TEER	x			x	[250]	2020
CFBE41o-wt or -F508del-CFTR HEK 293T cells	IF WB Wound healing	↑ N-cadherin vimentin and collagen I (protein) ↑ ZO-1 and CX31, while ↓ claudin-1 and Desmoplakin I/II Multilayered organisation for CF versus monolayer for control EMT markers localisation differences ↑ Ki-67-positive cells in basal cell layers ↑ time to close the wounds ↓ TEER ↑ TWIST1 (protein) +CFTR modulator treatments ↓ N-cadherin and vimentin ↓ TWIST1 +response to TGF-β1: No difference in response ↓ CK18 and ↑ N-cadherin (protein) WT more resistant to EMT induction than CF TWIST1 shRNA knockdown in HEK293T cells: vimentin inhibition	x	x		x	[250]	2020
CF versus control lung tissues	IF RT-qPCR	↑ occludin, tight junction protein 1/zonula occludens-1, connexin 43, connexin 26 and cytokeratin 18 (mRNA) ↑ vimentin (mRNA) Polygonal flat cells on the CF epithelial layer surface Fewer cylindrical-shaped columnar cells and several cell layers in CF tissue ↑ TWIST1 and ZEB1 (mRNA) Positive staining for SNAIL1 + Slug and ZEB1 Partial EMT	x			x	[250]	2020
CF versus control lung tissue	IHC	↑ MUC5AC (protein) ↓ β-tubulin (protein) ↑ vimentin-positive spindle-shaped Thickening of the RBM	x			x	[251]	2021
ALI HBECs (CF and controls lung tissue 2 weeks)	TEER WB ELISA RT-qPCR IHC	↑ MUC5AC (protein) ↓ MCIDAS, MYB and FOXJ1 (mRNA) ↓ TEER ↓ E-cadherin and occludin Spindle-shaped cells and thickness of epithelium +CFTR inhibition or *Pseudomonas* infection in control culture No EMT induction	x			x		

Table 4. Cont.

Models	Techniques	EMT Program	AR	EMT Signal	EMT TFs	EMT Mark	Refs.	Year
IB3-1 cells F508del/W1282X C38 cells as control	Cell migration assay WB IHC RT-qPCR ELISA	↗ fibronectin, N-cadherin and the transcription repressor Slug in IB3 cells compared to the C38 cells ↗ migratory phenotype ↗ TGF-β1 (mRNA and protein) +TGF-β receptor inhibitor ↘ fibronectin (protein) +TG2 inhibitors in IB3-1 cells Inhibition of cell migration ↘ TGF-β1 (protein)	x	x	x	x	[252]	2016
ALI HBECs 14 days	WB	+TGF-β1 3 ng/mL 48 h: ↗ TG2 ↗ Fibronectin and N-cadherin ↘ E-cadherin (protein) +TG2 overexpression: ↗ fibronectin, N-cadherin, Slug (protein) ↘ E-cadherin (protein) ↘ TEER				x	[252]	2016

AR, airway remodelling; CF, cystic fibrosis; CFBE, CF Bronchial Epithelial Cell Line; CFTR, cystic fibrosis transmembrane conductance regulator; EMT, epithelial–mesenchymal transition; FOXJ, Forkhead box J protein; HEK, human embryonic kidney; IF, immunofluorescence staining; IHC, immunohistochemistry; Mark, EMT-markers; MCIDAS, Multiciliate Differentiation And DNA Synthesis-associated Cell Cycle Protein; MUC, mucin; RBM, reticular basement membrane; RT-qPCR, quantitative real-time PCR; Signal, EMT-inducing signals; TEER, transepithelial electrical resistance; TFs, EMT transcription factors; TG, transglutaminase; TGF, Transforming Growth Factor; ZEB, Zinc finger E-box binding homeobox; ZO, Zonula occludens. Lung tissues were obtained from subjects undergoing lung resection for appropriate clinical indications or from CF patients at transplantation.

Collin et al. observed an alteration in cell differentiation in the lung explants of pwCF, with an increase in mucin 5AC labelling (goblet cells) and a decrease in β-tubulin-positive cells (ciliated cells). These alterations were associated with an increase in vimentin-positive cells, suggesting EMT-related remodelling in CF lungs [251]. A transcriptome meta-analysis also revealed an EMT signature in CF epithelium and the identified protein tyrosine phosphatase *PTP4A1/2* as being potentially involved [253].

Since the identification of the *CFTR* gene in 1989 [254,255], it has been shown that genotype alone cannot explain the phenotypic variations in CF [256]. In addition to the socio-environmental impact [257,258], the presence of modifier genes can indeed affect the pathophysiological features of pwCF and contribute to the severity of lung disease [259–265]. TGF-β is one of the major modifier genes and plays a critical role into myofibroblast differentiation [266]. Harris et al. showed an increase in myofibroblasts in CF lung explants that correlated with TGF-β [267]. In CF genome, three single-nucleotide polymorphisms in the *TGF-β* gene at position −509 (C or T; promoter region), +869 (T or C; codon 10 leucine or proline) and +915 (G or C; codon 15 arginine or proline) were identified [268–270]. The presence of proline at both sites induces low TGF-β1 expression [271], while leucine on codon 10 is associated with a high protein level [271–273]. TGF-β1 overexpression in pwCF has been associated with more a rapid deterioration of lung function [270]. Similarly, an increase in TGF-β1 in BALF, sputum, serum or plasma is associated with increased inflammation, some bacterial infections and the severity of pulmonary manifestations [271,274–279]. CFTR knock-out mice express more TGF-β protein in their lung tissue compared to wild-type mice [242]. Furthermore, these mice developed a more pronounced fibrotic signature with expression of ECM proteins in response to chronic

LPS exposure [242]. All these observations suggest a link between CFTR dysfunction and TGF-β upregulation. RBM thickness has been shown to be positively correlated with the levels of TGF-β1 in BALF in children with CF [227].

In addition to TGF-β, other TGF-β-related players are deregulated in CF. Recently tissue transglutaminase (TG2) was identified in vitro as a regulator of TGF-β1 [252]. TG2 inhibition leads indeed to a significant reduction in fibronectin, N-cadherin, SNAIL and TGF-β1 expression, resulting in a reversal of EMT. In human BEC line (16HBE14o-), miR1343 directly represses the activity of both TGFBR1 and TGFBR2 receptors by binding to the 3'UTR region and leading to the inhibition of canonical TGF-β signalling pathways [280].

FAM13A has been suggested to be a modifier gene for CF lung phenotype; the expression of FAM13A is downregulated in CF human BEC, and this decrease is associated with a decrease in E-cadherin [281], suggesting that FAM13A could be involved in EMT modulation in CF epithelial cells.

Recently, Quaresma et al. studied the presence of EMT features in CF tissue, primary cultures of human BEC and cells lines expressing mutant CFTR [250]. In CF tissue/cells defective of CFTR, mesenchymal markers, such as vimentin and N-cadherin, were upregulated, but most epithelial markers were not repressed, suggesting that partial EMT was active. Other features of active EMT were observed, including impaired wound healing, destructured epithelial proteins and defective cell junctions, as well as the upregulation of EMT-associated transcription factors. This study also shows that the observed EMT features were mediated by the EMT-associated transcription factor TWIST1. Interestingly, mutant CFTR has also been associated with increased WNT/β-catenin signalling and an exaggerated TGF-β secretion. Most recently, the same authors identified the Hippo-associated protein Yes-associated protein 1 as a potential driver of EMT and fibrosis in CF by using a multi-omics systems biology approach [282]. This transcription factor appears to be upregulated in CF as opposed to non-CF cells, and it has been shown to impair F508del CFTR trafficking, through the interaction with this protein. Furthermore, in the same study, five potential pathways were suggested to be involved in the link between mutant CFTR and EMT; these pathways (the Hippo, WNT, TGF-β, p53 pathways and MYC signalling) would need to be further investigated as potential therapeutic targets of EMT in CF.

Sousa et al. studied the protein Kruppel-like factors (KLF4), which is a transcription factor involved in the regulation of the proliferation, differentiation and wound-healing processes that are altered in CF [283]. The depletion of KLF4 in wild-type or F508del BEC had a different impact on epithelial integrity: in wild-type KLF4 KO cells, a decrease in transepithelial electrical resistance and no effect on wound closure were observed, while in F508del-KLF4 KO cells, higher levels of transepithelial electrical resistance were observed associated with a decrease in wound closure. Furthermore, the expression of EMT biomarkers and EMT-associated TFs were also differently affected. KLF4 depletion induced a switch of epithelial protein expression with a decrease in E-cadherin and cytokeratin 18 (epithelial markers) and an increase in N-cadherin and vimentin (mesenchymal proteins), that was associated with a marked decrease in TWIST1 in F508del-CFTR cells, but not in wild-type cells.

All these observations suggest that the pathways involved in the EMT process in CF are multiple and complex, and need further investigations.

5.3. Potential Effect of Novel Pharmacological Treatments on EMT and CF Airway Remodelling

Until the introduction of CFTR modulators in 2012 in the USA, the pharmacological treatment of CF lung disease was symptomatic and based on anti-inflammatory drugs, antibiotics and mucolytics/airway surface liquid hydrators. These treatments have significantly improved the life expectancy and quality of life of pwCF; however, the recent introduction of CFTR modulators had an impressive impact on clinical outcomes and the lives of pwCF [284]. This novel pharmacological approach aims at correcting the basic CFTR defect, in particular at correcting the expression (correctors) and/or improving the function (potentiators) of the CFTR protein in the respiratory epithelium. Several combinations of modulators have been developed. Among these, the most recently introduced in clinical practice was Elexacaftor–Tezacaftor–Ivacaftor (ETI), which led to an impressive improvement in CF clinical outcomes.

Some recent studies suggest a potential effect of CFTR modulators on CF airway remodelling, although the data are still scarce and need to be confirmed. Adam et al., in an in vitro study, have shown that Lumacaftor–Ivacaftor (the first CFTR modulators association commercialised) accelerated airway epithelial wound repair and improved transepithelial electrical resistance in the absence and even in the presence of *Pseudomonas aeruginosa* exoproducts [285]. Interestingly, more recently, a similar effect was reported in vitro in a CF cell line overexpressing CFTR-F508del treated with ETI. In fact, this drug not only restored F508del CFTR maturation and function but also enhanced wound repair in these cells [286]. These findings seem to indicate the potential of CFTR modulators in enhancing reparative and regenerative processes into the airways. Bec et al. studied adult pwCF treated with ETI for one year and documented that the overall CT score decreased as a consequence of decreased mucus plugging and peribronchial thickening, while bronchial, parenchymal, and hyperinflation scores remained unchanged [287]. Although some few and indirect evidence concerning a potential effect of modulators on mechanisms and features of airways remodelling in CF seem encouraging, further studies need to address this topic; longitudinal studies with CFTR modulators could provide further insight into their effects as well as on mechanisms involved in the EMT process associated with CF.

5.4. Inflammation and EMT in NCFB

NCFB are muco-obstructive diseases, including a heterogenous group of diseases with a number of underlying conditions, that affect patients of different age. The prevalence of the disease is increasing worldwide [288,289]; however, in 38.1% of cases of NCFB, the aetiology is not yet identified [290]. Similarly to CF lung disease, pathophysiology of bronchiectasis is linked to a vicious vortex of impaired mucociliary clearance, airway inflammation and infection, ultimately leading to structural damage. In particular, an aberrant epithelial remodelling with impaired mucociliary escalator architecture is present in both large and small airways. As in CF, in NCFB patients, their mucus is dehydrated and more viscous, favouring the development of infections. A marked neutrophilic inflammation is observed in the airways of patients with (pw)NCFB, and similarly to CF, a protease/antiprotease imbalance occurs with the release of NE and the overexpression of MMPs that correlate with impaired lung function and the increased neutrophilic inflammation [291]. Furthermore, severe disease was shown to be associated with an upregulation of neutrophils and NET [292].

Eosinophilic inflammation has also been observed in approximately 20% of NCFB patients and may be involved in disease exacerbations. Moreover, in pwNCFB polymorphism at position −1607 (1G or 2G) of the MMP-1 promoter induces the upregulation of MMP-1 activity, which is associated with post-inflammatory lung destruction and fibrogenesis. In a small cohort of pwNCFB, it has been shown that 1G/1G or 1G/2G genotypes are associated with a more severe phenotype and higher serum levels of MMP-1 and TGF-β1 [293].

Although airway remodelling and structural damage are key components of NCFB, so far, there are no data relating these features to the presence of EMT in these diseases, but this is certainly a field that needs active investigations as it could potentially suggest novel treatment approaches.

6. Conclusions

Chronic obstructive airway diseases are characterised by an exuberant and persistent inflammatory process into the airways as the result of repeated aggression by different triggers and noxious agents, such as cigarette smoking, pollutants, allergens and pathogens. In the context of this chronic inflammatory environment, tissue repair becomes excessive and altered, leading to the development of structural alterations, such as remodelling and fibrotic changes with different degrees of reversibility over time; these changes ultimately lead to organ dysfunction and the progressive impairment of lung function.

Although the role of EMT in the pathogenesis of these diseases is not yet well defined, several studies seem to suggest that this process is involved in pathological changes and airway remodelling, leading to chronic airflow obstruction and progressive lung function impairment.

Even if common signalling pathways are implicated in structural changes observed in these diseases, some specific triggers and a different regulation may be operating. Thus, it would be important to better define, according to the guidelines for research on EMT [1], the different mechanisms and pathways involved in the EMT process in these different airway diseases in order to identify potential specific therapeutic targets to prevent or limit the remodelling process and the irreversible structural changes.

Author Contributions: Conceptualisation, V.D.R.; investigation, A.M., L.R., A.F., S.S., S.G. and V.D.R.; writing—original draft preparation, A.M., L.R., A.F., S.S. and V.D.R.; writing—review and editing, A.M., L.R., S.G. and V.D.R.; supervision: S.G. and V.D.R. All authors have read and agreed to the published version of the manuscript.

Funding: This research received no external funding.

Institutional Review Board Statement: Not applicable.

Informed Consent Statement: Not applicable.

Data Availability Statement: Not applicable.

Acknowledgments: The authors thank Alessia Qiu for the English review.

Conflicts of Interest: A.M. received a grant, the CF Research Innovation Awards from Vertex Pharmaceutics. All other authors declare no conflict of interest.

References

1. Yang, J.; Antin, P.; Berx, G.; Blanpain, C.; Brabletz, T.; Bronner, M.; Campbell, K.; Cano, A.; Casanova, J.; Christofori, G.; et al. Guidelines and definitions for research on epithelial–mesenchymal transition. *Nat. Rev. Mol. Cell Biol.* **2020**, *21*, 341–352, Erratum in *Nat. Rev. Mol. Cell Biol.* **2021**, *22*, 834. [CrossRef]
2. Hay, E.D. An Overview of Epithelio-Mesenchymal Transformation. *Acta Anat.* **1995**, *154*, 8–20. [CrossRef]
3. Nieto, M.A.; Huang, R.Y.; Jackson, R.A.; Thiery, J.P. Emt: 2016. *Cell* **2016**, *166*, 21–45. [CrossRef] [PubMed]
4. Thiery, J.P. EMT: An Update. *Methods Mol. Biol.* **2021**, *2179*, 35–39. [PubMed]
5. Wu, B.; Sodji, Q.H.; Oyelere, A.K. Inflammation, Fibrosis and Cancer: Mechanisms, Therapeutic Options and Challenges. *Cancers* **2022**, *14*, 552. [CrossRef] [PubMed]
6. Mahmood, M.Q.; Ward, C.; Muller, H.K.; Sohal, S.S.; Walters, E.H. Epithelial mesenchymal transition (EMT) and non-small cell lung cancer (NSCLC): A mutual association with airway disease. *Med. Oncol.* **2017**, *34*, 45. [CrossRef] [PubMed]
7. Menju, T.; Date, H. Lung cancer and epithelial-mesenchymal transition. *Gen. Thorac. Cardiovasc. Surg.* **2021**, *69*, 781–789. [CrossRef]
8. Nam, M.-W.; Kim, C.-W.; Choi, K.-C. Epithelial-Mesenchymal Transition-Inducing Factors Involved in the Progression of Lung Cancers. *Biomol. Ther.* **2022**, *30*, 213–220. [CrossRef]
9. Kage, H.; Borok, Z. EMT and interstitial lung disease: A mysterious relationship. *Curr. Opin. Pulm. Med.* **2012**, *18*, 517–523. [CrossRef]
10. Hackett, T.-L. Epithelial–mesenchymal transition in the pathophysiology of airway remodelling in asthma. *Curr. Opin. Allergy Clin. Immunol.* **2012**, *12*, 53–59. [CrossRef]
11. Su, X.; Wu, W.; Zhu, Z.; Lin, X.; Zeng, Y. The effects of epithelial–mesenchymal transitions in COPD induced by cigarette smoke: An update. *Respir. Res.* **2022**, *23*, 225. [CrossRef]
12. Rout-Pitt, N.; Farrow, N.; Parsons, D.; Donnelley, M. Epithelial mesenchymal transition (EMT): A universal process in lung diseases with implications for cystic fibrosis pathophysiology. *Respir. Res.* **2018**, *19*, 136. [CrossRef]
13. James, A.L.; Wenzel, S. Clinical relevance of airway remodelling in airway diseases. *Eur. Respir. J.* **2007**, *30*, 134–155. [CrossRef] [PubMed]
14. Martinez, F.J.; Collard, H.R.; Pardo, A.; Raghu, G.; Richeldi, L.; Selman, M.; Swigris, J.J.; Taniguchi, H.; Wells, A.U. Idiopathic pulmonary fibrosis. *Nat. Rev. Dis. Primers* **2017**, *3*, 17074. [CrossRef] [PubMed]
15. Kalluri, R.; Weinberg, R.A. The basics of epithelial-mesenchymal transition. *J. Clin. Investig.* **2009**, *119*, 1420–1428. [CrossRef] [PubMed]
16. Zeisberg, M.; Neilson, E.G. Biomarkers for epithelial-mesenchymal transitions. *J. Clin. Investig.* **2009**, *119*, 1429–1437. [CrossRef]

17. Acloque, H.; Adams, M.S.; Fishwick, K.; Bronner-Fraser, M.; Nieto, M.A. Epithelial-mesenchymal transitions: The importance of changing cell state in development and disease. *J. Clin. Investig.* 2009, *119*, 1438–1449. [CrossRef]
18. Thiery, J.P.; Acloque, H.; Huang, R.Y.; Nieto, M.A. Epithelial-Mesenchymal Transitions in Development and Disease. *Cell* 2009, *139*, 871–890. [CrossRef]
19. Dongre, A.; Weinberg, R.A. New insights into the mechanisms of epithelial-mesenchymal transition and implications for cancer. *Nat. Rev. Mol. Cell Biol.* 2019, *20*, 69–84. [CrossRef]
20. Klingberg, F.; Hinz, B.; White, E.S. The myofibroblast matrix: Implications for tissue repair and fibrosis. *J. Pathol.* 2013, *229*, 298–309. [CrossRef]
21. Moulin, V.; Castilloux, G.; Auger, F.A.; Garrel, D.; O'Connor-McCourt, M.D.; Germain, L. Modulated response to cytokines of human wound healing myofibroblasts compared to dermal fibroblasts. *Exp. Cell Res.* 1998, *238*, 283–293. [CrossRef] [PubMed]
22. Ortiz-Zapater, E.; Signes-Costa, J.; Montero, P.; Roger, I. Lung Fibrosis and Fibrosis in the Lungs: Is It All about Myofibroblasts? *Biomedicines* 2022, *10*, 1423. [CrossRef]
23. Lamouille, S.; Xu, J.; Derynck, R. Molecular mechanisms of epithelial–mesenchymal transition. *Nat. Rev. Mol. Cell Biol.* 2014, *15*, 178–196. [CrossRef] [PubMed]
24. Cano, A.; Pérez-Moreno, M.A.; Rodrigo, I.; Locascio, A.; Blanco, M.J.; Del Barrio, M.G.; Portillo, F.; Nieto, M.A. The transcription factor Snail controls epithelial–mesenchymal transitions by repressing E-cadherin expression. *Nat. Cell Biol.* 2000, *2*, 76–83. [CrossRef] [PubMed]
25. Derynck, R.; Budi, E.H. Specificity, versatility, and control of TGF-beta family signaling. *Sci. Signal.* 2019, *12*, 570. [CrossRef]
26. Lee, J.H.; Massague, J. TGF-beta in developmental and fibrogenic EMTs. *Semin. Cancer Biol.* 2022, *86*, 136–145. [CrossRef]
27. Newfeld, S.J.; O'Connor, M.B. New aspects of TGF-beta superfamily signaling in development and disease (2022 FASEB meeting review). *Fac. Rev.* 2022, *11*, 36. [CrossRef]
28. Morikawa, M.; Derynck, R.; Miyazono, K. TGF-beta and the TGF-beta Family: Context-Dependent Roles in Cell and Tissue Physiology. *Cold Spring Harb. Perspect. Biol.* 2016, *8*. [CrossRef]
29. Batlle, E.; Massague, J. Transforming Growth Factor-beta Signaling in Immunity and Cancer. *Immunity* 2019, *50*, 924–940. [CrossRef]
30. Aschner, Y.; Downey, G.P. Transforming Growth Factor-beta: Master Regulator of the Respiratory System in Health and Disease. *Am. J. Respir. Cell Mol. Biol.* 2016, *54*, 647–655. [CrossRef]
31. Saito, A.; Horie, M.; Nagase, T. TGF-beta Signaling in Lung Health and Disease. *Int. J. Mol. Sci.* 2018, *19*, 2460. [CrossRef]
32. Sime, P.J.; Xing, Z.; Graham, F.L.; Csaky, K.G.; Gauldie, J. Adenovector-mediated gene transfer of active transforming growth factor-beta1 induces prolonged severe fibrosis in rat lung. *J. Clin. Investig.* 1997, *100*, 768–776. [CrossRef]
33. Wei, P.; Xie, Y.; Abel, P.W.; Huang, Y.; Ma, Q.; Li, L.; Hao, J.; Wolff, D.W.; Wei, T.; Tu, Y. Transforming growth factor (TGF)-beta1-induced miR-133a inhibits myofibroblast differentiation and pulmonary fibrosis. *Cell Death Dis.* 2019, *10*, 670. [CrossRef] [PubMed]
34. Ye, Z.; Hu, Y. TGF-beta1: Gentlemanly orchestrator in idiopathic pulmonary fibrosis (Review). *Int. J. Mol. Med.* 2021, *48*. [CrossRef] [PubMed]
35. Sisto, M.; Ribatti, D.; Lisi, S. Organ Fibrosis and Autoimmunity: The Role of Inflammation in TGFbeta-Dependent EMT. *Biomolecules* 2021, *11*, 310. [CrossRef] [PubMed]
36. Ong, C.H.; Tham, C.L.; Harith, H.H.; Firdaus, N.; Israf, D.A. TGF-beta-induced fibrosis: A review on the underlying mechanism and potential therapeutic strategies. *Eur. J. Pharmacol.* 2021, *911*, 174510. [CrossRef]
37. Lodyga, M.; Hinz, B. TGF-beta1—A truly transforming growth factor in fibrosis and immunity. *Semin. Cell Dev. Biol.* 2020, *101*, 123–139. [CrossRef]
38. Leask, A.; Abraham, D.J. TGF-beta signaling and the fibrotic response. *FASEB J.* 2004, *18*, 816–827. [CrossRef]
39. Brown, R.L.; Ormsby, I.; Doetschman, T.C.; Greenhalgh, D.G. Wound healing in the transforming growth factor-beta1-deficient mouse. *Wound Repair Regen.* 1995, *3*, 25–36. [CrossRef]
40. Camara, J.; Jarai, G. Epithelial-mesenchymal transition in primary human bronchial epithelial cells is Smad-dependent and enhanced by fibronectin and TNF-alpha. *Fibrogenesis Tissue Repair* 2010, *3*, 2. [CrossRef]
41. Kasai, H.; Allen, J.T.; Mason, R.M.; Kamimura, T.; Zhang, Z. TGF-beta1 induces human alveolar epithelial to mesenchymal cell transition (EMT). *Respir. Res.* 2005, *6*, 56. [CrossRef] [PubMed]
42. Shi, M.; Zhu, J.; Wang, R.; Chen, X.; Mi, L.; Walz, T.; Springer, T.A. Latent TGF-beta structure and activation. *Nature* 2011, *474*, 343–349. [CrossRef] [PubMed]
43. Tzavlaki, K.; Moustakas, A. TGF-beta Signaling. *Biomolecules* 2020, *10*, 487. [CrossRef]
44. Burgy, O.; Königshoff, M. The WNT signaling pathways in wound healing and fibrosis. *Matrix Biol.* 2018, *68–69*, 67–80. [CrossRef]
45. Cao, H.; Wang, C.; Chen, X.; Hou, J.; Xiang, Z.; Shen, Y.; Han, X. Inhibition of Wnt/beta-catenin signaling suppresses myofibroblast differentiation of lung resident mesenchymal stem cells and pulmonary fibrosis. *Sci. Rep.* 2018, *8*, 13644. [CrossRef] [PubMed]
46. Effendi, W.I.; Nagano, T. The Hedgehog Signaling Pathway in Idiopathic Pulmonary Fibrosis: Resurrection Time. *Int. J. Mol. Sci.* 2021, *23*, 171. [CrossRef] [PubMed]
47. Chanda, D.; Otoupalova, E.; Smith, S.R.; Volckaert, T.; De Langhe, S.P.; Thannickal, V.J. Developmental pathways in the pathogenesis of lung fibrosis. *Mol. Asp. Med.* 2019, *65*, 56–69. [CrossRef] [PubMed]

48. Hu, B.; Liu, J.; Wu, Z.; Liu, T.; Ullenbruch, M.R.; Ding, L.; Henke, C.A.; Bitterman, P.B.; Phan, S.H. Reemergence of Hedgehog Mediates Epithelial–Mesenchymal Crosstalk in Pulmonary Fibrosis. *Am. J. Respir. Cell Mol. Biol.* **2015**, *52*, 418–428. [CrossRef]
49. Cigna, N.; Farrokhi Moshai, E.; Brayer, S.; Marchal-Somme, J.; Wemeau-Stervinou, L.; Fabre, A.; Mal, H.; Leseche, G.; Dehoux, M.; Soler, P.; et al. The hedgehog system machinery controls transforming growth factor-beta-dependent myofibroblastic differentiation in humans: Involvement in idiopathic pulmonary fibrosis. *Am. J. Pathol.* **2012**, *181*, 2126–2137. [CrossRef]
50. Heldin, C.-H.; Lennartsson, J.; Westermark, B. Involvement of platelet-derived growth factor ligands and receptors in tumorigenesis. *J. Intern. Med.* **2018**, *283*, 16–44. [CrossRef]
51. Qin, W.; Pan, Y.; Zheng, X.; Li, D.; Bu, J.; Xu, C.; Tang, J.; Cui, R.; Lin, P.; Yu, X. MicroRNA-124 regulates TGF-alpha-induced epithelial-mesenchymal transition in human prostate cancer cells. *Int. J. Oncol.* **2014**, *45*, 1225–1231. [CrossRef]
52. Li, Y.; Li, H.; Duan, Y.; Cai, X.; You, D.; Zhou, F.; Yang, C.; Tuo, X.; Liu, Z. Blockage of TGF-alpha Induced by Spherical Silica Nanoparticles Inhibits Epithelial-Mesenchymal Transition and Proliferation of Human Lung Epithelial Cells. *BioMed Res. Int.* **2019**, *2019*, 8231267. [CrossRef] [PubMed]
53. Tang, J.; Xiao, L.; Cui, R.; Li, D.; Zheng, X.; Zhu, L.; Sun, H.; Pan, Y.; Du, Y.; Yu, X. CX3CL1 increases invasiveness and metastasis by promoting epithelial-to-mesenchymal transition through the TACE/TGF-alpha/EGFR pathway in hypoxic androgen-independent prostate cancer cells. *Oncol. Rep.* **2016**, *35*, 1153–1162. [CrossRef] [PubMed]
54. Saitoh, M. Epithelial-Mesenchymal Transition by Synergy between Transforming Growth Factor-beta and Growth Factors in Cancer Progression. *Diagnostics* **2022**, *12*, 2127. [CrossRef]
55. Shu, D.Y.; Lovicu, F.J. Enhanced EGF receptor-signaling potentiates TGFbeta-induced lens epithelial-mesenchymal transition. *Exp. Eye Res.* **2019**, *185*, 107693. [CrossRef]
56. Schelch, K.; Wagner, C.; Hager, S.; Pirker, C.; Siess, K.; Lang, E.; Lin, R.; Kirschner, M.B.; Mohr, T.; Brcic, L.; et al. FGF2 and EGF induce epithelial–mesenchymal transition in malignant pleural mesothelioma cells via a MAPKinase/MMP1 signal. *Carcinog.* **2018**, *39*, 534–545. [CrossRef]
57. Chen, R.; Jin, G.; Li, W.; McIntyre, T.M. Epidermal Growth Factor (EGF) Autocrine Activation of Human Platelets Promotes EGF Receptor-Dependent Oral Squamous Cell Carcinoma Invasion, Migration, and Epithelial Mesenchymal Transition. *J. Immunol.* **2018**, *201*, 2154–2164. [CrossRef] [PubMed]
58. Jia, C.; Wang, G.; Wang, T.; Fu, B.; Zhang, Y.; Huang, L.; Deng, Y.; Chen, G.; Wu, X.; Chen, J.; et al. Cancer-associated Fibroblasts induce epithelial-mesenchymal transition via the Transglutaminase 2-dependent IL-6/IL6R/STAT3 axis in Hepatocellular Carcinoma. *Int. J. Biol. Sci.* **2020**, *16*, 2542–2558. [CrossRef]
59. Park, M.K.; You, H.J.; Lee, H.J.; Kang, J.H.; Oh, S.H.; Kim, S.Y.; Lee, C.H. Transglutaminase-2 induces N-cadherin expression in TGF-beta1-induced epithelial mesenchymal transition via c-Jun-N-terminal kinase activation by protein phosphatase 2A down-regulation. *Eur. J. Cancer* **2013**, *49*, 1692–1705. [CrossRef]
60. Shafiq, A.; Suwakulsiri, W.; Rai, A.; Chen, M.; Greening, D.W.; Zhu, H.J.; Xu, R.; Simpson, R.J. Transglutaminase-2, RNA-binding proteins and mitochondrial proteins selectively traffic to MDCK cell-derived microvesicles following H-Ras-induced epithelial-mesenchymal transition. *Proteomics* **2021**, *21*, e2000221. [CrossRef]
61. Milton, A.V.; Konrad, D.B. Epithelial-mesenchymal transition and H(2)O(2) signaling—A driver of disease progression and a vulnerability in cancers. *Biol. Chem.* **2022**, *403*, 377–390. [CrossRef] [PubMed]
62. Ma, M.; Shi, F.; Zhai, R.; Wang, H.; Li, K.; Xu, C.; Yao, W.; Zhou, F. TGF-beta promote epithelial-mesenchymal transition via NF-kappaB/NOX4/ROS signal pathway in lung cancer cells. *Mol. Biol. Rep.* **2021**, *48*, 2365–2375. [CrossRef]
63. Kolaczkowska, E.; Kubes, P. Neutrophil recruitment and function in health and inflammation. *Nat. Rev. Immunol.* **2013**, *13*, 159–175. [CrossRef]
64. Plikus, M.V.; Wang, X.; Sinha, S.; Forte, E.; Thompson, S.M.; Herzog, E.L.; Driskell, R.R.; Rosenthal, N.; Biernaskie, J.; Horsley, V. Fibroblasts: Origins, definitions, and functions in health and disease. *Cell* **2021**, *184*, 3852–3872. [CrossRef]
65. Porsbjerg, C.; Melen, E.; Lehtimaki, L.; Shaw, D. Asthma. *Lancet* **2023**, *401*, 858–873. [CrossRef] [PubMed]
66. Ritchie, A.I.; Jackson, D.J.; Edwards, M.R.; Johnston, S.L. Airway Epithelial Orchestration of Innate Immune Function in Response to Virus Infection. A Focus on Asthma. *Ann. Am. Thorac. Soc.* **2016**, *13*, S55–S63. [CrossRef]
67. Noble, P.B.; Pascoe, C.D.; Lan, B.; Ito, S.; Kistemaker, L.E.; Tatler, A.L.; Pera, T.; Brook, B.S.; Gosens, R.; West, A.R. Airway smooth muscle in asthma: Linking contraction and mechanotransduction to disease pathogenesis and remodelling. *Pulm. Pharmacol. Ther.* **2014**, *29*, 96–107. [CrossRef] [PubMed]
68. King, G.G.; Noble, P.B. Airway remodelling in asthma: It's not going away. *Respirology* **2016**, *21*, 203–204. [CrossRef]
69. Girodet, P.O.; Allard, B.; Thumerel, M.; Begueret, H.; Dupin, I.; Ousova, O.; Lassalle, R.; Maurat, E.; Ozier, A.; Trian, T.; et al. Bronchial Smooth Muscle Remodeling in Nonsevere Asthma. *Am. J. Respir. Crit. Care Med.* **2016**, *193*, 627–633. [CrossRef]
70. Prakash, Y.S. Emerging concepts in smooth muscle contributions to airway structure and function: Implications for health and disease. *Am. J. Physiol. Lung Cell. Mol. Physiol.* **2016**, *311*, L1113–L1140. [CrossRef]
71. Holgate, S.T. Pathogenesis of asthma. *Clin. Exp. Allergy* **2008**, *38*, 872–897. [CrossRef] [PubMed]
72. Mostaço-Guidolin, L.B.; Osei, E.T.; Ullah, J.; Hajimohammadi, S.; Fouadi, M.; Li, X.; Li, V.; Shaheen, F.; Yang, C.X.; Chu, F.; et al. Defective Fibrillar Collagen Organization by Fibroblasts Contributes to Airway Remodeling in Asthma. *Am. J. Respir. Crit. Care Med.* **2019**, *200*, 431–443. [CrossRef] [PubMed]

73. Osei, E.T.; Mostaço-Guidolin, L.B.; Hsieh, A.; Warner, S.M.; Al-Fouadi, M.; Wang, M.; Cole, D.J.; Maksym, G.N.; Hallstrand, T.S.; Timens, W.; et al. Epithelial-interleukin-1 inhibits collagen formation by airway fibroblasts: Implications for asthma. *Sci. Rep.* **2020**, *10*, 8721. [CrossRef] [PubMed]
74. Mosmann, T.R.; Cherwinski, H.; Bond, M.W.; Giedlin, M.A.; Coffman, R.L. Two types of murine helper T cell clone. I. Definition according to profiles of lymphokine activities and secreted proteins. *J. Immunol.* **1986**, *136*, 2348–2357.
75. Le Gros, G.; Ben-Sasson, S.Z.; Seder, R.; Finkelman, F.D.; Paul, W.E. Generation of interleukin 4 (IL-4)-producing cells in vivo and in vitro: IL-2 and IL-4 are required for in vitro generation of IL-4-producing cells. *J. Exp. Med.* **1990**, *172*, 921–929. [CrossRef]
76. Metcalfe, D.D.; Baram, D.; Mekori, Y.A. Mast cells. *Physiol. Rev.* **1997**, *77*, 1033–1079. [CrossRef]
77. Tsartsali, L.; Hislop, A.A.; McKay, K.; James, A.L.; Elliot, J.; Zhu, J.; Rosenthal, M.; Payne, D.N.; Jeffery, P.K.; Bush, A.; et al. Development of the bronchial epithelial reticular basement membrane: Relationship to epithelial height and age. *Thorax* **2011**, *66*, 280–285. [CrossRef]
78. Dunnill, M.S. The pathology of asthma, with special reference to changes in the bronchial mucosa. *J. Clin. Pathol.* **1960**, *13*, 27–33. [CrossRef]
79. Roche, W.R.; Beasley, R.; Williams, J.H.; Holgate, S.T. Subepithelial fibrosis in the bronchi of asthmatics. *Lancet* **1989**, *333*, 520–524. [CrossRef]
80. Broekema, M.; Timens, W.; Vonk, J.M.; Volbeda, F.; Lodewijk, M.E.; Hylkema, M.N.; Ten Hacken, N.H.; Postma, D.S. Persisting Remodeling and Less Airway Wall Eosinophil Activation in Complete Remission of Asthma. *Am. J. Respir. Crit. Care Med.* **2011**, *183*, 310–316. [CrossRef]
81. Pain, M.; Bermudez, O.; Lacoste, P.; Royer, P.-J.; Botturi, K.; Tissot, A.; Brouard, S.; Eickelberg, O.; Magnan, A. Tissue remodelling in chronic bronchial diseases: From the epithelial to mesenchymal phenotype. *Eur. Respir. Rev.* **2014**, *23*, 118–130. [CrossRef] [PubMed]
82. Iwano, M.; Plieth, D.; Danoff, T.M.; Xue, C.; Okada, H.; Neilson, E.G. Evidence that fibroblasts derive from epithelium during tissue fibrosis. *J. Clin. Investig.* **2002**, *110*, 341–350. [CrossRef] [PubMed]
83. Royce, S.G.; Cheng, V.; Samuel, C.S.; Tang, M.L. The regulation of fibrosis in airway remodeling in asthma. *Mol. Cell. Endocrinol.* **2012**, *351*, 167–175. [CrossRef] [PubMed]
84. Loffredo, L.F.; Abdala-Valencia, H.; Anekalla, K.R.; Cuervo-Pardo, L.; Gottardi, C.J.; Berdnikovs, S. Beyond epithelial-to-mesenchymal transition: Common suppression of differentiation programs underlies epithelial barrier dysfunction in mild, moderate, and severe asthma. *Allergy* **2017**, *72*, 1988–2004. [CrossRef]
85. Nawijn, M.C.; Hackett, T.L.; Postma, D.S.; van Oosterhout, A.J.; Heijink, I.H. E-cadherin: Gatekeeper of airway mucosa and allergic sensitization. *Trends Immunol.* **2011**, *32*, 248–255. [CrossRef]
86. Magnan, A.; Frachon, I.; Rain, B.; Peuchmaur, M.; Monti, G.; Lenot, B.; Fattal, M.; Simonneau, G.; Galanaud, P.; Emilie, D. Transforming growth factor beta in normal human lung: Preferential location in bronchial epithelial cells. *Thorax* **1994**, *49*, 789–792. [CrossRef]
87. Hackett, T.L.; Warner, S.M.; Stefanowicz, D.; Shaheen, F.; Pechkovsky, D.V.; Murray, L.A.; Argentieri, R.; Kicic, A.; Stick, S.M.; Bai, T.R.; et al. Induction of epithelial-mesenchymal transition in primary airway epithelial cells from patients with asthma by transforming growth factor-beta1. *Am. J. Respir. Crit. Care Med.* **2009**, *180*, 122–133. [CrossRef]
88. Halwani, R.; Al-Muhsen, S.; Al-Jahdali, H.; Hamid, Q. Role of transforming growth factor-beta in airway remodeling in asthma. *Am. J. Respir. Cell Mol. Biol.* **2011**, *44*, 127–133. [CrossRef]
89. Redington, A.E.; Madden, J.; Frew, A.J.; Djukanovic, R.; Roche, W.R.; Holgate, S.T.; Howarth, P.H. Transforming growth factor-beta 1 in asthma. Measurement in bronchoalveolar lavage fluid. *Am. J. Respir. Crit. Care Med.* **1997**, *156*, 642–647. [CrossRef]
90. Ji, X.; Li, J.; Xu, L.; Wang, W.; Luo, M.; Luo, S.; Ma, L.; Li, K.; Gong, S.; He, L.; et al. IL4 and IL-17A provide a Th2/Th17-polarized inflammatory milieu in favor of TGF-beta1 to induce bronchial epithelial-mesenchymal transition (EMT). *Int. J. Clin. Exp. Pathol.* **2013**, *6*, 1481–1492.
91. Evasovic, J.M.; Singer, C.A. Regulation of IL-17A and implications for TGF-beta1 comodulation of airway smooth muscle remodeling in severe asthma. *Am. J. Physiol. Lung Cell. Mol. Physiol.* **2019**, *316*, L843–L868. [CrossRef] [PubMed]
92. Yang, H.W.; Lee, S.A.; Shin, J.M.; Park, I.H.; Lee, H.M. Glucocorticoids ameliorate TGF-beta1-mediated epithelial-to-mesenchymal transition of airway epithelium through MAPK and Snail/Slug signaling pathways. *Sci. Rep.* **2017**, *7*, 3486. [CrossRef] [PubMed]
93. Heijink, I.H.; Postma, D.S.; Noordhoek, J.A.; Broekema, M.; Kapus, A. House Dust Mite–Promoted Epithelial-to-Mesenchymal Transition in Human Bronchial Epithelium. *Am. J. Respir. Cell Mol. Biol.* **2010**, *42*, 69–79. [CrossRef]
94. Hackett, T.-L.; de Bruin, H.G.; Shaheen, F.; van den Berge, M.; van Oosterhout, A.J.; Postma, D.S.; Heijink, I.H. Caveolin-1 Controls Airway Epithelial Barrier Function. Implications for Asthma. *Am. J. Respir. Cell Mol. Biol.* **2013**, *49*, 662–671. [CrossRef]
95. Johnson, J.R.; Roos, A.; Berg, T.; Nord, M.; Fuxe, J. Chronic Respiratory Aeroallergen Exposure in Mice Induces Epithelial-Mesenchymal Transition in the Large Airways. *PLoS ONE* **2011**, *6*, e16175. [CrossRef]
96. Minshall, E.M.; Leung, D.Y.; Martin, R.J.; Song, Y.L.; Cameron, L.; Ernst, P.; Hamid, Q. Eosinophil-associated TGF-beta1 mRNA expression and airways fibrosis in bronchial asthma. *Am. J. Respir. Cell Mol. Biol.* **1997**, *17*, 326–333. [CrossRef]
97. Vignola, A.M.; Chanez, P.; Chiappara, G.; Merendino, A.; Pace, E.; Rizzo, A.; la Rocca, A.M.; Bellia, V.; Bonsignore, G.; Bousquet, J. Transforming growth factor-beta expression in mucosal biopsies in asthma and chronic bronchitis. *Am. J. Respir. Crit. Care Med.* **1997**, *156*, 591–599. [CrossRef]

98. Yasukawa, A.; Hosoki, K.; Toda, M.; Miyake, Y.; Matsushima, Y.; Matsumoto, T.; Boveda-Ruiz, D.; Gil-Bernabe, P.; Nagao, M.; Sugimoto, M.; et al. Eosinophils promote epithelial to mesenchymal transition of bronchial epithelial cells. *PLoS ONE* **2013**, *8*, e64281. [CrossRef]
99. Doerner, A.M.; Zuraw, B.L. TGF-beta1 induced epithelial to mesenchymal transition (EMT) in human bronchial epithelial cells is enhanced by IL-1beta but not abrogated by corticosteroids. *Respir. Res.* **2009**, *10*, 100. [CrossRef]
100. Commins, S.; Steinke, J.W.; Borish, L. The extended IL-10 superfamily: IL-10, IL-19, IL-20, IL-22, IL-24, IL-26, IL-28, and IL-29. *J. Allergy Clin. Immunol.* **2008**, *121*, 1108–1111. [CrossRef]
101. Zissler, U.M.; Ulrich, M.; Jakwerth, C.A.; Rothkirch, S.; Guerth, F.; Weckmann, M.; Schiemann, M.; Haller, B.; Schmidt-Weber, C.B.; Chaker, A.M. Biomatrix for upper and lower airway biomarkers in patients with allergic asthma. *J. Allergy Clin. Immunol.* **2018**, *142*, 1980–1983. [CrossRef] [PubMed]
102. Feng, K.-N.; Meng, P.; Zou, X.-L.; Zhang, M.; Li, H.-K.; Yang, H.-L.; Li, H.-T.; Zhang, T.-T. IL-37 protects against airway remodeling by reversing bronchial epithelial–mesenchymal transition via IL-24 signaling pathway in chronic asthma. *Respir. Res.* **2022**, *23*, 244. [CrossRef] [PubMed]
103. Roberts, C.R. Is asthma a fibrotic disease? *Chest* **1995**, *107*, 111S–117S. [CrossRef]
104. Laitinen, A.; Altraja, A.; Kämpe, M.; Linden, M.; Virtanen, I.; Laitinen, L.A. Tenascin Is Increased in Airway Basement Membrane of Asthmatics and Decreased by an Inhaled Steroid. *Am. J. Respir. Crit. Care Med.* **1997**, *156*, 951–958. [CrossRef]
105. Elias, J.A.; Zhu, Z.; Chupp, G.; Homer, R.J. Airway remodeling in asthma. *J. Clin. Investig.* **1999**, *104*, 1001–1006. [CrossRef]
106. Zeiger, R.S.; Dawson, C.; Weiss, S. Relationships between duration of asthma and asthma severity among children in the Childhood Asthma Management Program (CAMP). *J. Allergy Clin. Immunol.* **1999**, *103*, 376–386. [CrossRef]
107. Boulet, L.-P.; Turcotte, H.; Laviolette, M.; Naud, F.; Bernier, M.-C.; Martel, S.; Chakir, J. Airway hyperresponsiveness, inflammation, and subepithelial collagen deposition in recently diagnosed versus long-standing mild asthma. Influence of inhaled corticosteroids. *Am. J. Respir. Crit. Care Med.* **2000**, *162*, 1308–1313. [CrossRef]
108. James, A.L.; Maxwell, P.S.; Pearce-Pinto, G.; Elliot, J.G.; Carroll, N.G. The Relationship of Reticular Basement Membrane Thickness to Airway Wall Remodeling in Asthma. *Am. J. Respir. Crit. Care Med.* **2002**, *166*, 1590–1595. [CrossRef]
109. Burgess, J.K.; Mauad, T.; Tjin, G.; Karlsson, J.C.; Westergren-Thorsson, G. The extracellular matrix - the under-recognized element in lung disease? *J. Pathol.* **2016**, *240*, 397–409. [CrossRef]
110. Carroll, N.G.; Perry, S.; Karkhanis, A.; Harji, S.; Butt, J.; James, A.L.; Green, F.H. The Airway Longitudinal Elastic Fiber Network and Mucosal Folding in Patients with Asthma. *Am. J. Respir. Crit. Care Med.* **2000**, *161*, 244–248. [CrossRef]
111. Boser, S.R.; Mauad, T.; Araújo-Paulino, B.B.; Mitchell, I.; Shrestha, G.; Chiu, A.; Butt, J.; Kelly, M.M.; Caldini, E.; James, A.; et al. Myofibroblasts are increased in the lung parenchyma in asthma. *PLoS ONE* **2017**, *12*, e0182378. [CrossRef] [PubMed]
112. Payne, D.N.; Rogers, A.V.; Ädelroth, E.; Bandi, V.; Guntupalli, K.K.; Bush, A.; Jeffery, P.K. Early Thickening of the Reticular Basement Membrane in Children with Difficult Asthma. *Am. J. Respir. Crit. Care Med.* **2003**, *167*, 78–82. [CrossRef] [PubMed]
113. Cokugras, H.; Akcakaya, N.; Seckin, ; Camcioglu, Y.; Sarimurat, N.; Aksoy, F. Ultrastructural examination of bronchial biopsy specimens from children with moderate asthma. *Thorax* **2001**, *56*, 25–29. [CrossRef] [PubMed]
114. Nihlberg, K.; Larsen, K.; Hultgardh-Nilsson, A.; Malmstrom, A.; Bjermer, L.; Westergren-Thorsson, G. Tissue fibrocytes in patients with mild asthma: A possible link to thickness of reticular basement membrane? *Respir. Res.* **2006**, *7*, 50. [CrossRef]
115. Snelgrove, R.J.; Patel, D.F. Zooming into the Matrix: Using Nonlinear Optical Microscopy to Visualize Collagen Remodeling in Asthmatic Airways. *Am. J. Respir. Crit. Care Med.* **2019**, *200*, 403–405. [CrossRef]
116. Ward, C.; Pais, M.; Bish, R.; Reid, D.; Feltis, B.; Johns, D.; Walters, E.H. Airway inflammation, basement membrane thickening and bronchial hyperresponsiveness in asthma. *Thorax* **2002**, *57*, 309–316. [CrossRef]
117. Tsurikisawa, N.; Oshikata, C.; Tsuburai, T.; Saito, H.; Sekiya, K.; Tanimoto, H.; Takeichi, S.; Mitomi, H.; Akiyama, K. Bronchial hyperresponsiveness to histamine correlates with airway remodelling in adults with asthma. *Respir. Med.* **2010**, *104*, 1271–1277. [CrossRef]
118. Mostaco-Guidolin, L.; Hajimohammadi, S.; Vasilescu, D.M.; Hackett, T.-L. Application of Euclidean distance mapping for assessment of basement membrane thickness distribution in asthma. *J. Appl. Physiol. (1985)* **2017**, *123*, 473–481. [CrossRef]
119. Kuo, C.; Lim, S.; King, N.J.; Johnston, S.L.; Burgess, J.K.; Black, J.L.; Oliver, B.G. Rhinovirus infection induces extracellular matrix protein deposition in asthmatic and nonasthmatic airway smooth muscle cells. *Am. J. Physiol. Lung Cell. Mol. Physiol.* **2011**, *300*, L951–L957. [CrossRef]
120. Carroll, N.; Elliot, J.; Morton, A.; James, A. The Structure of Large and Small Airways in Nonfatal and Fatal Asthma. *Am. Rev. Respir. Dis.* **1993**, *147*, 405–410. [CrossRef]
121. Hirst, S.J.; Barnes, P.J.; Twort, C.H. PDGF isoform-induced proliferation and receptor expression in human cultured airway smooth muscle cells. *Am. J. Physiol.* **1996**, *270*, L415–L428. [CrossRef] [PubMed]
122. Panettieri, R.A., Jr.; Goldie, R.G.; Rigby, P.J.; Eszterhas, A.J.; Hay, D.W. Endothelin-1-induced potentiation of human airway smooth muscle proliferation: An ETA receptor-mediated phenomenon. *Br. J. Pharmacol.* **1996**, *118*, 191–197. [CrossRef] [PubMed]
123. Cohen, P.; Rajah, R.; Rosenbloom, J.; Herrick, D.J. IGFBP-3 mediates TGF-beta1-induced cell growth in human airway smooth muscle cells. *Am. J. Physiol. Lung Cell. Mol. Physiol.* **2000**, *278*, L545–L551. [CrossRef]
124. Chen, G.; Khalil, N. TGF-beta1 increases proliferation of airway smooth muscle cells by phosphorylation of map kinases. *Respir. Res.* **2006**, *7*, 2. [CrossRef] [PubMed]

125. Hakonarson, H.; Maskeri, N.; Carter, C.; Chuang, S.; Grunstein, M.M. Autocrine interaction between IL-5 and IL-1beta mediates altered responsiveness of atopic asthmatic sensitized airway smooth muscle. *J. Clin. Investig.* **1999**, *104*, 657–667. [CrossRef] [PubMed]
126. Coutts, A.; Chen, G.; Stephens, N.; Hirst, S.; Douglas, D.; Eichholtz, T.; Khalil, N. Release of biologically active TGF-beta from airway smooth muscle cells induces autocrine synthesis of collagen. *Am. J. Physiol. Lung Cell. Mol. Physiol.* **2001**, *280*, L999–L1008. [CrossRef]
127. Moynihan, B.J.; Tolloczko, B.; El Bassam, S.; Ferraro, P.; Michoud, M.-C.; Martin, J.G.; Laberge, S. IFN-gamma, IL-4 and IL-13 modulate responsiveness of human airway smooth muscle cells to IL-13. *Respir. Res.* **2008**, *9*, 84. [CrossRef] [PubMed]
128. Grainge, C.L.; Lau, L.C.; Ward, J.A.; Dulay, V.; Lahiff, G.; Wilson, S.; Holgate, S.; Davies, D.E.; Howarth, P.H. Effect of bronchoconstriction on airway remodeling in asthma. *N. Engl. J. Med.* **2011**, *364*, 2006–2015. [CrossRef]
129. Camoretti-Mercado, B.; Lockey, R.F. Airway smooth muscle pathophysiology in asthma. *J. Allergy Clin. Immunol.* **2021**, *147*, 1983–1995. [CrossRef]
130. Kuyper, L.M.; Paré, P.D.; Hogg, J.C.; Lambert, R.K.; Ionescu, D.; Woods, R.; Bai, T.R. Characterization of airway plugging in fatal asthma. *Am. J. Med.* **2003**, *115*, 6–11. [CrossRef]
131. Wenzel, S.E.; Vitari, C.A.; Shende, M.; Strollo, D.C.; Larkin, A.; Yousem, S.A. Asthmatic granulomatosis: A novel disease with asthmatic and granulomatous features. *Am. J. Respir. Crit. Care Med.* **2012**, *186*, 501–507. [CrossRef]
132. Malmström, K.; Lohi, J.; Sajantila, A.; Jahnsen, F.L.; Kajosaari, M.; Sarna, S.; Mäkelä, M.J. Immunohistology and remodeling in fatal pediatric and adolescent asthma. *Respir. Res.* **2017**, *18*, 94. [CrossRef]
133. Treho Bittar, H.E.; Doberer, D.; Mehrad, M.; Strollo, D.C.; Leader, J.K.; Wenzel, S.; Yousem, S.A. Histologic Findings of Severe/Therapy-Resistant Asthma from Video-assisted Thoracoscopic Surgery Biopsies. *Am. J. Surg. Pathol.* **2017**, *41*, 182–188. [CrossRef]
134. Elliot, J.G.; Noble, P.B.; Mauad, T.; Bai, T.R.; Abramson, M.J.; McKay, K.O.; Green, F.H.Y.; James, A.L. Inflammation-dependent and independent airway remodelling in asthma. *Respirology* **2018**, *23*, 1138–1145. [CrossRef] [PubMed]
135. Curran, D.R.; Cohn, L. Advances in mucous cell metaplasia: A plug for mucus as a therapeutic focus in chronic airway disease. *Am. J. Respir. Cell Mol. Biol.* **2010**, *42*, 268–275. [CrossRef]
136. Sun, Z.; Ji, N.; Ma, Q.; Zhu, R.; Chen, Z.; Wang, Z.; Qian, Y.; Wu, C.; Hu, F.; Huang, M.; et al. Epithelial-Mesenchymal Transition in Asthma Airway Remodeling Is Regulated by the IL-33/CD146 Axis. *Front. Immunol.* **2020**, *11*, 1598. [CrossRef] [PubMed]
137. Haddad, A.; Gaudet, M.; Plesa, M.; Allakhverdi, Z.; Mogas, A.K.; Audusseau, S.; Baglole, C.J.; Eidelman, D.H.; Olivenstein, R.; Ludwig, M.S.; et al. Neutrophils from severe asthmatic patients induce epithelial to mesenchymal transition in healthy bronchial epithelial cells. *Respir. Res.* **2019**, *20*, 234. [CrossRef] [PubMed]
138. Zhao, J.; Jiang, T.; Li, P.; Dai, L.; Shi, G.; Jing, X.; Gao, S.; Jia, L.; Wu, S.; Wang, Y.; et al. Tissue factor promotes airway pathological features through epithelial-mesenchymal transition of bronchial epithelial cells in mice with house dust mite-induced asthma. *Int. Immunopharmacol.* **2021**, *97*, 107690. [CrossRef]
139. Yang, Z.C.; Qu, Z.H.; Yi, M.J.; Shan, Y.C.; Ran, N.; Xu, L.; Liu, X.J. MiR-448-5p inhibits TGF-beta1-induced epithelial-mesenchymal transition and pulmonary fibrosis by targeting Six1 in asthma. *J. Cell. Physiol.* **2019**, *234*, 8804–8814. [CrossRef]
140. Gong, J.-H.; Cho, I.-H.; Shin, D.; Han, S.-Y.; Park, S.-H.; Kang, Y.-H. Inhibition of airway epithelial-to-mesenchymal transition and fibrosis by kaempferol in endotoxin-induced epithelial cells and ovalbumin-sensitized mice. *Lab. Investig.* **2014**, *94*, 297–308. [CrossRef] [PubMed]
141. Mattos, W.; Lim, S.; Russell, R.; Jatakanon, A.; Chung, K.F.; Barnes, P.J. Matrix metalloproteinase-9 expression in asthma: Effect of asthma severity, allergen challenge, and inhaled corticosteroids. *Chest* **2002**, *122*, 1543–1552. [CrossRef] [PubMed]
142. Jeffery, P.K.; Godfrey, R.W.; Ädelroth, E.; Nelson, F.; Rogers, A.; Johansson, S.-A. Effects of Treatment on Airway Inflammation and Thickening of Basement Membrane Reticular Collagen in Asthma: A Quantitative Light and Electron Microscopic Study. *Am. Rev. Respir. Dis.* **1992**, *145*, 890–899. [CrossRef] [PubMed]
143. Bergeron, C.; Hauber, H.P.; Gotfried, M.; Newman, K.; Dhanda, R.; Servi, R.J.; Ludwig, M.S.; Hamid, Q. Evidence of remodeling in peripheral airways of patients with mild to moderate asthma: Effect of hydrofluoroalkane-flunisolide. *J. Allergy Clin. Immunol.* **2005**, *116*, 983–989. [CrossRef] [PubMed]
144. Lundgren, R.; Soderberg, M.; Horstedt, P.; Stenling, R. Morphological studies of bronchial mucosal biopsies from asthmatics before and after ten years of treatment with inhaled steroids. *Eur. Respir. J.* **1988**, *1*, 883–889. [CrossRef] [PubMed]
145. Phelan, P.D.; Robertson, C.F.; Olinsky, A. The Melbourne Asthma Study: 1964–1999. *J. Allergy Clin. Immunol.* **2002**, *109*, 189–194. [CrossRef]
146. Sears, M.R.; Greene, J.M.; Willan, A.R.; Wiecek, E.M.; Taylor, D.R.; Flannery, E.M.; Cowan, J.O.; Herbison, G.P.; Silva, P.A.; Poulton, R. A Longitudinal, Population-Based, Cohort Study of Childhood Asthma Followed to Adulthood. *N. Engl. J. Med.* **2003**, *349*, 1414–1422. [CrossRef]
147. Chen, W.-J.; Liaw, S.-F.; Lin, C.-C.; Lin, M.-W.; Chang, F.-T. Effects of Zileuton on Airway Smooth Muscle Remodeling after Repeated Allergen Challenge in Brown Norway Rats. *Respiration* **2013**, *86*, 421–429. [CrossRef]
148. Hur, J.; Kang, J.Y.; Rhee, C.K.; Kim, Y.K.; Lee, S.Y. The leukotriene receptor antagonist pranlukast attenuates airway remodeling by suppressing TGF-beta signaling. *Pulm. Pharmacol. Ther.* **2018**, *48*, 5–14. [CrossRef]
149. Janssen, E.M.; Dy, S.M.; Meara, A.S.; Kneuertz, P.J.; Presley, C.J.; Bridges, J.F.P. Analysis of Patient Preferences in Lung Cancer – Estimating Acceptable Tradeoffs Between Treatment Benefit and Side Effects. *Patient Prefer. Adherence* **2020**, *14*, 927–937. [CrossRef]

150. Donovan, G.M.; Wang, K.C.W.; Shamsuddin, D.; Mann, T.S.; Henry, P.J.; Larcombe, A.N.; Noble, P.B. Pharmacological ablation of the airway smooth muscle layer—Mathematical predictions of functional improvement in asthma. *Physiol. Rep.* **2020**, *8*, e14451. [CrossRef]
151. Clifford, R.L.; Knox, A.J. Vitamin D—A new treatment for airway remodelling in asthma? *Br. J. Pharmacol.* **2009**, *158*, 1426–1428. [CrossRef] [PubMed]
152. Britt, R.D., Jr.; Thompson, M.A.; Freeman, M.R.; Stewart, A.L.; Pabelick, C.M.; Prakash, Y.S. Vitamin D Reduces Inflammation-induced Contractility and Remodeling of Asthmatic Human Airway Smooth Muscle. *Ann. Am. Thorac. Soc.* **2016**, *13*, S97–S98. [CrossRef] [PubMed]
153. Britt, R.D., Jr.; Faksh, A.; Vogel, E.R.; Thompson, M.A.; Chu, V.; Pandya, H.C.; Amrani, Y.; Martin, R.J.; Pabelick, C.M.; Prakash, Y.S. Vitamin D Attenuates Cytokine-Induced Remodeling in Human Fetal Airway Smooth Muscle Cells. *J. Cell. Physiol.* **2015**, *230*, 1189–1198. [CrossRef] [PubMed]
154. Roth, M.; Zhao, F.; Zhong, J.; Lardinois, D.; Tamm, M. Serum IgE Induced Airway Smooth Muscle Cell Remodeling Is Independent of Allergens and Is Prevented by Omalizumab. *PLoS ONE* **2015**, *10*, e0136549. [CrossRef]
155. Hoshino, M.; Ohtawa, J. Effects of Adding Omalizumab, an Anti-Immunoglobulin E Antibody, on Airway Wall Thickening in Asthma. *Respiration* **2012**, *83*, 520–528. [CrossRef]
156. Tajiri, T.; Niimi, A.; Matsumoto, H.; Ito, I.; Oguma, T.; Otsuka, K.; Takeda, T.; Nakaji, H.; Inoue, H.; Iwata, T.; et al. Comprehensive efficacy of omalizumab for severe refractory asthma: A time-series observational study. *Ann. Allergy Asthma Immunol.* **2014**, *113*, 470–475.e2. [CrossRef]
157. Przybyszowski, M.; Paciorek, K.; Zastrzeżyńska, W.; Gawlewicz-Mroczka, A.; Trojan-Królikowska, A.; Orłowska, A.; Soja, J.; Pawlik, W.; Sładek, K. Influence of Omalizumab Therapy on Airway Remodeling Assessed with High-Resolution Computed Tomography (HRCT) in Severe Allergic Asthma Patients. *Adv. Respir. Med.* **2018**, *86*, 282–290. [CrossRef]
158. Haldar, P.; Brightling, C.E.; Hargadon, B.; Gupta, S.; Monteiro, W.; Sousa, A.; Marshall, R.P.; Bradding, P.; Green, R.H.; Wardlaw, A.J.; et al. Mepolizumab and Exacerbations of Refractory Eosinophilic Asthma. *N. Engl. J. Med.* **2009**, *360*, 973–984. [CrossRef]
159. Flood-Page, P.; Menzies-Gow, A.; Phipps, S.; Ying, S.; Wangoo, A.; Ludwig, M.S.; Barnes, N.; Robinson, D.; Kay, A.B. Anti-IL-5 treatment reduces deposition of ECM proteins in the bronchial subepithelial basement membrane of mild atopic asthmatics. *J. Clin. Investig.* **2003**, *112*, 1029–1036. [CrossRef]
160. Chachi, L.; Diver, S.; Kaul, H.; Rebelatto, M.C.; Boutrin, A.; Nisa, P.; Newbold, P.; Brightling, C. Computational modelling prediction and clinical validation of impact of benralizumab on airway smooth muscle mass in asthma. *Eur. Respir. J.* **2019**, *54*, 1900930. [CrossRef]
161. Stolz, D.; Mkorombindo, T.; Schumann, D.M.; Agusti, A.; Ash, S.Y.; Bafadhel, M.; Bai, C.; Chalmers, J.D.; Criner, G.J.; Dharmage, S.C.; et al. Towards the elimination of chronic obstructive pulmonary disease: A Lancet Commission. *Lancet* **2022**, *400*, 921–972. [CrossRef] [PubMed]
162. Agustí, A.; Hogg, J.C. Update on the Pathogenesis of Chronic Obstructive Pulmonary Disease. *N. Engl. J. Med.* **2019**, *381*, 1248–1256. [CrossRef] [PubMed]
163. Sohal, S.S.; Ward, C.; Danial, W.; Wood-Baker, R.; Walters, E.H. Recent advances in understanding inflammation and remodeling in the airways in chronic obstructive pulmonary disease. *Expert Rev. Respir. Med.* **2013**, *7*, 275–288. [CrossRef] [PubMed]
164. Hogg, J.C.; Chu, F.; Utokaparch, S.; Woods, R.; Elliott, W.M.; Buzatu, L.; Cherniack, R.M.; Rogers, R.M.; Sciurba, F.C.; Coxson, H.O.; et al. The Nature of Small-Airway Obstruction in Chronic Obstructive Pulmonary Disease. *N. Engl. J. Med.* **2004**, *350*, 2645–2653. [CrossRef] [PubMed]
165. Nowrin, K.; Sohal, S.S.; Peterson, G.; Patel, R.; Walters, E.H. Epithelial-mesenchymal transition as a fundamental underlying pathogenic process in COPD airways: Fibrosis, remodeling and cancer. *Expert Rev. Respir. Med.* **2014**, *8*, 547–559. [CrossRef]
166. Kheradmand, F.; Shan, M.; Xu, C.; Corry, D.B. Autoimmunity in chronic obstructive pulmonary disease: Clinical and experimental evidence. *Expert Rev. Clin. Immunol.* **2012**, *8*, 285–292. [CrossRef]
167. Norrby, K. Mast cells and angiogenesis. *Apmis* **2002**, *110*, 355–371. [CrossRef]
168. Soltani, A.; Ewe, Y.P.; Lim, Z.S.; Sohal, S.S.; Reid, D.; Weston, S.; Wood-Baker, R.; Walters, E.H. Mast cells in COPD airways: Relationship to bronchodilator responsiveness and angiogenesis. *Eur. Respir. J.* **2012**, *39*, 1361–1367. [CrossRef]
169. Mahmood, M.Q.; Walters, E.H.; Shukla, S.D.; Weston, S.; Muller, H.K.; Ward, C.; Sohal, S.S. Beta-catenin, Twist and Snail: Transcriptional regulation of EMT in smokers and COPD, and relation to airflow obstruction. *Sci. Rep.* **2017**, *7*, 10832. [CrossRef]
170. Koczulla, A.-R.; Jonigk, D.; Wolf, T.; Herr, C.; Noeske, S.; Klepetko, W.; Vogelmeier, C.; von Neuhoff, N.; Rische, J.; Wrenger, S.; et al. Krüppel-like zinc finger proteins in end-stage COPD lungs with and without severe alpha1-antitrypsin deficiency. *Orphanet J. Rare Dis.* **2012**, *7*, 29. [CrossRef]
171. Gohy, S.T.; Hupin, C.; Fregimilicka, C.; Detry, B.R.; Bouzin, C.; Gaide Chevronay, H.; Lecocq, M.; Weynand, B.; Ladjemi, M.Z.; Pierreux, C.E.; et al. Imprinting of the COPD airway epithelium for dedifferentiation and mesenchymal transition. *Eur. Respir. J.* **2015**, *45*, 1258–1272. [CrossRef] [PubMed]
172. Zheng, L.; Jiang, Y.-L.; Fei, J.; Cao, P.; Zhang, C.; Xie, G.-F.; Wang, L.-X.; Cao, W.; Fu, L.; Zhao, H. Circulatory cadmium positively correlates with epithelial-mesenchymal transition in patients with chronic obstructive pulmonary disease. *Ecotoxicol. Environ. Saf.* **2021**, *215*, 112164. [CrossRef] [PubMed]

173. Shirahata, T.; Nakamura, H.; Nakajima, T.; Nakamura, M.; Chubachi, S.; Yoshida, S.; Tsuduki, K.; Mashimo, S.; Takahashi, S.; Minematsu, N.; et al. Plasma sE-cadherin and the plasma sE-cadherin/sVE-cadherin ratio are potential biomarkers for chronic obstructive pulmonary disease. *Biomarkers* **2018**, *23*, 414–421. [CrossRef] [PubMed]
174. Milara, J.; Peiro, T.; Serrano, A.; Cortijo, J. Epithelial to mesenchymal transition is increased in patients with COPD and induced by cigarette smoke. *Thorax* **2013**, *68*, 410–420. [CrossRef]
175. Sohal, S.S.; Reid, D.; Soltani, A.; Ward, C.; Weston, S.; Muller, H.K.; Wood-Baker, R.; Walters, E.H. Reticular basement membrane fragmentation and potential epithelial mesenchymal transition is exaggerated in the airways of smokers with chronic obstructive pulmonary disease. *Respirology* **2010**, *15*, 930–938. [CrossRef]
176. Soltani, A.; Reid, D.W.; Sohal, S.S.; Wood-Baker, R.; Weston, S.; Muller, H.K.; Walters, E.H. Basement membrane and vascular remodelling in smokers and chronic obstructive pulmonary disease: A cross-sectional study. *Respir. Res.* **2010**, *11*, 105. [CrossRef]
177. Sohal, S.S.; Reid, D.; Soltani, A.; Ward, C.; Weston, S.; Muller, H.K.; Wood-Baker, R.; Walters, E.H. Evaluation of epithelial mesenchymal transition in patients with chronic obstructive pulmonary disease. *Respir. Res.* **2011**, *12*, 130. [CrossRef]
178. Gonzalez, D.M.; Medici, D. Signaling mechanisms of the epithelial-mesenchymal transition. *Sci. Signal.* **2014**, *7*, re8. [CrossRef]
179. Zuo, H.; Cattani-Cavalieri, I.; Valença, S.S.; Musheshe, N.; Schmidt, M. Function of cAMP scaffolds in obstructive lung disease: Focus on epithelial-to-mesenchymal transition and oxidative stress. *Br. J. Pharmacol.* **2019**, *176*, 2402–2415. [CrossRef]
180. Takizawa, H.; Tanaka, M.; Takami, K.; Ohtoshi, T.; Ito, K.; Satoh, M.; Okada, Y.; Yamasawa, F.; Nakahara, K.; Umeda, A. Increased expression of transforming growth factor-beta1 in small airway epithelium from tobacco smokers and patients with chronic obstructive pulmonary disease (COPD). *Am. J. Respir. Crit. Care Med.* **2001**, *163*, 1476–1483. [CrossRef]
181. Schupp, J.C.; Binder, H.; Jäger, B.; Cillis, G.; Zissel, G.; Müller-Quernheim, J.; Prasse, A. Macrophage Activation in Acute Exacerbation of Idiopathic Pulmonary Fibrosis. *PLoS ONE* **2015**, *10*, e0116775. [CrossRef] [PubMed]
182. Wang, S.; Meng, X.M.; Ng, Y.Y.; Ma, F.Y.; Zhou, S.; Zhang, Y.; Yang, C.; Huang, X.R.; Xiao, J.; Wang, Y.Y.; et al. TGF-beta/Smad3 signalling regulates the transition of bone marrow-derived macrophages into myofibroblasts during tissue fibrosis. *Oncotarget* **2016**, *7*, 8809–8822. [CrossRef] [PubMed]
183. Yang, F.; Chang, Y.; Zhang, C.; Xiong, Y.; Wang, X.; Ma, X.; Wang, Z.; Li, H.; Shimosawa, T.; Pei, L.; et al. UUO induces lung fibrosis with macrophage-myofibroblast transition in rats. *Int. Immunopharmacol.* **2021**, *93*, 107396. [CrossRef]
184. Hallgren, O.; Rolandsson, S.; Andersson-Sjöland, A.; Nihlberg, K.; Wieslander, E.; Kvist-Reimer, M.; Dahlbäck, M.; Eriksson, L.; Bjermer, L.; Erjefält, J.S.; et al. Enhanced ROCK1 dependent contractility in fibroblast from chronic obstructive pulmonary disease patients. *J. Transl. Med.* **2012**, *10*, 171. [CrossRef] [PubMed]
185. Sohal, S.S.; Soltani, A.; Reid, D.; Ward, C.; Wills, K.E.; Muller, H.K.; Walters, E.H. A randomized controlled trial of inhaled corticosteroids (ICS) on markers of epithelial–mesenchymal transition (EMT) in large airway samples in COPD: An exploratory proof of concept study. *Int. J. Chronic Obstruct. Pulmon. Dis.* **2014**, *9*, 533–542. [CrossRef]
186. Eapen, M.S.; Lu, W.; Hackett, T.L.; Singhera, G.K.; Mahmood, M.Q.; Hardikar, A.; Ward, C.; Walters, E.H.; Sohal, S.S. Increased myofibroblasts in the small airways, and relationship to remodelling and functional changes in smokers and COPD patients: Potential role of epithelial-mesenchymal transition. *ERJ Open Res.* **2021**, *7*, 00876-2020. [CrossRef]
187. Togo, S.; Holz, O.; Liu, X.; Sugiura, H.; Kamio, K.; Wang, X.; Kawasaki, S.; Ahn, Y.; Fredriksson, K.; Skold, C.M.; et al. Lung Fibroblast Repair Functions in Patients with Chronic Obstructive Pulmonary Disease Are Altered by Multiple Mechanisms. *Am. J. Respir. Crit. Care Med.* **2008**, *178*, 248–260. [CrossRef]
188. Barnes, P.J. Small airway fibrosis in COPD. *Int. J. Biochem. Cell Biol.* **2019**, *116*, 105598. [CrossRef]
189. Lynch, M.D.; Watt, F.M. Fibroblast heterogeneity: Implications for human disease. *J. Clin. Investig.* **2018**, *128*, 26–35. [CrossRef]
190. Marginean, C.; Popescu, M.S.; Vladaia, M.; Tudorascu, D.; Pirvu, D.C.; Petrescu, F. Involvement of Oxidative Stress in COPD. *Curr. Health Sci. J.* **2018**, *44*, 48–55.
191. Antar, S.A.; Ashour, N.A.; Marawan, M.E.; Al-Karmalawy, A.A. Fibrosis: Types, Effects, Markers, Mechanisms for Disease Progression, and Its Relation with Oxidative Stress, Immunity, and Inflammation. *Int. J. Mol. Sci.* **2023**, *24*, 4004. [CrossRef] [PubMed]
192. Willis, B.C.; Borok, Z. TGF-beta-induced EMT: Mechanisms and implications for fibrotic lung disease. *Am. J. Physiol. Lung Cell. Mol. Physiol.* **2007**, *293*, L525–L534. [CrossRef] [PubMed]
193. Carlier, F.M.; Dupasquier, S.; Ambroise, J.; Detry, B.; Lecocq, M.; Biétry–Claudet, C.; Boukala, Y.; Gala, J.-L.; Bouzin, C.; Verleden, S.E.; et al. Canonical WNT pathway is activated in the airway epithelium in chronic obstructive pulmonary disease. *Ebiomedicine* **2020**, *61*, 103034. [CrossRef] [PubMed]
194. Zou, W.; Zou, Y.; Zhao, Z.; Li, B.; Ran, P. Nicotine-induced epithelial-mesenchymal transition via Wnt/beta-catenin signaling in human airway epithelial cells. *Am. J. Physiol. Lung Cell. Mol. Physiol.* **2013**, *304*, L199–L209. [CrossRef]
195. Kneidinger, N.; Yildirim, A.O.; Callegari, J.; Takenaka, S.; Stein, M.M.; Dumitrascu, R.; Bohla, A.; Bracke, K.R.; Morty, R.E.; Brusselle, G.G.; et al. Activation of the WNT/beta-catenin pathway attenuates experimental emphysema. *Am. J. Respir. Crit. Care Med.* **2011**, *183*, 723–733. [CrossRef]
196. Milara, J.; Peiro, T.; Serrano, A.; Artigues, E.; Aparicio, J.; Tenor, H.; Sanz, C.; Cortijo, J. Simvastatin Increases the Ability of Roflumilast N-oxide to Inhibit Cigarette Smoke-Induced Epithelial to Mesenchymal Transition in Well-differentiated Human Bronchial Epithelial Cells in vitro. *COPD* **2015**, *12*, 320–331. [CrossRef]
197. Jiang, Z.; Zhang, Y.; Zhu, Y.; Li, C.; Zhou, L.; Li, X.; Zhang, F.; Qiu, X.; Qu, Y. Cathelicidin induces epithelial-mesenchymal transition to promote airway remodeling in smoking-related chronic obstructive pulmonary disease. *Ann. Transl. Med.* **2021**, *9*, 223. [CrossRef]

198. Zhang, Y.; Jiang, Y.; Sun, C.; Wang, Q.; Yang, Z.; Pan, X.; Zhu, M.; Xiao, W. The human cathelicidin LL-37 enhances airway mucus production in chronic obstructive pulmonary disease. *Biochem. Biophys. Res. Commun.* **2014**, *443*, 103–109. [CrossRef]
199. Sun, C.; Zhu, M.; Yang, Z.; Pan, X.; Zhang, Y.; Wang, Q.; Xiao, W. LL-37 secreted by epithelium promotes fibroblast collagen production: A potential mechanism of small airway remodeling in chronic obstructive pulmonary disease. *Lab. Investig.* **2014**, *94*, 991–1002. [CrossRef]
200. Chu, S.; Ma, L.; Wu, Y.; Zhao, X.; Xiao, B.; Pan, Q. C-EBPbeta mediates in cigarette/IL-17A-induced bronchial epithelial-mesenchymal transition in COPD mice. *BMC Pulm. Med.* **2021**, *21*, 376. [CrossRef]
201. Wang, Q.; Wang, Y.; Zhang, Y.; Zhang, Y.; Xiao, W. The role of uPAR in epithelial-mesenchymal transition in small airway epithelium of patients with chronic obstructive pulmonary disease. *Respir. Res.* **2013**, *14*, 67. [CrossRef] [PubMed]
202. Mahmood, M.Q.; Sohal, S.S.; Shukla, S.D.; Hardikar, A.; Noor, W.D.; Muller, H.K.; Knight, D.A.; Walters, E.H.; Ward, C. Epithelial mesenchymal transition in smokers: Large versus small airways and relation to airflow obstruction. *Int. J. Chronic Obstruct. Pulmon. Dis.* **2015**, *10*, 1515–1524. [CrossRef] [PubMed]
203. Zhang, Y.; Xiao, W.; Jiang, Y.; Wang, H.; Xu, X.; Ma, D.; Chen, H.; Wang, X. Levels of Components of the Urokinase-Type Plasminogen Activator System are Related to Chronic Obstructive Pulmonary Disease Parenchymal Destruction and Airway Remodelling. *J. Int. Med. Res.* **2012**, *40*, 976–985. [CrossRef] [PubMed]
204. Young, R.P.; Whittington, C.F.; Hopkins, R.J.; Hay, B.A.; Epton, M.J.; Black, P.N.; Gamble, G.D. Chromosome 4q31 locus in COPD is also associated with lung cancer. *Eur. Respir. J.* **2010**, *36*, 1375–1382. [CrossRef]
205. Wu, N.; Wu, Z.; Sun, J.; Yan, M.; Wang, B.; Du, X.; Liu, Y. Small airway remodeling in diabetic and smoking chronic obstructive pulmonary disease patients. *Aging* **2020**, *12*, 7927–7944. [CrossRef]
206. Pan, K.; Lu, J.; Song, Y. Artesunate ameliorates cigarette smoke-induced airway remodelling via PPAR-gamma/TGF-beta1/Smad2/3 signalling pathway. *Respir. Res.* **2021**, *22*, 91. [CrossRef]
207. Wang, W.; Zha, G.; Zou, J.J.; Wang, X.; Li, C.N.; Wu, X.J. Berberine Attenuates Cigarette Smoke Extract-induced Airway Inflammation in Mice: Involvement of TGF-beta1/Smads Signaling Pathway. *Curr. Med. Sci.* **2019**, *39*, 748–753. [CrossRef]
208. Mahmood, M.Q.; Reid, D.; Ward, C.; Muller, H.K.; Knight, D.A.; Sohal, S.S.; Walters, E.H. Transforming growth factor (TGF) β(1) and Smad signalling pathways: A likely key to EMT-associated COPD pathogenesis. *Respirology* **2017**, *22*, 133–140. [CrossRef]
209. de Boer, W.I.; van Schadewijk, A.; Sont, J.K.; Sharma, H.S.; Stolk, J.; Hiemstra, P.S.; van Krieken, J.H. Transforming growth factor beta1 and recruitment of macrophages and mast cells in airways in chronic obstructive pulmonary disease. *Am. J. Respir. Crit. Care Med.* **1998**, *158*, 1951–1957. [CrossRef]
210. Ge, F.; Feng, Y.; Huo, Z.; Li, C.; Wang, R.; Wen, Y.; Gao, S.; Peng, H.; Wu, X.; Liang, H.; et al. Inhaled corticosteroids and risk of lung cancer among chronic obstructive pulmonary disease patients: A comprehensive analysis of nine prospective cohorts. *Transl. Lung Cancer Res.* **2021**, *10*, 1266–1276. [CrossRef]
211. Zhu, L.; Xu, F.; Kang, X.; Zhou, J.; Yao, Q.; Lin, Y.; Zhang, W. The antioxidant N-acetylcysteine promotes immune response and inhibits epithelial-mesenchymal transition to alleviate pulmonary fibrosis in chronic obstructive pulmonary disease by suppressing the VWF/p38 MAPK axis. *Mol. Med.* **2021**, *27*, 97. [CrossRef] [PubMed]
212. Martorana, P.A.; Beume, R.; Lucattelli, M.; Wollin, L.; Lungarella, G. Roflumilast fully prevents emphysema in mice chronically exposed to cigarette smoke. *Am. J. Respir. Crit. Care Med.* **2005**, *172*, 848–853. [CrossRef]
213. Milara, J.; Peiró, T.; Serrano, A.; Guijarro, R.; Zaragozá, C.; Tenor, H.; Cortijo, J. Roflumilast N-oxide inhibits bronchial epithelial to mesenchymal transition induced by cigarette smoke in smokers with COPD. *Pulm. Pharmacol. Ther.* **2014**, *28*, 138–148. [CrossRef] [PubMed]
214. Liu, X.; Wu, Y.; Zhou, Z.; Huang, M.; Deng, W.; Wang, Y.; Zhou, X.; Chen, L.; Li, Y.; Zeng, T.; et al. Celecoxib inhibits the epithelial-to-mesenchymal transition in bladder cancer via the miRNA-145/TGFBR2/Smad3 axis. *Int. J. Mol. Med.* **2019**, *44*, 683–693. [CrossRef] [PubMed]
215. Watanabe, Y.; Imanishi, Y.; Ozawa, H.; Sakamoto, K.; Fujii, R.; Shigetomi, S.; Habu, N.; Otsuka, K.; Sato, Y.; Sekimizu, M.; et al. Selective EP2 and Cox-2 inhibition suppresses cell migration by reversing epithelial-to-mesenchymal transition and Cox-2 overexpression and E-cadherin downregulation are implicated in neck metastasis of hypopharyngeal cancer. *Am. J. Transl. Res.* **2020**, *12*, 1096–1113. [PubMed]
216. Paller, C.; Pu, H.; Begemann, D.E.; Wade, C.A.; Hensley, P.J.; Kyprianou, N. TGF-beta receptor I inhibitor enhances response to enzalutamide in a pre-clinical model of advanced prostate cancer. *Prostate* **2019**, *79*, 31–43. [CrossRef] [PubMed]
217. Bobal, P.; Lastovickova, M.; Bobalova, J. The Role of ATRA, Natural Ligand of Retinoic Acid Receptors, on EMT-Related Proteins in Breast Cancer: Minireview. *Int. J. Mol. Sci.* **2021**, *22*, 13345. [CrossRef]
218. Blaschuk, O.W. Potential Therapeutic Applications of N-Cadherin Antagonists and Agonists. *Front. Cell Dev. Biol.* **2022**, *10*, 866200. [CrossRef]
219. Lv, Q.; Wang, J.; Xu, C.; Huang, X.; Ruan, Z.; Dai, Y. Pirfenidone alleviates pulmonary fibrosis in vitro and in vivo through regulating Wnt/GSK-3beta/beta-catenin and TGF-beta1/Smad2/3 signaling pathways. *Mol. Med.* **2020**, *26*, 49. [CrossRef]
220. Ihara, H.; Mitsuishi, Y.; Kato, M.; Takahashi, F.; Tajima, K.; Hayashi, T.; Hidayat, M.; Winardi, W.; Wirawan, A.; Hayakawa, D.; et al. Nintedanib inhibits epithelial-mesenchymal transition in A549 alveolar epithelial cells through regulation of the TGF-beta/Smad pathway. *Respir. Investig.* **2020**, *58*, 275–284. [CrossRef]
221. Shteinberg, M.; Haq, I.J.; Polineni, D.; Davies, J.C. Cystic fibrosis. *Lancet* **2021**, *397*, 2195–2211. [CrossRef] [PubMed]
222. Françoise, A.; Héry-Arnaud, G. The Microbiome in Cystic Fibrosis Pulmonary Disease. *Genes* **2020**, *11*, 536. [CrossRef] [PubMed]

223. Bedrossian, C.W.; Greenberg, S.D.; Singer, D.B.; Hansen, J.J.; Rosenberg, H.S. The lung in cystic fibrosis. A quantitative study including prevalence of pathologic findings among different age groups. *Hum. Pathol.* **1976**, *7*, 195–204. [CrossRef] [PubMed]
224. Durieu, I.; Peyrol, S.; Gindre, D.; Bellon, G.; Durand, D.V.; Pacheco, Y. Subepithelial Fibrosis and Degradation of the Bronchial Extracellular Matrix in Cystic Fibrosis. *Am. J. Respir. Crit. Care Med.* **1998**, *158*, 580–588. [CrossRef]
225. Burgel, P.-R.; Montani, D.; Danel, C.; Dusser, D.J.; Nadel, J.A. A morphometric study of mucins and small airway plugging in cystic fibrosis. *Thorax* **2007**, *62*, 153–161. [CrossRef]
226. Hays, S.R.; Ferrando, R.E.; Carter, R.; Wong, H.H.; Woodruff, P.G. Structural changes to airway smooth muscle in cystic fibrosis. *Thorax* **2005**, *60*, 226–228. [CrossRef]
227. Hilliard, T.N.; Regamey, N.; Shute, J.K.; Nicholson, A.G.; Alton, E.W.; Bush, A.; Davies, J.C. Airway remodelling in children with cystic fibrosis. *Thorax* **2007**, *62*, 1074–1080. [CrossRef]
228. Boucher, R.C. Muco-Obstructive Lung Diseases. Reply. *N. Engl. J. Med.* **2019**, *381*, e20.
229. De Rose, V. Mechanisms and markers of airway inflammation in cystic fibrosis. *Eur. Respir. J.* **2002**, *19*, 333–340. [CrossRef]
230. Sagel, S.D.; Wagner, B.D.; Anthony, M.M.; Emmett, P.; Zemanick, E.T. Sputum Biomarkers of Inflammation and Lung Function Decline in Children with Cystic Fibrosis. *Am. J. Respir. Crit. Care Med.* **2012**, *186*, 857–865. [CrossRef] [PubMed]
231. Sly, P.D.; Gangell, C.L.; Chen, L.; Ware, R.S.; Ranganathan, S.; Mott, L.S.; Murray, C.P.; Stick, S.M.; Investigators, A.C. Risk factors for bronchiectasis in children with cystic fibrosis. *N. Engl. J. Med.* **2013**, *368*, 1963–1970. [CrossRef] [PubMed]
232. Garratt, L.W.; Sutanto, E.N.; Ling, K.-M.; Looi, K.; Iosifidis, T.; Martinovich, K.M.; Shaw, N.C.; Buckley, A.G.; Kicic-Starcevich, E.; Lannigan, F.J.; et al. Alpha-1 Antitrypsin Mitigates the Inhibition of Airway Epithelial Cell Repair by Neutrophil Elastase. *Am. J. Respir. Cell Mol. Biol.* **2016**, *54*, 341–349. [CrossRef] [PubMed]
233. Doumas, S.; Kolokotronis, A.; Stefanopoulos, P. Anti-Inflammatory and Antimicrobial Roles of Secretory Leukocyte Protease Inhibitor. *Infect. Immun.* **2005**, *73*, 1271–1274. [CrossRef] [PubMed]
234. Twigg, M.S.; Brockbank, S.; Lowry, P.; FitzGerald, S.P.; Taggart, C.; Weldon, S. The Role of Serine Proteases and Antiproteases in the Cystic Fibrosis Lung. *Mediat. Inflamm.* **2015**, *2015*, 293053. [CrossRef] [PubMed]
235. McMahon, M.; Ye, S.; Pedrina, J.; Dlugolenski, D.; Stambas, J. Extracellular Matrix Enzymes and Immune Cell Biology. *Front. Mol. Biosci.* **2021**, *8*. [CrossRef] [PubMed]
236. Devereux, G.; Steele, S.; Jagelman, T.; Fielding, S.; Muirhead, R.; Brady, J.; Grierson, C.; Brooker, R.; Winter, J.; Fardon, T.; et al. An observational study of matrix metalloproteinase (MMP)-9 in cystic fibrosis. *J. Cyst. Fibros.* **2014**, *13*, 557–563. [CrossRef]
237. Sagel, S.D.; Kapsner, R.K.; Osberg, I. Induced sputum matrix metalloproteinase-9 correlates with lung function and airway inflammation in children with cystic fibrosis. *Pediatr. Pulmonol.* **2005**, *39*, 224–232. [CrossRef]
238. Esposito, R.; Mirra, D.; Spaziano, G.; Panico, F.; Gallelli, L.; D'Agostino, B. The Role of MMPs in the Era of CFTR Modulators: An Additional Target for Cystic Fibrosis Patients? *Biomolecules* **2023**, *13*, 350. [CrossRef]
239. Garratt, L.W.; Sutanto, E.N.; Ling, K.-M.; Looi, K.; Iosifidis, T.; Martinovich, K.M.; Shaw, N.C.; Kicic-Starcevich, E.; Knight, D.A.; Ranganathan, S.; et al. Matrix metalloproteinase activation by free neutrophil elastase contributes to bronchiectasis progression in early cystic fibrosis. *Eur. Respir. J.* **2015**, *46*, 384–394. [CrossRef]
240. Duszyk, M.; Shu, Y.; Sawicki, G.; Radomski, A.; Man, S.F.; Radomski, M.W. Inhibition of matrix metalloproteinase MMP-2 activates chloride current in human airway epithelial cells. *Can. J. Physiol. Pharmacol.* **1999**, *77*, 529–535. [CrossRef]
241. Adam, D.; Roux-Delrieu, J.; Luczka, E.; Bonnomet, A.; Lesage, J.; Mérol, J.-C.; Polette, M.; Abély, M.; Coraux, C. Cystic fibrosis airway epithelium remodelling: Involvement of inflammation. *J. Pathol.* **2015**, *235*, 408–419. [CrossRef]
242. Bruscia, E.M.; Zhang, P.X.; Barone, C.; Scholte, B.J.; Homer, R.; Krause, D.S.; Egan, M.E. Increased susceptibility of Cftr$^{-/-}$ mice to LPS-induced lung remodeling. *Am. J. Physiol. Lung Cell. Mol. Physiol.* **2016**, *310*, L711–L719. [CrossRef] [PubMed]
243. Conese, M.; Di Gioia, S. Pathophysiology of Lung Disease and Wound Repair in Cystic Fibrosis. *Pathophysiology* **2021**, *28*, 155–188. [CrossRef]
244. Southey, M.C.; Batten, L.; Andersen, C.R.; McCredie, M.R.; Giles, G.G.; Dite, G.; Hopper, J.L.; Venter, D.J. CFTR deltaF508 carrier status, risk of breast cancer before the age of 40 and histological grading in a population-based case-control study. *Int. J. Cancer* **1998**, *79*, 487–489. [CrossRef]
245. Zhang, J.T.; Jiang, X.H.; Xie, C.; Cheng, H.; Da Dong, J.; Wang, Y.; Fok, K.L.; Zhang, X.H.; Sun, T.T.; Tsang, L.L.; et al. Downregulation of CFTR promotes epithelial-to-mesenchymal transition and is associated with poor prognosis of breast cancer. *Biochim. Biophys. Acta* **2013**, *1833*, 2961–2969. [CrossRef] [PubMed]
246. Liu, K.; Dong, F.; Gao, H.; Guo, Y.; Li, H.; Yang, F.; Zhao, P.; Dai, Y.; Wang, J.; Zhou, W.; et al. Promoter hypermethylation of the CFTR gene as a novel diagnostic and prognostic marker of breast cancer. *Cell Biol. Int.* **2020**, *44*, 603–609. [CrossRef]
247. Amaral, M.D.; Quaresma, M.C.; Pankonien, I. What Role Does CFTR Play in Development, Differentiation, Regeneration and Cancer? *Int. J. Mol. Sci.* **2020**, *21*, 3133. [CrossRef]
248. Cohen, J.C.; Larson, J.E.; Killeen, E.; Love, D.; Takemaru, K. CFTR and Wnt/beta-catenin signaling in lung development. *BMC Dev. Biol.* **2008**, *8*, 70. [CrossRef]
249. Sun, H.; Wang, Y.; Zhang, J.; Chen, Y.; Liu, Y.; Lin, Z.; Liu, M.; Sheng, K.; Liao, H.; Tsang, K.S.; et al. CFTR mutation enhances Dishevelled degradation and results in impairment of Wnt-dependent hematopoiesis. *Cell Death Dis.* **2018**, *9*, 275. [CrossRef]
250. Quaresma, M.C.; Pankonien, I.; Clarke, L.A.; Sousa, L.S.; Silva, I.A.L.; Railean, V.; Doušová, T.; Fuxe, J.; Amaral, M.D. Mutant CFTR Drives TWIST1 mediated epithelial–mesenchymal transition. *Cell Death Dis.* **2020**, *11*, 920. [CrossRef]

251. Collin, A.M.; Lecocq, M.; Detry, B.; Carlier, F.M.; Bouzin, C.; de Sany, P.; Hoton, D.; Verleden, S.; Froidure, A.; Pilette, C.; et al. Loss of ciliated cells and altered airway epithelial integrity in cystic fibrosis. *J. Cyst. Fibros.* **2021**, *20*, e129–e139. [CrossRef] [PubMed]
252. Nyabam, S.; Wang, Z.; Thibault, T.; Oluseyi, A.; Basar, R.; Marshall, L.; Griffin, M. A novel regulatory role for tissue transglutaminase in epithelial-mesenchymal transition in cystic fibrosis. *Biochim. Biophys. Acta* **2016**, *1863*, 2234–2244. [CrossRef] [PubMed]
253. Clarke, L.A.; Botelho, H.M.; Sousa, L.; Falcao, A.O.; Amaral, M.D. Transcriptome meta-analysis reveals common differential and global gene expression profiles in cystic fibrosis and other respiratory disorders and identifies CFTR regulators. *Genomics* **2015**, *106*, 268–277. [CrossRef] [PubMed]
254. Kerem, B.; Rommens, J.M.; Buchanan, J.A.; Markiewicz, D.; Cox, T.K.; Chakravarti, A.; Buchwald, M.; Tsui, L.C. Identification of the cystic fibrosis gene: Genetic analysis. *Science* **1989**, *245*, 1073–1080. [CrossRef]
255. Riordan, J.R.; Rommens, J.M.; Kerem, B.; Alon, N.; Rozmahel, R.; Grzelczak, Z.; Zielenski, J.; Lok, S.; Plavsic, N.; Chou, J.-L.; et al. Identification of the Cystic Fibrosis Gene: Cloning and Characterization of Complementary DNA. *Science* **1989**, *245*, 1066–1073. [CrossRef]
256. Bombieri, C.; Seia, M.; Castellani, C. Genotypes and Phenotypes in Cystic Fibrosis and Cystic Fibrosis Transmembrane Regulator-Related Disorders. *Semin. Respir. Crit. Care Med.* **2015**, *36*, 180–193. [CrossRef]
257. Bustamante, A.E.; Fernández, L.T.; Rivas, L.C.; Mercado-Longoria, R. Disparities in cystic fibrosis survival in Mexico: Impact of socioeconomic status. *Pediatr. Pulmonol.* **2021**, *56*, 1566–1572. [CrossRef]
258. Oates, G.R.; Baker, E.; Rowe, S.M.; Gutierrez, H.H.; Schechter, M.S.; Morgan, W.; Harris, W.T. Tobacco smoke exposure and socioeconomic factors are independent predictors of pulmonary decline in pediatric cystic fibrosis. *J. Cyst. Fibros.* **2020**, *19*, 783–790. [CrossRef]
259. Merlo, C.A.; Boyle, M.P. Modifier genes in cystic fibrosis lung disease. *J. Lab. Clin. Med.* **2003**, *141*, 237–241. [CrossRef]
260. Drumm, M.L.; Konstan, M.W.; Schluchter, M.D.; Handler, A.; Pace, R.; Zou, F.; Zariwala, M.; Fargo, D.; Xu, A.; Dunn, J.M.; et al. Genetic Modifiers of Lung Disease in Cystic Fibrosis. *N. Engl. J. Med.* **2005**, *353*, 1443–1453. [CrossRef]
261. Taylor-Cousar, J.L.; Zariwala, M.A.; Burch, L.H.; Pace, R.G.; Drumm, M.L.; Calloway, H.; Fan, H.; Weston, B.W.; Wright, F.A.; Knowles, M.R.; et al. Histo-Blood Group Gene Polymorphisms as Potential Genetic Modifiers of Infection and Cystic Fibrosis Lung Disease Severity. *PLoS ONE* **2009**, *4*, e4270. [CrossRef] [PubMed]
262. Guillot, L.; Beucher, J.; Tabary, O.; Le Rouzic, P.; Clement, A.; Corvol, H. Lung disease modifier genes in cystic fibrosis. *Int. J. Biochem. Cell Biol.* **2014**, *52*, 83–93. [CrossRef] [PubMed]
263. He, G.; Panjwani, N.; Avolio, J.; Ouyang, H.; Keshavjee, S.; Rommens, J.M.; Gonska, T.; Moraes, T.J.; Strug, L.J. Expression of cystic fibrosis lung disease modifier genes in human airway models. *J. Cyst. Fibros.* **2022**, *21*, 616–622. [CrossRef]
264. Mésinèle, J.; Ruffin, M.; Guillot, L.; Corvol, H. Modifier Factors of Cystic Fibrosis Phenotypes: A Focus on Modifier Genes. *Int. J. Mol. Sci.* **2022**, *23*, 14205. [CrossRef] [PubMed]
265. Sepahzad, A.; Morris-Rosendahl, D.J.; Davies, J.C. Cystic Fibrosis Lung Disease Modifiers and Their Relevance in the New Era of Precision Medicine. *Genes* **2021**, *12*, 562. [CrossRef]
266. Hinz, B.; McCulloch, C.A.; Coelho, N.M. Mechanical regulation of myofibroblast phenoconversion and collagen contraction. *Exp. Cell Res.* **2019**, *379*, 119–128. [CrossRef]
267. Harris, W.T.; Kelly, D.R.; Zhou, Y.; Wang, D.; MacEwen, M.; Hagood, J.S.; Clancy, J.P.; Ambalavanan, N.; Sorscher, E.J. Myofibroblast differentiation and enhanced TGF-B signaling in cystic fibrosis lung disease. *PLoS ONE* **2013**, *8*, e70196. [CrossRef]
268. Brazova, J.; Sismova, K.; Vavrova, V.; Bartosova, J.; Macek, M., Jr.; Lauschman, H.; Sediva, A. Polymorphisms of TGF-beta1 in cystic fibrosis patients. *Clin. Immunol.* **2006**, *121*, 350–357. [CrossRef]
269. Faria, E.J.; Faria, I.C.; Ribeiro, J.D.; Ribeiro, A.F.; Hessel, G.; Bertuzzo, C.S. Association of MBL2, TGF-beta1 and CD14 gene polymorphisms with lung disease severity in cystic fibrosis. *J. Bras. Pneumol.* **2009**, *35*, 334–342. [CrossRef]
270. Trojan, T.; Alejandre Alcazar, M.A.; Fink, G.; Thomassen, J.C.; Maessenhausen, M.V.; Rietschel, E.; Schneider, P.M.; van Koningsbruggen-Rietschel, S. The effect of TGF-β(1) polymorphisms on pulmonary disease progression in patients with cystic fibrosis. *BMC Pulm. Med.* **2022**, *22*, 183. [CrossRef]
271. Corvol, H.; Boelle, P.Y.; Brouard, J.; Knauer, N.; Chadelat, K.; Henrion-Caude, A.; Flamant, C.; Muselet-Charlier, C.; Boule, M.; Fauroux, B.; et al. Genetic variations in inflammatory mediators influence lung disease progression in cystic fibrosis. *Pediatr. Pulmonol.* **2008**, *43*, 1224–1232. [CrossRef]
272. Arkwright, P.D.; Laurie, S.; Super, M.; Pravica, V.; Schwarz, M.J.; Webb, A.K.; Hutchinson, I.V. TGF-β(1) genotype and accelerated decline in lung function of patients with cystic fibrosis. *Thorax* **2000**, *55*, 459–462. [CrossRef] [PubMed]
273. Ghigo, A.; De Santi, C.; Hart, M.; Mitash, N.; Swiatecka-Urban, A. Cell signaling and regulation of CFTR expression in cystic fibrosis cells in the era of high efficiency modulator therapy. *J. Cyst. Fibros.* **2023**, *22*, S12–S16. [CrossRef]
274. Harris, W.T.; Muhlebach, M.S.; Oster, R.A.; Knowles, M.R.; Noah, T.L. Transforming growth factor-β(1) in bronchoalveolar lavage fluid from children with cystic fibrosis. *Pediatr. Pulmonol.* **2009**, *44*, 1057–1064. [CrossRef] [PubMed]
275. Harris, W.T.; Muhlebach, M.S.; Oster, R.A.; Knowles, M.R.; Clancy, J.P.; Noah, T.L. Plasma TGF-β(1) in pediatric cystic fibrosis: Potential biomarker of lung disease and response to therapy. *Pediatr. Pulmonol.* **2011**, *46*, 688–695. [CrossRef]
276. Thomassen, J.C.; Trojan, T.; Walz, M.; Vohlen, C.; Fink, G.; Rietschel, E.; Alejandre Alcazar, M.A.; van Koningsbruggen-Rietschel, S. Reduced neutrophil elastase inhibitor elafin and elevated transforming growth factor-β(1) are linked to inflammatory response in sputum of cystic fibrosis patients with *Pseudomonas aeruginosa*. *ERJ Open Res.* **2021**, *7*, 00636-2020. [CrossRef]

277. Sagwal, S.; Chauhan, A.; Kaur, J.; Prasad, R.; Singh, M.; Singh, M. Association of Serum TGF-beta1 Levels with Different Clinical Phenotypes of Cystic Fibrosis Exacerbation. *Lung* **2020**, *198*, 377–383. [CrossRef]
278. Eickmeier, O.; Boom, L.; Schreiner, F.; Lentze, M.J.; NGampolo, D.; Schubert, R.; Zielen, S.; Schmitt-Grohé, S. Transforming Growth Factorβ1 Genotypes in Relation to TGFbeta1, Interleukin-8, and Tumor Necrosis Factor Alpha in Induced Sputum and Blood in Cystic Fibrosis. *Mediat. Inflamm.* **2013**, *2013*, 913135. [CrossRef] [PubMed]
279. Peterson-Carmichael, S.L.; Harris, W.T.; Goel, R.; Noah, T.L.; Johnson, R.; Leigh, M.W.; Davis, S.D. Association of lower airway inflammation with physiologic findings in young children with cystic fibrosis. *Pediatr. Pulmonol.* **2009**, *44*, 503–511. [CrossRef]
280. Stolzenburg, L.R.; Wachtel, S.; Dang, H.; Harris, A. miR-1343 attenuates pathways of fibrosis by targeting the TGF-beta receptors. *Biochem. J.* **2016**, *473*, 245–256. [CrossRef]
281. Corvol, H.; Rousselet, N.; Thompson, K.E.; Berdah, L.; Cottin, G.; Foussigniere, T.; Longchampt, E.; Fiette, L.; Sage, E.; Prunier, C.; et al. FAM13A is a modifier gene of cystic fibrosis lung phenotype regulating rhoa activity, actin cytoskeleton dynamics and epithelial-mesenchymal transition. *J. Cyst. Fibros.* **2018**, *17*, 190–203. [CrossRef]
282. Quaresma, M.C.; Botelho, H.M.; Pankonien, I.; Rodrigues, C.S.; Pinto, M.C.; Costa, P.R.; Duarte, A.; Amaral, M.D. Exploring YAP1-centered networks linking dysfunctional CFTR to epithelial–mesenchymal transition. *Life Sci. Alliance* **2022**, *5*, e202101326. [CrossRef] [PubMed]
283. Sousa, L.; Pankonien, I.; Simões, F.B.; Chanson, M.; Amaral, M.D. Impact of KLF4 on Cell Proliferation and Epithelial Differentiation in the Context of Cystic Fibrosis. *Int. J. Mol. Sci.* **2020**, *21*, 6717. [CrossRef] [PubMed]
284. McBennett, K.A.; Davis, P.B.; Konstan, M.W. Increasing life expectancy in cystic fibrosis: Advances and challenges. *Pediatr. Pulmonol.* **2022**, *57*, S5–S12. [CrossRef] [PubMed]
285. Adam, D.; Bilodeau, C.; Sognigbé, L.; Maillé, E.; Ruffin, M.; Brochiero, E. CFTR rescue with VX-809 and VX-770 favors the repair of primary airway epithelial cell cultures from patients with class II mutations in the presence of *Pseudomonas aeruginosa* exoproducts. *J. Cyst. Fibros.* **2018**, *17*, 705–714. [CrossRef] [PubMed]
286. Laselva, O.; Conese, M. Elexacaftor/Tezacaftor/Ivacaftor Accelerates Wound Repair in Cystic Fibrosis Airway Epithelium. *J. Pers. Med.* **2022**, *12*, 1577. [CrossRef] [PubMed]
287. Bec, R.; Reynaud-Gaubert, M.; Arnaud, F.; Naud, R.; Dufeu, N.; Di Bisceglie, M.; Coiffard, B.; Gaubert, J.-Y.; Bermudez, J.; Habert, P. Chest computed tomography improvement in patients with cystic fibrosis treated with elexacaftor-tezacaftor-ivacaftor: Early report. *Eur. J. Radiol.* **2022**, *154*. [CrossRef]
288. Maselli, D.J.; Amalakuhan, B.; Keyt, H.; Diaz, A.A. Suspecting non-cystic fibrosis bronchiectasis: What the busy primary care clinician needs to know. *Int. J. Clin. Pract.* **2017**, *71*, e12924. [CrossRef]
289. Bergin, D.A.; Hurley, K.; Mehta, A.; Cox, S.; Ryan, D.; O'Neill, S.J.; Reeves, E.P.; McElvaney, N.G. Airway inflammatory markers in individuals with cystic fibrosis and non-cystic fibrosis bronchiectasis. *J. Inflamm. Res.* **2013**, *6*, 1–11. [CrossRef]
290. Chalmers, J.D.; Polverino, E.; Crichton, M.L.; Ringshausen, F.C.; De Soyza, A.; Vendrell, M.; Burgel, P.R.; Haworth, C.S.; Loebinger, M.R.; Dimakou, K.; et al. Bronchiectasis in Europe: Data on disease characteristics from the European Bronchiectasis registry (EMBARC). *Lancet Respir. Med.* **2023**, *11*, 637–649. [CrossRef]
291. Taylor, S.L.; Rogers, G.B.; Chen, A.C.; Burr, L.D.; McGuckin, M.A.; Serisier, D.J. Matrix Metalloproteinases Vary with Airway Microbiota Composition and Lung Function in Non–Cystic Fibrosis Bronchiectasis. *Ann. Am. Thorac. Soc.* **2015**, *12*, 701–707. [CrossRef] [PubMed]
292. Keir, H.R.; Shoemark, A.; Dicker, A.J.; Perea, L.; Pollock, J.; Giam, Y.H.; Suarez-Curtin, G.; Crichton, M.L.; Lonergan, M.; Oriano, M.; et al. Neutrophil extracellular traps, disease severity, and antibiotic response in bronchiectasis: An international, observational, multicohort study. *Lancet Respir. Med.* **2021**, *9*, 873–884. [CrossRef] [PubMed]
293. Hsieh, M.-H.; Chou, P.-C.; Chou, C.-L.; Ho, S.-C.; Joa, W.-C.; Chen, L.-F.; Sheng, T.-F.; Lin, H.-C.; Wang, T.-Y.; Chang, P.-J.; et al. Matrix Metalloproteinase-1 Polymorphism (-1607G) and Disease Severity in Non-Cystic Fibrosis Bronchiectasis in Taiwan. *PLoS ONE* **2013**, *8*, e66265. [CrossRef] [PubMed]

Disclaimer/Publisher's Note: The statements, opinions and data contained in all publications are solely those of the individual author(s) and contributor(s) and not of MDPI and/or the editor(s). MDPI and/or the editor(s) disclaim responsibility for any injury to people or property resulting from any ideas, methods, instructions or products referred to in the content.

Review

Experimental Models to Study Epithelial-Mesenchymal Transition in Proliferative Vitreoretinopathy

Azine Datlibagi [1,2], Anna Zein-El-Din [1], Maxime Frohly [1,2], François Willermain [3], Christine Delporte [1,*] and Elie Motulsky [2]

[1] Laboratory of Pathophysiological and Nutritional Biochemistry, Université Libre de Bruxelles, 1070 Brussels, Belgium
[2] Department of Ophthalmology, Erasme Hospital, Hôpital Universitaire de Bruxelles, Université Libre de Bruxelles, 1070 Brussels, Belgium
[3] Department of Ophthalmology, CHU St Pierre and Brugmann, Université Libre de Bruxelles, 1000 Brussels, Belgium
* Correspondence: christine.delporte@ulb.be

Abstract: Proliferative vitreoretinal diseases (PVDs) encompass proliferative vitreoretinopathy (PVR), epiretinal membranes, and proliferative diabetic retinopathy. These vision-threatening diseases are characterized by the development of proliferative membranes above, within and/or below the retina following epithelial-mesenchymal transition (EMT) of the retinal pigment epithelium (RPE) and/or endothelial-mesenchymal transition of endothelial cells. As surgical peeling of PVD membranes remains the sole therapeutic option for patients, development of in vitro and in vivo models has become essential to better understand PVD pathogenesis and identify potential therapeutic targets. The in vitro models range from immortalized cell lines to human pluripotent stem-cell-derived RPE and primary cells subjected to various treatments to induce EMT and mimic PVD. In vivo PVR animal models using rabbit, mouse, rat, and swine have mainly been obtained through surgical means to mimic ocular trauma and retinal detachment, and through intravitreal injection of cells or enzymes to induce EMT and investigate cell proliferation and invasion. This review offers a comprehensive overview of the usefulness, advantages, and limitations of the current models available to investigate EMT in PVD.

Keywords: proliferative vitreoretinal diseases (PVDs); experimental models; epithelial-mesenchymal transition (EMT)

Citation: Datlibagi, A.; Zein-El-Din, A.; Frohly, M.; Willermain, F.; Delporte, C.; Motulsky, E. Experimental Models to Study Epithelial-Mesenchymal Transition in Proliferative Vitreoretinopathy. *Int. J. Mol. Sci.* **2023**, *24*, 4509. https://doi.org/10.3390/ijms24054509

Academic Editors: Margherita Sisto and Sabrina Lisi

Received: 31 January 2023
Revised: 22 February 2023
Accepted: 24 February 2023
Published: 24 February 2023

Copyright: © 2023 by the authors. Licensee MDPI, Basel, Switzerland. This article is an open access article distributed under the terms and conditions of the Creative Commons Attribution (CC BY) license (https://creativecommons.org/licenses/by/4.0/).

1. Introduction

Proliferative vitreoretinal diseases (PVDs) are a vision-threatening group of pathologies that comprise proliferative vitreoretinopathy (PVR), epiretinal membranes (ERM), and proliferative diabetic retinopathy (PDR). Similarly, membranes found in neovascular age-related macular degeneration (nAMD) share common pathological pathways with PVD [1]. PVDs are characterized by avascular or fibrovascular membranes developing above, inside and/or beneath the retina. While PVR usually occurs after retinal detachment or ocular trauma, due to an excessive wound healing response, membranes in nAMD and PDR are triggered by local inflammation or oxidative stress and develop when the diseases are left unchecked or fail to respond to treatment [2–7]. Clinically, PVD membranes exert a vitreoretinal traction which may lead to retinal detachment and are responsible for most of secondary retinal detachments following initial surgical repair. The incidence of PVD is expected to rise in the coming decades due to the increase of diseases and risk factors responsible for PVD development [4,6–12]. However, surgical peeling of these membranes by specialized surgical teams remains the sole therapeutic option to this day, limiting patients' access to treatment and burdening healthcare systems. Commonly, peeling of these membranes involves the use of dyes and drugs to stain the internal limiting membrane

(ILM) on the neuroretina and/or vitreous to help visualizing these transparent structures and avoid iatrogenic lesions of the retina. However, compounds frequently used to stain the ILM and vitreous, such as Brilliant Blue G and triamcinolone acetonide, respectively, may diffuse through the ILM and exert a cytotoxic effect on the neuroretina and retinal pigment epithelium (RPE), potentially limiting patients' post-operative visual prognosis [13,14].

RPE and Müller cells have been identified as the main cell types involved in PVD. Commonly, following disruption of the blood retinal barrier caused by chronic pathologies, retinal tear, retinal detachment, or penetrating ocular trauma, RPE cells acquire myofibroblast characteristics, allowing them to migrate and form the contractile membranes found in PVD [3,5,15–19]. This process, named epithelial-to-mesenchymal transition (EMT), can occur in both physiological conditions such as embryogenesis and wound healing, and pathological conditions such as cancers and tissue fibrosis. EMT is characterized by a loss of apical-basal polarity, a switch in the expression of cytokeratins to vimentin, and increased cellular motility and invasive ability [18–22]. Similarly, endothelial cells can also undergo a process called endothelial–mesenchymal transition, as seen in embryogenesis, cardiac fibrosis, and fibrovascular membranes found in PDR and nAMD [5,15,16,20]. Müller cells also play a significant role in PVD development, mainly through the secretion of cytokines and growth factors, leading to gliosis and proliferation [23]. The detailed role of EMT in PVD has been summarized in Figure 1.

Despite sharing similar mechanisms during their development, members of the PVD spectrum have mainly been investigated separately. Therefore, it is of paramount importance to study EMT, a process occurring in all PVD, to better understand the pathogenesis of these diseases. Since proliferative membranes in PVR often develop within a few weeks following RPE layer disruption, experimental models mimicking acute or subacute development of membranes to study EMT as a key process of PVD will be useful to identify alternative and/or complementary treatments to improve patients' visual prognostic [2]. A review published in 2017 sorted and described the animal models that have been used until now to study PVR and to perform pharmaceutical investigations [24]. However, the use of these in vivo models to study EMT in PVR has not been described. Furthermore, to the best of our knowledge, no review describing currently used in vitro PVR models has ever been published.

This review provides an overview of the currently existing in vitro and in vivo PVR models, as PVR membranes are mainly characterized by EMT of retinal cells. The review will also highlight the use of PVR models for research purposes as well as their advantages and limitations to study the EMT process involved in the pathogenesis of PVD.

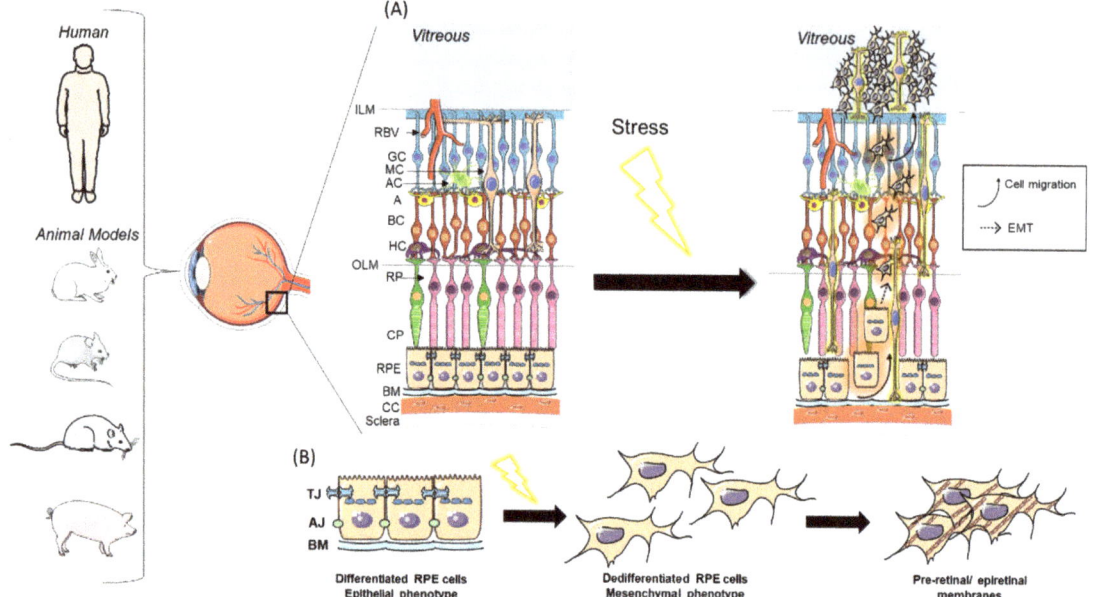

Figure 1. Role of EMT in PVR. (**A**) Depiction of PVR formation after stress factors' impact on retinal cells: EMT and migration of RPE cells through retinal cellular layers (highlighted in red) as well as gliosis of Müller cells (highlighted in yellow). (**B**) EMT of RPE cells due to stress, leading to preretinal and epiretinal membranes formation: Following disruption of BRB or compromised retinal architecture due to stress (ageing, hypoxia, inflammation, or traumatism), normal cobblestone-shaped RPE cells lose their tight and adherent junctions, their apical basal polarity, and obtain a mesenchymal phenotype, which increases their migrative and proliferative abilities. These RPE cells undergoing EMT then migrate through the different retinal layers to form preretinal and epiretinal membranes. A: Amacrine cell; AC: Astrocyte; AJ: Adherens Junctions; BC: Bipolar cell; BM: Bruch's Membrane; CC: Choroidal Capillaries; CP: Cone photoreceptors; EMT: Epithelial–Mesenchymal Transition; GC: Ganglion Cell; HC: Horizontal Cell; ILM: Inner Limiting Membrane; OLM: Outer Limiting Membrane; RBV: Retinal Blood Vessels; RP: Rod photoreceptor; RPE: Retinal Pigmented Epithelium; TJ: Tight Junctions. The Figure was partly generated using Servier Medical Art, provided by Servier, licensed under a Creative Commons Attribution 3.0 Unported license.

2. In Vitro Models of PVR

Most in vitro models of PVR rely on immortalized cell lines, pluripotent stem cells, and primary cells, mainly RPE and Müller cells, which represent the most abundant cells found in PVR membranes. However, only a few published reports have attempted to use cells directly isolated from PVR membranes.

In vitro models provide several benefits over in vivo models such as the ease of access to cell lines, a lower cost, and the possibility to obtain highly reproducible models. However, in vitro models also possess limitations compared to in vivo models, and the most widely used cell lines often possess abnormal karyotypes, which may restrain the conclusions drawn from such models.

Research groups studying PVD in vitro have often had recourse to inducing EMT in cell lines or primary culture by exogenous adjunction of transforming growth factor β (TGF-β), tumor necrosis factor α (TNF-α) or other cytokines and growth factors. The proliferative and contractile properties of RPE cells after wound healing or in presence of animal vitreous have also been evaluated. Hereafter we describe the different in vitro models that have been used to this day to study PVR, with an emphasis on their advantages and drawbacks.

2.1. EMT Induction by Growth Factors or Cytokines

EMT induction by addition of TGF-β to the culture medium of RPE cells, first described in 2001, has become the most often used method to study EMT in PVR in vitro models [25]. Both TGF-β1 and TGF-β2, used in different cell lines and primary cells, seem to induce EMT mainly through the activation of the Smad signaling pathway [26–28]. The vast majority of studies treating RPE with TGF-β showed increased expression of EMT markers, modification of cell morphology towards a mesenchymal state as well as increased migration, proliferative and contractile abilities using wound healing, invasion, and collagen contraction assays [26–30]. Interestingly, a recent study showed that exosomes produced by TGF-β2-treated ARPE-19 cells induced EMT in normal ARPE-19 cells, which underlines the importance of the microenvironment to initiate EMT in RPE cells [31]. However, TGF-β alone does not suffice to induce EMT efficiently but requires a loss of cell–cell contact to initiate this process [32].

In 2010, a combination of TGF-β and TNF-α, added for the first time in RPE cell culture medium, revealed their synergistic effect to induce EMT [33]. Ever since, this combined treatment has only been used thrice on ARPE-19 and primary human RPE cells despite showing promising results to induce EMT in RPE cells, leading to the formation of membranes and fibrotic deposits [34–36].

Other growth factors and cytokines such as epidermal growth factor (EGF), TNF-α, interleukin 6 (IL-6), fibroblast growth factor 2 (FGF2), Gremlin or Factor Xa have also been used to induce EMT and study EMT markers, proliferation, migration, and morphology of RPE cells [29,30,37,38]. These models could provide alternatives to the TGF-β-induced EMT model but remain to be more extensively studied.

2.2. EMT Induction through Mechanical Stimulation

Few groups have studied the behavior of human induced pluripotent stem cell (hiPSC)-derived RPE cells, primary RPE cells, and immortalized human RPE cells in reaction to a wound healing assay mainly to investigate their proliferative ability and the effect of potential antiproliferative drugs. These models have also been used to investigate the cell contractile properties using collagen matrix contraction assays [27,39]. Surprisingly, it has been reported that exposition to normal vitreous fluid during wound healing tends to increase the fibrotic response of hiPSC-derived RPE cells [40].

Recently, two studies have revealed that low-density cell culture for an extended period potentiated EMT in TGF-β1-treated ARPE-19 cells and spontaneously induced EMT in human embryonic stem cell-derived RPE [30,41].

2.3. Advantages and Limitations of In Vitro PVR Models

To establish in vitro PVR models, research groups have mostly used human cell lines that have spontaneously developed from primary RPE and Müller cells (such as ARPE-19 and MIO-M1, respectively), as well as human, rabbit, mouse, rat, and porcine primary RPE cells. In the last decade, differentiated RPE cells derived from human pluripotent stem cells have also been used to explore the EMT process involved in PVD pathogenesis, as they share functional and mature characteristics of native human RPE cells [27,28,36,42–44].

ARPE-19, a spontaneously arising human RPE cell line, has been most widely used to investigate EMT in PVR and other retinal disorders. ARPE-19 have been successfully used in all the aforementioned in vitro PVR models and offer an easily accessible source of RPE cells [45,46]. However, the use of these cells does not come without any drawback, as they show an abnormal karyotype and a loss of key characteristics of differentiated RPE including the cobblestone morphology of RPE cells favoring a mesenchymal cell morphology, the apicobasal polarity and the expression of some RPE markers [47,48]. Therefore, ARPE-19 may have already undergone partial EMT and do not represent an ideal cell line to study the initiation of EMT in PVD models [48,49]. This limitation can however be overcome by proper differentiation of ARPE-19 into mature RPE cells through addition of pyruvate in the culture medium for three to four months [50]. Recently, a rapid differentiation protocol using culture medium supplemented with nicotinamide has been reported to allow the cells to form a polarized epithelium with cobblestone appearance but lacking pigmentation within two to four weeks and to regain RPE functions [51]. Therefore, the use of differentiated ARPE-19 may represent an affordable and easy-to-handle in vitro model to study EMT induction in mature RPE cells and to mimic PVD pathogenesis.

Other spontaneously immortalized human Müller and RPE cell lines, such as MIO-M1 and D407, respectively, have also been used to investigate PVR [42,52]. However, they have been seldom used to study the EMT process in PVD. Furthermore, the D407 cell line shows similar limitations to the ARPE-19, such as an abnormal karyotype and lack of differentiated RPE characteristics [49].

Primary RPE cells originating from human, rabbit, mouse, and swine have been isolated from ocular globes and used for in vitro PVR studies. As for the ARPE-19, EMT induction in primary cells has been achieved by treatment with EGF, TGF-β and/or TNF-α, by mechanical wound healing or by cultivating the cells in presence of vitreous fluid [27,35,53,54]. The use of primary cells allows researchers to establish RPE sheets possessing in situ RPE features before inducing EMT, without biases of potential abnormal karyotypes of cell lines. However, all research teams do not have access to human donors shortly after their death or to animal eyes, nor possess the expertise to perform RPE isolation, which limits studies relying on primary cells.

Pluripotent stem-cell-derived RPE represents an alternative to obtain fully mature cells exhibiting all characteristics of native RPE and may therefore be the ideal cell type to develop in vitro PVR models. Furthermore, hiPSC can be obtained from human with minimally invasive techniques, such as skin biopsy or blood sampling [27,41]. However, hiPSC culture and differentiation into RPE cells is costly and time-consuming, whether the differentiation is spontaneous or guided, therefore limiting its use in routine research [40,49].

Cells isolated from human PVD membranes have been maintained in culture or subjected to TNF-α treatment to explore EMT and their proliferative and contractile properties [55–58]. This model allows to investigate the characteristics of proliferating cells composing the pathological membranes but has been very scarcely used due to the limited access to human samples. Furthermore, this model cannot be considered a true in vitro PVD model since the cells are already in an end-stage EMT prior to being isolated, even though their invasive properties increased after exposition to TNF-α and/or TGF-β [56,57].

3. Animal Models

Several animal models have been developed to study EMT in PVD. Animals used for this purpose mainly consist of rabbits, mice, and rats. In vivo animal models are valuable for PVD evaluation as they can be used to mimic human disease by inducing post-traumatic PVR or other pathological PVD and investigate novel therapeutics. Nonetheless, animal PVDs' pathophysiology and clinical expression can be different from the human disease which limits the extrapolation of animal studies on human PVR. Table 1 summarizes the characteristics of animal PVR models.

Table 1. Characteristics of rabbit, mouse, rat and swine PVR models.

Model Type	Methodology	Strengths	Limitations	Therapeutic Investigations	Pathogenesis Studies
Rabbit Models					
Cell-induced models	*Intravitreal injection of:* • 50,000–200,000 dermal, corneal or conjunctival fibroblasts [59,60] • 250,000 cultured human RPE cells [61–63] • 250,000 ARPE-19 cells [64] • 200,000 primary homologous RPE cells [65] • 50,000 Müller cells [66] • 70,000–800,000 Macrophages [67]	• Avoid major surgical side effects and anterior chamber lesions • Evaluate proliferation in PVR • Rapid onset models (3–4-day onset, ERM around day 28)	• Not suitable for study of chronic PVR • Injected exogenous cells induce inflammatory host reaction • Absence of the traumatic component of PVR • Absence of blood and plasma components' activation	[59,64,65,68–71]	[66,72]
	• Intravitreal injection of PRP (10⁷ platelets in a volume of 30µL) or autologous blood [73–75]	• Stimulate growth factors and cytokine secretion by platelets • Induce proliferation • Well-established, efficient, and cost-effective models • Mimic high-risk human PVR	• Absence of retinal detachment if not associated with traumatic lesions (PRP) • Rapid PVR onset (2 weeks)	[74]	[75]
	• Intravitreal or subretinal injection of 0.05–0.07 UI dispase [76,77]	• Recruitment of endogenous cells • Eased access to dispase • Histological changes close to human PVR • Avoid surgical side effects and anterior chamber lesions • Induction of high stages PVR	• Inconsistency of study results due to uncertain purity of dispase solutions • Prolonged exposure to dispase may induce cataract or lens subluxation • Absence of the traumatic component of PVR	[78]	[79]
Biologically induced models	• Intravitreal injection of 10–20 µg of VEGF [80,81]	• Mimics the neovascular proliferative aspect of PVR after 7 days	• Rabbit retinal vascularization pattern different from humans	[80]	
	• Injection of solutions containing 40 nmol Xanthine and 0.001 UI Xanthine oxidase [82]	• Representative of the inflammatory PVR aspect • ERM and retinal detachment 28 days after injection	• Inflammatory host reaction in anterior and posterior chambers	[82]	

Table 1. Cont.

Model Type	Methodology	Strengths	Limitations	Therapeutic Investigations	Pathogenesis Studies
		Rabbit Models			
Surgically induced models	• Unilateral surgical vitrectomy, retinotomy or cryopexy [83,84]	• Representative of traumatic PVR aspect • Disease onset 4 weeks post-surgery • Stimulation of inflammatory response with cryotherapy	• Variable extent of different surgeries • Risk of being non-reproducible • Risk of hemorrhage and excessive exudation of active components	[83]	[84]
	• Open or closed-globe injury by scleral incision and fluid percussion injury device (FPI) on the center of the cornea at a 65° angle [37]	• PVR developed in 2 weeks–6 months • Mimics post-traumatic human PVR	• Operator-dependent procedure • Risk of anterior segment injury in closed-globe injury model		[37]
Association of cell- and biologically induced models	*Intravitreal injection of:* • 10⁶ ARPE-19 cells treated with 10 ng/mL TGF-β2 [70] • 250,000 RPE cells and PRP [52]	• Intense cellular proliferation and preretinal neovascularization • Induction of high stages PVR	• Rapid induction of high grade PVR (1–2 days)	[52,70]	
	• Unilateral surgical intervention followed by fibroblast [85], PRP [86] or cytokine injections [87,88]	• Representative of traumatic and inflammatory/proliferative PVR aspects • PVR onset closer to human disease • Evaluation of acute PVR phases	• Rabbit retinal vascularization pattern different to humans	[85,86,88]	
Association of cell-, biologically and Surgically induced models	• Gas compression by intravitreal injection of C3F8 or SF6 gas followed by cell injection (RPE, PRP and/or fibroblasts) [89–91]	• Eased application due to small lens size • Eased vitreous manipulations without damage to the lens or retina • Liquification of vitreous avoiding anterior segment lesions • Posterior detachment of vitreous and preretinal membrane development • Emptying of vitreous chamber before other injections and subsequent lowering of IOP	• Rapid onset as soon as day 3	[61,91]	

Table 1. *Cont.*

Model Type	Methodology	Strengths	Limitations	Therapeutic Investigations	Pathogenesis Studies
Mouse Models					
Cell-induced models	• Injection of 50,000–160,000 ARPE-19 cells [92,93]	• Injection of RPE cells involved in retinal remodeling • Minimal operative complications	• Injection of exogenous cells which may induce inflammatory host reaction	[92]	
Biologically induced models	• Intravitreal injection of 0.1 U/µL–0.4 U/µL dispase [37,39,94–99] • Injection of dispase/collagenase solution (0.02–0.4 UI/µL) [38,99]	• Recruitment of endogenous RPE cells • Eased access to dispase • Reproducible technique • Minimal operative complications	• Prolonged exposure to dispase may induce cataract or lens subluxation • Risk of intravitreal hemorrhage	[38,94,98]	[37,39,95,97,99]
Surgically induced models	• Surgical retinal detachment with forceps without damaging the RPE layer [100–102] • Lesion of peripheral retina using silicone rubber needle [26] • Intravitreal injection of 0.5 µL of 100% SF6 gas followed by injection of 10,000 RPE cells [103]	• Representative of traumatic PVR aspect • Mimic key steps of human PVR	• Operator-dependent model which limits reproducibility • EMT occurs on the Bruch's membrane	[26,100–103]	
Transgenic models	• Homozygous rho/PDGF-B mice [104,105] • Lama1 deletion [106]	• Spontaneous development of proliferative membranes • Investigation of neovascular and proliferative processes of PVD	• Costly • Limited accessibility to transgenic mouse species	[105]	

Table 1. *Cont.*

Model Type	Methodology	Strengths	Limitations	Therapeutic Investigations	Pathogenesis Studies
Rat Models					
Cell-induced models	• Intravitreal injection of 106 ARPE-19 cells transfected with TP53BP2-specific siRNA [107] • Intravitreal injection of 2.4 × 106 RPE-J cells and/or PRP containing 2 × 1010 platelets [108,109] • Intravitreal injection of PRP containing 2.4 × 106 ARPE-19 cells [110] • Intravitreal injection of 250,000 macrophages [111]	• Minimal operative complications • Induction of high-stage PVR • Involvement of several cell types (RPE cells, glial cells, macrophages, and fibroblasts) as in human PVR	• Injection of exogenous cells • Risk of intravitreal hemorrhage	[108–110]	[108–111]
Biologically induced models	• Intravitreal injection of 0,03 IU/μL of dispase [112] • Subretinal injection of 3 μg of dispase [41]	• Recruitment of endogenous RPE cells • Stable and reproducible model	• Prolonged exposure to dispase may induce cataract or lens subluxation • Risk of intravitreal hemorrhage	[41,112]	
Swine Models					
Association of cell- and surgically induced models	• Surgical retinal detachment induced by subretinal BSS injection followed by intravitreal injection of 8.108 RPE cells with or without PRP [54,113,114]	• Best alternative to non-human primates • Tractional retinal detachment 2 weeks after surgery and injection	• Operator-dependent model which limits reproducibility • Costly	[54]	[114]

ARPE-19: Spontaneously arising RPE cell line; C3F8: Perfluoropropane; CTGF: Connective Tissue Growth Factor; EMT: Epithelial-Mesenchymal Transition; ERM: Epiretinal Membrane; PRP: Platelet-Rich-Plasma; PVR: Proliferative Vitreoretinopathy; RPE cells: Retinal Pigment Epithelial cells; SF6: Sulfur hexafluoride; IOP: Intraocular Pressure; TGF: Transforming Growth Factor; VEGF: Vascular Endothelial Growth Factor; siRNA: silencing ribonucleic acid; TP53BP2: Tumor Suppressor P53-Binding Protein 2.

3.1. Rabbit PVR Models

Rabbits are some of the most common animals used for in vivo experimental PVR research. This species presents many advantages, such as its ease of acquisition and handling, a small lens size and a voluminous posterior chamber close to the human vitreous' volume, which allows an easy access to the retina and better visualization of developing PVR [80].

PVR has been mainly induced in rabbits by intravitreal injections of cells or biologically active compounds and by ocular trauma through surgical means. These PVR models have been developed to study various components involved in PVR pathogenesis. Indeed, the proliferative and inflammatory vitreous reaction following the injection of cells, cytokines, growth factors and/or other blood components reproduce different stages of PVR [73,115,116]. In addition, surgical techniques mimicking ocular trauma reproduce the most frequent cause of PVR, disrupting the blood–retinal barrier (BRB), and allowing the subsequent activation and recruitment of macrophages, fibroblasts, and glial cells [37,117]. Interestingly, PVR has also been induced by associating ocular trauma with intravitreal injections or the simultaneous injection of different cell types and/or active compounds [52,70,117,118]. Such combination allows a concomitant representation of various aspects of PVR occurring at different stages, leading to better mimicking the human disease.

The first described injection model reproducing PVR in rabbits was described in 1984, based on the induction of a "fibroplasia" by injecting dermal connective tissue into the vitreous [119]. Since then, the intravitreal injection of cells or other compounds has been widely used in rabbit PVR models [37,59,61,70].

3.1.1. Cell-Induced Rabbit PVR Models

Homologous or heterologous intravitreal injections of fibroblasts of various origins [59,60,120], human RPE cells [61–63], transfected ARPE-19 cells [64,121], primary RPE cultures [66], Müller cells [66], macrophages [67,121], or platelet-rich plasma (PRP) [122] have been used to develop PVR model. The most frequently studied model is the intravitreal injection of fibroblasts [59,60,120]. Such injections can trigger the development of epiretinal and intravitreal proliferative membranes within a few days after injection, due to the host reaction to these exogenous fibroblasts, ultimately leading to retinal detachment [115,123]. The injection of fibroblasts triggers an inflammatory infiltration, migration of RPE cells from the subretinal space, and loss of the initial RPE hexagonal shape towards a fibroblast-like appearance [124].

PRP and blood derivatives' injection stimulates EMT, mainly through the secretion of growth factors and active mediators provided by platelets, such as platelet-derived growth factor (PDGF) and vascular endothelial growth factor (VEGF) [122,125]. Retinal blood vessels' occlusion by platelets further stimulates neovascularization and proliferation [126]. Furthermore, co-injection of PRP and fibroblasts leads to the development of higher stage PVR with intense intraocular proliferation and preretinal vascularization [127,128].

3.1.2. Biologically Induced Rabbit PVR Models

Several groups have also performed intravitreal injections of biologically active compounds such as dispase [76,129], TGF-β [70], VEGF [80,81], recombinant human IL-1β [87], xanthine and xanthine oxidase [82], to induce inflammatory and proliferative reactions and prompt PVR development. Dispase is an easily accessible enzyme that induces histological changes in the retina with very few side effects [76]. As a metalloprotease, dispase dissociates cells from their surrounding matrix, leading to RPE cells' exposition and disrupting the vitreoretinal continuity [130]. Leakage and recruitment of endogenous cells such as fibroblasts, macrophages, and glial cells then follow, driving the expression of growth factors and cytokines which will stimulate the cells giving rise to PVR [77].

3.1.3. Cell- and Biologically Induced Rabbit PVR Models

Some of the compounds and cells previously described have also been used in conjunction to induce PVR. For instance, some cells (RPE cells, human fibroblasts, heterologous fibroblasts etc.) were injected along with cytokines or other cells (PRP rich in trophic factors and cytokines) [52,118]. The co-injection of cytokines and PRP leads to dissociation and migration of RPE cells, mainly stimulating the proliferative process with moderate inflammation and subsequent development of PVR exhibiting thicker ERM compared to cytokine injection alone [118].

3.1.4. Surgically Induced Rabbit PVR Models

Since ocular trauma is the main cause of PVR in humans, several research groups have attempted to mimic trauma through surgical techniques to induce PVR in rabbits. These models involve performing open-globe injury or vitrectomy followed by retinotomy or cryopexy [37,83,84]. Multiple features of human PVR can be represented in these models, such as retinal tear and BRB disruption that follow retinal detachments or traumatisms in human PVR. Epiretinal scarring as well as proliferation processes involving endogenous cells' recruitment reproduce more accurately PVR pathogenesis compared to models where injections of exogenous cells and agents are performed [131]. Furthermore, PVR onset time in mechanical models is around 4–12 weeks, similarly to observations made in humans, which allows the evaluation of drugs and long-run interventions [132]. However, surgical/traumatic models might exacerbate proliferation by excessive vitreous hemorrhage related to the experience of the surgeon, which renders these models less reproducible.

3.1.5. Cell- or Biologically Induced Rabbit PVR Models following Surgery

Surgical techniques combined with injection of one or several cell types and/or cytokine have been used to obtain a more reliable PVR model and to better investigate the physiopathology of human PVR where a retinal tear is often the main precursor of the disease [85–87]. Retinal defects, potentialized by the injected cytokines, allow the migration and proliferation of various cells into the vitreous through the retina or the interaction of injected cells with leaking cytokines, cells, and growth factors. These cells will then induce subsequent epiretinal membranes, surface wrinkling retinopathy and star-fold-like configurations 4 weeks after surgery [87]. Nonetheless, as rabbit retinas are less vascularized compared to human, these models are not ideal to study the actual impact of BRB disruption and subsequent ERM formation [65,88]. In fact, rabbit retinal vascularization pattern (merangiotic) is different compared to humans (euangiotic/holangiotic) with PVR in rabbits beginning on or around the medullary rays where retinal vasculature is present, the rest of the rabbit retina being avascular [133]. Retinal vasculature in humans plays an important role in PVR development as the anatomical disturbance of the retina and BRB disruption play a significant role in subsequent migration of inflammatory cells and proteins. Furthermore, newly formed retinal vessels following retinal detachment are common in PVR and may be sources of growth factors and inflammatory cells leading to ERM formation [134].

Cell injection following gas vitrectomy, using mainly perfluoropropane (C3F8), leads to PVR development within 7 to 28 days [128]. This technique allows emptying the vitreous chamber before cell injection, lowering the intraocular pressure, and softening the ocular globe easing intravitreal manipulations and subsequent injections rendering the model more reproducible [128]. This model is particularly interesting in rabbits due to the smaller lens size compared with the eyeball which allows vitreous manipulations to be performed without damage to the lens or retina [89]. Such a procedure allows the posterior detachment of vitreous and preretinal membranes development, the latter attributable to the break of retinal cell-to-cell contact and the disruption of the BRB, happening occasionally with posterior vitreous detachment, leading to RPE cells, collagen fibers, myofibroblasts, growth factors and cytokine leakage into the vitreous [61,90].

3.1.6. Advantages and Limitations of Rabbit PVR Models

Briefly, rabbit injection models can be preferred to traumatic and surgical ones due to their ease of manipulation and less traumatic application avoiding non-naturally occurring side lesions [76]. More specifically, intravitreal injection of cells allows the study of the proliferative stage of PVR due to the reaction of local cells to the injected ones while the injection of blood derivatives and active inflammatory components mimic the inflammatory reaction that leads to EMT and tractional membrane development [80,115,116,118]. However, the injection of exogenous cells bearing foreign antigens as well as the rapid disease development must be taken in account before selecting this type of model [76]. Some proliferative retinal diseases such as macular PVR and post-traumatic PVR can have a rapid onset and be simulated by the injection models [135]. However, other proliferative retinal diseases linked to chronic pathologies such as proliferative diabetic retinopathy and exudative age-related macular degeneration can take years for proliferation to be clinically apparent and thus cannot be accurately reproduced by these models.

On the other hand, traumatic and surgical models offer the advantage to stimulate locally available cells without foreign agent injection and with an onset time close to clinical PVR [132]. Different manipulations can also be facilitated by the lowered intraocular pressure and softening of the ocular globe occurring after gas injection [128]. Nonetheless, surgical techniques performed by different manipulators may not be perfectly reproducible without proper surgical expertise, leading to variable degrees of cell liberation and local reaction [86,117]. Furthermore, such procedures hold a high risk of severe vitreous hemorrhage that does not reproduce the clinical situation [136].

The use of PVR models combining surgical and injection techniques represents an alternative to the use of a sole technique. These models offer the advantages to induce PVR in a timely manner closer to human disease onset with changes in vitreous consistent with clinical PVR pathogenesis. These phenomena are attributable to the surgical BRB disruption and its subsequent effect on cell leakage and local reactions, as well as the inflammatory and proliferative benefits of the injected compounds [87]. Nonetheless, the vascularization of rabbits' retina differs from humans, making these results less extrapolatable to human PVR [88].

3.2. Mouse PVR Models

Mice represent the second most frequently used in vivo PVR models due to their ease of handling, accessibility, and similarities to human physiology and anatomy. Furthermore, they offer the possibility to induce the disease by various means such as injections, surgery or genetic modifications. Most mouse models of PVR are mainly derived from the C57BL/6 strain which has been used for over 50 years and allows the generation of reliable transgenic models [38,97,104,106,137,138]. Mouse PVR models using intravitreal injections range from active compounds' injection such as dispase or coagulation factors to cell injections of ARPE-19, preceded or not by surgical discontinuation of the RPE layer [37,92,93,103]. Transgenic models however are capable of developing PVR spontaneously and allow the evaluation of cell proliferation and ERM formation [105,106].

3.2.1. Biologically Induced Mouse PVR Models

Intravitreal injection of dispase has been mostly used to induce PVR development in mice. The low cost of dispase and high reproducibility of the model, makes it ideal for PVR pathophysiological investigations without involving specific immune response [96,98]. In addition, the severity of induced PVR can be tuned by adjusting the concentration of dispase [94]. PVR severity can also be increased by the simultaneous injection of a coagulation factor (FXa) [37]. Furthermore, the inflammatory response of RPE cells can be triggered in such model by the simultaneous intravitreal hemorrhage mixed with vitreous profibrotic factors [37,99]. The RPE, macrophages, and glial cells enter an inflammatory phase participating in the creation of a sub- and epi-retinal membrane, with only the subretinal membrane containing RPE cells [94,99].

3.2.2. Cell-Induced Mouse PVR Models

To the best of our knowledge, PVR induction by intravitreal injection of cells in mice has only been performed in two studies using ARPE-19 cells [92,93]. Such PVR model was characterized by the formation of an ERM resembling those observed in patients with PVR [92]. Injection of large quantities of exogenous cells may lead to significant inflammation due to the host reaction, making this model less representative of human PVR.

3.2.3. Surgically Induced Mouse PVR Models

Gentle retinal detachment using forceps or silicon rubber needle has been performed to induce PVR in mice [26,100–102]. Such methods induce EMT by detaching the retina without damaging the underlying RPE. PVR development mostly occurs on the Bruch's membrane and not on the surface of the detached retina [26,102]. However, released vitreous cytokines and BRB rupture lead to immunological reaction, making the experimental conditions difficult to control [100].

3.2.4. Cell-Induced Mouse PVR Models following Surgery

ARPE-19 cell injection following intravitreal gas injection has been recently performed to induce PVR development in mice. Interestingly, intravitreal gas injection leads to posterior vitreous detachment and increases the severity of subsequent PVR formation possibly by facilitating cell migration [103]. This model mimics key pathological aspects of human PVR without compromising retinal integrity and represents a valuable model as it allows therapeutic and pathophysiological studies. However, injection of human RPE cells may induce excessive inflammation related to the use of foreign cells, which may bias results' interpretations [103].

3.2.5. Transgenic Mouse PVR Models

Transgenic mice models have been generated to spontaneously induce the development of PVR [104–106]. The transgenic specific overexpression of Rho-PDGF A and B in photoreceptor cells results in vascular and glial cell proliferation [104]. Rho-PDGF A transgenic mice allow glial cell proliferation, formation of an ERM made of astrocytes and RPE cells, and superficial vascularization of the retina. Rho-PDGF B transgenic mice are more interesting for studying vascular proliferative retinopathies as they develop deep retinal vascularization with an epiretinal membrane containing glial cells, endothelial cells, and pericytes. Considering the involvement of these growth factors in the pathogenesis of PVR, specific aptamers have been developed and used in therapeutic trials to modulate PVR [105]. Mice showing mutated Laminin Subunit Alpha 1 (LAMA1) retain a fetal vitreous vascularization and a pre-retinal glial membrane, and the ERM present is very similar to PVR [106]. Overall, these transgenic mice models present severe proliferative aspects allowing further study of the pathogenesis of proliferative retinopathies.

3.2.6. Advantages and Limitations of Mouse PVR Models

Mice represent interesting and valuable models to study PVR due to their short reproductive cycle, and ease of handling and housing. However, compared to rabbits, murine eyes are anatomically smaller and possess a large lens and small vitreous volume, making them more technically challenging for injection, surgery, and observation [103,139]. Furthermore, mouse retina is not completely comparable to the human's due to the absence of macula [140]. Nevertheless, their accessibility and ease of genetic modification make them a particularly interesting model for studying pathophysiological mechanisms involved in PVR [38,94,98,104,106].

As in rabbits, PVR can be induced in mice by intravitreal injection of cells or active compounds, or by inducing trauma, to mimic some of the key steps of PVR development in humans, such as proliferation and inflammation. These models have been mainly used to study EMT initiation occurring in PVR [37–39,94–98]. Transgenic mice models are particularly interesting to study the inflammatory stage of PVR as the pathophysiological hypotheses suspected of being involved can be isolated and studied separately [38,94,98]. Since the different modalities of PVR induction do not allow pharmacological study of ERM and neovascularization, some transgenic models that spontaneously develop PVR are valuable to study the aggressive proliferative aspects and therefore the late stages of PVR.

Despite technical challenges related to large lens and small vitreous volumes, intravitreal injections remain interesting models of PVR as they minimize operative complications compared to PVR model induced by surgery. Nonetheless, caution must be taken during cell injections as they may prone an excessive inflammation that does not reproduce human PVR.

3.3. Rat PVR Models

A wide variety of rat strains (such as Long Evans, Wistar, Sprague Dawley, Brown Norway) has been used to develop PVR models, which could lead to problems of reproducibility. These models mainly rely on intravitreal injection of cells or active compounds such as dispase or PRP. Although rats possess large globes and smaller lenses, they have not been as frequently used as mice and rabbits to study PVR pathogenesis [41,112,141,142].

3.3.1. Cell-Induced Rat PVR Models

Cells such as RPE, macrophages, or a combination of RPE and PRP have been used to induce PVR in rats [107–111]. PVR induction by intravitreal injection of ARPE19 has also been shown to induce vitreoretinal fibrosis, similarly to rabbits [107]. During EMT induction, the key step of RPE migration and proliferation in the intravitreal space can be enhanced by PRP co-injection with either ARPE-19 or primary RPE cells from 7-day-old Long-Evans rats (RPE-J) [108,110]. These co-injection models constitute valuable in vivo PVR models as they adequately mimic human PVR through the involvement of RPE cells, glial cells, macrophages, and fibroblasts in the fibrocellular membranes [111].

3.3.2. Biologically Induced Rat PVR Model

Intravitreal or subretinal injection of dispase has been performed in rats to induce PVR [41,112]. Rat PVR models using dispase injection leads to disruption of retinal integrity and to EMT initiation. These models show similar benefits to mice models as they also allow investigation of EMT initiation and the testing of different therapeutic agents [41,112].

3.3.3. Advantages and Limitations of Rat PVR Models

Compared to mice, rats possess larger ocular globes with proportionally smaller lens, a larger vitreous volume and present very similar advantages and disadvantages. However, the vast heterogeneity of rat strains used to investigate PVR in vivo limits the reproducibility of these models. Furthermore, rats have not been as extensively used to investigate PVR pathogenesis when compared to rabbits and mice [41,112,141,142].

3.4. Swine Models

Pigs have not been widely used to investigate the pathophysiology of PVR due to their heavy housing and maintenance cost, although pigs share several physiological and anatomical similarities with humans. Indeed, pigs possess a retinal structure characterized with a high density of photoreceptors, a holangiotic vascularization and a vitreous composition similar to humans. Moreover, due to the resemblance to human's anatomy, the results obtained are the closest to human pathophysiology, making the pigs the best alternative to non-human primates for PVR studies. Furthermore, the use of pigs to develop PVR models has been motivated by the failure to translate all results found in other models to humans [113]. PVR induction in the porcine models is commonly performed through retinal detachment induced by subretinal balanced salt solution injection during vitrectomy, followed by an intravitreal injection of RPE associated or not with PRP [54,113,114]. PVR development in pigs is similar to humans, starting with the development of fibrotic membranes followed by tractional RD at day 14 [113,114]. Recently, a PVR model was obtained in minipigs by scraping the endogenous RPE layer following vitrectomy with induction of bleb retinal detachment [113].

4. Conclusions

This review provides a summary of existing in vitro and in vivo PVR models that allow investigation of the EMT process occurring in PVD, along with their advantages and limitations. Most in vitro models rely on the use of immortalized cell lines such as ARPE-19 cells due to their ease of access and lower cost compared to primary cells and hiPSCs. However, ARPE-19 cells lack several key features of mature RPE, which may limit correlation of experimental results to human PVD. A recent simple protocol allowing to rapidly differentiate ARPE-19 into mature RPE cells using nicotinamide would be very useful to study EMT initiation in PVD pathogenesis. Other in vitro models using primary and human pluripotent stem-cell-derived RPE cells offer the benefits of possessing most characteristics and features of native RPE cells, making them the best in vitro models for experimental investigation. However, the limited access to primary human RPE cells as well as the heavy cost and required expertise for primary RPE cells' isolation and hiPSCs' differentiation greatly limit their use in many research laboratories. Therefore, nicotinamide-induced differentiated ARPE-19 cells subjected to EMT through adjunction of TGF-β with or without TNF-α in the culture medium represents an attractive and relevant model to study PVR and PVD pathogenesis.

Among in vivo PVR models, rabbits and mice have been widely used to mimic PVR pathogenesis, mainly through intravitreal injection of dispase, fibroblasts, or RPE cells, or through surgical means mimicking penetrating ocular trauma. Mice and rabbits represent accessible animals to investigate PVD pathogenesis and therapeutic agents in preclinical models. Furthermore, mice offer the possibility of performing genetic modifications to study specific pathways that may be involved in PVD and identify potential therapeutic targets. Despite sharing many similarities with human retinal physiology and anatomy, making them the ideal in vivo model for PVR pathogenesis investigation, pigs have been rarely used, mainly due to their heavy housing and maintenance costs. Minipigs may represent an alternative to classical swine research as they need smaller facilities and offer similar benefits. However, the use of minipigs to study PVR needs further validation, as this model has so far only been used once.

Overall, both in vitro and in vivo PVR models present advantages and limitations. In vitro models provide a controlled environment for analyzing specific cellular and molecular processes involved in PVR pathogenesis. In vivo animal PVR models offer a more realistic representation of the disease but are limited by the difficulty in controlling variables and extrapolating findings to humans. These models represent complementary valuable tools to deepen our current understanding of PVD and to develop effective treatments for patients. Furthermore, PVR induction in 3D models such as organs-on-a-chip or 3D bioprinted outer retina may also provide innovating and interesting alternatives to study molecular mechanisms of PVD.

Author Contributions: Conceptualization, A.D., C.D. and E.M.; writing—original draft preparation, A.D., A.Z.-E.-D., M.F. and C.D.; writing—review and editing, A.D., F.W., C.D., E.M.; visualization, A.D., A.Z.-E.-D. and M.F.; supervision, A.D., F.W., C.D. and E.M.; project administration, A.D., F.W., C.D. and E.M.; funding acquisition, A.D., F.W., C.D. and E.M. All authors have read and agreed to the published version of the manuscript.

Funding: This work was supported by a Research Fellowship and research credit from Fonds Erasme pour la recherche médicale (A.D.), research credit (CDR J.0128.23) from the Fund for Scientific Research (F.R.S.–FNRS) (E.M.) and Fund IRIS-Research managed by the King Baudouin Foundation (F.W., E.M., A.D., C.D.).

Institutional Review Board Statement: Not applicable.

Informed Consent Statement: Not applicable.

Data Availability Statement: Not applicable.

Conflicts of Interest: The authors declare no conflict of interest.

References

1. Yoshida, S.; Nakama, T.; Ishikawa, K.; Nakao, S.; Sonoda, K.-H.; Ishibashi, T. Periostin in vitreoretinal diseases. *Cell. Mol. Life Sci.* **2017**, *74*, 4329–4337. [CrossRef]
2. Idrees, S.; Sridhar, J.; Kuriyan, A.E. Proliferative Vitreoretinopathy: A Review. *Int. Ophthalmol. Clin.* **2019**, *59*, 221–240. [CrossRef]
3. Ishikawa, K.; Kannan, R.; Hinton, D.R. Molecular mechanisms of subretinal fibrosis in age-related macular degeneration. *Exp. Eye Res.* **2016**, *142*, 19–25. [CrossRef]
4. Al-Zamil, W.M.; Yassin, S.A. Recent developments in age-related macular degeneration: A review. *Clin. Interv. Aging* **2017**, *12*, 1313–1330. [CrossRef]
5. Shu, D.Y.; Butcher, E.; Saint-Geniez, M. EMT and EndMT: Emerging Roles in Age-Related Macular Degeneration. *Int. J. Mol. Sci.* **2020**, *21*, 4271. [CrossRef] [PubMed]
6. Nentwich, M.M.; Ulbig, M.W. Diabetic retinopathy—Ocular complications of diabetes mellitus. *World J. Diabetes* **2015**, *6*, 489–499. [CrossRef] [PubMed]
7. Sabanayagam, C.; Banu, R.; Chee, M.L.; Lee, R.; Wang, Y.X.; Tan, G.; Jonas, J.B.; Lamoureux, E.L.; Cheng, C.-Y.; Klein, B.E.K.; et al. Incidence and progression of diabetic retinopathy: A systematic review. *Lancet Diabetes Endocrinol.* **2019**, *7*, 140–149. [CrossRef] [PubMed]
8. Ogurtsova, K.; da Rocha Fernandes, J.D.; Huang, Y.; Linnenkamp, U.; Guariguata, L.; Cho, N.H.; Cavan, D.; Shaw, J.E.; Makaroff, L.E. IDF Diabetes Atlas: Global estimates for the prevalence of diabetes for 2015 and 2040. *Diabetes Res. Clin. Pract.* **2017**, *128*, 40–50. [CrossRef] [PubMed]
9. Sardarinia, M.; Asgari, S.; Hizomi Arani, R.; Eskandari, F.; Azizi, F.; Khalili, D.; Hadaegh, F. Incidence and risk factors of severe non-proliferative/proliferative diabetic retinopathy: More than a decade follow up in the Tehran Lipids and Glucose Study. *J. Diabetes Investig.* **2022**, *13*, 317–327. [CrossRef]
10. van Leeuwen, R.; Haarman, A.E.G.; van de Put, M.A.J.; Klaver, C.C.W.; Los, L.I. Dutch Rhegmatogenous Retinal Detachment Study Group Association of Rhegmatogenous Retinal Detachment Incidence With Myopia Prevalence in The Netherlands. *JAMA Ophthalmol.* **2021**, *139*, 85–92. [CrossRef] [PubMed]
11. Fricke, T.R.; Jong, M.; Naidoo, K.S.; Sankaridurg, P.; Naduvilath, T.J.; Ho, S.M.; Wong, T.Y.; Resnikoff, S. Global prevalence of visual impairment associated with myopic macular degeneration and temporal trends from 2000 through 2050: Systematic review, meta-analysis and modelling. *Br. J. Ophthalmol.* **2018**, *102*, 855–862. [CrossRef] [PubMed]
12. Holden, B.A.; Fricke, T.R.; Wilson, D.A.; Jong, M.; Naidoo, K.S.; Sankaridurg, P.; Wong, T.Y.; Naduvilath, T.J.; Resnikoff, S. Global Prevalence of Myopia and High Myopia and Temporal Trends from 2000 through 2050. *Ophthalmology* **2016**, *123*, 1036–1042. [CrossRef]
13. Lazzara, F.; Conti, F.; Ferrara, M.; Lippera, M.; Coppola, M.; Rossi, S.; Drago, F.; Bucolo, C.; Romano, M.R. Safety Profile of Lutein- Versus Triamcinolone Acetonide-Based Vitreous Staining. *Transl. Vis. Sci. Technol.* **2023**, *12*, 5. [CrossRef]
14. Spadaro, A.; Rao, M.; Lorenti, M.; Romano, M.R.; Augello, A.; Eandi, C.M.; Platania, C.B.M.; Drago, F.; Bucolo, C. New Brilliant Blue G Derivative as Pharmacological Tool in Retinal Surgery. *Front. Pharmacol.* **2020**, *11*, 708. [CrossRef] [PubMed]
15. Shu, D.Y.; Lovicu, F.J. Myofibroblast transdifferentiation: The dark force in ocular wound healing and fibrosis. *Prog. Retin. Eye Res.* **2017**, *60*, 44–65. [CrossRef] [PubMed]
16. Little, K.; Ma, J.H.; Yang, N.; Chen, M.; Xu, H. Myofibroblasts in macular fibrosis secondary to neovascular age-related macular degeneration—The potential sources and molecular cues for their recruitment and activation. *EBioMedicine* **2018**, *38*, 283–291. [CrossRef]
17. Tenbrock, L.; Wolf, J.; Boneva, S.; Schlecht, A.; Agostini, H.; Wieghofer, P.; Schlunck, G.; Lange, C. Subretinal fibrosis in neovascular age-related macular degeneration: Current concepts, therapeutic avenues, and future perspectives. *Cell Tissue Res.* **2022**, *387*, 361–375. [CrossRef]
18. Yang, S.; Li, H.; Li, M.; Wang, F. Mechanisms of epithelial-mesenchymal transition in proliferative vitreoretinopathy. *Discov. Med.* **2015**, *20*, 207–217. [PubMed]
19. Tamiya, S.; Kaplan, H.J. Role of epithelial-mesenchymal transition in proliferative vitreoretinopathy. *Exp. Eye Res.* **2016**, *142*, 26–31. [CrossRef]
20. Yang, J.; Antin, P.; Berx, G.; Blanpain, C.; Brabletz, T.; Bronner, M.; Campbell, K.; Cano, A.; Casanova, J.; Christofori, G.; et al. Guidelines and definitions for research on epithelial-mesenchymal transition. *Nat. Rev. Mol. Cell Biol.* **2020**, *21*, 341–352. [CrossRef] [PubMed]
21. Dongre, A.; Weinberg, R.A. New insights into the mechanisms of epithelial-mesenchymal transition and implications for cancer. *Nat. Rev. Mol. Cell Biol.* **2019**, *20*, 69–84. [CrossRef]
22. Ribatti, D.; Tamma, R.; Annese, T. Epithelial-Mesenchymal Transition in Cancer: A Historical Overview. *Transl. Oncol.* **2020**, *13*, 100773. [CrossRef]

23. Eastlake, K.; Banerjee, P.J.; Angbohang, A.; Charteris, D.G.; Khaw, P.T.; Limb, G.A. Müller glia as an important source of cytokines and inflammatory factors present in the gliotic retina during proliferative vitreoretinopathy. *Glia* **2016**, *64*, 495–506. [CrossRef] [PubMed]
24. Hou, H.; Nudleman, E.; Weinreb, R.N. Animal Models of Proliferative Vitreoretinopathy and Their Use in Pharmaceutical Investigations. *Ophthalmic Res.* **2018**, *60*, 195–204. [CrossRef] [PubMed]
25. Lee, S.C.; Kim, S.H.; Koh, H.J.; Kwon, O.W. TGF-betas synthesized by RPE cells have autocrine activity on mesenchymal transformation and cell proliferation. *Yonsei Med. J.* **2001**, *42*, 271–277. [CrossRef] [PubMed]
26. Saika, S.; Kono-Saika, S.; Tanaka, T.; Yamanaka, O.; Ohnishi, Y.; Sato, M.; Muragaki, Y.; Ooshima, A.; Yoo, J.; Flanders, K.C.; et al. Smad3 is required for dedifferentiation of retinal pigment epithelium following retinal detachment in mice. *Lab. Investig.* **2004**, *84*, 1245–1258. [CrossRef] [PubMed]
27. Por, E.D.; Greene, W.A.; Burke, T.A.; Wang, H.-C. Trichostatin A Inhibits Retinal Pigmented Epithelium Activation in an In Vitro Model of Proliferative Vitreoretinopathy. *J. Ocul. Pharmacol. Ther.* **2016**, *32*, 415–424. [CrossRef]
28. Wei, J.; Wu, L.; Yang, S.; Zhang, C.; Feng, L.; Wang, M.; Li, H.; Wang, F. E-cadherin to N-cadherin switching in the TGF-β1 mediated retinal pigment epithelial to mesenchymal transition. *Exp. Eye Res.* **2022**, *220*, 109085. [CrossRef]
29. Lee, H.; O'Meara, S.J.; O'Brien, C.; Kane, R. The role of gremlin, a BMP antagonist, and epithelial-to-mesenchymal transition in proliferative vitreoretinopathy. *Investig. Ophthalmol. Vis. Sci.* **2007**, *48*, 4291–4299. [CrossRef]
30. He, H.; Kuriyan, A.E.; Su, C.-W.; Mahabole, M.; Zhang, Y.; Zhu, Y.-T.; Flynn, H.W.; Parel, J.-M.; Tseng, S.C.G. Inhibition of Proliferation and Epithelial Mesenchymal Transition in Retinal Pigment Epithelial Cells by Heavy Chain-Hyaluronan/Pentraxin 3. *Sci. Rep.* **2017**, *7*, 43736. [CrossRef]
31. Zhang, Y.; Wang, K.; Pan, J.; Yang, S.; Yao, H.; Li, M.; Li, H.; Lei, H.; Jin, H.; Wang, F. Exosomes mediate an epithelial-mesenchymal transition cascade in retinal pigment epithelial cells: Implications for proliferative vitreoretinopathy. *J. Cell. Mol. Med.* **2020**, *24*, 13324–13335. [CrossRef] [PubMed]
32. Tamiya, S.; Liu, L.; Kaplan, H.J. Epithelial-mesenchymal transition and proliferation of retinal pigment epithelial cells initiated upon loss of cell-cell contact. *Investig. Ophthalmol. Vis. Sci.* **2010**, *51*, 2755–2763. [CrossRef]
33. Takahashi, E.; Nagano, O.; Ishimoto, T.; Yae, T.; Suzuki, Y.; Shinoda, T.; Nakamura, S.; Niwa, S.; Ikeda, S.; Koga, H.; et al. Tumor necrosis factor-alpha regulates transforming growth factor-beta-dependent epithelial-mesenchymal transition by promoting hyaluronan-CD44-moesin interaction. *J. Biol. Chem.* **2010**, *285*, 4060–4073. [CrossRef] [PubMed]
34. Matoba, R.; Morizane, Y.; Shiode, Y.; Hirano, M.; Doi, S.; Toshima, S.; Araki, R.; Hosogi, M.; Yonezawa, T.; Shiraga, F. Suppressive effect of AMP-activated protein kinase on the epithelial-mesenchymal transition in retinal pigment epithelial cells. *PLoS ONE* **2017**, *12*, e0181481. [CrossRef] [PubMed]
35. Schiff, L.; Boles, N.C.; Fernandes, M.; Nachmani, B.; Gentile, R.; Blenkinsop, T.A. P38 inhibition reverses TGFβ1 and TNFα-induced contraction in a model of proliferative vitreoretinopathy. *Commun. Biol.* **2019**, *2*, 162. [CrossRef] [PubMed]
36. Boles, N.C.; Fernandes, M.; Swigut, T.; Srinivasan, R.; Schiff, L.; Rada-Iglesias, A.; Wang, Q.; Saini, J.S.; Kiehl, T.; Stern, J.H.; et al. Epigenomic and Transcriptomic Changes During Human RPE EMT in a Stem Cell Model of Epiretinal Membrane Pathogenesis and Prevention by Nicotinamide. *Stem Cell Rep.* **2020**, *14*, 631–647. [CrossRef] [PubMed]
37. Han, H.; Zhao, X.; Liao, M.; Song, Y.; You, C.; Dong, X.; Yang, X.; Wang, X.; Huang, B.; Du, M.; et al. Activated Blood Coagulation Factor X (FXa) Contributes to the Development of Traumatic PVR Through Promoting RPE Epithelial-Mesenchymal Transition. *Investig. Ophthalmol. Vis. Sci.* **2021**, *62*, 29. [CrossRef] [PubMed]
38. Chen, X.; Yang, W.; Deng, X.; Ye, S.; Xiao, W. Interleukin-6 promotes proliferative vitreoretinopathy by inducing epithelial-mesenchymal transition via the JAK1/STAT3 signaling pathway. *Mol. Vis.* **2020**, *26*, 517–529.
39. Zhang, W.; Han, H. Targeting matrix stiffness-induced activation of retinal pigment epithelial cells through the RhoA/YAP pathway ameliorates proliferative vitreoretinopathy. *Exp. Eye Res.* **2021**, *209*, 108677. [CrossRef]
40. Greene, W.A.; Kaini, R.R.; Wang, H.-C. Utility of Induced Pluripotent Stem Cell-Derived Retinal Pigment Epithelium for an In Vitro Model of Proliferative Vitreoretinopathy. *Adv. Exp. Med. Biol.* **2019**, *1186*, 33–53. [CrossRef]
41. Chen, Y.; Wu, B.; He, J.F.; Chen, J.; Kang, Z.W.; Liu, D.; Luo, J.; Fang, K.; Leng, X.; Tian, H.; et al. Effectively Intervening Epithelial-Mesenchymal Transition of Retinal Pigment Epithelial Cells With a Combination of ROCK and TGF-β Signaling Inhibitors. *Investig. Ophthalmol. Vis. Sci.* **2021**, *62*, 21. [CrossRef] [PubMed]
42. Oki, K.; Miyata, Y.; Shimada, A.; Nagase, T.; Katsura, Y.; Kosano, H. Cell-mediated contraction of vitreous explants from chicken embryo: Possibility of screening for therapeutic agents against proliferative vitreoretinal diseases. *Mol. Vis.* **2013**, *19*, 2374–2384. [PubMed]
43. Chen, H.; Wang, H.; An, J.; Shang, Q.; Ma, J. Inhibitory Effects of Plumbagin on Retinal Pigment Epithelial Cell Epithelial-Mesenchymal Transition In Vitro and In Vivo. *Med. Sci. Monit.* **2018**, *24*, 1502–1510. [CrossRef] [PubMed]
44. Pacheco-Domínguez, R.L.; Palma-Nicolas, J.P.; López, E.; López-Colomé, A.M. The activation of MEK-ERK1/2 by glutamate receptor-stimulation is involved in the regulation of RPE proliferation and morphologic transformation. *Exp. Eye Res.* **2008**, *86*, 207–219. [CrossRef]
45. Davis, A.A.; Bernstein, P.S.; Bok, D.; Turner, J.; Nachtigal, M.; Hunt, R.C. A human retinal pigment epithelial cell line that retains epithelial characteristics after prolonged culture. *Investig. Ophthalmol. Vis. Sci.* **1995**, *36*, 955–964.
46. Kozlowski, M.R. The ARPE-19 cell line: Mortality status and utility in macular degeneration research. *Curr. Eye Res.* **2015**, *40*, 501–509. [CrossRef]

47. Fasler-Kan, E.; Aliu, N.; Wunderlich, K.; Ketterer, S.; Ruggiero, S.; Berger, S.; Meyer, P. The Retinal Pigment Epithelial Cell Line (ARPE-19) Displays Mosaic Structural Chromosomal Aberrations. *Methods Mol. Biol.* **2018**, *1745*, 305–314. [CrossRef]
48. Lehmann, G.L.; Benedicto, I.; Philp, N.J.; Rodriguez-Boulan, E. Plasma membrane protein polarity and trafficking in RPE cells: Past, present and future. *Exp. Eye Res.* **2014**, *126*, 5–15. [CrossRef]
49. Bharti, K.; den Hollander, A.I.; Lakkaraju, A.; Sinha, D.; Williams, D.S.; Finnemann, S.C.; Bowes-Rickman, C.; Malek, G.; D'Amore, P.A. Cell culture models to study retinal pigment epithelium-related pathogenesis in age-related macular degeneration. *Exp. Eye Res.* **2022**, *222*, 109170. [CrossRef]
50. Ahmado, A.; Carr, A.-J.; Vugler, A.A.; Semo, M.; Gias, C.; Lawrence, J.M.; Chen, L.L.; Chen, F.K.; Turowski, P.; da Cruz, L.; et al. Induction of differentiation by pyruvate and DMEM in the human retinal pigment epithelium cell line ARPE-19. *Investig. Ophthalmol. Vis. Sci.* **2011**, *52*, 7148–7159. [CrossRef]
51. Hazim, R.A.; Volland, S.; Yen, A.; Burgess, B.L.; Williams, D.S. Rapid differentiation of the human RPE cell line, ARPE-19, induced by nicotinamide. *Exp. Eye Res.* **2019**, *179*, 18–24. [CrossRef] [PubMed]
52. Wang, Y.; Yuan, Z.; You, C.; Han, J.; Li, H.; Zhang, Z.; Yan, H. Overexpression p21WAF1/CIP1 in suppressing retinal pigment epithelial cells and progression of proliferative vitreoretinopathy via inhibition CDK2 and cyclin E. *BMC Ophthalmol.* **2014**, *14*, 144. [CrossRef] [PubMed]
53. Yan, F.; Hui, Y.; Li, Y.; Guo, C.; Meng, H. Epidermal growth factor receptor in cultured human retinal pigment epithelial cells. *Ophthalmologica* **2007**, *221*, 244–250. [CrossRef] [PubMed]
54. Umazume, K.; Liu, L.; Scott, P.A.; de Castro, J.P.F.; McDonald, K.; Kaplan, H.J.; Tamiya, S. Inhibition of PVR with a tyrosine kinase inhibitor, dasatinib, in the swine. *Investig. Ophthalmol. Vis. Sci.* **2013**, *54*, 1150–1159. [CrossRef]
55. Wu, W.-C.; Hu, D.-N.; Mehta, S.; Chang, Y.-C. Effects of retinoic acid on retinal pigment epithelium from excised membranes from proliferative vitreoretinopathy. *J. Ocul. Pharmacol. Ther.* **2005**, *21*, 44–54. [CrossRef]
56. Amarnani, D.; Machuca-Parra, A.I.; Wong, L.L.; Marko, C.K.; Stefater, J.A.; Stryjewski, T.P.; Eliott, D.; Arboleda-Velasquez, J.F.; Kim, L.A. Effect of Methotrexate on an In Vitro Patient-Derived Model of Proliferative Vitreoretinopathy. *Investig. Ophthalmol. Vis. Sci.* **2017**, *58*, 3940–3949. [CrossRef]
57. Delgado-Tirado, S.; Amarnani, D.; Zhao, G.; Rossin, E.J.; Eliott, D.; Miller, J.B.; Greene, W.A.; Ramos, L.; Arevalo-Alquichire, S.; Leyton-Cifuentes, D.; et al. Topical delivery of a small molecule RUNX1 transcription factor inhibitor for the treatment of proliferative vitreoretinopathy. *Sci. Rep.* **2020**, *10*, 20554. [CrossRef] [PubMed]
58. Kim, L.A.; Wong, L.L.; Amarnani, D.S.; Bigger-Allen, A.A.; Hu, Y.; Marko, C.K.; Eliott, D.; Shah, V.A.; McGuone, D.; Stemmer-Rachamimov, A.O.; et al. Characterization of cells from patient-derived fibrovascular membranes in proliferative diabetic retinopathy. *Mol. Vis.* **2015**, *21*, 673–687.
59. Oshima, Y.; Sakamoto, T.; Hisatomi, T.; Tsutsumi, C.; Ueno, H.; Ishibashi, T. Gene transfer of soluble TGF-beta type II receptor inhibits experimental proliferative vitreoretinopathy. *Gene Ther.* **2002**, *9*, 1214–1220. [CrossRef]
60. Dong, X.; Chen, N.; Xie, L.; Wang, S. Prevention of experimental proliferative vitreoretinopathy with a biodegradable intravitreal drug delivery system of all-trans retinoic acid. *Retina* **2006**, *26*, 210–213. [CrossRef]
61. Daftarian, N.; Baigy, O.; Suri, F.; Kanavi, M.R.; Balagholi, S.; Afsar Aski, S.; Moghaddasi, A.; Nourinia, R.; Abtahi, S.-H.; Ahmadieh, H. Intravitreal connective tissue growth factor neutralizing antibody or bevacizumab alone or in combination for prevention of proliferative vitreoretinopathy in an experimental model. *Exp. Eye Res.* **2021**, *208*, 108622. [CrossRef]
62. Zhou, P.; Zhao, M.-W.; Li, X.-X.; Yu, W.-Z.; Bian, Z.-M. siRNA targeting mammalian target of rapamycin (mTOR) attenuates experimental proliferative vitreoretinopathy. *Curr. Eye Res.* **2007**, *32*, 973–984. [CrossRef] [PubMed]
63. Zhang, L.; Li, X.; Zhao, M.; He, P.; Yu, W.; Dong, J.; Liu, G.; Li, C.; Shi, X. Antisense oligonucleotide targeting c-fos mRNA limits retinal pigment epithelial cell proliferation: A key step in the progression of proliferative vitreoretinopathy. *Exp. Eye Res.* **2006**, *83*, 1405–1411. [CrossRef]
64. Pao, S.-I.; Lin, L.-T.; Chen, Y.-H.; Chen, C.-L.; Chen, J.-T. MicroRNA-4516 suppresses proliferative vitreoretinopathy development via negatively regulating OTX1. *PLoS ONE* **2022**, *17*, e0270526. [CrossRef]
65. Chen, M.; Hou, J.; Tan, G.; Xie, P.; Freeman, W.R.; Beadle, J.R.; Hostetler, K.Y.; Cheng, L. A novel lipid prodrug strategy for sustained delivery of hexadecyloxypropyl 9-[2-(phosphonomethoxy)ethyl]guanine (HDP-PMEG) on unwanted ocular proliferation. *Drug Deliv.* **2017**, *24*, 1703–1712. [CrossRef]
66. McGillem, G.S.; Dacheux, R.F. Rabbit retinal Müller cells undergo antigenic changes in response to experimentally induced proliferative vitreoretinopathy. *Exp. Eye Res.* **1999**, *68*, 617–627. [CrossRef]
67. Zhang, X.; Wei, J.; Ma, P.; Mu, H.; Wang, A.; Zhang, L.; Wu, Z.; Sun, K. Preparation and evaluation of a novel biodegradable long-acting intravitreal implant containing ligustrazine for the treatment of proliferative vitreoretinopathy. *J. Pharm. Pharmacol.* **2015**, *67*, 160–169. [CrossRef]
68. Hou, H.; Huffman, K.; Rios, S.; Freeman, W.R.; Sailor, M.J.; Cheng, L. A Novel Approach of Daunorubicin Application on Formation of Proliferative Retinopathy Using a Porous Silicon Controlled Delivery System: Pharmacodynamics. *Investig. Ophthalmol. Vis. Sci.* **2015**, *56*, 2755–2763. [CrossRef]
69. Wong, C.A.; Potter, M.J.; Cui, J.Z.; Chang, T.S.; Ma, P.; Maberley, A.L.; Ross, W.H.; White, V.A.; Samad, A.; Jia, W.; et al. Induction of proliferative vitreoretinopathy by a unique line of human retinal pigment epithelial cells. *Can. J. Ophthalmol.* **2002**, *37*, 211–220. [CrossRef] [PubMed]

70. Pao, S.-I.; Lin, L.-T.; Chen, Y.-H.; Chen, C.-L.; Chen, J.-T. Repression of Smad4 by MicroRNA-1285 moderates TGF-β-induced epithelial-mesenchymal transition in proliferative vitreoretinopathy. *PLoS ONE* **2021**, *16*, e0254873. [CrossRef] [PubMed]
71. Wu, W.; Yang, X.; Lu, B. [Intravitreally injectable poly (D, L-Lactide) microspheres containing dexamethasone acetate for sustained release]. *Yao Xue Xue Bao* **2001**, *36*, 766–770. [PubMed]
72. Lei, H.; Rheaume, M.-A.; Kazlauskas, A. Recent developments in our understanding of how platelet-derived growth factor (PDGF) and its receptors contribute to proliferative vitreoretinopathy. *Exp. Eye Res.* **2010**, *90*, 376–381. [CrossRef]
73. Baudouin, C.; Khosravi, E.; Pisella, P.J.; Ettaiche, M.; Elena, P.P. Inflammation measurement and immunocharacterization of cell proliferation in an experimental model of proliferative vitreoretinopathy. *Ophthalmic Res.* **1998**, *30*, 340–350. [CrossRef]
74. Ozer, M.A.; Polat, N.; Ozen, S.; Ogurel, T.; Parlakpinar, H.; Vardi, N. Histopathological and ophthalmoscopic evaluation of apocynin on experimental proliferative vitreoretinopathy in rabbit eyes. *Int. Ophthalmol.* **2017**, *37*, 599–605. [CrossRef]
75. Cheng, L.; Hostetler, K.; Valiaeva, N.; Tammewar, A.; Freeman, W.R.; Beadle, J.; Bartsch, D.-U.; Aldern, K.; Falkenstein, I. Intravitreal crystalline drug delivery for intraocular proliferation diseases. *Investig. Ophthalmol. Vis. Sci.* **2010**, *51*, 474–481. [CrossRef]
76. Kralinger, M.T.; Kieselbach, G.F.; Voigt, M.; Hayden, B.; Hernandez, E.; Fernandez, V.; Parel, J.-M. Experimental model for proliferative vitreoretinopathy by intravitreal dispase: Limited by zonulolysis and cataract. *Ophthalmologica* **2006**, *220*, 211–216. [CrossRef]
77. Goczalik, I.; Ulbricht, E.; Hollborn, M.; Raap, M.; Uhlmann, S.; Weick, M.; Pannicke, T.; Wiedemann, P.; Bringmann, A.; Reichenbach, A.; et al. Expression of CXCL8, CXCR1, and CXCR2 in neurons and glial cells of the human and rabbit retina. *Investig. Ophthalmol. Vis. Sci.* **2008**, *49*, 4578–4589. [CrossRef]
78. Mandava, N.; Blackburn, P.; Paul, D.B.; Wilson, M.W.; Read, S.B.; Alspaugh, E.; Tritz, R.; Barber, J.R.; Robbins, J.M.; Kruse, C.A. Ribozyme to proliferating cell nuclear antigen to treat proliferative vitreoretinopathy. *Investig. Ophthalmol. Vis. Sci.* **2002**, *43*, 3338–3348.
79. Isiksoy, S.; Basmak, H.; Kasapoglu Dundar, E.; Ozer, A. Expression of proteins associated with cell-matrix adhesion in proliferative vitreoretinopathy designed by Dispase model. *Eur. J. Ophthalmol.* **2007**, *17*, 89–103. [CrossRef]
80. Moon, S.W.; Sun, Y.; Warther, D.; Huffman, K.; Freeman, W.R.; Sailor, M.J.; Cheng, L. New model of proliferative vitreoretinopathy in rabbit for drug delivery and pharmacodynamic studies. *Drug Deliv.* **2018**, *25*, 600–610. [CrossRef] [PubMed]
81. Badaro, E.; Novais, E.A.; Abdala, K.; Chun, M.; Urias, M.; de Arruda Melo Filho, P.A.; Farah, M.E.; Rodrigues, E.B. Development of an experimental model of proliferative retinopathy by intravitreal injection of VEGF165. *J. Ocul. Pharmacol. Ther.* **2014**, *30*, 752–756. [CrossRef] [PubMed]
82. Baudouin, C.; Pisella, P.J.; Ettaiche, M.; Goldschild, M.; Becquet, F.; Gastaud, P.; Droy-Lefaix, M.T. Effects of EGb761 and superoxide dismutase in an experimental model of retinopathy generated by intravitreal production of superoxide anion radical. *Graefes Arch. Clin. Exp. Ophthalmol.* **1999**, *237*, 58–66. [CrossRef] [PubMed]
83. Nassar, K.; Lüke, J.; Lüke, M.; Kamal, M.; Abd El-Nabi, E.; Soliman, M.; Rohrbach, M.; Grisanti, S. The novel use of decorin in prevention of the development of proliferative vitreoretinopathy (PVR). *Graefes Arch. Clin. Exp. Ophthalmol.* **2011**, *249*, 1649–1660. [CrossRef]
84. Hoerster, R.; Muether, P.S.; Vierkotten, S.; Hermann, M.M.; Kirchhof, B.; Fauser, S. Upregulation of TGF-ß1 in experimental proliferative vitreoretinopathy is accompanied by epithelial to mesenchymal transition. *Graefes Arch. Clin. Exp. Ophthalmol.* **2014**, *252*, 11–16. [CrossRef] [PubMed]
85. Tahara, Y.R.; Sakamoto, T.R.; Oshima, Y.R.; Ishibashi, T.R.; Inomata, H.R.; Murata, T.R.; Hinton, D.R.; Ryan, S.J. The antidepressant hypericin inhibits progression of experimental proliferative vitreoretinopathy. *Curr. Eye Res.* **1999**, *19*, 323–329. [CrossRef]
86. Pastor, J.C.; Rodríguez, E.; Marcos, M.A.; Lopez, M.I. Combined pharmacologic therapy in a rabbit model of proliferative vitreoretinopathy (PVR). *Ophthalmic Res.* **2000**, *32*, 25–29. [CrossRef]
87. Liou, G.I.; Pakalnis, V.A.; Matragoon, S.; Samuel, S.; Behzadian, M.A.; Baker, J.; Khalil, I.E.; Roon, P.; Caldwell, R.B.; Hunt, R.C.; et al. HGF regulation of RPE proliferation in an IL-1beta/retinal hole-induced rabbit model of PVR. *Mol. Vis.* **2002**, *8*, 494–501.
88. Kosnosky, W.; Li, T.H.; Pakalnis, V.A.; Fox, A.; Hunt, R.C. Interleukin-1-beta changes the expression of metalloproteinases in the vitreous humor and induces membrane formation in eyes containing preexisting retinal holes. *Investig. Ophthalmol. Vis. Sci.* **1994**, *35*, 4260–4267.
89. Zhou, G.; Duan, Y.; Ma, G.; Wu, W.; Hu, Z.; Chen, N.; Chee, Y.; Cui, J.; Samad, A.; Matsubara, J.A.; et al. Introduction of the MDM2 T309G Mutation in Primary Human Retinal Epithelial Cells Enhances Experimental Proliferative Vitreoretinopathy. *Investig. Ophthalmol. Vis. Sci.* **2017**, *58*, 5361–5367. [CrossRef] [PubMed]
90. Lee, J.J.; Park, J.K.; Kim, Y.-T.; Kwon, B.-M.; Kang, S.G.; Yoo, Y.D.; Yu, Y.S.; Chung, H. Effect of 2'-benzoyl-oxycinnamaldehyde on RPE cells in vitro and in an experimental proliferative vitreoretinopathy model. *Investig. Ophthalmol. Vis. Sci.* **2002**, *43*, 3117–3124.
91. Lei, H.; Velez, G.; Cui, J.; Samad, A.; Maberley, D.; Matsubara, J.; Kazlauskas, A. N-acetylcysteine suppresses retinal detachment in an experimental model of proliferative vitreoretinopathy. *Am. J. Pathol.* **2010**, *177*, 132–140. [CrossRef] [PubMed]
92. Chen, S.-H.; Lin, Y.-J.; Wang, L.-C.; Tsai, H.-Y.; Yang, C.-H.; Teng, Y.-T.; Hsu, S.-M. Doxycycline Ameliorates the Severity of Experimental Proliferative Vitreoretinopathy in Mice. *Int. J. Mol. Sci.* **2021**, *22*, 11670. [CrossRef]
93. Zhang, J.; Zhou, Q.; Yuan, G.; Dong, M.; Shi, W. Notch signaling regulates M2 type macrophage polarization during the development of proliferative vitreoretinopathy. *Cell. Immunol.* **2015**, *298*, 77–82. [CrossRef] [PubMed]

94. Szczesniak, A.-M.; Porter, R.F.; Toguri, J.T.; Borowska-Fielding, J.; Gebremeskel, S.; Siwakoti, A.; Johnston, B.; Lehmann, C.; Kelly, M.E.M. Cannabinoid 2 receptor is a novel anti-inflammatory target in experimental proliferative vitreoretinopathy. *Neuropharmacology* **2017**, *113*, 627–638. [CrossRef] [PubMed]
95. Iribarne, M.; Ogawa, L.; Torbidoni, V.; Dodds, C.M.; Dodds, R.A.; Suburo, A.M. Blockade of endothelinergic receptors prevents development of proliferative vitreoretinopathy in mice. *Am. J. Pathol.* **2008**, *172*, 1030–1042. [CrossRef] [PubMed]
96. Gao, Q.; Wang, W.; Lan, Y.; Chen, X.; Yang, W.; Yuan, Y.; Tan, J.; Zong, Y.; Jiang, Z. The inhibitory effect of small interference RNA protein kinase C-alpha on the experimental proliferative vitreoretinopathy induced by dispase in mice. *Int. J. Nanomed.* **2013**, *8*, 1563–1572. [CrossRef]
97. Yoo, K.; Son, B.K.; Kim, S.; Son, Y.; Yu, S.-Y.; Hong, H.S. Substance P prevents development of proliferative vitreoretinopathy in mice by modulating TNF-α. *Mol. Vis.* **2017**, *23*, 933–943. [PubMed]
98. Zhang, W.; Tan, J.; Liu, Y.; Li, W.; Gao, Q.; Lehmann, P.V. Assessment of the innate and adaptive immune system in proliferative vitreoretinopathy. *Eye (Lond.)* **2012**, *26*, 872–881. [CrossRef]
99. Cantó Soler, M.V.; Gallo, J.E.; Dodds, R.A.; Suburo, A.M. A mouse model of proliferative vitreoretinopathy induced by dispase. *Exp. Eye Res.* **2002**, *75*, 491–504. [CrossRef]
100. Liang, C.-M.; Tai, M.-C.; Chang, Y.-H.; Chen, Y.-H.; Chen, C.-L.; Lu, D.-W.; Chen, J.-T. Glucosamine inhibits epithelial-to-mesenchymal transition and migration of retinal pigment epithelium cells in culture and morphologic changes in a mouse model of proliferative vitreoretinopathy. *Acta Ophthalmol.* **2011**, *89*, e505–e514. [CrossRef]
101. Saika, S.; Yamanaka, O.; Ikeda, K.; Kim-Mitsuyama, S.; Flanders, K.C.; Yoo, J.; Roberts, A.B.; Nishikawa-Ishida, I.; Ohnishi, Y.; Muragaki, Y.; et al. Inhibition of p38MAP kinase suppresses fibrotic reaction of retinal pigment epithelial cells. *Lab. Investig.* **2005**, *85*, 838–850. [CrossRef] [PubMed]
102. Saika, S.; Yamanaka, O.; Nishikawa-Ishida, I.; Kitano, A.; Flanders, K.C.; Okada, Y.; Ohnishi, Y.; Nakajima, Y.; Ikeda, K. Effect of Smad7 gene overexpression on transforming growth factor beta-induced retinal pigment fibrosis in a proliferative vitreoretinopathy mouse model. *Arch. Ophthalmol.* **2007**, *125*, 647–654. [CrossRef] [PubMed]
103. Heffer, A.; Wang, V.; Sridhar, J.; Feldon, S.E.; Libby, R.T.; Woeller, C.F.; Kuriyan, A.E. A Mouse Model of Proliferative Vitreoretinopathy Induced by Intravitreal Injection of Gas and RPE Cells. *Transl. Vis. Sci. Technol.* **2020**, *9*, 9. [CrossRef]
104. Mori, K.; Gehlbach, P.; Ando, A.; Dyer, G.; Lipinsky, E.; Chaudhry, A.G.; Hackett, S.F.; Campochiaro, P.A. Retina-specific expression of PDGF-B versus PDGF-A: Vascular versus nonvascular proliferative retinopathy. *Investig. Ophthalmol. Vis. Sci.* **2002**, *43*, 2001–2006.
105. Akiyama, H.; Kachi, S.; Silva, R.L.E.; Umeda, N.; Hackett, S.F.; McCauley, D.; McCauley, T.; Zoltoski, A.; Epstein, D.M.; Campochiaro, P.A. Intraocular injection of an aptamer that binds PDGF-B: A potential treatment for proliferative retinopathies. *J. Cell. Physiol.* **2006**, *207*, 407–412. [CrossRef]
106. Edwards, M.M.; McLeod, D.S.; Grebe, R.; Heng, C.; Lefebvre, O.; Lutty, G.A. Lama1 mutations lead to vitreoretinal blood vessel formation, persistence of fetal vasculature, and epiretinal membrane formation in mice. *BMC Dev. Biol.* **2011**, *11*, 60. [CrossRef]
107. Chen, X.-L.; Bai, Y.-J.; Hu, Q.-R.; Li, S.-S.; Huang, L.-Z.; Li, X.-X. Small Interfering RNA Targeted to ASPP2 Promotes Progression of Experimental Proliferative Vitreoretinopathy. *Mediators Inflamm.* **2016**, *2016*, 7920631. [CrossRef]
108. Zheng, X.-Z.; Du, L.-F.; Wang, H.-P. An immunohistochemical analysis of a rat model of proliferative vitreoretinopathy and a comparison of the expression of TGF-β and PDGF among the induction methods. *Bosn. J. Basic Med. Sci.* **2010**, *10*, 204–209. [CrossRef]
109. Zheng, X.; Du, L.; Wang, H.; Gu, Q. A novel approach to attenuate proliferative vitreoretinopathy using ultrasound-targeted microbubble destruction and recombinant adeno-associated virus-mediated RNA interference targeting transforming growth factor-β2 and platelet-derived growth factor-B. *J. Gene Med.* **2012**, *14*, 339–347. [CrossRef]
110. Lyu, Y.; Xu, W.; Zhang, J.; Li, M.; Xiang, Q.; Li, Y.; Tan, T.; Ou, Q.; Zhang, J.; Tian, H.; et al. Protein Kinase A Inhibitor H89 Attenuates Experimental Proliferative Vitreoretinopathy. *Investig. Ophthalmol. Vis. Sci.* **2020**, *61*, 1. [CrossRef]
111. Lin, M.; Li, Y.; Li, Z.; Lin, J.; Zhou, X.; Liang, D. Macrophages acquire fibroblast characteristics in a rat model of proliferative vitreoretinopathy. *Ophthalmic Res.* **2011**, *45*, 180–190. [CrossRef] [PubMed]
112. Uslubas, I.; Kanli, A.; Kasap, M.; Akpinar, G.; Karabas, L. Effect of aflibercept on proliferative vitreoretinopathy: Proteomic analysis in an experimental animal model. *Exp. Eye Res.* **2021**, *203*, 108425. [CrossRef]
113. Wong, C.W.; Busoy, J.M.F.; Cheung, N.; Barathi, V.A.; Storm, G.; Wong, T.T. Endogenous or Exogenous Retinal Pigment Epithelial Cells: A Comparison of Two Experimental Animal Models of Proliferative Vitreoretinopathy. *Transl. Vis. Sci. Technol.* **2020**, *9*, 46. [CrossRef]
114. Umazume, K.; Barak, Y.; McDonald, K.; Liu, L.; Kaplan, H.J.; Tamiya, S. Proliferative vitreoretinopathy in the Swine-a new model. *Investig. Ophthalmol. Vis. Sci.* **2012**, *53*, 4910–4916. [CrossRef]
115. Velikay, M.; Stolba, U.; Wedrich, A.; Datlinger, P.; Akramian, J.; Binder, S. The antiproliferative effect of fractionized radiation therapy: Optimization of dosage. *Doc. Ophthalmol.* **1994**, *87*, 265–269. [CrossRef] [PubMed]
116. Lei, H.; Velez, G.; Hovland, P.; Hirose, T.; Gilbertson, D.; Kazlauskas, A. Growth factors outside the PDGF family drive experimental PVR. *Investig. Ophthalmol. Vis. Sci.* **2009**, *50*, 3394–3403. [CrossRef] [PubMed]
117. Wong, C.W.; Cheung, N.; Ho, C.; Barathi, V.; Storm, G.; Wong, T.T. Characterisation of the inflammatory cytokine and growth factor profile in a rabbit model of proliferative vitreoretinopathy. *Sci. Rep.* **2019**, *9*, 15419. [CrossRef]

118. Khoroshilova-Maslova, I.P.; Leparskaya, N.L.; Nabieva, M.M.; Andreeva, L.D. Experimental Modeling of Proliferative Vitreoretinopathy. An Experimental Morphological Study. *Bull. Exp. Biol. Med.* **2015**, *159*, 100–102. [CrossRef]
119. Wiedemann, P.; Sorgente, N.; Ryan, S.J. Proliferative vitreoretinopathy: The rabbit cell injection model for screening of antiproliferative drugs. *J. Pharmacol. Methods* **1984**, *12*, 69–78. [CrossRef]
120. Kawahara, S.; Hata, Y.; Kita, T.; Arita, R.; Miura, M.; Nakao, S.; Mochizuki, Y.; Enaida, H.; Kagimoto, T.; Goto, Y.; et al. Potent inhibition of cicatricial contraction in proliferative vitreoretinal diseases by statins. *Diabetes* **2008**, *57*, 2784–2793. [CrossRef]
121. Guo, L.; Yu, W.; Li, X.; Zhao, G.; Liang, J.; He, P.; Wang, K.; Zhou, P.; Jiang, Y.; Zhao, M. Targeting of integrin-linked kinase with a small interfering RNA suppresses progression of experimental proliferative vitreoretinopathy. *Exp. Eye Res.* **2008**, *87*, 551–560. [CrossRef]
122. Agrawal, R.N.; He, S.; Spee, C.; Cui, J.Z.; Ryan, S.J.; Hinton, D.R. In vivo models of proliferative vitreoretinopathy. *Nat. Protoc.* **2007**, *2*, 67–77. [CrossRef] [PubMed]
123. Bali, E.; Willermain, F.; Caspers-Velu, L.; Dubois, C.; Dehou, M.-F.; Velu, T.; Libert, J.; Bruyns, C. IL-10 in vivo gene expression in a cell-induced animal model of proliferative vitreoretinopathy. *Int. J. Mol. Med.* **2003**, *12*, 305–310. [CrossRef] [PubMed]
124. Khoroshilova-Maslova, I.P.; Leparskaya, N.L.; Vorotelyak, E.A.; Vasiliev, A.V. The significance of fibroblasts in experimental modeling of proliferative vitreoretinopathy. *Vestn. Oftalmol.* **2017**, *133*, 4–10. [CrossRef] [PubMed]
125. Pennock, S.; Rheaume, M.-A.; Mukai, S.; Kazlauskas, A. A novel strategy to develop therapeutic approaches to prevent proliferative vitreoretinopathy. *Am. J. Pathol.* **2011**, *179*, 2931–2940. [CrossRef]
126. Ramirez, M.; Davidson, E.A.; Luttenauer, L.; Elena, P.P.; Cumin, F.; Mathis, G.A.; De Gasparo, M. The renin-angiotensin system in the rabbit eye. *J. Ocul. Pharmacol. Ther.* **1996**, *12*, 299–312. [CrossRef]
127. Nakagawa, M.; Refojo, M.F.; Marin, J.F.; Doi, M.; Tolentino, F.I. Retinoic acid in silicone and silicone-fluorosilicone copolymer oils in a rabbit model of proliferative vitreoretinopathy. *Investig. Ophthalmol. Vis. Sci.* **1995**, *36*, 2388–2395.
128. Zheng, Y.; Ikuno, Y.; Ohj, M.; Kusaka, S.; Jiang, R.; Cekiç, O.; Sawa, M.; Tano, Y. Platelet-derived growth factor receptor kinase inhibitor AG1295 and inhibition of experimental proliferative vitreoretinopathy. *Jpn. J. Ophthalmol.* **2003**, *47*, 158–165. [CrossRef]
129. Yu, Z.; Ma, S.; Wu, M.; Cui, H.; Wu, R.; Chen, S.; Xu, C.; Lu, X.; Feng, S. Self-assembling hydrogel loaded with 5-FU PLGA microspheres as a novel vitreous substitute for proliferative vitreoretinopathy. *J. Biomed. Mater. Res. A* **2020**, *108*, 2435–2446. [CrossRef]
130. Frenzel, E.M.; Neely, K.A.; Walsh, A.W.; Cameron, J.D.; Gregerson, D.S. A new model of proliferative vitreoretinopathy. *Investig. Ophthalmol. Vis. Sci.* **1998**, *39*, 2157–2164.
131. Zhao, X.; Han, H.; Song, Y.; Du, M.; Liao, M.; Dong, X.; Wang, X.; Kuhn, F.; Hoskin, A.; Xu, H.; et al. The Role of Intravitreal Anti-VEGF Agents in Rabbit Eye Model of Open-Globe Injury. *J. Ophthalmol.* **2021**, *2021*, 5565178. [CrossRef] [PubMed]
132. Yang, C.S.; Khawly, J.A.; Hainsworth, D.P.; Chen, S.N.; Ashton, P.; Guo, H.; Jaffe, G.J. An intravitreal sustained-release triamcinolone and 5-fluorouracil codrug in the treatment of experimental proliferative vitreoretinopathy. *Arch. Ophthalmol.* **1998**, *116*, 69–77. [CrossRef] [PubMed]
133. Lossi, L.; D'Angelo, L.; De Girolamo, P.; Merighi, A. Anatomical features for an adequate choice of experimental animal model in biomedicine: II. Small laboratory rodents, rabbit, and pig. *Ann. Anat.* **2016**, *204*, 11–28. [CrossRef]
134. Tosi, G.M.; Marigliani, D.; Romeo, N.; Toti, P. Disease pathways in proliferative vitreoretinopathy: An ongoing challenge. *J. Cell. Physiol.* **2014**, *229*, 1577–1583. [CrossRef]
135. Khateb, S.; Aweidah, H.; Halpert, M.; Jaouni, T. Postoperative Macular Proliferative Vitreoretinopathy: A Case Series and Literature Review. *Case Rep. Ophthalmol.* **2021**, *12*, 464–472. [CrossRef]
136. Men, G.; Peyman, G.A.; Kuo, P.-C.; Bezerra, Y.; Ghahramani, F.; Naaman, G.; Livir-Rallatos, C.; Lee, P.J. The role of scleral buckle in experimental posterior penetrating eye injury. *Retina* **2003**, *23*, 202–208. [CrossRef] [PubMed]
137. Heffer, A.M.; Wang, V.; Libby, R.T.; Feldon, S.E.; Woeller, C.F.; Kuriyan, A.E. Salinomycin inhibits proliferative vitreoretinopathy formation in a mouse model. *PLoS ONE* **2020**, *15*, e0243626. [CrossRef]
138. Zhang, W.; Li, J. Yes-associated protein is essential for proliferative vitreoretinopathy development via the epithelial-mesenchymal transition in retinal pigment epithelial fibrosis. *J. Cell. Mol. Med.* **2021**, *25*, 10213–10223. [CrossRef]
139. Peirson, S.N.; Brown, L.A.; Pothecary, C.A.; Benson, L.A.; Fisk, A.S. Light and the laboratory mouse. *J. Neurosci. Methods* **2018**, *300*, 26–36. [CrossRef]
140. Chang, B. Mouse models for studies of retinal degeneration and diseases. *Methods Mol. Biol.* **2013**, *935*, 27–39. [CrossRef]
141. Zhu, W.; Wu, Y.; Cui, C.; Zhao, H.-M.; Ba, J.; Chen, H.; Yu, J. Expression of IGFBP-6 in proliferative vitreoretinopathy rat models and its effects on retinal pigment epithelial-J cells. *Mol. Med. Rep.* **2014**, *9*, 33–38. [CrossRef] [PubMed]
142. Yue, Y.-K.; Chen, X.-L.; Liu, S.; Liu, W. Upregulation of ASPP2 expression alleviates the development of proliferative vitreoretinopathy in a rat model. *Int. J. Ophthalmol.* **2021**, *14*, 1813–1819. [CrossRef] [PubMed]

Disclaimer/Publisher's Note: The statements, opinions and data contained in all publications are solely those of the individual author(s) and contributor(s) and not of MDPI and/or the editor(s). MDPI and/or the editor(s) disclaim responsibility for any injury to people or property resulting from any ideas, methods, instructions or products referred to in the content.

Review

The Nexus of Inflammation-Induced Epithelial-Mesenchymal Transition and Lung Cancer Progression: A Roadmap to Pentacyclic Triterpenoid-Based Therapies

Kirill V. Odarenko [1,2], Marina A. Zenkova [1] and Andrey V. Markov [1,*]

[1] Institute of Chemical Biology and Fundamental Medicine, Siberian Branch of the Russian Academy of Sciences, 630090 Novosibirsk, Russia; k.odarenko@yandex.ru (K.V.O.); marzen@niboch.nsc.ru (M.A.Z.)
[2] Faculty of Natural Sciences, Novosibirsk State University, 630090 Novosibirsk, Russia
* Correspondence: andmrkv@gmail.com or markov_av@niboch.nsc.ru; Tel.: +7-383-363-51-61

Abstract: Lung cancer is the leading cause of cancer-related death worldwide. Its high mortality is partly due to chronic inflammation that accompanies the disease and stimulates cancer progression. In this review, we analyzed recent studies and highlighted the role of the epithelial–mesenchymal transition (EMT) as a link between inflammation and lung cancer. In the inflammatory tumor microenvironment (iTME), fibroblasts, macrophages, granulocytes, and lymphocytes produce inflammatory mediators, some of which can induce EMT. This leads to increased invasiveness of tumor cells and self-renewal of cancer stem cells (CSCs), which are associated with metastasis and tumor recurrence, respectively. Based on published data, we propose that inflammation-induced EMT may be a potential therapeutic target for the treatment of lung cancer. This prospect is partially realized in the development of EMT inhibitors based on pentacyclic triterpenoids (PTs), described in the second part of our study. PTs reduce the metastatic potential and stemness of tumor cells, making PTs promising candidates for lung cancer therapy. We emphasize that the high diversity of molecular mechanisms underlying inflammation-induced EMT far exceeds those that have been implicated in drug development. Therefore, analysis of information on the relationship between the iTME and EMT is of great interest and may provide ideas for novel treatment approaches for lung cancer.

Keywords: inflammation; pulmonary malignancy; epithelial-to-mesenchymal transition; natural products; aggressiveness; tumor stem cells; mechanism of action

1. Introduction

Lung cancer is one of the most commonly diagnosed cancers and the leading cause of cancer-related deaths worldwide, with an estimated 2,207,000 new cases and 1,796,000 new deaths in 2020 [1]. In Russia, lung cancer ranks second in overall incidence and first in males, with a 5-year survival rate of 15–20% [2]. Despite significant advances in diagnostic and surgical techniques, as well as targeted drug development, the mortality of patients with lung cancer is high and steadily increasing. One of the major reasons for this failure is the lack of therapeutics that are able to effectively control the crosstalk between cancer and the tumor microenvironment (TME), leading to the enhancement of tumor aggressiveness and metastasis.

Numerous studies have shown that stromal and immune cells surrounding tumor tissue are capable of producing various growth factors and cytokines that induce the epithelial–mesenchymal transition (EMT) in cancer cells [3]. Malignant cells undergoing EMT are characterized by the disruption of adherens junctions, the cadherin switch from E-cadherin to N-cadherin, and cytoskeletal reorganization, leading to the acquisition of a motile and invasive phenotype [3]. Moreover, a range of evidence has been collected that suggests that the EMT of lung cancer cells mediates not only the enhancement of their metastatic potential but also their evasion of the immune system and the development of drug resistance [4], clearly indicating an aggravating role of EMT in lung cancer pathology. Indeed, a significant positive correlation between low E-cadherin/high vimentin expression (an important EMT feature) and worse overall survival of lung cancer patients has been reported [5].

It is known that inflammatory airway injury induced by various stressors, including tobacco smoke, industrial dust, allergens, and pulmonary infections, not only creates a milieu conducive to lung carcinogenesis and is associated with a high risk of lung cancer [6] but also promotes the EMT in the lung [7]. For example, an active EMT has been identified in the airway epithelium of smokers and patients with chronic obstructive pulmonary disease (COPD) [8] as well as in severe COVID-19 patients [9]. Because of the established links between chronic lung inflammation, the EMT, and lung cancer aggressiveness, therapeutic modulation of an inflammation-driven EMT can be considered a promising approach in the treatment of pulmonary malignancies.

One of the most important sources of bioactive compounds capable of suppressing tumor progression is pentacyclic triterpenoids (PTs), multitarget plant metabolites that effectively suppress the proliferative, migratory, and invasive capacities of tumor cells [10]. Moreover, these molecules have been shown to significantly block the EMT of tumor cells of different origins driven by various EMT inducers, including pro-inflammatory mediators [11–13]. Given this fact, as well as the demonstrated ability of PTs to ameliorate pulmonary inflammation in vivo [14], these compounds can be considered a promising platform for the development of novel blockers of inflammatory-associated EMT in lung cancer.

In this review, we summarize recent insights into the relationship between the inflammatory background and EMT-driven enhancement of the malignant traits in lung cancer cells and shed light on the possibility of using PTs as potential inhibitors of this interconnection. In view of the pandemic spread of SARS-CoV-2 and COVID-19-associated severe pulmonary pathologies, the knowledge gained in this review will be useful for both lung oncology and medicinal chemistry researchers.

2. Inflammatory-Driven EMT of Lung Cancer Cells: Key Players and Regulators

2.1. EMT-Associated Changes in Cancer Cells

The epithelial–mesenchymal transition (EMT) is a process of epithelial cell phenotype switching that plays a crucial role in a plethora of normal physiological processes, including embryonic development, wound healing, and tissue regeneration, as well as pathological processes such as fibrosis, neoplastic transformation, and cancer metastasis [15].

Various stimuli from TME, such as growth factors [16], hypoxia [17], low pH [18], or changes in the extracellular matrix (ECM) [19], upregulate EMT-associated transcription factors (EMT-TFs) of the Snail (Snail, Slug), ZEB (ZEB1, ZEB2), and Twist (Twist1, Twist2) families, which in turn shift the balance of EMT-associated cell markers toward mesenchymal ones (N-cadherin, vimentin, fibronectin, etc.) [20] (Figure 1).

The main feature of the EMT, an increase in the migratory and invasive abilities of tumor cells, results from a complex process of transdifferentiation involving many cellular components. First, E-cadherin, an epithelial cell–cell adhesion protein, is replaced by N-cadherin, which leads to the formation of relatively weak adhesion junctions [21]. This process leads to the release of β-catenin from the cadherin complex, followed by its translocation to the nucleus (Figure 1). There, β-catenin triggers the expression of EMT-TFs through the engagement of LEF/TCF transcription factors, thus forming a positive feedback loop to maintain the mesenchymal phenotype [21]. In addition to changes in surface markers, cytoskeletal rearrangements occur during EMT and enhance cell motility. The replacement of cytokeratin filaments with vimentin filaments increases the mechanical strength of the cytoskeleton by promoting microtubule polarization, stress fiber formation, and focal adhesion stabilization [22].

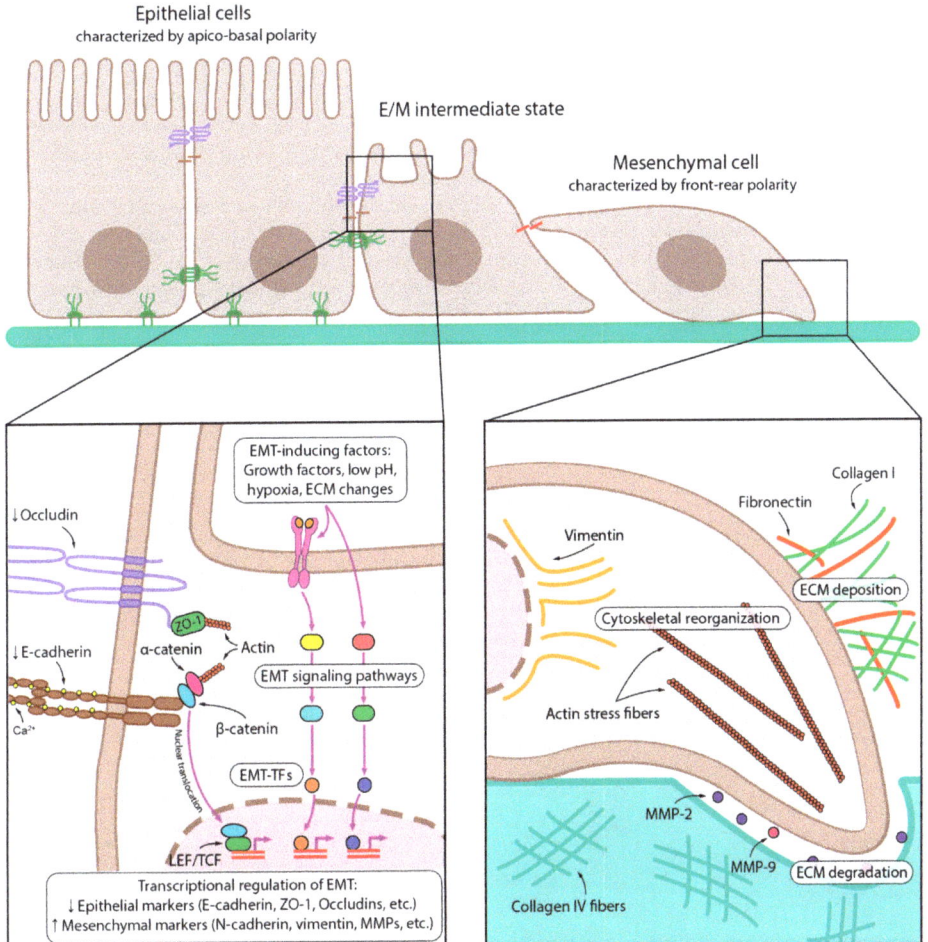

Figure 1. Key molecular events of EMT. Epithelial tumor cells receive EMT-inducing signals from the tumor microenvironment, such as growth factors, low pH, hypoxia, and ECM modifications. EMT-inducing factors trigger various intracellular signaling pathways and activate EMT-TFs, which then downregulate and upregulate epithelial and mesenchymal markers, respectively. EMT is further sustained by autocrine loops: for example, inhibition of E-cadherin leads to its dissociation from β-catenin, which translocates to the nucleus and transactivates EMT-associated genes. Later stages of EMT involve transformation of both intra- and extracellular compartments. The formation of actin stress fibers and vimentin intermediate filaments provides the mechanical force for migration. The cells that have undergone EMT have multidirectional effects on the ECM: they disrupt the basement membrane by degrading collagen IV with matrix metalloproteinases (MMPs) 2 and 9, but at the same time produce other ECM components such as collagen I and fibronectin, which further maintain EMT. Downward (↓) and upward (↑) arrows indicate downregulation and upregulation of expression, respectively.

Alterations in the cytoskeleton and cell–cell adhesion together lead to a change in cell morphology: the typical apical-basal polarity of epithelial cells is replaced by a front-rear polarity of rapidly migrating mesenchymal cells [23]. In addition to changes in the cellular architecture, the EMT is accompanied by extensive ECM rearrangements, which in turn provide molecular support to cells to complete their transition. For example, fibronectin fibers secreted by mesenchymal-like cells bind EMT-inducing growth factors and facilitate their delivery to receptors [24]. Collagen I stabilizes Snail by activating the discoidin domain receptor 2 (DDR2) and the downstream Src/ERK2 signaling pathway [25]. Simultaneous to the formation of the new ECM composed of fibronectin and collagen I, cells undergoing an EMT cleave collagen IV and laminin of the basal lamina by secreting proteolytic enzymes such as matrix metalloproteinases 2 and 9 (MMP-2 and MMP-9, respectively) and thus invade surrounding tissues [26] (Figure 1).

Altogether, the EMT leads to the disruption of cell–cell interactions, loss of apical-basal cell polarity, cytoskeletal rearrangements, and the acquisition of a motile invasive phenotype by tumor cells. These changes ultimately cause cells to migrate from the epithelium, enter the circulation, and disseminate to distant sites, where they can undergo a reverse transition, giving rise to micro- and macrometastases [27].

2.2. Inflammatory TME and Its Role in EMT Induction in Lung Cancer Cells

Inflammation is known to play a crucial role in the initiation and maintenance of tumor growth by providing pro-tumorigenic components to the TME [28]. Given that patients with diseases such as COPD and pulmonary fibrosis that progress to chronic inflammatory conditions have a higher risk of developing lung cancer [29,30], pulmonary inflammation can be considered an important participant in the malignant transformation of lung tissue. Moreover, pro-inflammatory cytokines produced in the microenvironment of established lung tumors have a pronounced pro-metastatic effect, including the stimulation of EMT [31]. To better understand the key aspects of inflammation-driven lung cancer progression, in this section, we focus on the cellular landscape that forms the inflammatory TME (iTME) and its impact on EMT induction in lung cancer cells (Figure 2).

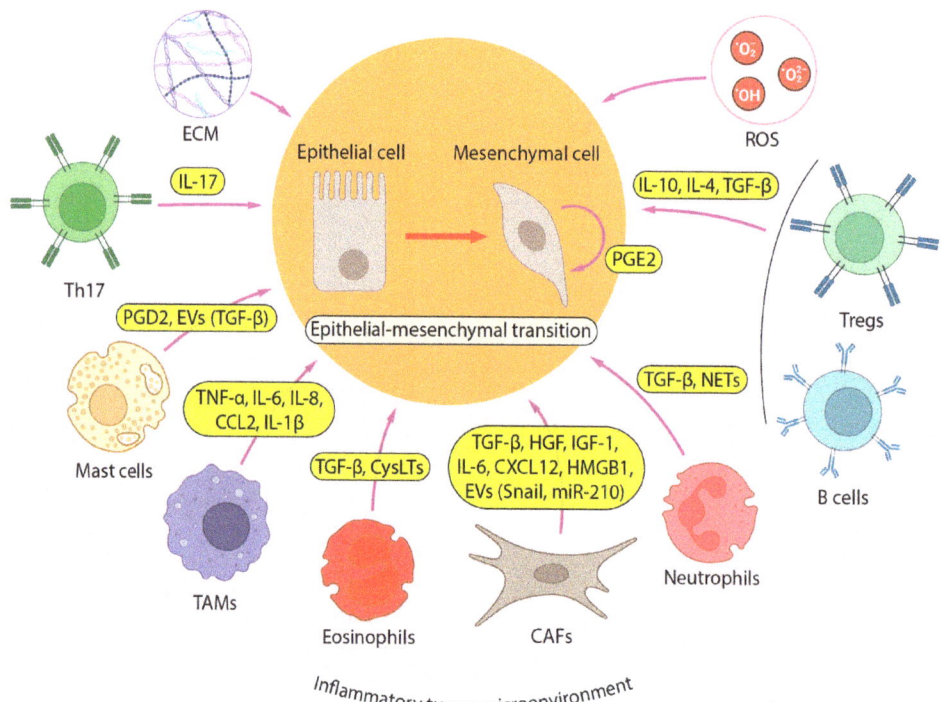

Figure 2. iTME as a driving source of EMT in lung cancer. A number of cell types from iTME, including cancer-associated fibroblasts (CAFs), tumor-associated macrophages (TAMs), neutrophils,

eosinophils, mast cells, T cells (T helper 17 (Th17) and regulatory T (Tregs) cells), and B cells, induce EMT in lung cancer cells by producing various cytokines. Among them, transforming growth factor (TGF-β) is the most studied, but EMT-inducing activity has also been reported for anti-inflammatory cytokines (interleukin 4 (IL-4), IL-10), pro-inflammatory cytokines (IL-1β, IL-6, IL-17, tumor necrosis factor α (TNF-α)), chemokines (C-C Motif Chemokine Ligand 2 (CCL2), C-X-C Motif Chemokine Ligand 12 (CXCL12), IL-8), prostaglandins D2 and E2 (PGD2, PGE2), and neutrophil extracellular traps (NETs). Along with secretion, some EMT-inducing factors are transported to tumor cells via extracellular vesicles (TGF-β, Snail, microRNA-210 (miR-210)). In addition, lung tumor cells produce PGE2 to stimulate their EMT in an autocrine manner.

2.2.1. Cancer-Associated Fibroblasts

There are two complementary pathways contributing to the formation of iTME, including the intrinsic pathway, when genetic changes in neoplastic cells drive the overproduction of pro-inflammatory mediators, and the extrinsic pathway, when iTME is induced by persistent infection, autoimmune disorders, or chronic exposure to an irritant [32]. Along with tumor cells, cancer-associated fibroblasts (CAFs), a heterogenous cell population originating from normal fibroblasts, adipocytes, epithelial and endothelial cells, bone marrow-derived mesenchymal stem cells, and some other cell types, play a key role in tumor stroma formation [33]. CAFs are able to produce ECM components and various growth factors, creating favorable conditions for the proliferation and survival of tumor cells, as well as cancer invasion and neoangiogenesis [34] (Figure 3).

To date, a large number of data have been accumulated confirming the ability of CAFs to induce the EMT of lung cancer cells [35–42]. TGF-β secreted by CAFs plays an important role in this process: co-incubation of lung adenocarcinoma A549 and NCI-H358 cells with CAF-derived conditioned medium and the selective TGF-β inhibitor SB431542 significantly suppressed CAFs-induced EMT progression [35]. In addition to TGF-β, the hepatocyte growth factor (HGF) and insulin-like growth factor 1 (IGF-1) secreted by CAFs participate in the control of EMT initiation in lung cancer cells by synergistically inducing ANXA2 expression and phosphorylation in tumor cells with subsequent cadherin switching and stimulation of vimentin expression, cell migration, and invasion [36].

In addition to growth factors, CAFs secrete a number of pro-inflammatory proteins with pronounced EMT-stimulating activity against lung cancer cells [37,38]. Shintani et al. found that the conditioned medium of CAFs is enriched in interleukin 6 (IL6), which induces the development of the EMT phenotype of non-small cell lung cancer (NSCLC) cells and mediates the development of cell resistance to cisplatin [37]. In an independent study by scientists from Tianjin Medical University General Hospital, the EMT-stimulating effect of IL6 produced by CAFs was shown to be based on the activation of the JAK2/STAT3 signaling pathway: it was found that the pretreatment of lung cancer cells with JAK2- and STAT3-specific inhibitors, AG490 and Stattic, respectively, significantly reversed the effect of the CAF-conditioned medium on the regulation of EMT- and metastasis-related genes and cell motility [38]. In addition to IL6, the promoting effect of CAFs on EMT programming in lung cancer cells is also mediated by the secretion of CXCL12 and high mobility group box 1 (HMGB1), which activate the CXCR4/β-catenin/PPARδ [39] and NFκB [40] signaling pathways, respectively.

A number of recent studies have demonstrated that CAFs are able to deliver EMT-stimulating signals to lung cancer cells via exosomes [41,42]. You et al. reported that the exosome-mediated secretion of Snail by CAFs resulted in the suppression of E-cadherin expression and upregulation of vimentin, accompanied by increased motility and invasiveness of A549 and H1299 cells [41]. Yang et al. found that CAFs isolated from patients with lung adenocarcinoma secreted exosomes containing miR-210, which significantly enhanced the migration and invasion abilities of NSCLC cells via the activation of the PTEN/PI3K/AKT signaling pathway [42]. Since both Snail and miR-210 are known to be associated not only with EMT but also with immune regulation [43,44], these molecules can also be considered important components of iTME that enhance the malignancy of lung cancer cells.

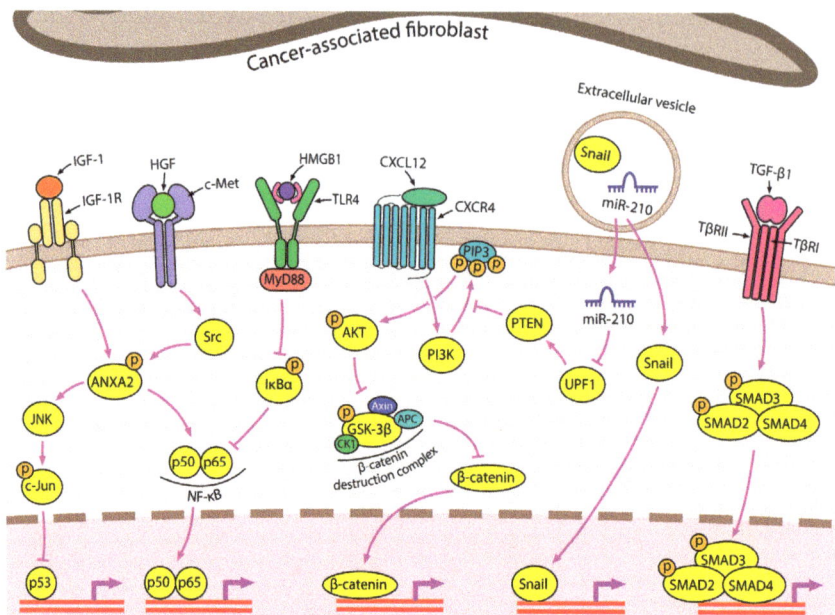

Figure 3. Contribution of CAFs to inflammation-induced EMT in lung cancer CAFs secrete a variety of EMT-inducing factors, including insulin-like growth factor 1 (IGF-1), hepatocyte growth factor (HGF), high mobility group box 1 (HMGB1), CXCL12, TGF-β, and extracellular vesicles (EVs). The IGF-1 and HGF signaling pathways are dependent on annexin A2 (ANXA2) phosphorylation. Autophagy-induced secretion of HMGB1 by CAFs induces EMT via the nuclear factor kappa B (NF-κB) pathway. CXCL12 triggers CXCR4/β-catenin to upregulate EMT-associated genes, among which peroxisome proliferator activated receptor delta (PPARδ) plays a specific role. CAF-derived EVs carry the EMT-TF Snail and miR-210, which induces the PTEN/PI3K/AKT pathway through UPF1 inhibition.

2.2.2. Immune Cells

Immune cells of both innate and adaptive immunity are another cellular component of the tumor stroma that has a multidirectional influence on cancer progression (Figure 2). Chae et al. analyzed the immune landscape of tissue samples from patients with lung adenocarcinoma and lung squamous cell carcinoma according to the "epithelial" or "mesenchymal" tumor phenotype, which differed in the expression of 16 EMT-associated genes [45]. The tissues with a "mesenchymal" phenotype, i.e., containing cells that have undergone EMT activation, show a lower infiltration of CD4+/CD8+ T cells and dendritic cells, which play a key role in immune surveillance against cancer, as well as increased infiltration of B cells and regulatory T cells (Tregs). The secretion of immunosuppressive cytokines such as IL-10, IL-4, and TGF-β by B cells and Tregs, in addition to those produced by tumor cells, allows them to escape immune surveillance and leads to subsequent tumor progression and metastasis [45]. These data agree well with the results of in vitro experiments using a co-culture system consisting of peripheral blood mononuclear cells from healthy donors with A549 cells treated with a combination of inflammation-related cytokines (TGF-β, TNF-α, and interferon γ (IFN-γ)) or the supernatant derived from a mixed lymphocyte reaction. A549 cells that have undergone EMT activation under these stimuli have been shown to acquire immunomodulatory properties, such as the ability to increase or decrease the proliferation of B cells and T cells, respectively, which is in good agreement with the levels of these cell types in "mesenchymal" type lung tumors [45,46].

Granulocytes

Some of the aforementioned immunomodulatory molecules exhibit pleiotropic effects, and in addition to their immunomodulatory activity, they directly or indirectly induce EMT in tumor cells (Figure 2). TGF-β is an anti-inflammatory cytokine with this mode of action. On the one hand, it induces the formation of a tumor-associated N2 neutrophil phenotype, which plays a role in supporting tumor growth and suppressing the antitumor immune response [47]. On the other hand, TGF-β triggers the EMT in lung cancer cells [48]. Interactions between these cell types have been demonstrated in co-culture experiments, revealing that N2 neutrophils contribute to the induction of EMT in A549 cells through the activation of the TGF-β/Smad signaling pathway [49] (Figure 4A). This signaling cascade is also activated in human bronchial epithelial BEAS-2B cells after incubation with primary human eosinophils or human eosinophilic leukemia Eol-1 cells [50,51]. Furthermore, cysteinyl leukotrienes (CysLTs), pro-inflammatory mediators contributing to the pathogenesis of chronic asthma [52], also play a crucial role in eosinophil-induced EMT. Treatment of BEAS-2B cells co-cultured with eosinophils using a selective CysLT inhibitor significantly suppressed TGF-β production, resulting in the blockade of the TGF-β/Smad3 signaling axis and, consequently, the inhibition of the expression of EMT markers and EMT-mediated changes in the morphology of BEAS-2B cells [51] (Figure 4A).

Tumor-Associated Macrophages

In addition to the previously discussed neutrophils and eosinophils, tumor-associated macrophages (TAMs) have been implicated in the formation of iTME. Macrophages are conventionally associated with tissue development and healing due to their capacity to remodel the extracellular matrix (ECM), generate growth and angiogenesis factors, and engulf apoptotic cells [53]. However, within the tumor microenvironment, these macrophage properties take on a pathological role, supporting cancer invasion and metastasis [54]. Several cytokines secreted by macrophages, including TNF-α, IL-6, and IL-8, have the potential to induce EMT in tumor cells (Figure 4B), thereby promoting metastasis [55,56].

Using a co-culture of human lung adenocarcinoma A549 and H1299 cells with THP-1-derived macrophages, Dehai et al. demonstrated a bidirectional interaction between neoplastic epithelial cells and macrophages [57]. Tumor cells stimulate macrophage differentiation toward the M2 phenotype, which is involved in immunosuppression, angiogenesis, and cancer invasion. Macrophages, in turn, enhance the invasive ability of tumor cells and induce an EMT, as evidenced by the switch in cadherin abundance (from E-cadherin to N-cadherin). The co-culture of THP-1-derived macrophages with A549 or H1299 cells resulted in the upregulation of IL-6 in co-cultured cell lines, and IL-6 was also found to induce the EMT in A549 and H1299 cells (Figure 4B). However, the blockade of the EMT by neutralizing anti-IL-6 antibody was less pronounced in the co-culture system compared to that observed in cells treated with recombinant IL-6, suggesting that some other molecules in the TME are involved in the EMT [57]. These results are in good agreement with a recent study by Qin et al., which showed that the treatment of co-cultured A549 and THP-1 cells with E. coli lipopolysaccharide (LPS) shifted the expression profile of EMT markers in tumor cells toward the mesenchymal state and increased the secretion of IL-6 and TGF-β by both cell types [58]. Furthermore, the IL-6-related JAK2/STAT3 pathway was shown to be involved in EMT induction in this experimental model, indirectly confirming the EMT-inducing effect of IL-6 in the co-culture system [58]. The EMT-inducing effect of IL-6 was also confirmed in the human bronchial epithelial BEAS-2B cell model, as cell treatment with IL-6 shifted the expression profile of EMT markers toward the mesenchymal state, increased the expression of EMT-TFs Snail, Twist, and ZEB1, and enhanced cell migration and invasion in BEAS-2B cells [58]. Chen et al. further demonstrated that another TME factor, monocyte chemoattractant protein 1 (MCP-1/CCL2), can enhance IL-6-mediated EMT by exerting a synergistic effect with IL-6 on STAT3 signaling, which causes the subsequent activation of Twist (Figure 4B). In addition, IL-6 and CCL2 have been shown to upregulate each other, forming positive feedback loops to maintain STAT3 signaling activity [56]. These findings are in good agreement with clinical data showing that in patients with NSCLC, serum IL-6 is negatively correlated with E-cadherin and positively correlated with the levels of N-cadherin and vimentin in tumor tissues and, consequently, with lymph node and distant metastasis [59].

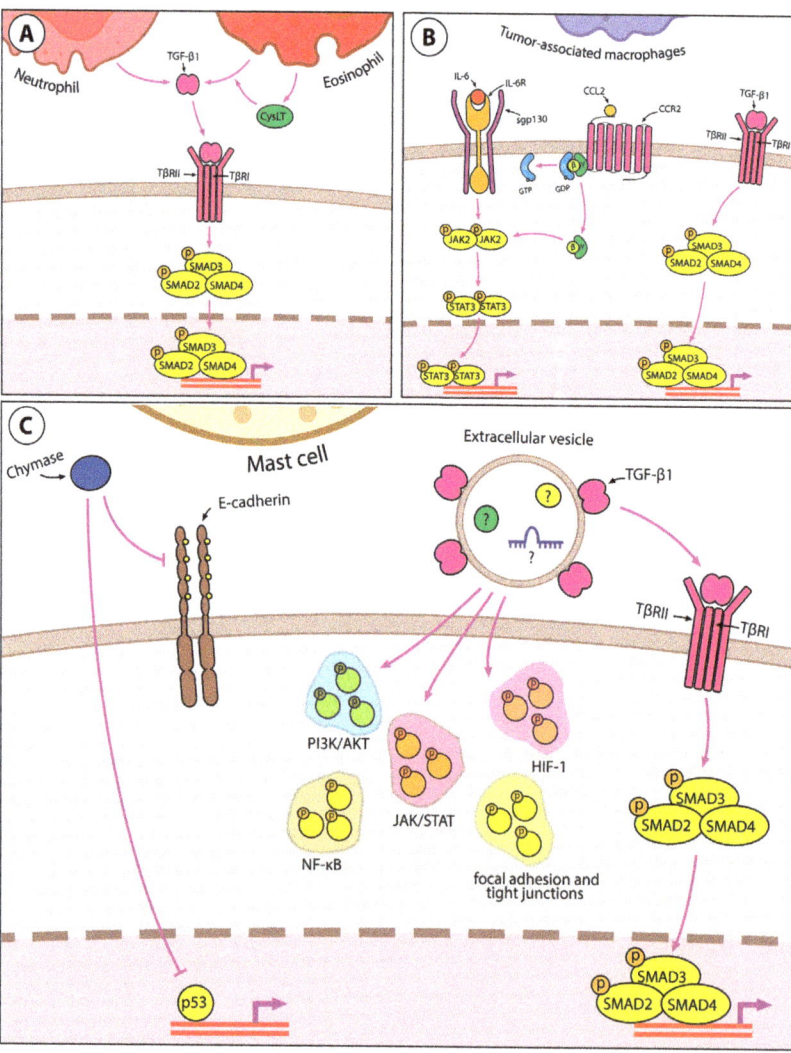

Figure 4. Immune component of inflammation-induced EMT in lung cancer. (**A**) Granulocytes, namely neutrophils and eosinophils, induce EMT through TGF-β/Smad signaling. CysLTs increase TGF-β production in eosinophils in an autocrine manner. (**B**) TAMs induce EMT in lung cancer by releasing IL-6, CCL2, and TGF-β. IL-6 and CCL2 share the same downstream JAK2/STAT3 signaling and mutually upregulate each other to induce EMT. (**C**) Mast cells release chymase, which reduces

cell–cell adhesion by cleaving E-cadherin. Inhibition of the p52 tumor suppressor can be mentioned as another process induced by chymase in lung cancer cells. Mast cells secrete TGF-β-coated EVs that induce the classical Smad-dependent pathway. However, EVs contain other EMT-inducing molecules as they activate the phosphorylation of many proteins in lung cancer cells involved in PI3K/AKT, JAK/STAT, NF-κB, and HIF-1 signaling pathways, as well as the formation of focal adhesions and tight junctions.

Mast Cells

Mast cells are another iTME component involved in EMT induction and the stimulation of metastasis (Figure 4C). Mast cells communicate with lung epithelial cells by extracellular vesicles (EVs) transporting proteins, RNA, and DNA. Yin et al. have shown that HMC-1 mast cell-derived EVs carry TGF-β1 on their surface and are able to induce EMT in A549 cells through the Smad pathway. However, the mechanism of EVs-induced EMT may involve some other signaling pathways, as incubation with mast cell-derived EVs induced the phosphorylation of proteins involved in PI3K/Akt, HIF-1, NF-κB, and JAK/STAT pathways, as well as proteins regulating focal adhesion and tight junctions in A549 cells [60] (Figure 4C). The migration of lung adenocarcinoma A549 and H520 cells induced by mast cells can be further enhanced by chymase secreted by mast cells. Chymase cleaves E-cadherin, attenuates cell–cell adhesion in the lung epithelium, and increases MMP-9 expression in the cells [61].

2.2.3. Other Pro-Inflammatory EMT-Inducing Mediators in iTME

In addition to the aforementioned TGF-β, IL-6, and IL-8, several other immunomodulatory proteins, such as TNF-α, IL-1β, IFN-γ, and IL-17, are capable of either directly inducing EMT or enhancing its induction by other cytokines [62,63]. For example, a mixture of TNF-α, IL-1β, and IFN-γ was shown to significantly enhance TGF-β-induced EMT induction in A549 cells via the upregulation of TGF-β receptor type I (TβRI) expression [64]. Recently, Li et al. conducted a thorough study on the mechanism of IL-1β-dependent EMT stimulation: they demonstrated a gradual change in the EMT-associated phenotype of A549 cells during long-term exposure to IL-1β, resulting in the retention of mesenchymal characteristics of tumor cells after removal of the inflammatory stimulus (a phenomenon called "EMT memory") [62]. It was shown that the initial step of EMT induction, which is dependent on the EMT-TF Slug, is transformed into a Slug-independent chronic step during which E-cadherin expression is controlled by epigenetic modifications, including DNA methylation and the repressive histone modifications H3K27Me3 and H3K9Me2/3, resulting in a prolonged mesenchymal phenotype in the absence of IL-1β. EMT memory has also been demonstrated when A549 are chronically treated with TGF-β or TNF-α [62].

IL-17 is a pro-inflammatory cytokine mainly produced by activated CD4+ T cells (Th17), which has been shown to stimulate lung cancer metastasis [65]. The analysis of human lung adenocarcinoma tissues performed by Huang et al. showed that IL-17 expression was positively correlated with the expression of N-cadherin, vimentin, Snail1, Snail2, and Twist1 and negatively correlated with E-cadherin expression [63]. This may indicate that IL-17 is involved in EMT induction in lung adenocarcinoma cells, which was further confirmed by in vitro studies. IL-17 induces EMT in A549 and Lewis lung carcinoma (LCC) cells through STAT3 signaling [63]. In addition, the NF-κB pathway also regulates IL-17-mediated EMT through the upregulation of ZEB1, as blocking NF-κB activity with a selective inhibitor abrogates EMT-associated changes in cancer cells [66].

TGF-β and TNF-α also implicate NF-κB in the induction of EMT: mesenchymal-related A549 spheroid cultures induced by these inflammatory cytokines showed high levels of phosphorylation of IKK and its downstream targets IκBα and RelA, as well as overexpression of the NF-κB-dependent genes *IL-8* and *BIRC3*. Furthermore, the inhibition of NF-κB resulted in the suppression of Twist, Slug, and ZEB expression and reduced the invasive activity of A549 cells, as well as their ability to metastasize to the lung in a mouse xenograft model [67].

Prostaglandin E2 (PGE2), produced from arachidonic acid, is another EMT-stimulating factor in the iTME [68,69]. Cyclooxygenase 2 (COX-2), the rate-limiting enzyme in PGE2 synthesis, is constitutively expressed in many tumor types, including lung adenocarcinoma [68]. PGE2 was shown to suppress E-cadherin expression via the upregulation of ZEB-1 and Snail in COX-2-overexpressing A549 cells. Consistent with this, immunohistochemical staining of human lung adenocarcinoma sections showed that COX-2 was positively and negatively correlated with ZEB2 and E-cadherin expression, respectively [69]. Interestingly, PGE2 had an opposite effect on TGF-β-treated A549 cells, as PGE2 or the PGE2 receptor agonists EP4 and EP2 have been shown to suppress TGF-β-mediated actin reorganization, cell migration, and fibronectin and collagen I expression in A549 cells, although they did not affect the switch from E-cadherin to N-cadherin [70]. Thus, the effect of PGE2 on EMT seems to be determined by interactions with other EMT-associated pathways.

Interestingly, prostaglandin D2 (PGD2), also a COX metabolite like PGE2, induced EMT in A549 cells by stimulating TGF-β expression, and this cytokine was found to be a key master regulator of PGD2-driven EMT. The inhibition of TβRI by selective inhibitor SB431542 or short hairpin RNA-mediated gene silencing effectively abrogated the EMT-stimulating effect of PGD2 in A549 cells [71].

2.2.4. ECM Components as EMT Regulators

Hyaluronan (HA), a polysaccharide component of the ECM, maintains tissue integrity due to its viscoelastic properties, but also regulates inflammation in a size-dependent manner: high-molecular-weight HA (~104 kDa) inhibits, whereas low-molecular-weight HA (~200 kDa) induces inflammation [72]. HA has been shown to be overproduced by tumor and stromal cells under inflammatory conditions [73].

TGF-β1 and IL-1β induce HA production by upregulating hyaluronan synthases (HAS) in lung adenocarcinoma cells [74]. The overexpression of HAS3 causes morphological changes, invasiveness, E-cadherin suppression, and vimentin induction in H358 cells, suggesting the role of HA as an autocrine factor in maintaining EMT [74]. When grown as 3D spheroids on HA-grafted chitosan membranes, A549 and H1299 cells exhibit higher levels of EMT markers and migration and invasion potential than their 2D counterparts [75]. Han et al. introduced mesenchymal stem cells (MSCs) into this system and demonstrated that HA provides structural support for their interaction with lung cancer cells. MSCs form a spheroid core and produce the EMT-inducing factors IL-10, TGF-β1, and CXCL12, which are thought to induce A549 cells from the spheroid margin to undergo EMT, as evidenced by molecular markers and increased motility in a zebrafish xenograft model [76]. HA interacts with hyaluronan mediated motility receptor (RHAMM) and CD44, and the inhibition of either of these proteins causes lung cancer cells to switch to an epithelial phenotype [77,78]. Furthermore, CD44+ primary lung adenocarcinoma cells show a clear EMT induction and have higher tumorigenic potential in nude mouse xenografts compared to CD44- cells [79]. Two molecular mechanisms of CD44-mediated EMT have been identified: (1) CD44 triggers the Wnt/β-catenin pathway, which causes FoxM1 to bind to the promoter and induce Twist [79]; (2) upon activation, CD44 colocalizes with epidermal growth factor receptor (EGFR) at the membrane and promotes the downstream activation of AKT and ERK pathways [80]. Interestingly, the disruption of CD44 expression inhibits EMT when A549 cells are treated with TGF-β1 alone, but not in combination with TNF-α, suggesting a close link between the HA/CD44 and TGF-β1 pathways [80,81]. The analysis of samples from patients with lung adenocarcinoma shows that CD44 levels are higher in tumors than in normal lung tissue [79], correlate with the expression of EMT markers (inversely with E-cadherin and directly with Snail and Twist) [77,82], and predict the likelihood of metastasis [79].

Collagens play an integrative role in both normal and tumor tissues, connecting different cell types [83,84]. In addition to their structural function, collagens I and XVII have been found to be induced during inflammation and induce EMT in lung cancer cells [85–88].

Collagen I provides structural stability to the bronchi, alveoli, and interstitium, but its abnormal deposition is associated with respiratory diseases, including asthma, pulmonary fibrosis, and lung cancer [89]. TGF-β upregulates type I collagen in lung tumor cells by inhibiting miR-200 and derepressing ZEB1, and collagen I then induces EMT through the integrin β1/FAK/Src pathway [85,88]. Only mature collagen fibers can induce EMT, as knockdown of the collagen cross-linking enzyme LOXL2, which is also regulated by ZEB1, abolishes FAK and Src activation and the metastasis of mouse lung cancer 344SQ cells in vitro and in vivo [85]. Another key player in this pathway is v-crk sarcoma virus CT10 oncogene homologue (avian)-like (CRKL), which serves as a scaffolding protein in integrin-based focal adhesions. Although CRKL knockdown does not affect EMT markers, it alters cell morphology to round-shaped, decreases migration and invasion, and dysregulates FAK and Src localization in human H157 lung cancer cells [88]. Shintani et al. showed that plating A549 cells on collagen I-coated dishes induces TGF-β3 expression via the PI3K/AKT pathway, thereby forming an autocrine loop to maintain EMT, as evidenced by the reversal of changes in morphology and molecular markers by neutralization with anti-TGF-β3 antibody or PI3K/ERK inhibitors [86]. The analysis of immunohistochemistry (IHC) data from 490 lung cancer samples by Peng et al. showed increased levels of collagen I in poorly differentiated tumors and its correlation with Zeb1 expression, which is consistent with the aforementioned in vitro results. In addition, increased collagen I (*COL1A1*) and *LOXL2* mRNA levels were associated with worse survival in lung adenocarcinoma patients from The Cancer Genome Atlas (TCGA) cohort [85].

Collagen XVII facilitates epithelial cell adhesion to the basement membrane but has also been associated with invasiveness, particularly in lung cancer [90]. The role of collagen XVII in cancer-associated lung inflammation is not fully understood. Autoantibodies to collagen XVII are elevated in the serum of lung cancer patients receiving anti-PD1/PD-L1 therapy and correlate with better responses to therapy and survival, suggesting an important role of collagen XVII expression for T cell effector function in the tumor [91]. On the other hand, collagen XVII induces EMT in lung cancer cells by inhibiting ubiquitin-mediated Snail degradation through the FAK/AKT/GSK3β pathway [87]. Important features of this pathway are the ADAM9/10-mediated shedding of collagen XVII and the subsequent stabilization of laminin-5, which was confirmed by the inhibition of EMT in A549 and CL1-1 cells by short hairpin RNA (shRNA)-mediated knockdown of the corresponding genes [87]. IHC analysis of surgically resected lung tumors from 98 patients showed a negative correlation between survival and positive staining for collagen XVII and laminin-5 [87].

2.2.5. Reactive Oxygen Species in the Regulation of EMT in Lung Cells

The production of reactive oxygen species (ROS) is induced in immune and epithelial cells during the inflammatory response as a defense mechanism against foreign cells such as bacteria or tumor cells. However, oxidative stress has been identified as another factor contributing to inflammation-induced EMT (Figure 2).

MSCs are an established cell type from the iTME that promotes the EMT in lung cancer cells by inducing ROS production. Luo et al. used a co-culture system to show that MSCs increase intracellular ROS in A549 cells, and this induces the EMT through an autophagy-dependent mechanism, as evidenced by autophagosome formation, increases in the autophagic markers Beclin-1 and LC3-II, and the sensitivity of EMT to autophagy modulators (rapamycin, 3-methyladenine, and bafilomycin) [92].

ROS have been recognized as important mediators of TGF-β-induced EMT. TGF-β triggers ROS production in lung cancer cells through the NF-κB-dependent induction of NADPH oxidase 4 (NOX4) expression [93,94]. Interestingly, in vitro studies suggest that this pathway is required for the induction of EMT by TGF-β, as pharmacological inhibition of NOX4 or NF-κB suppresses ROS production and reverses changes in motility and EMT markers in A549 cells [93,94]. ShRNA-mediated knockdown of the antioxidant transcription factor NRF2 enhances the response of A549 cells to TGF-β, increasing cell motility, p-SMAD2/3 and NOX4 expression, and intracellular ROS levels, identifying NRF2 as a negative regulator of the TGF-β/NF-κB/NOX4/ROS axis [94]. Another pathway associated with TGF-β-induced ROS production involves increasing the cellular pool of labile iron through ferritin heavy chain (FHS) suppression by transmembrane prostate androgen-induced protein (TMEPAI). ROS then inhibit the EMT suppressor insulin receptor substrate-1 (IRS-1) [95]. Hu et al. showed that TGF-β-induced EMT was inhibited by TMEPAI depletion in A549 cells and could be rescued by either hydrogen peroxide (H_2O_2) treatment or small-interfering RNA (siRNA)-mediated knockdown of insulin receptor substrate 1 (IRS-1). Increased expression of TMEPAI in tumors compared to adjacent lung tissue found in 30 patients with squamous cell lung cancer supports the clinical relevance of the above mechanism [95].

2.2.6. COVID-19-Associated Inducers of EMT in Lung Tissue

Given the unprecedented spread and severe consequences of severe acute respiratory syndrome coronavirus 2 (SARS-CoV-2) infection on global public health, the relationship between COVID-19, lung cancer progression, and the EMT requires brief discussion. To date, available clinical studies have clearly demonstrated the high susceptibility of lung cancer patients to SARS-CoV-2 infection [96]. This effect may be related to the overexpression of angiotensin converting enzyme 2 (ACE2) in lung cancer compared to normal tissue, which is a key target for SARS-CoV-2 cell entry [97], and the suppressed immunity of cancer patients [98]. In turn, viral infection may create a microenvironment that enhances the aggressive properties of lung cancer cells, including the induction of the EMT. For example, recent work by Saygideger et al. found that serum samples from COVID-19 patients increased the motility of A549 cells, their loss of intercellular junctions, and the switch in the expression of EMT-related markers to mesenchymal-type markers [99], which is consistent with significant EMT-associated changes in lung lesions in COVID-19 patients who died of the disease reported by Falleni et al. [100]. In addition, a retrospective analysis showed that cancer patients had increased pulmonary metastatic lesions 6 months after SARS-CoV-2 infection [99], while no direct oncogenic effect of this virus has been reported to date [101]. The major inducers of EMT produced by COVID-19 are TGF-β, pro-inflammatory cytokines including IL-6 and IL-1β [102], and the urokinase plasminogen activator (uPA), the latter of which is involved in the production of plasmin, which degrades components of the ECM and activates TGF-β [103]. Furthermore, neutrophil extracellular traps (NETs) produced by neutrophils recruited into SARS-CoV-2-infected lungs have also been implicated in EMT induction in lung cancer cells [9]. Thus, the inflammatory background caused by viral infection may also be involved in the induction of the EMT in lung cancer. More details on the relationship between SARS-CoV-2 infection, tumor progression, and the EMT can be found in recent comprehensive reviews [31,103–105].

3. Inflammation-Induced EMT as a Source of Cancer Stem Cells

The efficacy of current chemotherapy in the treatment of lung cancer is limited by the acquisition of drug resistance by cancer cells. According to the latest concept, a key role in this process is played by cancer stem cells (CSCs), a population of cancer cells that survive cytotoxic exposure in a quiescent state and then differentiate into heterogeneous types of tumor cells, leading to tumor relapse [106]. The origin of CSCs is still debated, but it is most likely that they arise from proliferating lung epithelial cells capable of transdifferentiation, i.e., the facultative stem cells (alveolar epithelial type II cells (AEC2s) and club cells in the distal lung regions and basal cells in the proximal lung regions) [107].

Cells from the tumor microenvironment, such as CAFs and TAMs, have been shown to provide a supportive niche for CSCs. Chronic inflammation that develops during tumor progression upregulates the stemness of CSCs through activation of the EMT by cytokines produced by inflammatory cells infiltrating the tumor stroma [107]. Macrophages with the M2 phenotype have been shown to play an important role in supporting CSCs through the secretion of anti-inflammatory cytokines TGF-β1 [108,109] and IL-10 [110], but the same activity has also been demonstrated for some pro-inflammatory cytokines such as TNF-α [111], IFN-γ [112], IL-6 [113,114], and IL-17 [115].

M1 and M2 macrophages stimulate the stemness of human H1299 and mouse D121 lung cancer cells by inducing the expression of the deubiquitinase ubiquitin-specific peptidase 17 (USP17), which disrupts the TNFR-associated factor (TRAF)2/TRAF3 complex and thereby inhibits the degradation of its downstream targets NIK, c-Rel, and IRF5, which are involved in the regulation of stemness-related genes [116]. Furthermore, the binding of USP17 to TRAF2/TRAF3 activates the NF-κB signaling pathway [116], which, in addition to regulating the inflammatory response, is involved in the induction of EMT and the acquisition of stem-like properties in NSCLC cells [117]. M2 macrophage-derived IL-10 has been shown to induce stemness in A549 and H460 NSCLC cell lines via the JAK1/STAT1/NF-κB/Notch1 pathway [110], and IL-10 expression by TAMs correlates closely with the NSCLC stage and survival of NSCLC patients [110,118]. Another cytokine secreted by M2 macrophages, TGF-β1, induces EMT in primary lung cancer cells through the downregulation of miR-138 with a subsequent increase in the colony-forming ability and expression of stem cell markers CD44 and CD90 in lung cancer cells [108]. The interaction between TGF-β receptor type II (TβRII) and epithelial membrane protein 3 (EMP3) plays a critical role in the TGF-β1-mediated induction of the EMT and stem-like properties, as evidenced by the reduced migratory ability and resistance to irradiation in EMP3 knockdown A549 cells and the positive correlation between high EMP3 levels and poor survival rates in patients with NSCLC [109].

IL-6 has been reported to increase proliferation and induce EMT in CSC-like A549 and H157 cells expressing the stem cell marker CD133 (CD133+ cells), but not in CD133- cells [113]. Increased CSC proliferation observed upon activation of the IL-6/JAK2/STAT3 signaling axis is associated with the upregulation of DNA methyltransferase I (DNMT1), which decreases the expression of p53 and p21 due to DNA hypermethylation [114]. IL-17 produced by Th17 cells has been shown to increase the migration, invasion, spheroid forming potential, and expression of mesenchymal and stem cell markers in A549 and H460 cells via the STAT3/NF-κB/Notch1 pathway [115]. The effect of IFN-γ on NSCLC cells is dose-dependent: low-dose IFN-γ induces the EMT and enhances the stemness of A549 and H460 cells both In vitro and in vivo by activating the intercellular adhesion molecule 1 (ICAM1)/PI3K/Akt/Notch1 signaling axis; high-dose IFN-γ induces apoptosis of NSCLC cells through the JAK1/STAT1/caspase-3,7 pathway. Analysis of tumor samples from NSCLC patients has shown that low, but not high, levels of IFN-γ and high expression of ICAM1 in the tumor microenvironment are positively correlated with enrichment in CD133+ cells and the highest expression of stemness-related genes [112].

Primary induction of the EMT by inflammatory cytokines stimulates NSCLC cells to secrete soluble factors that act in an autocrine manner to maintain tumor stemness. Wamsley et al. reported that the stimulation of A549 cells with TGF-β and TNF-α induces NF-κB-mediated production of inhibin subunit beta A (INHBA)/Activin, a member of the TGF-β superfamily of growth factors, which further maintains the expression of the EMT TFs Snail, Slug, and ZEB2, as well as the CSC markers N-Myc, SRY-box transcription factor 2 (SOX2), Krüppel-like factor 4 (KLF4), and high mobility group AT-hook 2 (HMGA2) [111]. Furthermore, Activin expression is elevated in tissue samples from patients with primary NSCLC tumors, including adenocarcinoma, squamous cell carcinoma, and large cell carcinoma, further supporting its role as an autocrine factor in maintaining CSC phenotypes [111]. In addition to Activin, the observed effect of TGF-β on the stemness of lung cancer cells appears to be dependent on chemokine (C-X-C motif) ligand 12 (CXCL12), as the upregulation of CXCL12 and its receptor chemokine (C-X-C motif) receptor 7 (CXCR7) is positively correlated with the increase in spheroid-forming potential and CSC marker expression in A549 cells treated with TGF-β. In clinical samples, the co-expression of TGF-β1 and CXCR7 positively correlates with CD44 levels in advanced lung adenocarcinoma, confirming the involvement of CXCL12 in maintaining stem-like properties of NSCLC cells [119]. Wnt3A is another key player in the autocrine/paracrine regulation of cancer cell stemness upon EMT activation. Interestingly, EMT induction alters the set of genes regulated by the Wnt3A/β-catenin pathway by switching from β-catenin/E-cadherin/Sox15 to β-catenin/Twist1/TCF4 complex formation [120]. In mesenchymal-like A549 cells, Twist1 stabilizes β-catenin and enhances the transcriptional activity of the β-catenin/TCF4 complex, leading to an increase in the spheroid-forming capacity and expression of CSC markers. The clinical significance of the Wnt/β-catenin pathway in promoting the CSC phenotype is supported by the analysis of human lung cancer samples, which shows that high-grade primary and metastatic cancers have higher levels of CD133 and nuclear fractions of β-catenin and Twist1, while E-cadherin and Sox15 levels are significantly decreased compared to low-grade primary cancers [120].

4. Clinical Trials of Drugs Targeting Key Regulators of Inflammation-Driven EMT

Rigorous study of the relationship between inflammation and tumor progression has led to the development of a wide range of drug candidates targeting key regulators of inflammation-driven EMT, some of which are now in various stages of clinical trials. Surprisingly, despite the encouraging tumor suppressive effects of TGF-β-targeted compounds in vitro and in vivo, their antitumor efficacy in oncology patients is mainly controversial with a small survival benefit [121]. For example, trabedersen (an antisense oligonucleotide targeting TGF-β2), Lucanix (a non-viral gene-based allogenic tumor cell vaccine targeting TGF-β2), and the combination of galunisertib (TβRI kinase inhibitor) and lomustine showed no significant antitumor activity in clinical trials in patients with brain tumors (phase IIb, NCT00431561) [122], NSCLC (phase III, NCT00676507) [123], and recurrent glioblastoma (phase II, NCT01582269) [124], respectively. Nevertheless, a number of studies still showed favorable survival outcomes in cancer patients after anti-TGF-β therapy, suggesting a possible relationship between the efficacy of TGF-β-targeting drugs and the tumor context and the level of their bioavailability. For example, galunisertib was shown to significantly enhance the antitumor effect of sorafenib in patients with hepatocellular carcinoma (phase II, NCT01246986) [125], and its combination with temozolomide-based chemoradiation improved overall survival in patients with glioblastoma (phase I/II, NCT01220271) [126]. In addition, the phase III clinical trial (NCT00761280) showed that trabedersen increased the two-year survival in patients with secondary glioblastoma and anaplastic astrocytoma to 35.7% compared with 23.1% for conventional chemotherapy [126]. The anti-TGF-β1-3 monoclonal antibody called fresolimumab (also known as GC1008) at 10 mg/kg in combination with radiation was found to cause a significantly longer median overall survival with better immunologic parameters in patients with metastatic breast cancer compared to fresolimumab administration at 1 mg/kg (Phase II, NCT01401062) [127].

In addition to anti-TGF-β therapy, the inhibition of a number of pro-inflammatory mediators has also conferred a favorable survival benefit in cancer patients. For example, the use of celecoxib (a selective COX-2 inhibitor) as an adjuvant to chemotherapy significantly improved the overall survival and disease-free survival in COX-2-positive gastric cancer patients [128], and its oral administration significantly improved responses in patients with recurrent ovarian cancer characterized by platinum-based drug resistance (phase II, NCT01124435) [129]. Another COX-2 inhibitor, rofecoxib, in combination with cyclophosphamide and vinblastine, showed a 30% clinical benefit in patients with advanced solid tumors [130], and low-dose aspirin prevented colorectal cancer in patients with familial adenomatous polyposis by suppressing the recurrence of colorectal polyps (UMIN000018736) [131], a process associated with EMT [132]. These results are consistent with the recent retrospective analysis by Cai et al. demonstrating that the use of non-steroidal anti-inflammatory drugs during radical resection in NSCLC patients correlates with outcomes [133]. In addition, the blockade of IL-1 signaling has demonstrated significant antitumor efficacy in clinical trials. Treatment of patients with advanced colorectal cancer with bermekimab (a monoclonal antibody against IL-1α) was found to significantly improve the clinical response rate (phase III, NCT01767857), and nadunolimab (a monoclonal antibody against IL-1 receptor accessory protein (IL1RAP)) in combination with nab-paclitaxel and gemcitabine demonstrated a clinical benefit rate of 74% (phase I/IIa, NCT03267316) [134]. Consistent with this, an additional analysis of the Canakinumab Anti-Inflammatory Thrombosis Outcomes Study (CANTOS) trial in 10,061 patients showed that the administration of canakinumab (a monoclonal antibody against IL-1β) dose-dependently reduced the incidence of lung cancer and lung cancer mortality (phase III, NCT01327846) [135].

Taken together, the described clinical data, although not directly indicative of blockade of inflammation-driven EMT in cancer patients, clearly confirm the stimulatory effect of iTME on tumor progression and metastasis. The demonstrated clinical benefit for cancer patients in response to anti-TGF-β and anti-inflammatory drugs suggests the prospect of further investigation of the "inflammation—tumor growth" relationship, including the induction of EMT in tumor cells by an inflammatory background.

5. Inhibitory Effect of Triterpenoids on Inflammation-Induced EMT in Lung Cancer Cells

Natural metabolites represent an important source of drug candidates with pronounced anti-inflammatory and anti-tumor properties [10]. These compounds can be considered effective inhibitors of the EMT in lung cancer cells, as has already been shown for lignans (arctigenin [136]), alkaloids (berberine [137]), flavonoids (luteolin [138]), phenolic compounds (eugenol [139]), resveratrol [140], curcumin [141]), and other molecules (more details on this topic can be found in the following recently published comprehensive reviews [142–144]). Given the lack of reviews analyzing in detail the bioactivity of pentacyclic triterpenoids (PTs) against the links between inflammation and EMT, this chapter will focus on this class of natural compounds.

PTs are terpenes whose carbon backbone consists of six isoprene units. They can be isolated from various sources such as fungi and marine organisms, but most bioactive PTs have been obtained from plants. In plants, the mevalonate and methylerythritol phosphate pathways synthesize isopentenyl pyrophosphate (IPP), some of which is isomerized to dimethylallyl pyrophosphate (DMAPP). Two IPP units are then added to DMAPP to form farnesyl pyrophosphate (FPP), which is dimerized to form squalene. Squalene epoxidase oxidizes squalene to 2,3-oxidosqualene, the common precursor of all PTs, after which the metabolic pathways diverge to produce different types of triterpenoid scaffolds (Figure 5) [145].

From a biological point of view, the most important types of PTs are lupane, oleanane, ursane, and friedelane. Lupane-type PTs have a five-carbon E-ring. In contrast, oleanane and ursane E-rings are six-membered and can be distinguished by the position of the methyl group attached to the C-20 and C-19 atoms, respectively. The friedelane-type backbone is derived from the oleanane type by methyl translocation [145] (Figure 5). Representative sources of PTs are *Betula pubescens* (white birch) for lupane [146], *Malus domestica* (apple), *Coffea arabica* (arabic coffee), and *Origanum vulgare* (oregano) for ursane [147], *Glycyrrhiza glabra* (licorice), *Olea europaea* (olive), and *Evodia rutaecarpa* (evodia) for oleanane [148–150], and *Tripterygium wilfordii* and *Celastrus orbiculatus* (oriental bittersweet) for friedelane [151,152]. However, the diversity of PTs is not limited to these species; they can be found in many food and medicinal plants.

PTs have a variety of biological activities, including lipid-lowering [153], anti-diabetic [154], antidepressant [155], antinociceptive [156], anti-inflammatory [157], and antitumor [158] effects. PTs inhibit the immune response by reducing the production of ROS, nitric oxide (NO), and proinflammatory cytokines by macrophages [157]. In tumors, PTs induce cell death via autophagy and apoptosis and inhibit metastasis and neoangiogenesis [158,159]. The ability to inhibit EMT has been reported for all types of PTs [160–163], but these studies have not paid special attention to inflammation-driven EMT. Therefore, the next part of our study focuses on the analysis of available published data dedicated to the modulating effect of PTs on EMT induced by inflammatory background in lung cancer. The results of this analysis are summarized in Figure 6 and Table 1.

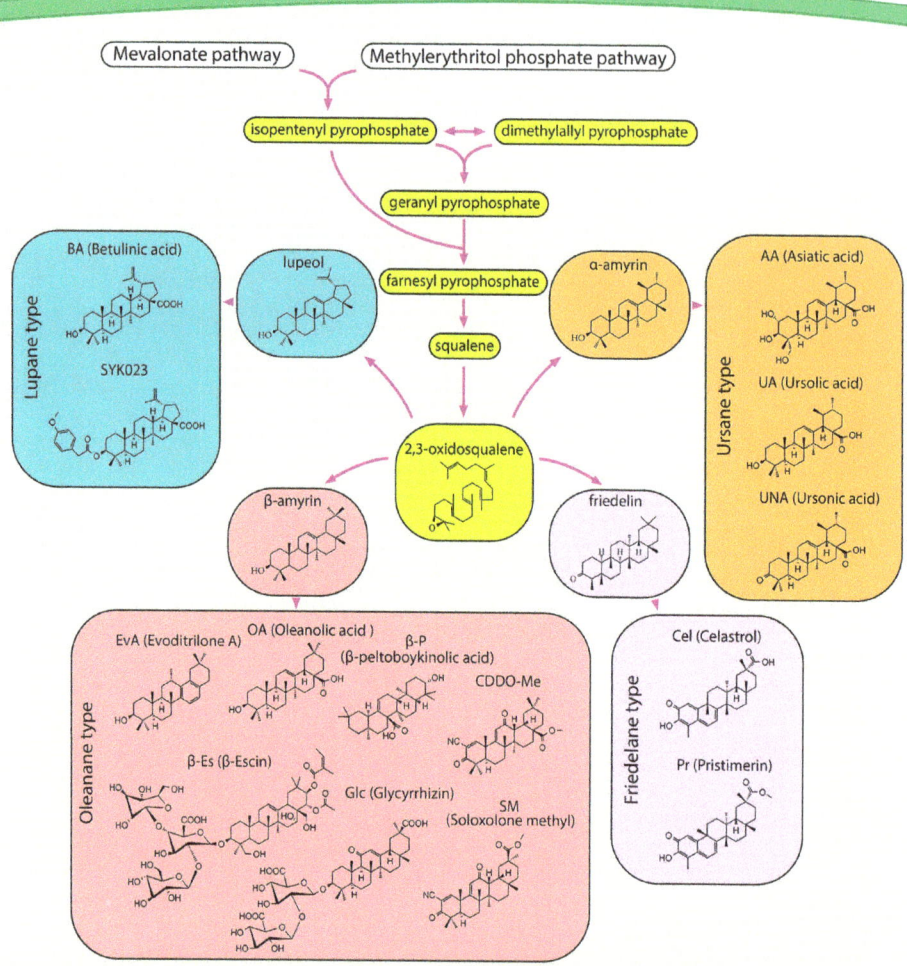

Figure 5. Biosynthetic pathway of PTs. In plant cells, the mevalonate and methylerythritol phosphate pathways provide isoprene units in the form of isopentenyl pyrophosphate (IPP) and dimethylallyl pyrophosphate (DMAPP), which are used as building blocks for the production of squalene. Oxidation of squalene yields 2,3-oxidosqualene, which serves as a precursor for the synthesis of all types of PT scaffolds, including lupane, ursane, oleanane, and friedelane. For simplicity, enzymes and most synthesis steps are omitted.

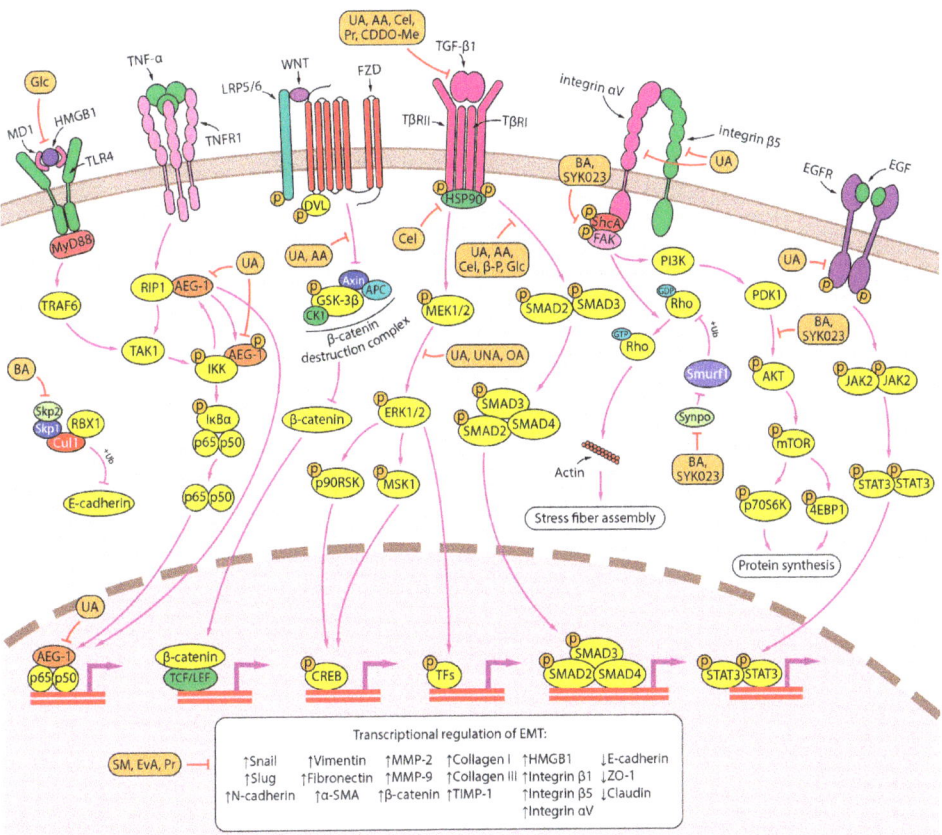

Figure 6. Inhibition of inflammation-induced EMT by PTs. Lupane-, oleanane-, and ursane-type PTs exhibit a variety of EMT inhibition mechanisms. PTs inhibit TGF-β signaling pathways (UA, AA, celastrol, pristimerin, CDDO-Me) and the downstream SMAD (UA, AA, celastrol, β-peltoboykinolic acid, glycyrrhizin) and ERK (UA, UNA, OA) signaling axes. PT-induced disruption of integrin signaling involves integrin αVβ5 expression inhibition (UA) and activation of FAK and AKT (BA, SYK023). The blockade of F-actin polymerization by the lupane-type PTs BA and SYK023 is mediated by the downregulation of Synpo, which presumably activates Smurf1-dependent ubiquitination of RhoA. In contrast, BA disrupts the Skp2-SCF E3 ligase complex, thereby protecting E-cadherin from degradation. The ursane-type PTs UA and AA inhibit β-catenin through a GSK-β-dependent mechanism. The effect of UA on EMT is also associated with repression of the AEG-1 oncogene, which regulates several steps in the NF-κB signaling pathway, and inhibition of the EGFR signaling pathway. Glycyrrhizin inhibits EMT induced by HMGB1, a nuclear protein released from tumor cells (TLR4 and RAGE are shown as possible HMGB1 receptors due to uncertainty in the downstream pathway). Some triterpenoids have an unknown mechanism of action but regulate EMT-associated genes (SM, evoditrilone A, pristimerin). Downward (↓) and upward (↑) arrows indicate downregulation and upregulation of expression, respectively.

Table 1. Studies on the inhibition of EMT and stemness by PTs.

PT		Experimental Model	Conditions		Biological Effects ***	Effects on Cell Signalings ****	Refs.
Type	Name		Conc. *	Time **			
Ursane	Asiatic acid (AA)	A549 treated with TGF-β1 (10 ng/mL)	10–40	24 h	↓ morphological changes (20 μM AA), ↓ migration ^{DD}, ↓ invasion ^{DD}	↓ β-catenin ^{DD}, ↓ p-GSK-3β ^{DD}, ↑ E-cadherin ^{DD}, ↓ N-cadherin ^{DD}, ↓ vimentin ^{DD}, ↓ Snail ^{DD}	[164]
		Pulmonary fibrosis in C57BL/6 mice was induced by single intratracheal administration of bleomycin (3 mg/kg)	5–20 mg/kg/day (intragastrically)	21 d	↓ collagen deposition ^{DD}, ↓ histopathological changes in the lungs ^{DD}, ↑ pulmonary function ^{DD}, ↓ macrophage, neutrophil and lymphocyte infiltration in BALF ^{DD}	↓ collagen I ^{DD}, ↓ collagen III ^{DD}, ↓ α-SMA ^{DD}, ↓ TIMP-1 ^{DD}, ↓ vimentin ^{DD}, ↑ E-cadherin ^{DD}, ↓ TGF-β1 ^{DD}, ↓ p-Smad2/3 ^{DD}, ↓ p-ERK1/2 ^{DD}, ↓ IL-1β ^{DD}, ↓ IL-18 ^{DD}, ↓ IL-6 ^{DD}, ↓ TNF-α ^{DD}, ↓ NLRP3 ^{DD}, ↓ ASC ^{DD}, ↓ pro-Caspase-1 ^{DD}, ↓ active Caspase-1 ^{DD}	[165]
		A549, H1975	5–30	24 h	↓ adhesion to Matrigel ^{DD} (A549), ↓ migration ^{DD} (A549, H1975), ↓ invasion ^{DD} (A549, H1975)	↑E-cadherin ^{DD} (A549), ↓ N-cadherin ^{DD} (A549), ↓ vimentin (A549, 20 μM UA); 84 genes regulated by UA (A549, 30 μM UA) were associated with the signaling pathways of TGF-β, ECM-receptors, adherens junctions, Wnt, VEGF, tight junctions, cell adhesion molecules; ↓ AEG-1 ^{DD,K} (A549)	[166]
Ursane	Ursolic acid (UA)	A549 treated with TNF-α (5 ng/mL)	5–20	12/24 h with UA, then 12/24 h with TNF-α		↓ NF-κB p65 subunit^K (20 μM UA, 12 h), ↓ AEG-1 ^{DD} (24 h)	
		H1975 treated with TGF-β1 (5 ng/mL)	0.02	24 h	↓ morphological changes, ↓ migration, ↓ invasion	↑ E-cadherin, ↓ N-cadherin, ↓ MMP-2 catalytic activity, ↓ MMP-9 catalytic activity, ↓ MMP-2, ↓ MMP-9, ↓ integrin αVβ5^K	[167]

Table 1. Cont.

Type	PT Name	Experimental Model	Conditions Conc. *	Time **	Biological Effects ***	Effects on Cell Signalings ****	Refs.
		HBE treated with 1% cigarette smoke extract (CSE)	10	2 h prior to CSE		↓ TGF-β1, ↓ p-Smad2/3, ↓ S100A4, ↑ IGF-1	[168]
		Emphysema in Wistar rats was induced by exposure to cigarette smoke for 30 min, two times a day, 6 days a week, for 3 months	10–40 mg/kg/day (intragastrically)	3 mos	↓ airway-vessel remodeling, ↓ collagen deposition, ↓ mucus secretion	↓ α-SMA ^{DD}, ↓ S100A4, ↓ TGF-β1, ↓ p-Smad2/3, ↑ IGF-1 ^{DD}	
		Emphysema in SD rats induced by exposure to cigarette smoke for 30 min, two times a day, 6 days a week, for 3 months	10–40 mg/kg/day (intragastrically)	3 mos		↓ p-IRE1, ↓ XBP1	[169]
Ursane	Ursolic acid (UA)	Emphysema in SD rats induced by intraperitoneal injection of CSE on days 1, 8, 15, 21	20 mg/kg/day (intragastrically)	2–4 wk	↓ airway remodeling	↓ p-Smad2/3, ↓ p-PERK, ↓ PERK, p-eIF-2α, ↓ e-IF-2α, ↓ ATF4, ↓ CHOP, ↓ p-IRE1, ↓ ATF6, ↓ active Caspase 12	
		A549, H460	10, 20	1–14 d	↓ migration ^{DD}, ↓ invasion ^{DD}, ↓ tumorsphere formation (20 μM UA; 7 d, 14 d)	↓ VEGF ^{DD} (24 h), ↓ NANOG (tumorpheres; 20 μM UA; 24 h), ↓ OCT4 (tumorpheres; 20 μM UA; 24 h), ↓ SOX2 (tumorpheres; 20 μM UA; 24 h), ↓ pEGFR (24 h), ↓ pJAK2 ^{DD} (24 h), ↓ pSTAT3 ^{DD} (24 h), ↓ PD-L1 ^{DD} (24 h), ↓ MMP2 ^{DD} (24 h), ↓ MMP3 ^{DD} (24 h), ↓ MMP9 ^{DD} (24 h), ↓ VEGF ^{DD} (24 h), ↓ the binding of STAT3 to MMP2 and PD-L1 promoters (20 μM UA, 24 h)	[170]
		A549 and H460 treated with EGF (25 ng/mL)	20	1 h with EGF, then 24 h with UA		↓ pEGFR	

Table 1. Cont.

Type	PT Name	Experimental Model	Conditions Conc. *	Time **	Biological Effects ***	Effects on Cell Signalings ****	Refs.
Ursane	Ursolic acid (UA)	H1975 harbouring L858R/T790M mutation	25	12–72 h	↓ migration	↓ CT45A2^K (12 h), ↓ the binding of TCF4 to CT45A2 promotor^K (12 h), ↓ TCF4 (48 h), ↓ p-β-catenin (48 h), ↑ p-GSK-3β (48 h), ↓ nuclear translocation of β-catenin	[171]
		subcutaneous injection of H1975 in athymic nude mice	25 mg/kg/day	18 d		↓ CT45A2, ↓ p-β-catenin, ↑ p-GSK-3β, ↓ TCF4	
Ursane	Ursonic acid	A549, H1299	2.5, 5	24 h, 48 h	↓ invasion (24 h)	↓ MMP-2 catalytic activity ^{DD} (A549, H1299; 48 h), ↓ MMP-9 catalytic activity (H1299, 48 h), ↓ MMP-2 ^{DD} (A549, H1299; 48 h), ↓ MMP-9 ^{DD} (H1299, 48 h), ↓ RECK ^{DD} (A549, 48 h), ↑ RECK (H1299, 48 h), ↓ p-ERK ^{DD} (A549, H1299; 24 h), ↓ p-CREB ^{DD} (A549, H1299; 24 h)	[172]
Oleanane	CDDO-Me	Radiation-induced lung inflammation in C57BL/6 mice was induced by thoracic irradiation with a single dose of 12.5 Gy	600 ng intragastrically on days -1, 1, 3 and 5	3 wk	↓ inflammatory cells infiltration in BALF, ↓ total protein in BALF, ↓ histopathological changes in the lungs	↓ IL-6, ↓ TGF-β, ↑ IL-10, ↓ fibronectin, ↓ α-SMA, ↓ collagen I	[173]
		Radiation-induced pulmonary fibrosis in C57BL/6 mice was induced by thoracic irradiation with a single dose of 22.5 Gy	600 ng intragastrically on days -1, 1, 3, 5, 7 and 9	12 wk	↓ collagen deposition	↓ fibronectin, ↓ α-SMA, ↓ collagen I	

Table 1. Cont.

Type	PT Name	Experimental Model	Conditions Conc. *	Conditions Time **	Biological Effects ***	Effects on Cell Signalings ****	Refs.
Friedelane	Celastrol	A549 treated with TGF-β1 (5 ng/mL)	1	30 min with celastrol, then 24 h–72 h with TGF-β1	↓ morphological changes (72 h), ↓ invasion (36 h)	↑ E-cadherin (72 h), ↓ Snail (24 h)	[152]
		A549 treated with TGF-β1 (5 ng/mL)	5	24 h		↑ E-cadherin, ↑ ZO-1, ↓ N-cadherin, ↓ vimentin, ↓ Snail, ↓ Slug	
		Pulmonary fibrosis in Wistar albino rats was induced by single intratracheal administration of bleomycin (3 U/kg)	5 mg/kg, twice a week	28 d		↓ TGF-β1, ↓ p-Smad2/3, ↑ E-cadherin, ↑ claudin, ↓ N-cadherin, ↓ Snail, ↓ Slug, ↓ β-catenin, ↓ Hsp90	[174]
Oleanane	Evoditrilone A	A549	1–2	48 h	↓ colony formation ability ^{DD}, ↓ migration ^{DD}	↑ E-cadherin ^{DD}, ↓ MMP-2 ^{DD}, ↓ N-cadherin ^{DD}	[148]
Oleanane	Glycyrrhizin	A549 and BEAS-2B treated with TGF-β1 (5 ng/mL)	50–200 (A549); 25–100 (BEAS-2B)	2 h with glycyrrhizin, then 24 h with TGF-β1	↓ migration ^{DD}	↓ HMGB1 secretion ^{DD}, ↓ HMGB1 ^{DD}, ↓ p-Smad2/3 ^{DD}, ↑ E-cadherin ^{DD}, ↓ vimentin ^{DD}	[149]
		A549 and BEAS-2B with lentivirus-mediated HMGB1 overexpression	100 (A549); 50 (BEAS-2B)	24 h		↓ HMGB1 secretion, ↓ HMGB1, ↓ TGF-β1, ↓ p-Smad2/3, ↑ E-cadherin, ↓ vimentin	
Oleanane	Oleanolic acid (OA), OA-loaded P105/TPGS mixed micelles	A549, PC-9	15, 30	24 h	↓ migration (OA-micelles > free OA), ↓ invasion (OA-micelles > free OA)	↑ E-cadherin (OA-micelles > free OA), ↓ N-cadherin (OA-micelles > free OA), ↓ p-ERK (OA-micelles > free OA)	[175]
Friedelane	Pristimerin (Pr)	H1299	0.9–3.6	24–72 h	↓ colony formation ability ^{TD} (1.8 μM Pr, 48 h), ↓ migration ^{TD,DD}, ↓ invasion ^{DD} (48 h)	↓ vimentin (3.6 μM Pr, 48 h), ↓ F-actin (0.9–3.6 μM Pr, 48 h), ↓ integrin β1 (3.6 μM Pr, 48 h), ↓ MMP-2 (0.9–3.6 μM Pr, 48 h), ↓ Snail (0.9–3.6 μM Pr, 48 h)	[151]

Table 1. Cont.

Type	PT Name	Experimental Model	Conditions Conc. *	Time **	Biological Effects ***	Effects on Cell Signalings ****	Refs.
Oleanane	Soloxolone methyl	A549 treated with TGF-β1 (50 ng/mL)	0.5	24, 48 h	↓ morphological changes, ↓ migration (24 h, 48 h), ↓ invasion (48 h)	↑ E-cadherin (48 h), ↑ ZO-1 (48 h), ↓ vimentin (48 h), ↓ fibronectin (48 h)	[176]
Oleanane	β-escin (β-Es)	H460	5–40	24 h		↓ ALDH$^+$ cell population (5–40 μM β-Es), ↑ p21 (20 μM β-Es; both in ALDH$^+$ and ALDH$^-$ cells)	[177]
		lung tumors in A/J mice were induced by single intraperitoneal injection of tobacco carcinogen NNK (10 μmol/mouse)	3 weeks after NNK treatment mice were fed with the diet containing 500 ppm β-Es	17, 34 wk	↓ lung tumor formation, ↓ progression of adenomas to adenocarcinomas	↑ p21 (34 wk), ↓ ALDH1A1 (34 wk), ↓ p-Akt (34 wk)	
Oleanane	β-peltoboykinolic acid (β-P)	A549 treated with TGF-β1 (2 ng/mL)	1–10 μg/mL	24–48 h	↓ morphological changes (5 μg/mL β-P, 10 μg/mL β-P; 48 h), ↓ migration ^{DD} (24 h, 36 h)	↑ E-cadherin ^{DD} (48 h), ↓ N-cadherin (10 μg/mL β-P, 48 h), ↓ vimentin ^{DD} (48 h), ↓ fibronectin ^{DD} (48 h), ↓ collagen I ^{DD} (48 h), ↓ p-Smad2 (10 μg/mL β-P, 48 h), ↓ Snail ^{DD} (48 h)	[178]
Lupane	Betulinic acid	293T, A549, H1299	10–30	4 h–7 d	↓ migration (A549, H1299; 20 μM BA > 10 μM BA; 24 h), ↓ invasion (A549, H1299; 20 μM BA > 10 μM BA; 24 h), ↓ the sphere-forming ability (A549, H1299; 20 μM BA > 10 μM BA; 7 d)	the direct binding to Skp2 by forming H-bonds with Lys145, ↓ Skp2-Skp1 interactions (exogenous Flag-Skp1 was transfected in 293T, endogenous Skp2-Skp1 interactions in H1299; 20, 30 μM BA; 4 h), ↓ Skp2-mediated ubiquitination of p27 (exogenous p27 in 293T, endogenous p27 in A549; 10, 20 μM BA; 24 h), ↓ Skp2-mediated ubiquitination of E-cadherin (exogenous E-cadherin in 293T, endogenous E-cadherin in A549; 10, 20 μM BA; 24 h), ↓ Skp2 ^{DD,TD} (A549, H1299), ↑ p27 ^{DD,TD} (A549, H1299), ↑ E-cadherin ^{DD,TD} (A549, H1299)	[179]

Table 1. Cont.

Type	PT Name	Experimental Model	Conditions Conc.*	Time**	Biological Effects ***	Effects on Cell Signalings ****	Refs.
Lupane	Betulinic acid	intravenous injection of A549 into BALB/C nude mice (metastasis model)	50 or 75 mg/kg BA was administered on the day 7 after LLC injection	2 mos	↓ metastasis [DD]		[179]
		A549 and H1299 treated with TGF-β1 (10 ng/mL)	10–20	24 h		↑ E-cadherin [DD], ↓ vimentin (A549: 20 μM; H1299 [DD]), ↓ N-cadherin [DD], ↓ Skp2 [DD]	
		LLC	10–20	24 h	↓ migration [DD], ↓ invasion [DD]	↓ Skp2 [DD], ↑ E-cadherin [DD]	
		subcutaneous injection of LLC into C57BL/6 mice (spontaneous metastasis model)	50 or 75 mg/kg each day after LLC injection	21 d	↓ primary tumor growth [DD], ↓ lung metastasis [DD]	↓ Skp2 [DD], ↑ E-cadherin [DD], ↑ p27 [DD]	
		intravenous injection of LLC into C57BL/6 mice (metastasis model)	50 or 75 mg/kg BA was administered on the day 7 after LLC injection	60 d	↓ lung metastasis [DD]	↓ Skp2, ↑ E-cadherin	
Lupane	Betulinic acid, SYK023	H1299	0.1–30	36 h	↓ migration (BA: 10 μM > 5 μM; SYK023 [DD]), ↓ invasion (BA: 10 μM > 5 μM; SYK023 [DD])	↓ F-actin polymerization (BA: 10 μM > 5 μM; SYK023 [DD]), ↓ p-FAK (SYK023: 1 μM, 5 μM), ↓ p-Src (BA: 5 μM > 1 μM; SYK023: 0.5–5 μM), ↓ p-Akt (BA: 5 μM; SYK023: 0.5–5 μM), ↓ p-mTOR (SYK023: 1 μM, 5 μM), ↓ N-cadherin (BA: 30 μM; SYK023 [DD]), ↓ β-catenin (BA: 30 μM; SYK023 [DD]), ↓ vimentin (SYK023: 20 μM, 30 μM), ↓ c-Myc (BA: 20 μM, 30 μM; SYK023: 20 μM, 30 μM), ↓ SYPD [K] (BA, SYK023: 20 μM)	[180]

* Concentrations are shown in μM unless otherwise stated. ** Time is given in hours (h), days (d), weeks (wk) and months (mos). *** arrows indicate suppression and activation of biological process, respectively. **** Downward (↓) and upward (↑) arrows indicate downregulation and upregulation of expression, respectively. [DD] Dose-dependent increase in effect. [TD] Time-dependent increase in effect. [K] Demonstrated as a key mechanism using pharmacological inhibition and/or genetic engineering methods.

5.1. Ursane-Type Triterpenoids

Ursolic acid (UA), a compound synthesized in a variety of plants, has been shown to inhibit basement membrane adhesion, migration, invasion, and EMT of A549 cells via NF-κB-dependent downregulation of astrocyte-elevated gene-1 (AEG-1) [166]. A study by Ruan et al. demonstrated that UA inhibited TGF-β-induced EMT in H1875 cells through the downregulation of the Smad-independent integrin αVβ5 pathway [167]. However, the canonical Smad-dependent pathway is also responsive to UA treatment, as evidenced by the correlation between reduced levels of TGF-β and phosphorylated Smad3 and the attenuation of EMT and airway-vessel remodeling in the lungs of UA-treated Wistar rats with cigarette smoke-induced emphysema. In addition, UA induced the expression of insulin-like growth factor 1 (IGF-1) [168], which has been suggested to play a key role in the regeneration of epithelial and muscle cells in patients with COPD [181]. Lin et al. also demonstrated that UA effectively blocks the unfolded protein response (UPR) signaling cascades that cause the alleviation of emphysema and airway remodeling in the lungs of Sprague–Dawley rats induced by the administration of cigarette smoke extract [169].

Another ursane-type triterpenoid, asiatic acid (AA), extracted from *Centella asiatica* (L.), also inhibited TGF-β-induced EMT of A549 cells [164]. This effect was further confirmed in a bleomycin-induced pulmonary fibrosis model in C57BL/6 mice; the administration of AA for 21 days resulted in the alleviation of pulmonary fibrosis by blocking EMT through the downregulation of TGF-β1/Smad2/3 and ERK1/2 pathways. Interestingly, along with the reversal of EMT, AA suppressed lung inflammation by reducing inflammatory cell infiltration in bronchoalveolar lavage fluid and the expression of pro-inflammatory cytokines (IL-1β, IL-18, IL-6, and TNF-α) in lung tissue as well as inhibiting IL-1β and IL-18 secretion via the inactivation of NLRP3 inflammasome [165]. Considering the above-mentioned EMT-inducing activity of these cytokines, AA-mediated inhibition of EMT in the lungs of mice with pulmonary fibrosis may be associated with the inhibition of TGF-β signaling pathways, but also with the inhibition of some other signaling axes susceptible to EMT-inducing cytokines.

5.2. Oleanane- and Friedelane-Type Triterpenoids

Oleanolic acid, a plant oleanane-type triterpenoid, can inhibit EMT in the human NSCLC cell lines A549 and PC-9 through the ERK pathway [175]. A number of other oleanane-type triterpenoids have been reported to inhibit the EMT in lung epithelial cells, primarily through the inhibition of the TGF-β1/Smad2/3 pathway. Celastrol isolated from *Tripterygium wilfordii Hook F.* inhibited TGF-β1-induced EMT in A549 cells [152]. These results are supported by in vivo studies showing that celastrol treatment suppressed the activation of the TGF-β1/Smad2/3 pathway in the lungs of Wistar rats with bleomycin-induced pulmonary fibrosis. As a result, the EMT was inhibited as evidenced by an increase in the expression of epithelial markers (E-cadherin and claudin) and suppression of the expression of mesenchymal markers (N-cadherin, β-catenin, Snail, and Slug) in rat lung tissues, respectively. These effects of celastrol were partly due to the downregulation of Hsp90 expression [174], which is a negative regulator of E-cadherin expression [182]. Like celastrol, β-peltoboykinolic acid isolated from *Astilbe rubra* also inhibits the EMT in A549 cells through the TGF-β1/Smad2/3 pathway [178].

Gui et al. reported that HMGB1 downregulation underlies glycyrrhizin-mediated reversal of the EMT in TGF-β1-treated A549 and BEAS-2B cells, as HMGB1 has been shown to promote TGF-β1-induced Smad2/3 phosphorylation [149]. Additionally, Ren et al. found that the HMGB1-blocking effect of glycyrrhizin determines its suppressive effect on the CAF-driven highly motile and invasive phenotype of A549 and H661 cells [40].

Glycyrrhizin is the parent compound in the synthesis of soloxolone methyl (SM), a triterpenoid with a cyano-enone pharmacophore group in its structure [183], which reduced migratory and invasive abilities and inhibited the EMT in TGF-β1-stimulated A549 cells. The analysis of the gene association network consisting of previously established EMT-associated genes and in silico-predicted molecular targets of SM with further verification by molecular docking approach revealed that SM-mediated EMT inhibition could be due to its interactions with MMP-2, MMP-9, and mitogen-activated protein kinase 8 (MAPK8), but further experimental verification is needed to prove the interaction between SM and its targets [176]. 2-cyano-3, 12-dioxooleana-1, 9-dien-28-oic acid (CDDO), a structural analog of SM synthesized from oleanolic acid, has been shown to attenuate radiation-induced lung inflammation and fibrosis in C57BL/6 mice. CDDO inhibited inflammatory cell infiltration, thereby reducing the levels of the EMT-inducing cytokines TGF-β and IL-6 produced by these cells in the lungs of X-ray-treated mice. In addition, CDDO inhibited the expression of the mesenchymal markers fibronectin, α-SMA, and collagen I, as well as the deposition of collagen fibers in the lungs of irradiated mice, suggesting that CDDO inhibits the EMT in lung epithelial cells in mice with radiation-induced lung fibrosis [173]. Pristimerin isolated from *Celastraceae* plants and evoditrilone A isolated from *Evodia rutaecarpa* were reported to inhibit migration ability and regulate EMT-associated markers in H1299 and A549 cells, respectively [148,151]. The exact mechanisms of the EMT-inhibitory activity of cyanoenone-bearing PTs, as well as pristimerin and evoditrilone A, remain to be elucidated.

5.3. Lupane-Type Triterpenoids

Among the lupane triterpenoids, betulinic acid (BA) inhibited TGF-β1-induced EMT in H1299 and A549 cells in vitro and metastasis in a variety of mouse models in vivo. He et al. demonstrated that BA inhibits NSCLC metastasis by direct interaction with Skp2, thereby blocking the assembly of the Skp2-SCF E3 ligase complex, which causes impairment of the ubiquitin-dependent degradation of its downstream target E-cadherin [179]. The introduction of a 4-methoxyphenylacetic group at the C-3 position of BA (a derivative named SYK023) significantly improved its inhibitory effect on the migration and invasive ability of H1299 in vitro and lung metastasis in nude mice bearing CL1-5 human lung cancer xenografts in vivo. An additional mechanism contributing to the antimetastatic activity of BA and SYK023 was reported by Hsu et al. It was shown that these compounds effectively blocked F-actin polymerization by inhibiting the Src/FAK and Akt/mTOR pathways and suppressing the expression of many actin polymerization genes, among which synaptopodin (Synpo) plays a key role, as its knockdown prevented the inhibitory effect of BA and SYK023 on cell motility and F-actin polymerization [180]. These effects may be related to the established role of Synpo in blocking Smurf1-mediated ubiquitination and degradation of RhoA, a key regulator of the actin cytoskeleton [184]. In addition, both BA and SYK023 were shown to inhibit the expression of N-cadherin, vimentin, and β-catenin in H1299 cells, which is in good agreement with their anti-EMT effect demonstrated by He et al. [179,180].

6. Inhibitory Effect of Triterpenoids on Stem-like Properties of Lung Cancer Cells

As previously described, chronic inflammation that develops during cancer progression induces the transformation of normal tissue stem cells into cancer stem cells (CSCs), which have the ability to self-renew and differentiate into cancer cells, thus contributing to tumor recurrence and drug resistance. Because EMT plays a critical role in maintaining the stem-like properties of CSCs, targeting EMT pathways could potentially alter the response of CSCs to anticancer drugs and reduce the probability of tumor relapse after chemotherapy.

UA has been shown to inhibit tumorsphere formation and expression of the CSCs markers NANOG, OCT4, and SOX2 in A549 and H460 cells by targeting the activation of EGFR and its downstream JAK2/STAT3 pathway [170] (Figure 7). A similar effect on the sphere-forming potential of A549 and H1299 cells was demonstrated for BA, but the mechanism of its inhibitory effect on the stem-like properties of NSCLC cells differs from that of UA and seems to be related to the above-mentioned direct interaction of BA with Skp2 and blocking the assembly of the Skp2-SCF complex, which inhibits the degradation of the cyclin-dependent kinase inhibitor p27 and thus compromises CSCs proliferation [179,185].

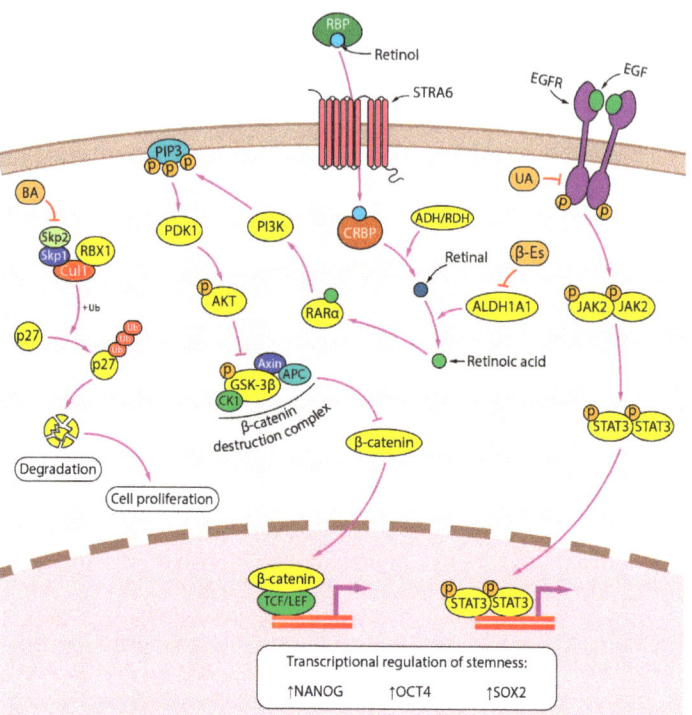

Figure 7. Inhibition of stemness by PTs. Lupane-type PT BA inhibits stem cell proliferation by interfering with Skp2-SCF-dependent degradation of the cyclin-dependent kinase inhibitor p27. Oleanane-type PT β-escin suppresses the activity of ALDH1A1 and the downstream AKT pathway, which is thought to regulate the CSC transcriptome through the β-catenin pathway. The anti-stem cell activity of ursane-type PT UA is mediated through the EGFR pathway. Downward (↓) and upward (↑) arrows indicate downregulation and upregulation of expression, respectively.

Aldehyde dehydrogenase 1A1 (ALDH1A1) has been shown to promote the stemness of CSCs by catalyzing the conversion of retinol to retinoic acid, which induces the expression of stem cell-related genes either through the classical pathway, in which retinoic acid binds to the nuclear retinoic acid receptor α (RARα)/retinoid X receptor (RXR) heterodimer, or through the alternative pathways, including activation of the PI3K/Akt pathway via binding to cytosolic RARα [186]. β-Escin, a triterpenoid saponin isolated from *Aesculus hippocastanum*, was shown to decrease ALDH activity and ALDH1A1 expression in H460 cells in vitro and eradicate the population of ALDH1A1+ cells in lung adenocarcinomas induced by the tobacco carcinogen 4-(methyl nitrosamino)-1-(3-pyridyl)-1-butanone (NNK) in A/J mice in vivo. As a result, there was a marked decrease in Akt activation in the lung tumors of β-Escin-treated mice, confirming the relationship between the inhibitory effect of BA on CSCs proliferation and the suppression of retinoic acid-activated signaling pathways [177].

7. Future Prospective and Limitations

An increasing number of studies indicate that the inflammatory microenvironment established during tumor progression promotes lung cancer metastasis through the induction of EMT in tumor cells. A number of inflammation-related cytokines produced by various cells in iTME have pleiotropic effects and, in addition to their immunomodulatory potency, are able to stimulate EMT in lung cancer cells. The invasive capacity of cancer cells increases as they undergo EMT due to cytoskeletal rearrangements and the production of enzymes that degrade ECM components. In addition, recent studies have shown a link between EMT induction and the formation of lung CSCs, a population of stem-like cancer cells involved in the acquisition of chemoresistance and tumor recurrence. Therefore, targeting iTME-mediated induction of EMT with chemotherapeutic agents may be a promising strategy to suppress metastasis and drug resistance in lung cancer, as evidenced by the clinical benefit for cancer patients in response to drugs targeting iTME components in clinical trials.

PTs of the oleanane-, ursane-, and lupane-type exhibit suppressive potency against inflammation-induced EMT in lung cancer cells. It should be emphasized that in the majority of published studies on this topic, researchers have described the effect of PTs mainly on the TGF-β signaling pathway, while pathways sensitive to other EMT-associated regulators remained outside their attention and require further detailed investigation. The demonstrated ability of some PTs to markedly inhibit the expression of inflammation-related cytokines in the lungs of mice with pulmonary fibrosis, accompanied by modulation of the expression of EMT markers, suggests the prospect of further, more detailed studies of the effect of PTs on cytokine-driven EMT in lung cancer cells.

Since the inflammatory process is characterized by the production of a large pool of cytokines acting synergistically, co-culture models of lung cancer cells with activated immune cells are of great interest in this direction. It should be noted that the majority of published in vitro experiments on the anti-EMT potency of PTs have been performed using very simple cell models activated by EMT inducers only in a monotherapy regimen. Co-culture studies would allow a more thorough comparison of the data obtained from in vitro and in vivo experiments and allow more objective conclusions about the mechanism of anti-EMT action of the studied compounds.

Despite the currently active research on the relationship between EMT and CSCs, there are only fragmentary data on the effect of PTs on the population of lung CSCs enriched in inflammatory conditions due to EMT induction, and therefore it is not possible to fully describe the molecular mechanism underlying stemness inhibition by PTs. However, the paucity of studies on this topic may not be due to the lack of CSC-inhibitory activity of PTs, but to the fact that researchers have overwhelmingly focused on the effect of compounds on the motility and invasiveness of EMT-derived lung cancer cells in the context of their antimetastatic activity, overlooking their effect on stem-like properties. Thus, the study of the effect of PTs with previously confirmed anti-EMT activity, as well as newly developed triterpenoids, on the stemness of lung cancer cells is of great interest for further characterization of the molecular mechanism of their antitumor activity.

8. Conclusions

Taken together, the data presented in this review suggest a clear link between inflammation and the induction of the EMT in tumor cells and, as a consequence, the enhancement of their malignancy, which opens new opportunities for the development of antitumor agents. PTs exhibiting pronounced anti-inflammatory and antitumor potential can be considered a promising source of effective blockers of inflammation-driven EMT in lung cancer cells, but research in this direction is currently at an early stage and requires further development. The information on key inflammation-related signaling pathways associated with the EMT in lung cells described in this work can be used as a kind of roadmap for researchers developing new phytochemical compounds targeting processes closely associated with the EMT, such as cancer and fibrosis.

Author Contributions: Conceptualization, A.V.M. and K.V.O.; methodology, K.V.O.; formal analysis, K.V.O.; investigation, K.V.O.; resources, A.V.M.; data curation, A.V.M.; writing—original draft preparation, K.V.O. and A.V.M.; writing—review and editing, M.A.Z.; visualization, K.V.O.; supervision, A.V.M.; project administration, M.A.Z.; funding acquisition, M.A.Z. All authors have read and agreed to the published version of the manuscript.

Funding: This research was funded by the Russian Science Foundation, grant number 19-74-30011.

Institutional Review Board Statement: Not applicable.

Informed Consent Statement: Not applicable.

Data Availability Statement: Not applicable.

Conflicts of Interest: The authors declare no conflict of interest.

References

1. Sung, H.; Ferlay, J.; Siegel, R.L.; Laversanne, M.; Soerjomataram, I.; Jemal, A.; Bray, F. Global Cancer Statistics 2020: GLOBOCAN Estimates of Incidence and Mortality Worldwide for 36 Cancers in 185 Countries. *CA Cancer J. Clin.* **2021**, *71*, 209–249. [CrossRef]
2. Arseniev, A.I.; Nefedov, A.O.; Novikov, S.N.; Barchuk, A.A.; Tarkov, S.A.; Kostitsin, K.A.; Nefedova, A.V.; Aristidov, N.Y.; Semiletova, Y.V.; Ryazankina, A.A. Algorithms of non-invasive, minimally-invasive and invasive diagnostics for lung cancer (review). *Prev. Clin. Med.* **2021**, *2*, 69–77. (In Russian)
3. Ribatti, D.; Tamma, R.; Annese, T. Epithelial-Mesenchymal Transition in Cancer: A Historical Overview. *Transl. Oncol.* **2020**, *13*, 100773. [CrossRef] [PubMed]
4. Menju, T.; Date, H. Lung cancer and epithelial-mesenchymal transition. *Gen. Thorac. Cardiovasc. Surg.* **2021**, *69*, 781–789. [CrossRef] [PubMed]
5. Tsoukalas, N.; Aravantinou-Fatorou, E.; Tolia, M.; Giaginis, C.; Galanopoulos, M.; Kiakou, M.; Kostakis, I.D.; Dana, E.; Vamvakaris, I.; Korogiannos, A.; et al. Epithelial-Mesenchymal Transition in Non Small-cell Lung Cancer. *Anticancer Res.* **2017**, *37*, 1773–1778. [CrossRef] [PubMed]
6. Trandafir, L.M.; Miron, O.; Afrasanie, V.-A.; Paduraru, M.-I.; Miron, L. The relationship between chronic lung diseases and lung cancer-a narrative review. *JBUON* **2020**, *25*, 1687–1692.
7. Szalontai, K.; Gémes, N.; Furák, J.; Varga, T.; Neuperger, P.; Balog, J.Á.; Puskás, L.G.; Szebeni, G.J. Chronic Obstructive Pulmonary Disease: Epidemiology, Biomarkers, and Paving the Way to Lung Cancer. *J. Clin. Med.* **2021**, *10*, 2889. [CrossRef] [PubMed]
8. Mahmood, M.Q.; Walters, E.H.; Shukla, S.D.; Weston, S.; Muller, H.K.; Ward, C.; Sohal, S.S. β-catenin, Twist and Snail: Transcriptional regulation of EMT in smokers and COPD, and relation to airflow obstruction. *Sci. Rep.* **2017**, *7*, 1–12. [CrossRef]
9. Pandolfi, L.; Bozzini, S.; Frangipane, V.; Percivalle, E.; De Luigi, A.; Violatto, M.B.; Lopez, G.; Gabanti, E.; Carsana, L.; D'Amato, M.; et al. Neutrophil Extracellular Traps Induce the Epithelial-Mesenchymal Transition: Implications in Post-COVID-19 Fibrosis. *Front. Immunol.* **2021**, *12*, 663303. [CrossRef]
10. Markov, A.V.; Zenkova, M.A.; Logashenko, E.B. Modulation of Tumour-Related Signaling Pathways by Natural Pentacyclic Triterpenoids and their Semisynthetic Derivatives. *Curr. Med. Chem.* **2017**, *24*, 1277–1320. [CrossRef]
11. Kozak, J.; Forma, A.; Czeczelewski, M.; Kozyra, P.; Sitarz, E.; Radzikowska-Büchner, E.; Sitarz, M.; Baj, J. Inhibition or Reversal of the Epithelial-Mesenchymal Transition in Gastric Cancer: Pharmacological Approaches. *Int. J. Mol. Sci.* **2021**, *22*, 277. [CrossRef] [PubMed]
12. Pearlman, R.L.; Montes de Oca, M.K.; Pal, H.C.; Afaq, F. Potential therapeutic targets of epithelial–mesenchymal transition in melanoma. *Cancer Lett.* **2017**, *391*, 125–140. [CrossRef] [PubMed]
13. Tang, Z.Y.; Li, Y.; Tang, Y.T.; Ma, X.D.; Tang, Z.Y. Anticancer activity of oleanolic acid and its derivatives: Recent advances in evidence, target profiling and mechanisms of action. *Biomed. Pharmacother.* **2022**, *145*, 112397. [CrossRef] [PubMed]
14. He, Y.Q.; Zhou, C.C.; Yu, L.Y.; Wang, L.; Deng, J.L.; Tao, Y.L.; Zhang, F.; Chen, W.S. Natural product derived phytochemicals in managing acute lung injury by multiple mechanisms. *Pharmacol. Res.* **2021**, *163*, 105224. [CrossRef] [PubMed]
15. Kalluri, R.; Weinberg, R.A. The basics of epithelial-mesenchymal transition. *J. Clin. Investig.* **2009**, *119*, 1420–1428. [CrossRef] [PubMed]
16. Montanari, M.; Rossetti, S.; Cavaliere, C.; D'Aniello, C.; Malzone, M.G.; Vanacore, D.; Di Franco, R.; La Mantia, E.; Iovane, G.; Piscitelli, R.; et al. Epithelial-mesenchymal transition in prostate cancer: An overview. *Oncotarget* **2017**, *8*, 35376–35389. [CrossRef] [PubMed]
17. Jiang, J.; Tang, Y.L.; Liang, X.H. EMT: A new vision of hypoxia promoting cancer progression. *Cancer Biol. Ther.* **2011**, *11*, 714–723. [CrossRef]
18. Suzuki, A.; Maeda, T.; Baba, Y.; Shimamura, K.; Kato, Y. Acidic extracellular pH promotes epithelial mesenchymal transition in lewis lung carcinoma model. *Cancer Cell Int.* **2014**, *14*, 129. [CrossRef]
19. Deng, Z.; Fear, M.W.; Suk Choi, Y.; Wood, F.M.; Allahham, A.; Mutsaers, S.E.; Prêle, C.M. The extracellular matrix and mechanotransduction in pulmonary fibrosis. *Int. J. Biochem. Cell Biol.* **2020**, *126*, 105802. [CrossRef]
20. Debnath, P.; Huirem, R.S.; Dutta, P.; Palchaudhuri, S. Epithelial–mesenchymal transition and its transcription factors. *Biosci. Rep.* **2021**, *42*, BSR20211754. [CrossRef]
21. Loh, C.Y.; Chai, J.Y.; Tang, T.F.; Wong, W.F.; Sethi, G.; Shanmugam, M.K.; Chong, P.P.; Looi, C.Y. The E-Cadherin and N-Cadherin Switch in Epithelial-to-Mesenchymal Transition: Signaling, Therapeutic Implications, and Challenges. *Cells* **2019**, *8*, 1118. [CrossRef] [PubMed]
22. Liu, C.-Y.Y.; Lin, H.-H.H.; Tang, M.-J.J.; Wang, Y.-K.K. Vimentin contributes to epithelial-mesenchymal transition ancer cell mechanics by mediating cytoskeletal organization and focal adhesion maturation. *Oncotarget* **2015**, *6*, 15966–15983. [CrossRef] [PubMed]

23. Lamouille, S.; Xu, J.; Derynck, R. Molecular mechanisms of epithelial-mesenchymal transition. *Nat. Rev. Mol. Cell Biol.* **2014**, *15*, 178–196. [CrossRef] [PubMed]
24. Griggs, L.A.; Hassan, N.T.; Malik, R.S.; Griffin, B.P.; Martinez, B.A.; Elmore, L.W.; Lemmon, C.A. Fibronectin fibrils regulate TGF-β1-induced Epithelial-Mesenchymal Transition. *Matrix Biol.* **2017**, *60–61*, 157–175. [CrossRef]
25. Zhang, K.; Corsa, C.A.; Ponik, S.M.; Prior, J.L.; Piwnica-Worms, D.; Eliceiri, K.W.; Keely, P.J.; Longmore, G.D. The collagen receptor discoidin domain receptor 2 stabilizes SNAIL1 to facilitate breast cancer metastasis. *Nat. Cell Biol.* **2013**, *15*, 677–687. [CrossRef]
26. Gonzalez, D.M.; Medici, D. Signaling mechanisms of the epithelial-mesenchymal transition. *Sci. Signal.* **2014**, *7*, re8. [CrossRef] [PubMed]
27. Dongre, A.; Weinberg, R.A. New insights into the mechanisms of epithelial–mesenchymal transition and implications for cancer. *Nat. Rev. Mol. Cell Biol.* **2019**, *20*, 69–84. [CrossRef]
28. Rajasegaran, T.; How, C.W.; Saud, A.; Ali, A.; Lim, J.C. Targeting Inflammation in Non-Small Cell Lung Cancer through Drug Repurposing. *Pharmaceuticals* **2023**, *16*, 451. [CrossRef]
29. Houghton, A.M.G. Mechanistic links between COPD and lung cancer. *Nat. Rev. Cancer* **2013**, *13*, 233–245. [CrossRef]
30. Samet, J.M. Does idiopathic pulmonary fibrosis increase lung cancer risk? *Am. J. Respir. Crit. Care Med.* **2000**, *161*, 1–2. [CrossRef]
31. Ahmad, S.; Manzoor, S.; Siddiqui, S.; Mariappan, N.; Zafar, I.; Ahmad, A.; Ahmad, A. Epigenetic underpinnings of inflammation: Connecting the dots between pulmonary diseases, lung cancer and COVID-19. *Semin. Cancer Biol.* **2022**, *83*, 384–398. [CrossRef]
32. Islam, M.S.; Morshed, M.R.; Babu, G.; Khan, M.A. The role of inflammations and EMT in carcinogenesis. *Adv. Cancer Biol.-Metastasis* **2022**, *5*, 100055. [CrossRef]
33. Ping, Q.; Yan, R.; Cheng, X.; Wang, W.; Zhong, Y.; Hou, Z.; Shi, Y.; Wang, C.; Li, R. Cancer-associated fibroblasts: Overview, progress, challenges, and directions. *Cancer Gene Ther.* **2021**, *28*, 984–999. [CrossRef] [PubMed]
34. Heinrich, E.L.; Walser, T.C.; Krysan, K.; Liclican, E.L.; Grant, J.L.; Rodriguez, N.L.; Dubinett, S.M. The Inflammatory Tumor Microenvironment, Epithelial Mesenchymal Transition and Lung Carcinogenesis. *Cancer Microenviron.* **2012**, *5*, 5–18. [CrossRef] [PubMed]
35. Shintani, Y.; Abulaiti, A.; Kimura, T.; Funaki, S.; Nakagiri, T.; Inoue, M.; Sawabata, N.; Minami, M.; Morii, E.; Okumura, M. Pulmonary Fibroblasts Induce Epithelial Mesenchymal Transition and Some Characteristics of Stem Cells in Non-Small Cell Lung Cancer. *Ann. Thorac. Surg.* **2013**, *96*, 425–433. [CrossRef] [PubMed]
36. Yi, Y.; Zeng, S.; Wang, Z.; Wu, M.; Ma, Y.; Ye, X.; Zhang, B.; Liu, H. Cancer-associated fibroblasts promote epithelial-mesenchymal transition and EGFR-TKI resistance of non-small cell lung cancers via HGF/IGF-1/ANXA2 signaling. *Biochim. Biophys. Acta-Mol. Basis Dis.* **2018**, *1864*, 793–803. [CrossRef] [PubMed]
37. Shintani, Y.; Fujiwara, A.; Kimura, T.; Kawamura, T.; Funaki, S.; Minami, M.; Okumura, M. IL-6 Secreted from Cancer-Associated Fibroblasts Mediates Chemoresistance in NSCLC by Increasing Epithelial-Mesenchymal Transition Signaling. *J. Thorac. Oncol.* **2016**, *11*, 1482–1492. [CrossRef]
38. Wang, L.; Cao, L.; Wang, H.; Liu, B.; Zhang, Q.; Meng, Z.; Wu, X.; Zhou, Q.; Xu, K. Cancer-associated fibroblasts enhance metastatic potential of lung cancer cells through IL-6/STAT3 signaling pathway. *Oncotarget* **2017**, *8*, 76116–76128. [CrossRef]
39. Wang, Y.; Lan, W.; Xu, M.; Song, J.; Mao, J.; Li, C.; Du, X.; Jiang, Y.; Li, E.; Zhang, R.; et al. Cancer-associated fibroblast-derived SDF-1 induces epithelial-mesenchymal transition of lung adenocarcinoma via CXCR4/β-catenin/PPARδ signalling. *Cell Death Dis.* **2021**, *12*, 214. [CrossRef]
40. Ren, Y.; Cao, L.; Wang, L.; Zheng, S.; Zhang, Q.; Guo, X.; Li, X.; Chen, M.; Wu, X.; Furlong, F.; et al. Autophagic secretion of HMGB1 from cancer-associated fibroblasts promotes metastatic potential of non-small cell lung cancer cells via NFκB signaling. *Cell Death Dis.* **2021**, *12*, 858. [CrossRef]
41. You, J.; Li, M.; Cao, L.M.; Gu, Q.H.; Deng, P.B.; Tan, Y.; Hu, C.P. Snail1-dependent cancer-associated fibroblasts induce epithelial-mesenchymal transition in lung cancer cells via exosomes. *QJM Int. J. Med.* **2019**, *112*, 581–590. [CrossRef]
42. Yang, F.; Yan, Y.; Yang, Y.; Hong, X.; Wang, M.; Yang, Z.; Liu, B.; Ye, L. MiR-210 in exosomes derived from CAFs promotes non-small cell lung cancer migration and invasion through PTEN/PI3K/AKT pathway. *Cell. Signal.* **2020**, *73*, 109675. [CrossRef]
43. Chattopadhyay, I.; Ambati, R.; Gundamaraju, R. Exploring the Crosstalk between Inflammation and Epithelial-Mesenchymal Transition in Cancer. *Mediat. Inflamm.* **2021**, *2021*, 9918379. [CrossRef]
44. Hui, X.; Al-Ward, H.; Shaher, F.; Liu, C.-Y.; Liu, N. The Role of miR-210 in the Biological System: A Current Overview. *Hum. Hered.* **2020**, *84*, 233–239. [CrossRef]
45. Chae, Y.K.; Chang, S.; Ko, T.; Anker, J.; Agte, S.; Choi, W.M.; Lee, K.; Cruz, M. Epithelial-mesenchymal transition (EMT) signature is inversely associated with T-cell infiltration in non-small cell lung cancer (NSCLC). *Sci. Rep.* **2018**, *8*, 2918. [CrossRef]
46. Ricciardi, M.; Zanotto, M.; Malpeli, G.; Bassi, G.; Perbellini, O.; Chilosi, M.; Bifari, F.; Krampera, M. Epithelial-to-mesenchymal transition (EMT) induced by inflammatory priming elicits mesenchymal stromal cell-like immune- modulatory properties in cancer cells. *Br. J. Cancer* **2015**, *116*, 1067–1075. [CrossRef] [PubMed]
47. Fridlender, Z.G.; Sun, J.; Kim, S.; Kapoor, V.; Cheng, G.; Worthen, G.S.; Albelda, S.M. Polarization of TAN phenotype by TGFb: "N1" versus "N2" TAN. *Cancer Cell* **2010**, *16*, 183–194. [CrossRef] [PubMed]
48. Massagué, J. TGFβ in Cancer. *Cell* **2008**, *134*, 215–230. [CrossRef] [PubMed]
49. Hu, P.; Shen, M.; Zhang, P.; Zheng, C.; Pang, Z.; Zhu, L.; Du, J. Intratumoral neutrophil granulocytes contribute to epithelial-mesenchymal transition in lung adenocarcinoma cells. *Tumor Biol.* **2015**, *36*, 7789–7796. [CrossRef] [PubMed]

50. Yasukawa, A.; Hosoki, K.; Toda, M.; Miyake, Y.; Matsushima, Y.; Matsumoto, T.; Boveda-Ruiz, D.; Gil-Bernabe, P.; Nagao, M.; Sugimoto, M.; et al. Eosinophils Promote Epithelial to Mesenchymal Transition of Bronchial Epithelial Cells. *PLoS ONE* **2013**, *8*, e64281. [CrossRef]
51. Hosoki, K.; Kainuma, K.; Toda, M.; Harada, E.; Chelakkot-Govindalayathila, A.L.; Roeen, Z.; Nagao, M.; D'Alessandro-Gabazza, C.N.; Fujisawa, T.; Gabazza, E.C. Montelukast suppresses epithelial to mesenchymal transition of bronchial epithelial cells induced by eosinophils. *Biochem. Biophys. Res. Commun.* **2014**, *449*, 351–356. [CrossRef] [PubMed]
52. Trinh, H.K.T.; Lee, S.-H.; Cao, T.B.T.; Park, H.-S. Asthma pharmacotherapy: An update on leukotriene treatments. *Expert Rev. Respir. Med.* **2019**, *13*, 1169–1178. [CrossRef]
53. Kim, S.Y.; Nair, M.G. Macrophages in wound healing: Activation and plasticity. *Immunol. Cell Biol.* **2019**, *97*, 258–267. [CrossRef] [PubMed]
54. Condeelis, J.; Pollard, J.W. Macrophages: Obligate partners for tumor cell migration, invasion, and metastasis. *Cell* **2006**, *124*, 263–266. [CrossRef] [PubMed]
55. Fernando, R.I.; Castillo, M.D.; Litzinger, M.; Hamilton, D.H.; Palena, C. IL-8 signaling plays a critical role in the epithelial-mesenchymal transition of human carcinoma cells. *Cancer Res.* **2011**, *71*, 5296–5306. [CrossRef] [PubMed]
56. Chen, W.; Gao, Q.; Han, S.; Pan, F.; Fan, W. The CCL2/CCR2 axis enhances IL-6-induced epithelial-mesenchymal transition by cooperatively activating STAT3-Twist signaling. *Tumor Biol.* **2014**, *36*, 973–981. [CrossRef] [PubMed]
57. Dehai, C.; Bo, P.; Qiang, T.; Lihua, S.; Fang, L.; Shi, J.; Jingyan, C.; Yan, Y.; Guangbin, W.; Zhenjun, Y. Enhanced invasion of lung adenocarcinoma cells after co-culture with THP-1-derived macrophages via the induction of EMT by IL-6. *Immunol. Lett.* **2014**, *160*, 1–10. [CrossRef]
58. Qin, Y.; Zhao, P.; Chen, Y.; Liu, X.; Dong, H.; Zheng, W.; Li, C.; Mao, X.; Li, J. Lipopolysaccharide induces epithelial–mesenchymal transition of alveolar epithelial cells cocultured with macrophages possibly via the JAK2/STAT3 signaling pathway. *Hum. Exp. Toxicol.* **2020**, *39*, 224–234. [CrossRef]
59. Shang, G.-S.; Liu, L.; Qin, Y.-W. IL-6 and TNF-α promote metastasis of lung cancer by inducing epithelial-mesenchymal transition. *Oncol. Lett.* **2017**, *13*, 4657–4660. [CrossRef]
60. Yin, Y.; Shelke, G.V.; Lässer, C.; Brismar, H.; Lötvall, J. Extracellular vesicles from mast cells induce mesenchymal transition in airway epithelial cells. *Respir. Res.* **2020**, *21*, 101. [CrossRef]
61. Jiang, Y.; Wu, Y.; Hardie, W.J.; Zhou, X. Mast cell chymase affects the proliferation and metastasis of lung carcinoma cells in vitro. *Oncol. Lett.* **2017**, *14*, 3193–3198. [CrossRef] [PubMed]
62. Li, R.; Ong, S.L.; Tran, L.M.; Jing, Z.; Liu, B.; Park, S.J.; Huang, Z.L.; Walser, T.C.; Heinrich, E.L.; Lee, G.; et al. Chronic IL-1β-induced inflammation regulates epithelial- to-mesenchymal transition memory phenotypes via epigenetic modifications in non-small cell lung cancer. *Sci. Rep.* **2020**, *10*, 377. [CrossRef] [PubMed]
63. Huang, Q.; Han, J.; Fan, J.; Duan, L.; Guo, M.; Lv, Z.; Hu, G.; Chen, L.; Wu, F.; Tao, X.; et al. IL-17 induces EMT via Stat3 in lung adenocarcinoma. *Am. J. Cancer Res.* **2016**, *6*, 440–451. [PubMed]
64. Liu, X. Inflammatory cytokines augments TGF-β1-induced epithelial-mesenchymal transition in A549 cells by up-regulating TβR-I. *Cell Motil. Cytoskelet.* **2008**, *65*, 935–944. [CrossRef]
65. Li, Q.; Han, Y.; Fei, G.; Guo, Z.; Ren, T.; Liu, Z. IL-17 promoted metastasis of non-small-cell lung cancer cells. *Immunol. Lett.* **2012**, *148*, 144–150. [CrossRef]
66. Gu, K.; Li, M.-M.; Shen, J.; Liu, F.; Cao, J.-Y.; Jin, S.; Yu, Y. Interleukin-17-induced EMT promotes lung cancer cell migration and invasion via NF-κB/ZEB1 signal pathway. *Am. J. Cancer Res.* **2015**, *5*, 1169–1179. [PubMed]
67. Kumar, M.; Allison, D.F.; Baranova, N.N.; Wamsley, J.J.; Katz, A.J.; Bekiranov, S.; Jones, D.R.; Mayo, M.W. NF-κB regulates mesenchymal transition for the induction of non-small cell lung cancer initiating cells. *PLoS ONE* **2013**, *8*, e68597. [CrossRef] [PubMed]
68. Hida, T.; Yatabe, Y.; Achiwa, H.; Muramatsu, H.; Kozaki, K.I.; Nakamura, S.; Ogawa, M.; Mitsudomi, T.; Sugiura, T.; Takahashi, T. Increased expression of cyclooxygenase 2 occurs frequently in human lung cancers, specifically in adenocarcinomas. *Cancer Res.* **1998**, *58*, 3761–3764.
69. Dohadwala, M.; Yang, S.C.; Luo, J.; Sharma, S.; Batra, R.K.; Huang, M.; Lin, Y.; Goodglick, L.; Krysan, K.; Fishbein, M.C.; et al. Cyclooxygenase-2-dependent regulation of E-cadherin: Prostaglandin E 2 induces transcriptional repressors ZEB1 and snail in non-small cell lung cancer. *Cancer Res.* **2006**, *66*, 5338–5345. [CrossRef]
70. Takai, E.; Tsukimoto, M.; Kojima, S. TGF-β1 Downregulates COX-2 Expression Leading to Decrease of PGE2 Production in Human Lung Cancer A549 Cells, Which Is Involved in Fibrotic Response to TGF-β1. *PLoS ONE* **2013**, *8*, e76346. [CrossRef]
71. Vafaeinik, F.; Kum, H.J.; Jin, S.Y.; Min, D.S.; Song, S.H.; Ha, H.K.; Kim, C.D.; Bae, S.S. Regulation of Epithelial-Mesenchymal Transition of A549 Cells by Prostaglandin D2. *Cell. Physiol. Biochem.* **2022**, *56*, 89–104. [CrossRef] [PubMed]
72. Petrey, A.; de la Motte, C. Hyaluronan, a Crucial Regulator of Inflammation. *Front. Immunol.* **2014**, *5*, 101. [CrossRef] [PubMed]
73. Marozzi, M.; Parnigoni, A.; Negri, A.; Viola, M.; Vigetti, D.; Passi, A.; Karousou, E.; Rizzi, F. Inflammation, Extracellular Matrix Remodeling, and Proteostasis in Tumor Microenvironment. *Int. J. Mol. Sci.* **2021**, *22*, 8102. [CrossRef]
74. Chow, G.; Tauler, J.; Mulshine, J.L. Cytokines and Growth Factors Stimulate Hyaluronan Production: Role of Hyaluronan in Epithelial to Mesenchymal-Like Transition in Non-Small Cell Lung Cancer. *J. Biomed. Biotechnol.* **2010**, *2010*, 485468. [CrossRef] [PubMed]

75. Huang, Y.-J.; Hsu, S. Acquisition of epithelial–mesenchymal transition and cancer stem-like phenotypes within chitosan-hyaluronan membrane-derived 3D tumor spheroids. *Biomaterials* **2014**, *35*, 10070–10079. [CrossRef] [PubMed]
76. Han, H.-W.; Hsu, S. Chitosan-hyaluronan based 3D co-culture platform for studying the crosstalk of lung cancer cells and mesenchymal stem cells. *Acta Biomater.* **2016**, *42*, 157–167. [CrossRef]
77. Suda, K.; Murakami, I.; Yu, H.; Kim, J.; Tan, A.-C.; Mizuuchi, H.; Rozeboom, L.; Ellison, K.; Rivard, C.J.; Mitsudomi, T.; et al. CD44 Facilitates Epithelial-to-Mesenchymal Transition Phenotypic Change at Acquisition of Resistance to EGFR Kinase Inhibitors in Lung Cancer. *Mol. Cancer Ther.* **2018**, *17*, 2257–2265. [CrossRef]
78. Chen, F.; Zhu, X.; Zheng, J.; Xu, T.; Wu, K.; Ru, C. RHAMM regulates the growth and migration of lung adenocarcinoma A549 cell line by regulating Cdc2/CyclinB1 and MMP9 genes. *Math. Biosci. Eng.* **2020**, *17*, 2150–2163. [CrossRef]
79. Su, J.; Wu, S.; Wu, H.; Li, L.; Guo, T. CD44 is functionally crucial for driving lung cancer stem cells metastasis through Wnt/β-catenin-FoxM1-Twist signaling. *Mol. Carcinog.* **2016**, *55*, 1962–1973. [CrossRef]
80. Li, L.; Qi, L.; Liang, Z.; Song, W.; Liu, Y.; Wang, Y.; Sun, B.; Zhang, B.; Cao, W. Transforming growth factor-β1 induces EMT by the transactivation of epidermal growth factor signaling through HA/CD44 in lung and breast cancer cells. *Int. J. Mol. Med.* **2015**, *36*, 113–122. [CrossRef] [PubMed]
81. Nurwidya, F.; Takahashi, F.; Kato, M.; Baskoro, H.; Hidayat, M.; Wirawan, A.; Takahashi, K. CD44 silencing decreases the expression of stem cell-related factors induced by transforming growth factor β1 and tumor necrosis factor α in lung cancer: Preliminary findings. *Bosn. J. Basic Med. Sci.* **2017**, *17*, 228–234. [CrossRef] [PubMed]
82. Wang, C.-Y.; Huang, C.-S.; Yang, Y.-P.; Liu, C.-Y.; Liu, Y.-Y.; Wu, W.-W.; Lu, K.-H.; Chen, K.-H.; Chang, Y.-L.; Lee, S.-D.; et al. The subpopulation of CD44-positive cells promoted tumorigenicity and metastatic ability in lung adenocarcinoma. *J. Chin. Med. Assoc.* **2019**, *82*, 196–201. [CrossRef]
83. Monleón-Guinot, I.; Milian, L.; Martínez-Vallejo, P.; Sancho-Tello, M.; Llop-Miguel, M.; Galbis, J.M.; Cremades, A.; Carda, C.; Mata, M. Morphological Characterization of Human Lung Cancer Organoids Cultured in Type I Collagen Hydrogels: A Histological Approach. *Int. J. Mol. Sci.* **2023**, *24*, 10131. [CrossRef]
84. Lotsberg, M.L.; Røsland, G.V.; Rayford, A.J.; Dyrstad, S.E.; Ekanger, C.T.; Lu, N.; Frantz, K.; Stuhr, L.E.B.; Ditzel, H.J.; Thiery, J.P.; et al. Intrinsic Differences in Spatiotemporal Organization and Stromal Cell Interactions Between Isogenic Lung Cancer Cells of Epithelial and Mesenchymal Phenotypes Revealed by High-Dimensional Single-Cell Analysis of Heterotypic 3D Spheroid Models. *Front. Oncol.* **2022**, *12*, 818437. [CrossRef] [PubMed]
85. Peng, D.H.; Ungewiss, C.; Tong, P.; Byers, L.A.; Wang, J.; Canales, J.R.; Villalobos, P.A.; Uraoka, N.; Mino, B.; Behrens, C.; et al. ZEB1 induces LOXL2-mediated collagen stabilization and deposition in the extracellular matrix to drive lung cancer invasion and metastasis. *Oncogene* **2017**, *36*, 1925–1938. [CrossRef] [PubMed]
86. Shintani, Y.; Maeda, M.; Chaika, N.; Johnson, K.R.; Wheelock, M.J. Collagen I Promotes Epithelial-to-Mesenchymal Transition in Lung Cancer Cells via Transforming Growth Factor–β Signaling. *Am. J. Respir. Cell Mol. Biol.* **2008**, *38*, 95–104. [CrossRef]
87. Liu, C.-C.; Lin, J.-H.; Hsu, T.-W.; Hsu, J.-W.; Chang, J.-W.; Su, K.; Hsu, H.-S.; Hung, S.-C. Collagen XVII/laminin-5 activates epithelial-to-mesenchymal transition and is associated with poor prognosis in lung cancer. *Oncotarget* **2018**, *9*, 1656–1672. [CrossRef]
88. Ungewiss, C.; Rizvi, Z.H.; Roybal, J.D.; Peng, D.H.; Gold, K.A.; Shin, D.-H.; Creighton, C.J.; Gibbons, D.L. The microRNA-200/Zeb1 axis regulates ECM-dependent β1-integrin/FAK signaling, cancer cell invasion and metastasis through CRKL. *Sci. Rep.* **2016**, *6*, 18652. [CrossRef]
89. Liu, L.; Stephens, B.; Bergman, M.; May, A.; Chiang, T. Role of Collagen in Airway Mechanics. *Bioengineering* **2021**, *8*, 13. [CrossRef]
90. Jones, V.A.; Patel, P.M.; Gibson, F.T.; Cordova, A.; Amber, K.T. The Role of Collagen XVII in Cancer: Squamous Cell Carcinoma and Beyond. *Front. Oncol.* **2020**, *10*, 353. [CrossRef]
91. Hasan Ali, O.; Bomze, D.; Ring, S.S.; Berner, F.; Fässler, M.; Diem, S.; Abdou, M.-T.; Hammers, C.; Emtenani, S.; Braun, A.; et al. BP180-specific IgG is associated with skin adverse events, therapy response, and overall survival in non-small cell lung cancer patients treated with checkpoint inhibitors. *J. Am. Acad. Dermatol.* **2020**, *82*, 854–861. [CrossRef] [PubMed]
92. Luo, D.; Hu, S.; Tang, C.; Liu, G. Mesenchymal stem cells promote cell invasion and migration and autophagy-induced epithelial-mesenchymal transition in A549 lung adenocarcinoma cells. *Cell Biochem. Funct.* **2018**, *36*, 88–94. [CrossRef] [PubMed]
93. Ma, M.; Shi, F.; Zhai, R.; Wang, H.; Li, K.; Xu, C.; Yao, W.; Zhou, F. TGF-β promote epithelial-mesenchymal transition via NF-κB/NOX4/ROS signal pathway in lung cancer cells. *Mol. Biol. Rep.* **2021**, *48*, 2365–2375. [CrossRef] [PubMed]
94. Ryu, D.; Lee, J.-H.; Kwak, M.-K. NRF2 level is negatively correlated with TGF-β1-induced lung cancer motility and migration via NOX4-ROS signaling. *Arch. Pharm. Res.* **2020**, *43*, 1297–1310. [CrossRef] [PubMed]
95. Hu, Y.; He, K.; Wang, D.; Yuan, X.; Liu, Y.; Ji, H.; Song, J. TMEPAI regulates EMT in lung cancer cells by modulating the ROS and IRS-1 signaling pathways. *Carcinogenesis* **2013**, *34*, 1764–1772. [CrossRef] [PubMed]
96. Kong, Q.; Xiang, Z.; Wu, Y.; Gu, Y.; Guo, J.; Geng, F. Analysis of the susceptibility of lung cancer patients to SARS-CoV-2 infection. *Mol. Cancer* **2020**, *19*, 80. [CrossRef]
97. Zhang, H.; Quek, K.; Chen, R.; Chen, J.; Chen, B. Expression of the SAR2-Cov-2 receptor ACE2 reveals the susceptibility of COVID-19 in non-small cell lung cancer. *J. Cancer* **2020**, *11*, 5289–5292. [CrossRef]
98. Aramini, B.; Masciale, V.; Samarelli, A.V.; Tonelli, R.; Cerri, S.; Clini, E.; Stella, F.; Dominici, M. Biological effects of COVID-19 on lung cancer: Can we drive our decisions. *Front. Oncol.* **2022**, *12*, 1029830. [CrossRef]

99. Saygideger, Y.; Sezan, A.; Candevir, A.; Saygıdeğer Demir, B.; Güzel, E.; Baydar, O.; Derinoz, E.; Komur, S.; Kuscu, F.; Ozyılmaz, E.; et al. COVID-19 patients' sera induce epithelial mesenchymal transition in cancer cells. *Cancer Treat. Res. Commun.* **2021**, *28*, 100406. [CrossRef]
100. Falleni, M.; Tosi, D.; Savi, F.; Chiumello, D.; Bulfamante, G. Endothelial-Mesenchymal Transition in COVID-19 lung lesions. *Pathol.-Res. Pract.* **2021**, *221*, 153419. [CrossRef]
101. Jahankhani, K.; Ahangari, F.; Adcock, I.M.; Mortaz, E. Possible cancer-causing capacity of COVID-19: Is SARS-CoV-2 an oncogenic agent? *Biochimie* **2023**, *213*, 130–138. [CrossRef] [PubMed]
102. Zhang, L.; Zhang, X.; Deng, X.; Wang, P.; Mo, Y.; Zhang, Y.; Tong, X. Cytokines as drivers: Unraveling the mechanisms of epithelial-mesenchymal transition in COVID-19 lung fibrosis. *Biochem. Biophys. Res. Commun.* **2023**, *686*, 149118. [CrossRef] [PubMed]
103. Pi, P.; Zeng, Z.; Zeng, L.; Han, B.; Bai, X.; Xu, S. Molecular mechanisms of COVID-19-induced pulmonary fibrosis and epithelial-mesenchymal transition. *Front. Pharmacol.* **2023**, *14*, 1218059. [CrossRef] [PubMed]
104. Liapis, I.; Baritaki, S. COVID-19 vs. Cancer Immunosurveillance: A Game of Thrones within an Inflamed Microenviroment. *Cancers* **2022**, *14*, 4330. [CrossRef]
105. du Plessis, M.; Fourie, C.; Riedemann, J.; de Villiers, W.J.S.; Engelbrecht, A.M. Cancer and Covid-19: Collectively catastrophic. *Cytokine Growth Factor Rev.* **2022**, *63*, 78–89. [CrossRef]
106. Leon, G.; MacDonagh, L.; Finn, S.P.; Cuffe, S.; Barr, M.P. Cancer stem cells in drug resistant lung cancer: Targeting cell surface markers and signaling pathways. *Pharmacol. Ther.* **2016**, *158*, 71–90. [CrossRef]
107. Heng, W.S.; Gosens, R.; Kruyt, F.A.E. Lung cancer stem cells: Origin, features, maintenance mechanisms and therapeutic targeting. *Biochem. Pharmacol.* **2019**, *160*, 121–133. [CrossRef]
108. Zhang, F.; Li, T.; Han, L.; Qin, P.; Wu, Z.; Xu, B.; Gao, Q.; Song, Y. TGFβ1-induced down-regulation of microRNA-138 contributes to epithelial-mesenchymal transition in primary lung cancer cells. *Biochem. Biophys. Res. Commun.* **2018**, *496*, 1169–1175. [CrossRef]
109. Kahm, Y.-J.; Kim, R.-K.; Jung, U.; Kim, I.-G. Epithelial membrane protein 3 regulates lung cancer stem cells via the TGF-β signaling pathway. *Int. J. Oncol.* **2021**, *59*, 80. [CrossRef]
110. Yang, L.; Dong, Y.; Li, Y.; Wang, D.; Liu, S.; Wang, D.; Gao, Q.; Ji, S.; Chen, X.; Lei, Q.; et al. IL-10 derived from M2 macrophage promotes cancer stemness via JAK1/STAT1/NF-κB/Notch1 pathway in non-small cell lung cancer. *Int. J. Cancer* **2019**, *145*, 1099–1110. [CrossRef]
111. Wamsley, J.J.; Kumar, M.; Allison, D.F.; Clift, S.H.; Holzknecht, C.M.; Szymura, S.J.; Hoang, S.A.; Xu, X.; Moskaluk, C.A.; Jones, D.R.; et al. Activin upregulation by NF-κB is required to maintain mesenchymal features of cancer stem-like cells in non-small cell lung cancer. *Cancer Res.* **2015**, *75*, 426–435. [CrossRef]
112. Song, M.; Ping, Y.; Zhang, K.; Yang, L.; Li, F.; Zhang, C.; Cheng, S.; Yue, D.; Maimela, N.R.; Qu, J.; et al. Low-Dose IFNγ Induces Tumor Cell Stemness in Tumor Microenvironment of Non–Small Cell Lung Cancer. *Cancer Res.* **2019**, *79*, 3737–3748. [CrossRef] [PubMed]
113. Lee, S.O.; Yang, X.; Duan, S.; Tsai, Y.; Strojny, L.R.; Keng, P.; Chen, Y. IL-6 promotes growth and epithelial-mesenchymal transition of CD133+ cells of non-small cell lung cancer. *Oncotarget* **2016**, *7*, 6626–6638. [CrossRef] [PubMed]
114. Liu, C.-C.; Lin, J.-H.; Hsu, T.-W.; Su, K.; Li, A.F.-Y.; Hsu, H.-S.; Hung, S.-C. IL-6 enriched lung cancer stem-like cell population by inhibition of cell cycle regulators via DNMT1 upregulation. *Int. J. Cancer* **2015**, *136*, 547–559. [CrossRef] [PubMed]
115. Wang, R.; Yang, L.; Zhang, C.; Wang, R.; Zhang, Z.; He, Q.; Chen, X.; Zhang, B.; Qin, Z.; Wang, L.; et al. Th17 cell-derived IL-17A promoted tumor progression via STAT3/NF-κB/Notch1 signaling in non-small cell lung cancer. *Oncoimmunology* **2018**, *7*, e1461303. [CrossRef]
116. Lu, H.; Yeh, D.-W.; Lai, C.-Y.; Liu, Y.-L.; Huang, L.-R.; Lee, A.Y.-L.; Jin, S.-L.C.; Chuang, T.-H. USP17 mediates macrophage-promoted inflammation and stemness in lung cancer cells by regulating TRAF2/TRAF3 complex formation. *Oncogene* **2018**, *37*, 6327–6340. [CrossRef]
117. Zakaria, N.; Mohd Yusoff, N.; Zakaria, Z.; Widera, D.; Yahaya, B.H. Inhibition of NF-κB signaling reduces the stemness characteristics of lung cancer stem cells. *Front. Oncol.* **2018**, *8*, 166. [CrossRef]
118. Zeni, E.; Mazzetti, L.; Miotto, D.; Lo Cascio, N.; Maestrelli, P.; Querzoli, P.; Pedriali, M.; De Rosa, E.; Fabbri, L.M.; Mapp, C.E.; et al. Macrophage expression of interleukin-10 is a prognostic factor in nonsmall cell lung cancer. *Eur. Respir. J.* **2007**, *30*, 627–632. [CrossRef] [PubMed]
119. Wu, Y.-C.; Tang, S.-J.; Sun, G.-H.; Sun, K.-H. CXCR7 mediates TGFβ1-promoted EMT and tumor-initiating features in lung cancer. *Oncogene* **2016**, *35*, 2123–2132. [CrossRef]
120. Chang, Y.-W.; Su, Y.-J.; Hsiao, M.; Wei, K.-C.; Lin, W.-H.; Liang, C.-J.; Chen, S.-C.; Lee, J.-L. Diverse targets of β-catenin during the epithelial–mesenchymal transition define cancer stem cells and predict disease relapse. *Cancer Res.* **2015**, *75*, 3398–3410. [CrossRef]
121. Peng, D.; Fu, M.; Wang, M.; Wei, Y.; Wei, X. Targeting TGF-β signal transduction for fibrosis and cancer therapy. *Mol. Cancer* **2022**, *21*, 104. [CrossRef] [PubMed]
122. Uckun, F.M.; Qazi, S.; Hwang, L.; Trieu, V.N. Recurrent or Refractory High-Grade Gliomas Treated by Convection-Enhanced Delivery of a TGFβ2-Targeting RNA Therapeutic: A Post-Hoc Analysis with Long-Term Follow-Up. *Cancers* **2019**, *11*, 1891. [CrossRef] [PubMed]

123. Giaccone, G.; Bazhenova, L.A.; Nemunaitis, J.; Tan, M.; Juhász, E.; Ramlau, R.; van den Heuvel, M.M.; Lal, R.; Kloecker, G.H.; Eaton, K.D.; et al. A phase III study of belagenpumatucel-L, an allogeneic tumour cell vaccine, as maintenance therapy for non-small cell lung cancer. *Eur. J. Cancer* **2015**, *51*, 2321–2329. [CrossRef] [PubMed]
124. Brandes, A.A.; Carpentier, A.F.; Kesari, S.; Sepulveda-Sanchez, J.M.; Wheeler, H.R.; Chinot, O.; Cher, L.; Steinbach, J.P.; Capper, D.; Specenier, P.; et al. A Phase II randomized study of galunisertib monotherapy or galunisertib plus lomustine compared with lomustine monotherapy in patients with recurrent glioblastoma. *Neuro. Oncol.* **2016**, *18*, 1146–1156. [CrossRef]
125. Giannelli, G.; Santoro, A.; Kelley, R.K.; Gane, E.; Paradis, V.; Cleverly, A.; Smith, C.; Estrem, S.T.; Man, M.; Wang, S.; et al. Biomarkers and overall survival in patients with advanced hepatocellular carcinoma treated with TGF-βRI inhibitor galunisertib. *PLoS ONE* **2020**, *15*, e0222259. [CrossRef]
126. Huang, C.-Y.; Chung, C.-L.; Hu, T.-H.; Chen, J.-J.; Liu, P.-F.; Chen, C.-L. Recent progress in TGF-β inhibitors for cancer therapy. *Biomed. Pharmacother.* **2021**, *134*, 111046. [CrossRef]
127. Formenti, S.C.; Lee, P.; Adams, S.; Goldberg, J.D.; Li, X.; Xie, M.W.; Ratikan, J.A.; Felix, C.; Hwang, L.; Faull, K.F.; et al. Focal Irradiation and Systemic TGFβ Blockade in Metastatic Breast Cancer. *Clin. Cancer Res.* **2018**, *24*, 2493–2504. [CrossRef]
128. Guo, Q.; Liu, X.; Lu, L.; Yuan, H.; Wang, Y.; Chen, Z.; Ji, R.; Zhou, Y. Comprehensive evaluation of clinical efficacy and safety of celecoxib combined with chemotherapy in management of gastric cancer. *Medicine* **2017**, *96*, e8857. [CrossRef]
129. Legge, F.; Paglia, A.; D'Asta, M.; Fuoco, G.; Scambia, G.; Ferrandina, G. Phase II study of the combination carboplatin plus celecoxib in heavily pre-treated recurrent ovarian cancer patients. *BMC Cancer* **2011**, *11*, 214. [CrossRef]
130. Young, S.D.; Whissell, M.; Noble, J.C.S.; Cano, P.O.; Lopez, P.G.; Germond, C.J. Phase II Clinical Trial Results Involving Treatment with Low-Dose Daily Oral Cyclophosphamide, Weekly Vinblastine, and Rofecoxib in Patients with Advanced Solid Tumors. *Clin. Cancer Res.* **2006**, *12*, 3092–3098. [CrossRef]
131. Ishikawa, H.; Mutoh, M.; Sato, Y.; Doyama, H.; Tajika, M.; Tanaka, S.; Horimatsu, T.; Takeuchi, Y.; Kashida, H.; Tashiro, J.; et al. Chemoprevention with low-dose aspirin, mesalazine, or both in patients with familial adenomatous polyposis without previous colectomy (J-FAPP Study IV): A multicentre, double-blind, randomised, two-by-two factorial design trial. *Lancet Gastroenterol. Hepatol.* **2021**, *6*, 474–481. [CrossRef] [PubMed]
132. Lu, J.; Kornmann, M.; Traub, B. Role of Epithelial to Mesenchymal Transition in Colorectal Cancer. *Int. J. Mol. Sci.* **2023**, *24*, 14815. [CrossRef] [PubMed]
133. Cai, R.; Liao, X.; Li, G.; Xiang, J.; Ye, Q.; Chen, M.; Feng, S. The use of non-steroid anti-inflammatory drugs during radical resection correlated with the outcome in non-small cell lung cancer. *World J. Surg. Oncol.* **2023**, *21*, 358. [CrossRef]
134. Pretre, V.; Papadopoulos, D.; Regard, J.; Pelletier, M.; Woo, J. Interleukin-1 (IL-1) and the inflammasome in cancer. *Cytokine* **2022**, *153*, 155850. [CrossRef]
135. Ridker, P.M.; MacFadyen, J.G.; Thuren, T.; Everett, B.M.; Libby, P.; Glynn, R.J.; Ridker, P.; Lorenzatti, A.; Krum, H.; Varigos, J.; et al. Effect of interleukin-1β inhibition with canakinumab on incident lung cancer in patients with atherosclerosis: Exploratory results from a randomised, double-blind, placebo-controlled trial. *Lancet* **2017**, *390*, 1833–1842. [CrossRef]
136. Xu, Y.; Lou, Z.; Lee, S.-H. Arctigenin represses TGF-β-induced epithelial mesenchymal transition in human lung cancer cells. *Biochem. Biophys. Res. Commun.* **2017**, *493*, 934–939. [CrossRef] [PubMed]
137. Malyla, V.; De Rubis, G.; Paudel, K.R.; Chellappan, D.K.; Hansbro, N.G.; Hansbro, P.M.; Dua, K. Berberine nanostructures attenuate ß-catenin, a key component of epithelial mesenchymal transition in lung adenocarcinoma. *Naunyn. Schmiedebergs. Arch. Pharmacol.* **2023**, *396*, 3595–3603. [CrossRef]
138. Hussain, Y.; Cui, J.H.; Khan, H.; Aschner, M.; Batiha, G.E.-S.; Jeandet, P. Luteolin and cancer metastasis suppression: Focus on the role of epithelial to mesenchymal transition. *Med. Oncol.* **2021**, *38*, 66. [CrossRef]
139. Choudhury, P.; Barua, A.; Roy, A.; Pattanayak, R.; Bhattacharyya, M.; Saha, P. Eugenol emerges as an elixir by targeting β-catenin, the central cancer stem cell regulator in lung carcinogenesis: An in vivo and in vitro rationale. *Food Funct.* **2021**, *12*, 1063–1078. [CrossRef]
140. Wang, H.; Zhang, H.; Tang, L.; Chen, H.; Wu, C.; Zhao, M.; Yang, Y.; Chen, X.; Liu, G. Resveratrol inhibits TGF-β1-induced epithelial-to-mesenchymal transition and suppresses lung cancer invasion and metastasis. *Toxicology* **2013**, *303*, 139–146. [CrossRef]
141. Bahrami, A.; Majeed, M.; Sahebkar, A. Curcumin: A potent agent to reverse epithelial-to-mesenchymal transition. *Cell. Oncol.* **2019**, *42*, 405–421. [CrossRef]
142. Ang, H.L.; Mohan, C.D.; Shanmugam, M.K.; Leong, H.C.; Makvandi, P.; Rangappa, K.S.; Bishayee, A.; Kumar, A.P.; Sethi, G. Mechanism of epithelial-mesenchymal transition in cancer and its regulation by natural compounds. *Med. Res. Rev.* **2023**, *43*, 1141–1200. [CrossRef]
143. Anwar, S.; Malik, J.A.; Ahmed, S.; Kameshwar, V.A.; Alanazi, J.; Alamri, A.; Ahemad, N. Can Natural Products Targeting EMT Serve as the Future Anticancer Therapeutics? *Molecules* **2022**, *27*, 7668. [CrossRef]
144. Nan, Y.; Su, H.; Zhou, B.; Liu, S. The function of natural compounds in important anticancer mechanisms. *Front. Oncol.* **2023**, *12*, 1049888. [CrossRef] [PubMed]
145. Li, Y.; Wang, J.; Li, L.; Song, W.; Li, M.; Hua, X.; Wang, Y.; Yuan, J.; Xue, Z. Natural products of pentacyclic triterpenoids: From discovery to heterologous biosynthesis. *Nat. Prod. Rep.* **2023**, *40*, 1303–1353. [CrossRef] [PubMed]
146. Lou, H.; Li, H.; Zhang, S.; Lu, H.; Chen, Q. A Review on Preparation of Betulinic Acid and Its Biological Activities. *Molecules* **2021**, *26*, 5583. [CrossRef]

147. Gudoityte, E.; Arandarcikaite, O.; Mazeikiene, I.; Bendokas, V.; Liobikas, J. Ursolic and oleanolic acids: Plant metabolites with neuroprotective potential. *Int. J. Mol. Sci.* **2021**, *22*, 4599. [CrossRef] [PubMed]
148. Shi, Y.-S.; Xia, H.-M.; Wu, C.-H.; Li, C.-B.; Duan, C.-C.; Che, C.; Zhang, X.-J.; Li, H.-T.; Zhang, Y.; Zhang, X.-F. Novel nortriterpenoids with new skeletons and limonoids from the fruits of Evodia rutaecarpa and their bioactivities. *Fitoterapia* **2020**, *142*, 104503. [CrossRef] [PubMed]
149. Gui, Y.; Sun, J.; You, W.; Wei, Y.; Tian, H.; Jiang, S. Glycyrrhizin suppresses epithelial-mesenchymal transition by inhibiting high-mobility group box1 via the TGF-β1/Smad2/3 pathway in lung epithelial cells. *PeerJ* **2020**, *8*, e8514. [CrossRef] [PubMed]
150. Ayeleso, T.B.; Matumba, M.G.; Mukwevho, E. Oleanolic Acid and Its Derivatives: Biological Activities and Therapeutic Potential in Chronic Diseases. *Molecules* **2017**, *22*, 1915. [CrossRef]
151. Li, J.; Guo, Q.; Lei, X.; Zhang, L.; Su, C.; Liu, Y.; Zhou, W.; Chen, H.; Wang, H.; Wang, F.; et al. Pristimerin induces apoptosis and inhibits proliferation, migration in H1299 Lung Cancer Cells. *J. Cancer* **2020**, *11*, 6348–6355. [CrossRef]
152. Kang, H.; Lee, M.; Jang, S.W. Celastrol inhibits TGF-β1-induced epithelial-mesenchymal transition by inhibiting Snail and regulating E-cadherin expression. *Biochem. Biophys. Res. Commun.* **2013**, *437*, 550–556. [CrossRef] [PubMed]
153. Mannino, G.; Iovino, P.; Lauria, A.; Genova, T.; Asteggiano, A.; Notarbartolo, M.; Porcu, A.; Serio, G.; Chinigò, G.; Occhipinti, A.; et al. Bioactive Triterpenes of Protium heptaphyllum Gum Resin Extract Display Cholesterol-Lowering Potential. *Int. J. Mol. Sci.* **2021**, *22*, 2664. [CrossRef] [PubMed]
154. Oboh, M.; Govender, L.; Siwela, M.; Mkhwanazi, B.N. Anti-Diabetic Potential of Plant-Based Pentacyclic Triterpene Derivatives: Progress Made to Improve Efficacy and Bioavailability. *Molecules* **2021**, *26*, 7243. [CrossRef] [PubMed]
155. Colla, A.R.S.; Rosa, J.M.; Cunha, M.P.; Rodrigues, A.L.S. Anxiolytic-like effects of ursolic acid in mice. *Eur. J. Pharmacol.* **2015**, *758*, 171–176. [CrossRef] [PubMed]
156. Maia, J.L.; Lima-Júnior, R.C.P.; David, J.P.; David, J.M.; Santos, F.A.; Rao, V.S. Oleanolic acid, a pentacyclic triterpene attenuates the mustard oil-induced colonic nociception in mice. *Biol. Pharm. Bull.* **2006**, *29*, 82–85. [CrossRef] [PubMed]
157. Miranda, R.D.S.; de Jesus, B.D.S.M.; da Silva Luiz, S.R.; Viana, C.B.; Adão Malafaia, C.R.; Figueiredo, F.D.S.; Carvalho, T.d.S.C.; Silva, M.L.; Londero, V.S.; da Costa-Silva, T.A.; et al. Antiinflammatory activity of natural triterpenes—An overview from 2006 to 2021. *Phyther. Res.* **2022**, *36*, 1459–1506. [CrossRef]
158. Özdemir, Z.; Wimmer, Z. Selected plant triterpenoids and their amide derivatives in cancer treatment: A review. *Phytochemistry* **2022**, *203*, 113340. [CrossRef]
159. Borella, R.; Forti, L.; Gibellini, L.; De Gaetano, A.; De Biasi, S.; Nasi, M.; Cossarizza, A.; Pinti, M. Synthesis and anticancer activity of CDDO and CDDO-me, two derivatives of natural triterpenoids. *Molecules* **2019**, *24*, 4097. [CrossRef]
160. Zhu, B.; Wei, Y. Antitumor activity of celastrol by inhibition of proliferation, invasion, and migration in cholangiocarcinoma via PTEN/PI3K/Akt pathway. *Cancer Med.* **2020**, *9*, 783–796. [CrossRef]
161. Wang, H.; Zhong, W.; Zhao, J.; Zhang, H.; Zhang, Q.; Liang, Y.; Chen, S.; Liu, H.; Zong, S.; Tian, Y.; et al. Oleanolic acid inhibits epithelial–mesenchymal transition of hepatocellular carcinoma by promoting iNOS dimerization. *Mol. Cancer Ther.* **2019**, *18*, 62–74. [CrossRef]
162. Cao, M.; Xiao, D.; Ding, X. The anti-tumor effect of ursolic acid on papillary thyroid carcinoma via suppressing Fibronectin-1. *Biosci. Biotechnol. Biochem.* **2020**, *84*, 2415–2424. [CrossRef] [PubMed]
163. Zheng, Y.; Liu, P.; Wang, N.; Wang, S.; Yang, B.; Li, M.; Chen, J.; Situ, H.; Xie, M.; Lin, Y.; et al. Betulinic Acid Suppresses Breast Cancer Metastasis by Targeting GRP78-Mediated Glycolysis and ER Stress Apoptotic Pathway. *Oxid. Med. Cell. Longev.* **2019**, *2019*, 8781690. [CrossRef]
164. Cui, Q.; Ren, J.; Zhou, Q.; Yang, Q.; Li, B. Effect of asiatic acid on epithelial-mesenchymal transition of human alveolar epithelium A549 cells induced by TGF-β1. *Oncol. Lett.* **2019**, *17*, 4285–4292. [CrossRef]
165. Dong, S.-H.H.; Liu, Y.-W.W.; Wei, F.; Tan, H.-Z.Z.; Han, Z.-D.D. Asiatic acid ameliorates pulmonary fibrosis induced by bleomycin (BLM) via suppressing pro-fibrotic and inflammatory signaling pathways. *Biomed. Pharmacother.* **2017**, *89*, 1297–1309. [CrossRef] [PubMed]
166. Liu, K.; Guo, L.; Miao, L.; Bao, W.; Yang, J.; Li, X.; Xi, T.; Zhao, W. Ursolic acid inhibits epithelial-mesenchymal transition by suppressing the expression of astrocyte-elevated gene-1 in human nonsmall cell lung cancer A549 cells. *Anticancer Drugs* **2013**, *24*, 494–503. [CrossRef] [PubMed]
167. Ruan, J.S.; Zhou, H.; Yang, L.; Wang, L.; Jiang, Z.S.; Sun, H.; Wang, S.M. Ursolic acid attenuates TGF-b1-induced epithelial-mesenchymal transition in NSCLC by targeting integrin Avb5/MMPs signaling. *Oncol. Res.* **2019**, *27*, 593–600. [CrossRef]
168. Lin, L.; Hou, G.; Han, D.; Yin, Y.; Kang, J.; Wang, Q. Ursolic acid alleviates airway-vessel remodeling and muscle consumption in cigarette smoke-induced emphysema rats. *BMC Pulm. Med.* **2019**, *19*, 103. [CrossRef]
169. Lin, L.; Hou, G.; Han, D.; Kang, J.; Wang, Q. Ursolic Acid Protected Lung of Rats From Damage Induced by Cigarette Smoke Extract. *Front. Pharmacol.* **2019**, *10*, 700. [CrossRef]
170. Kang, D.Y.; Sp, N.; Lee, J.-M.; Jang, K.-J. Antitumor effects of ursolic acid through mediating the inhibition of STAT3/PD-L1 signaling in non-small cell lung cancer cells. *Biomedicines* **2021**, *9*, 297. [CrossRef]
171. Yang, K.; Chen, Y.; Zhou, J.; Ma, L.; Shan, Y.; Cheng, X.; Wang, Y.; Zhang, Z.; Ji, X.; Chen, L.; et al. Ursolic acid promotes apoptosis and mediates transcriptional suppression of CT45A2 gene expression in non-small-cell lung carcinoma harbouring EGFR T790M mutations. *Br. J. Pharmacol.* **2019**, *176*, 4609–4624. [CrossRef]

172. Son, J.; Lee, S.Y. Ursonic acid exerts inhibitory effects on matrix metalloproteinases via ERK signaling pathway. *Chem. Biol. Interact.* **2020**, *315*, 108910. [CrossRef]
173. Wang, Y.-Y.; Zhang, C.-Y.; Ma, Y.-Q.; He, Z.-X.; Zhe, H.; Zhou, S.-F. Therapeutic effects of C-28 methyl ester of 2-cyano-3,12-dioxoolean-1,9-dien-28-oic acid (CDDO-Me; bardoxolone methyl) on radiation-induced lung inflammation and fibrosis in mice. *Drug Des. Dev. Ther.* **2015**, *9*, 3163–3178. [CrossRef]
174. Divya, T.; Velavan, B.; Sudhandiran, G. Regulation of Transforming Growth Factor-β/Smad-mediated Epithelial–Mesenchymal Transition by Celastrol Provides Protection against Bleomycin-induced Pulmonary Fibrosis. *Basic Clin. Pharmacol. Toxicol.* **2018**, *123*, 122–129. [CrossRef] [PubMed]
175. Wu, H.; Zhong, Q.; Zhong, R.; Huang, H.; Xia, Z.; Ke, Z.; Zhang, Z.; Song, J.; Jia, X. Preparation and antitumor evaluation of self-assembling oleanolic acid-loaded Pluronic P105/d-α-tocopheryl polyethylene glycol succinate mixed micelles for non-small-cell lung cancer treatment. *Int. J. Nanomed.* **2016**, *11*, 6337–6352. [CrossRef]
176. Markov, A.V.; Odarenko, K.V.; Sen'kova, A.V.; Salomatina, O.V.; Salakhutdinov, N.F.; Zenkova, M.A. Cyano enone-bearing triterpenoid soloxolone methyl inhibits epithelial-mesenchymal transition of human lung adenocarcinoma cells in vitro and metastasis of murine melanoma in vivo. *Molecules* **2020**, *25*, 5925. [CrossRef]
177. Patlolla, J.M.R.; Qian, L.; Biddick, L.; Zhang, Y.; Desai, D.; Amin, S.; Lightfoot, S.; Rao, C. V β-Escin inhibits NNK-induced lung adenocarcinoma and ALDH1A1 and RhoA/Rock expression in A/J mice and growth of H460 human lung cancer cells. *Cancer Prev. Res.* **2013**, *6*, 1140–1149. [CrossRef]
178. Bang, I.J.; Kim, H.R.; Jeon, Y.; Jeong, M.H.; Park, Y.J.; Kwak, J.H.; Chung, K.H. β-Peltoboykinolic Acid from Astilbe rubra Attenuates TGF-β1-Induced Epithelial-to-Mesenchymal Transitions in Lung Alveolar Epithelial Cells. *Molecules* **2019**, *24*, 2573. [CrossRef] [PubMed]
179. He, D.-H.; Chen, Y.-F.; Zhou, Y.-L.; Zhang, S.-B.; Hong, M.; Yu, X.; Wei, S.-F.; Fan, X.-Z.; Li, S.-Y.; Wang, Q.; et al. Phytochemical library screening reveals betulinic acid as a novel Skp2-SCF E3 ligase inhibitor in non–small cell lung cancer. *Cancer Sci.* **2021**, *112*, 3218–3232. [CrossRef]
180. Hsu, T.-I.I.; Chen, Y.-J.J.; Hung, C.-Y.Y.; Wang, Y.-C.C.; Lin, S.-J.J.; Su, W.-C.C.; Lai, M.-D.D.; Kim, S.-Y.Y.; Wang, Q.; Qian, K.; et al. A novel derivative of betulinic acid, SYK023, suppresses lung cancer growth and malignancy. *Oncotarget* **2015**, *6*, 13671–13687. [CrossRef]
181. Pansters, N.A.; Langen, R.C.; Wouters, E.F.; Schols, A.M. Synergistic stimulation of myogenesis by glucocorticoid and IGF-I signaling. *J. Appl. Physiol.* **2013**, *114*, 1329–1339. [CrossRef] [PubMed]
182. Hance, M.W.; Dole, K.; Gopal, U.; Bohonowych, J.E.; Jezierska-Drutel, A.; Neumann, C.A.; Liu, H.; Garraway, I.P.; Isaacs, J.S. Secreted Hsp90 is a novel regulator of the epithelial to mesenchymal transition (EMT) in prostate cancer. *J. Biol. Chem.* **2012**, *287*, 37732–37744. [CrossRef] [PubMed]
183. Salomatina, O.V.; Markov, A.V.; Logashenko, E.B.; Korchagina, D.V.; Zenkova, M.A.; Salakhutdinov, N.F.; Vlassov, V.V.; Tolstikov, G.A. Synthesis of novel 2-cyano substituted glycyrrhetinic acid derivatives as inhibitors of cancer cells growth and NO production in LPS-activated J-774 cells. *Bioorg. Med. Chem.* **2014**, *22*, 585–593. [CrossRef] [PubMed]
184. Asanuma, K.; Yanagida-Asanuma, E.; Faul, C.; Tomino, Y.; Kim, K.; Mundel, P. Synaptopodin orchestrates actin organization and cell motility via regulation of RhoA signalling. *Nat. Cell Biol.* **2006**, *8*, 485–491. [CrossRef]
185. Zhang, W.; Ren, Z.; Jia, L.; Li, X.; Jia, X.; Han, Y. Fbxw7 and Skp2 regulate stem cell switch between quiescence and mitotic division in lung adenocarcinoma. *BioMed Res. Int.* **2019**, *2019*, 9648269. [CrossRef] [PubMed]
186. Vassalli, G. Aldehyde dehydrogenases: Not just markers, but functional regulators of stem cells. *Stem Cells Int.* **2019**, *2019*, 3904645. [CrossRef]

Disclaimer/Publisher's Note: The statements, opinions and data contained in all publications are solely those of the individual author(s) and contributor(s) and not of MDPI and/or the editor(s). MDPI and/or the editor(s) disclaim responsibility for any injury to people or property resulting from any ideas, methods, instructions or products referred to in the content.

Article

Machine Learning and Single-Cell Analysis Identify Molecular Features of IPF-Associated Fibroblast Subtypes and Their Implications on IPF Prognosis

Jiwei Hou *, Yanru Yang and Xin Han *

Department of Biochemistry and Molecular Biology, School of Medicine & Holistic Integrative Medicine, Jiangsu Collaborative Innovation Canter of Chinese Medicinal Resources Industrialization, Nanjing University of Chinese Medicine, Nanjing 210023, China; yrr@njucm.edu.cn
* Correspondence: houjw@njucm.edu.cn (J.H.); xhan0220@njucm.edu.cn (X.H.)

Abstract: Idiopathic pulmonary fibrosis (IPF) is a devastating lung disease of unknown cause, and the involvement of fibroblasts in its pathogenesis is well recognized. However, a comprehensive understanding of fibroblasts' heterogeneity, their molecular characteristics, and their clinical relevance in IPF is lacking. In this study, we aimed to systematically classify fibroblast populations, uncover the molecular and biological features of fibroblast subtypes in fibrotic lung tissue, and establish an IPF-associated, fibroblast-related predictive model for IPF. Herein, a meticulous analysis of scRNA-seq data obtained from lung tissues of both normal and IPF patients was conducted to identify fibroblast subpopulations in fibrotic lung tissues. In addition, hdWGCNA was utilized to identify co-expressed gene modules associated with IPF-related fibroblasts. Furthermore, we explored the prognostic utility of signature genes for these IPF-related fibroblast subtypes using a machine learning-based approach. Two predominant fibroblast subpopulations, termed IPF-related fibroblasts, were identified in fibrotic lung tissues. Additionally, we identified co-expressed gene modules that are closely associated with IPF-fibroblasts by utilizing hdWGCNA. We identified gene signatures that hold promise as prognostic markers in IPF. Moreover, we constructed a predictive model specifically focused on IPF-fibroblasts which can be utilized to assess disease prognosis in IPF patients. These findings have the potential to improve disease prediction and facilitate targeted interventions for patients with IPF.

Citation: Hou, J.; Yang, Y.; Han, X. Machine Learning and Single-Cell Analysis Identify Molecular Features of IPF-Associated Fibroblast Subtypes and Their Implications on IPF Prognosis. *Int. J. Mol. Sci.* **2024**, 25, 94. https://doi.org/10.3390/ijms25010094

Academic Editors: Sabrina Lisi and Margherita Sisto

Received: 30 October 2023
Revised: 14 December 2023
Accepted: 16 December 2023
Published: 20 December 2023

Copyright: © 2023 by the authors. Licensee MDPI, Basel, Switzerland. This article is an open access article distributed under the terms and conditions of the Creative Commons Attribution (CC BY) license (https://creativecommons.org/licenses/by/4.0/).

Keywords: idiopathic pulmonary fibrosis; fibroblast; bioinformatics; heterogeneity; predictive model

1. Introduction

Idiopathic pulmonary fibrosis (IPF) is a parenchymal lung disease characterized by fibroblast proliferation and excessive accumulation of extracellular matrix (ECM) [1,2]. Unfortunately, the prognosis for patients with IPF is poor, with a median survival of approximately 3 years after diagnosis and limited treatment options available [3,4]. The underlying causes and mechanisms of fibrotic lung diseases, including IPF, are still not fully understood, and effective radical therapies are yet to be developed.

Numerous cell types, including alveolar epithelial cells, endothelial cells, immune cells, and fibroblasts, have been identified as contributors to fibrosis [2,5,6]. Among these, fibroblasts play a central role in the process of fibrogenesis, leading to the accumulation of extracellular matrix (ECM) and compromising lung structure and function [7]. In fibrotic lung tissue, fibroblasts demonstrate enhanced proliferative potential, increased migration, resistance to apoptosis, and invasive capacity, as well as leading to heightened deposition of ECM [8–10]. These characteristics significantly contribute to the pathogenesis of fibrosis, highlighting their potential value as both prognostic factors and therapeutic targets [11]. Pulmonary fibroblasts exhibit functional heterogeneity in lung homeostasis and disease [12]. Growing evidence suggests that specific subsets of fibroblasts actively contribute to lung pathophysiology by modulating the local immune microenvironment and producing ECM proteins [13–15]. The identification of these pathogenic fibroblast subsets presents new therapeutic possibilities for various fibrotic diseases.

In recent years, the application of single-cell RNA-sequencing (scRNA-seq) technology has significantly advanced our understanding of cellular heterogeneity in various pathological tissues [16,17]. Through this high-throughput analysis technique, the transcriptomic characteristics of fibroblasts in both normal and fibrotic lung tissues have been described [18]. However, there is still a lack of comprehensive understanding regarding the composition of fibroblast subsets, their gene expression profiling, and their specific functions in fibrotic lung tissue. In addition, the clinical association of fibroblast subtypes and their prognostic value for fibrogenesis remains to be illustrated. Here, we hypothesize that an integrated scRNA-seq analysis of lung fibroblasts can offer a more thorough characterization of fibroblast subtypes, novel insights into their biological characteristics, and signaling pathways they may activate in fibrotic lung tissue, which in turn may have an impact on clinical outcome.

Machine learning, a data analysis method that automatically constructs analytical models, has been widely utilized in clinical medicine [19]. Previous studies have demonstrated its potential in designing drugs, identifying pathologies, and developing predictive models [20–22]. In recent years, machine learning has been applied to diagnose and treat various diseases, including pulmonary fibrosis [23–25]. In these applications, least absolute shrinkage and selection operator (LASSO) logistic regression analysis, a linear regression method with regularization, is commonly utilized for high-dimensional analysis. Additionally, support vector machine–recursive feature elimination (SVM-RFE) can be employed to select optimal combinations of variables by leveraging its non-linear discrimination capabilities and ability to model different variable quantities [26]. Hence, the identification of biomarkers for idiopathic pulmonary fibrosis and the construction of prediction models using machine learning algorithms are of significant importance.

In the present study, we systematically classified fibroblast populations and revealed the molecular and biological characteristics of fibroblast subtypes in fibrotic lung tissue. Using hdWGCNA, we further identified the co-expressed gene modules associated with IPF-fibroblasts and addressed the valuable prognostic utility of signature genes for these IPF-related subtypes through a machine learning-based approach, providing valuable assistance for disease prediction and intervention.

2. Results

2.1. Single-Cell RNA Sequencing Reveals the Cellular Heterogeneity of Fibroblasts in Normal and Fibrotic Lung Tissues

To explore the cellular composition and diversity of fibroblasts in both normal and fibrotic lung tissues, we collected and analyzed scRNA-seq data from patients with IPF. Specifically, we selected sequencing data from both normal and lower lobe samples, as pulmonary fibrosis often initiates in the lower lobe in clinical practice. Uniform manifold approximation and projection (UMAP) analyses identified eight major cell populations, including endothelial cells, epithelial cells, macrophages, monocytes, NK cells, fibroblasts, T cells, and tissue stem cells (Figure 1A,B). Figure 1C illustrates the expression of known lineage markers in the eight major cell clusters in both the normal and IPF groups. We next investigated the proportion of each cell population in the different sample sets (Figure 1D). As expected, we observed a significant increase in the fraction of the fibroblast cluster in the IPF group compared with the normal group.

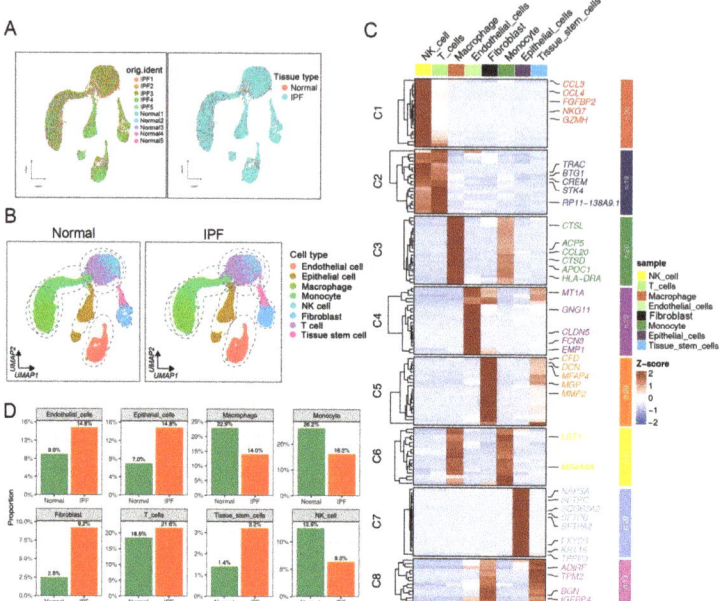

Figure 1. Integrated scRNA-seq analysis reveals heterogeneity of normal and fibrotic lung tissues, according to dataset GSE128033. (**A**) Cells on the UMAP plot of all 10 samples were colored as originating from normal and IPF patients. (**B**) Unbiased clustering of 26,129 cells reveals eight cellular clusters. Clusters are distinguished by different colors. (**C**) Heatmap showing representative differentially expressed genes between each cell population. (**D**) Cell proportions of eight cell types originating from normal and fibrotic lung tissues.

Then, we repeated the UMAP analysis to hierarchically cluster the fibroblasts. As shown in Figure S1, subclustering of fibrotic and normal lung fibroblasts further identified 10 distinct subtypes. We demonstrated that fibroblasts of cluster 4 and cluster 5 were significantly increased in fibrotic samples compared to normal lung samples, which were defined as IPF-fibroblast (Figure 2A,B). To more precisely characterize the distinctive patterns of differentially expressed gene signatures in these cell subtypes, a score was assigned to each gene based on its relative expression in each individual cell. These genes were then subjected to unsupervised clustering, resulting in the formation of distinct gene clusters (Figure 2C). In addition, we grouped genes with similar expression trends, resulting in the identification of nine distinct trends with implications in various biological functions, as revealed by the clustering results. It is noteworthy that genes in cluster 4, which exhibited high expression levels in IPF-fibroblasts, were found to be highly enriched for biological processes related to fibrogenesis, including extracellular matrix (ECM) organization, extracellular structure organization, and cellular response to transforming growth factor beta stimulus. (Figure 2C). Consistently, gene set enrichment analysis (GSEA) also suggested that IPF-fibroblasts expressed high levels of genes involved in the deposition of ECM, such as collagen fibril organization, collagen metabolic process, and collagen binding (Figure 2D). Moreover, the GSEA result showed a negative correlation between IPF-fibroblasts and pathways involving activation of immune response (Figure 2E). These findings suggest that IPF-fibroblasts play a critical role in the synthesis and production of extracellular matrix components within alveolar structures, indicating their potential culpability in the development and progression of IPF.

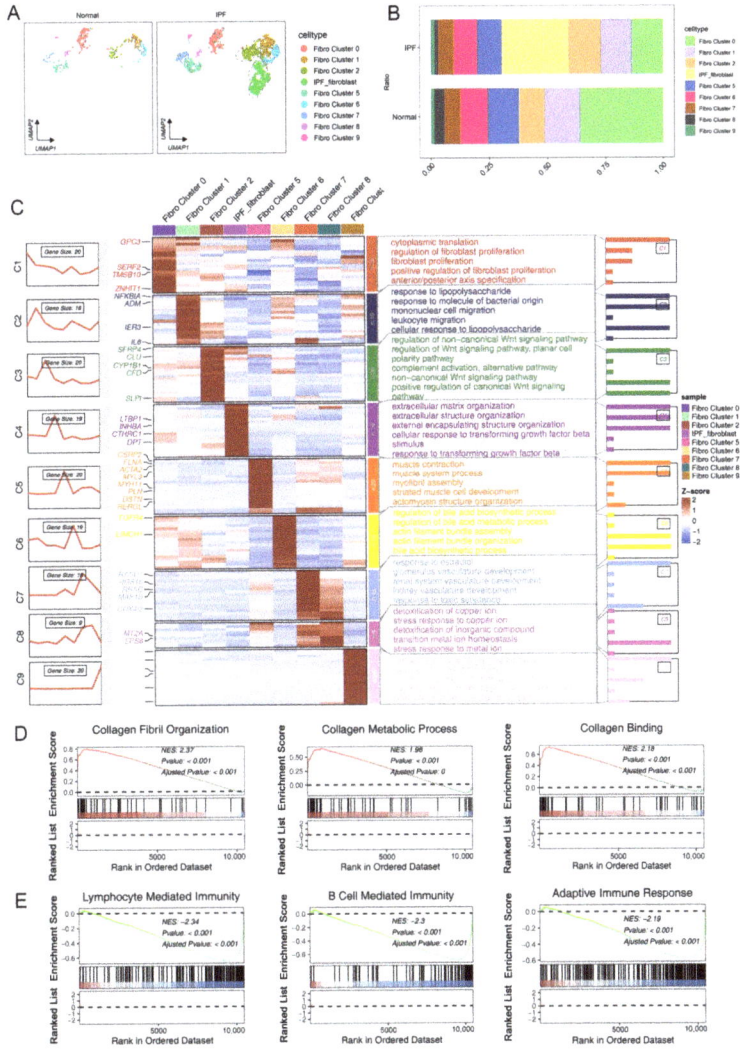

Figure 2. scRNA-seq analysis reveals heterogeneity of fibroblast subtypes in lung fibrosis. (**A**) Subclustering of fibrotic and normal lung fibroblasts further identified nine distinct subtypes. Each fibroblast subcluster is visualized via a color-coded UMAP plot. (**B**) Cell proportions of fibroblast subclusters in the lung tissues of normal and IPF patients. (**C**) Left panels: The series of diagrams illustrates the patterns of dynamic changes in representative differentially expressed genes (DEGs) in each fibroblast population. Middle panels: heatmap showing representative DEGs between each cell population. Right panels: representative enriched gene ontology (GO) terms for each cluster. (**D**,**E**) GSEA enrichment plots for representative signaling pathways upregulated (**D**) and downregulated (**E**) in IPF-related fibroblasts compared to other fibroblasts.

2.2. The Pseudotime Trajectory Analysis of Pathogenic Fibroblast Subtypes during Fibrogenesis

To investigate the origins of IPF-fibroblasts in the development of IPF, pseudotime trajectory analysis of fibroblasts was further performed. Fibroblast clusters 5 and 7 were observed at the start of the trajectory, whereas IPF-fibroblasts were found at both ends of trajectory branches 1 and 2 (Figure 3A–C). We conducted further analysis of the dynamic expression changes of genes along the trajectory to identify genes that are critical for fibrogenesis. We identified several genes that exhibited the most significant changes in expression during pseudotime progression in IPF-fibroblasts, including collagen triple helix repeat-containing protein 1 (CTHRC1), dermatopontin (DPT), inhibitor beta A chain (INHBA), and latent-transforming growth factor beta-binding protein 1 (LTBP1) (Figure 3D–F). Furthermore, we conducted clustering of the top 100 genes based on their pseudotemporal expression patterns and subsequently analyzed the functional enrichments of each resulting cluster. As a result, we identified five distinct patterns of gene expression changes over pseudotime (Figure 3G). The genes assigned to cluster 2 exhibited high expression levels during the end stage and were primarily associated with extracellular matrix (ECM) organization. Conversely, genes assigned to cluster 4 displayed high expression levels during the beginning stage and were mainly enriched in processes related to mesenchymal migration and differentiation of glomerular mesangial cells (Figure S2A). Subsequently, we attempted to uncover the molecular mechanisms that distinguished the two branches. Our analysis of the gene expression dynamics revealed that, in conjunction with the fate 2 branch, the genes assigned to cluster 2 that were activated towards the end of the transition were primarily associated with the gene ontology (GO) terms "response to cytokine", "negative regulation of apoptotic process", and "cell proliferation", all of which align with the characteristics of fibrotic differentiation (Figure S2B). Thus, these distinct gene expression patterns defined a successful IPF-fibroblast transition trajectory and highlighted a functional discrepancy within the pre-IPF-fibroblast subcluster.

2.3. Cell–Cell Communications Analyses in Lung Fibroblast Subpopulations

The availability of a single-cell dataset presented us with an exceptional opportunity to investigate cell–cell communication facilitated by ligand-receptor interactions. In order to elucidate the cell–cell communication network between fibroblast subpopulations and other cell types in fibrotic and normal lung tissues, we performed an analysis using CellChat, which was based on known ligand–receptor pairs and their cofactors [21]. Overall, IPF-fibroblasts exhibited strong communication abilities with other non-fibrotic cell types during the fibrogenesis process (Figure 4A). The results suggested that IPF-related fibroblasts exhibit a stronger secretory ability, as indicated by their higher levels of outgoing interaction strength compared to other fibroblasts (Figures 4B,C and S3). Notably, our study revealed that IPF-fibroblasts are capable of directly interacting with other fibroblasts through the adhesive ligand–receptor pairs CCL11/ACKR4 and CTGF/ITGA5 (Figure 4D,E). We further demonstrated at the single-cell level that NK cells transmit PARs and ADGRE5-dependent signaling to IPF-fibroblasts (Figure 4F,G). Additionally, our CellChat analysis revealed an upregulation of pro-fibrosis signaling (such as COLLAGEN and ANGPTL) in the communication between IPF-fibroblasts and other fibroblasts, as well as an increase in GDF signaling in the communication between epithelial cells and IPF-fibroblasts (Figure 4H–J). Collectively, our findings suggest that IPF-fibroblasts and other cell types establish an interaction network that supports each other's maintenance and function.

2.4. Identification of the Co-Expressed Gene Modules Associated with IPF-Fibroblasts by Using hdWGCNA

We utilized high-dimensional weighted gene co-expression network analysis (hdWGCNA), a comprehensive framework for co-expression network analysis in single-cell RNA sequencing data [22], to identify co-expressed gene modules and elucidate their functional roles within IPF-related fibroblasts. A scale-free co-expression network was established using an optimal soft thresholding power of 8 (Figure 5A). We identified a total of 12 distinct gene co-expression modules, among which the blue, purple, and magenta modules were highly activated, primarily in IPF-fibroblasts (clusters 3 and 4, as shown in Figure 5B,C). The correlation between each module was further investigated (Figure 5D–F). Figures 5G and S4 presented the top 10 hub genes of the 12 modules and the protein-protein interaction (PPI) network of the identified hub genes in each module, respectively.

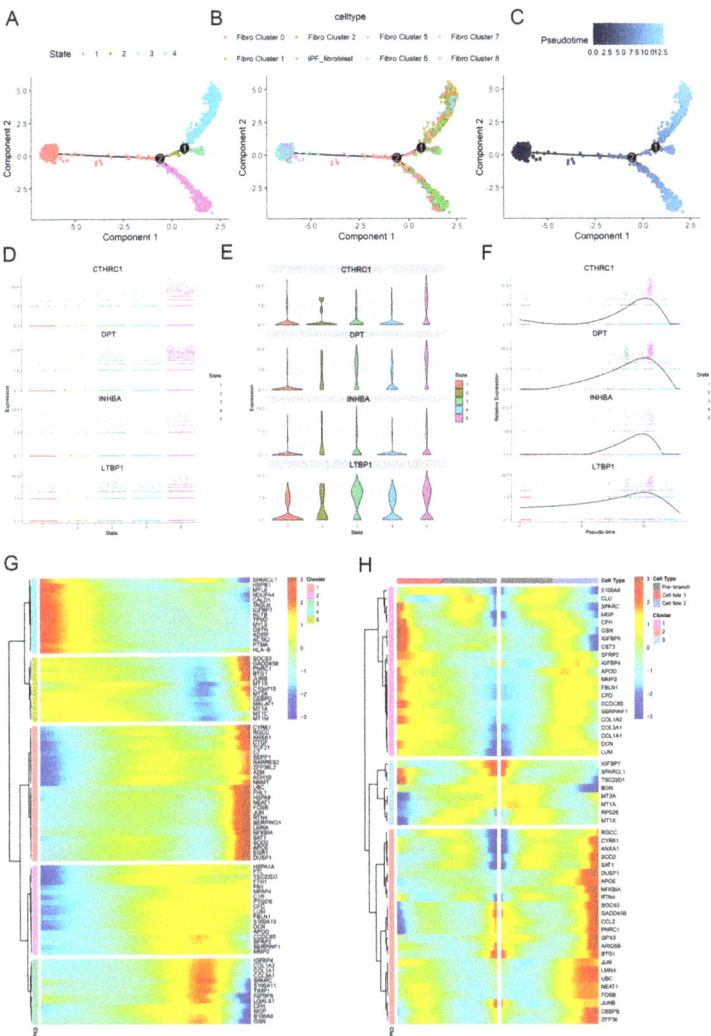

Figure 3. Trajectory analysis of fibroblast populations. (**A–C**) Monocle analysis for trajectory inference of the fibroblast subclusters, colored by cell cluster. The developmental trajectory of fibroblasts, color-coded by states (**A**), the associated cell subpopulations (**B**), and pseudotime (**C**). Scatter plots (**D**), violin plots (**E**), and pseudotime trajectories (**F**) show the expression of selected genes in different cell states as the pseudotime progresses. (**G**) A heatmap showing the dynamic changes in gene expression of the different cell clusters (**H**) The pseudotime heatmap shows the changes in selected genes after the changes in pseudotime.

2.5. Machine Learning-Based Construction of the IPF-Fibroblast-Related Predictive Model for IPF

Then, we focused on predicting the onset and progression of IPF using a predictive model based on hub genes from three IPF-fibroblast-related modules, which could differentiate IPF-related fibroblasts from other fibroblasts due to their gene significance. We utilized two bulk RNA-seq datasets for further analysis.

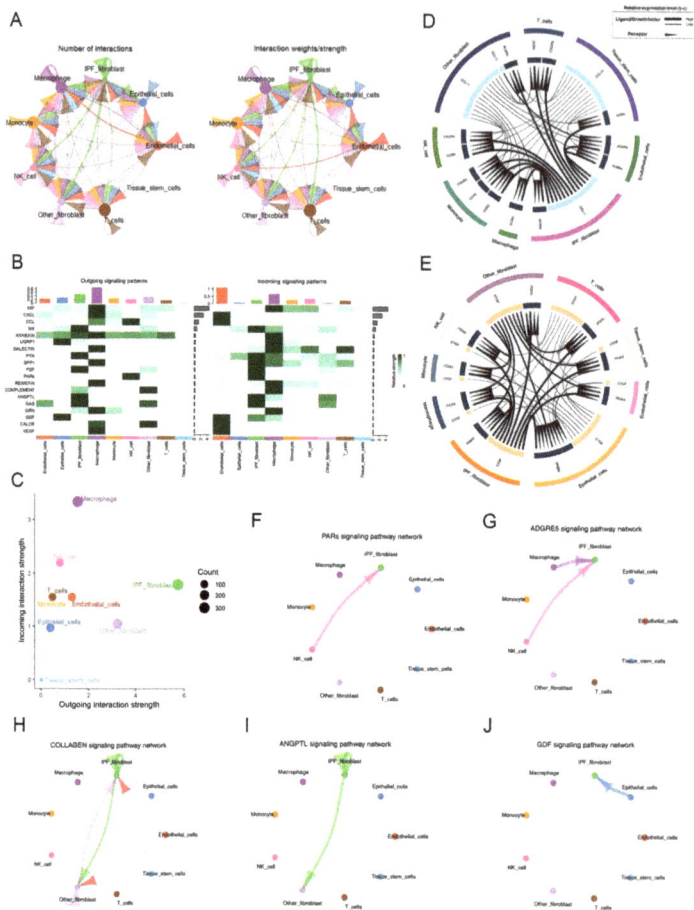

Figure 4. Cell communication analysis in fibroblast subpopulations. (**A**) Circle plots depict the number and strength of ligand–receptor interactions between pairs of cell populations. (**B**) The outgoing and incoming signaling patterns of fibroblasts and other cell populations. (**C**) A scatter plot reveals the variations in incoming and outgoing interaction strengths across all cell types. (**D**,**E**) Cell–cell ligand–receptor (LR) and cytokine-related pathway network in which fibroblasts interact with other cell populations in normal and fibrotic lung tissues. (**F–J**) Circle plots showing selected inferred differential signaling networks. The direction of the arrow indicates the direction of cell communication. The edge width represents the communication probability.

The GSE32537 dataset served as the training cohort, while the GSE14407 dataset was used to evaluate the predictive power of the final model developed. Initially, we conducted an analysis on the GSE32537 dataset, which consisted of 119 IPF patients. By employing the LASSO regression algorithm, we were able to identify 24 critical genes that exhibited a strong association with prognosis (Figure 6A,B). Subsequently, these identified genes underwent stepwise Cox proportional hazards regression, resulting in the final selection of 14 genes (Figure 6C). Next, we employed seven distinct machine learning algorithms and performed parameter optimization for each model using five repetitions of tenfold cross-validation. Subsequently, we evaluated the area under the curve (AUC) values of these models in the validation cohort. Through these rigorous mathematical procedures, we ultimately selected the "svm" machine learning algorithm model, which exhibited the highest AUC of 0.93 (Figure 6D,E). In addition, we constructed linear regression models to investigate the relationship between the expression levels of the 14 identified genes and lung function. Notably, Figure 6F demonstrates a significant negative correlation between the increased expression of CCDC80, COL6A1, CTHRC1, FBLN2, FSTL1, and GSN and a decline in the percent predicted diffusing capacity for lung carbon monoxide (% DLCO). The above results indicated the ideal predictive value of the IPF-fibroblast-related predictive model for IPF patients.

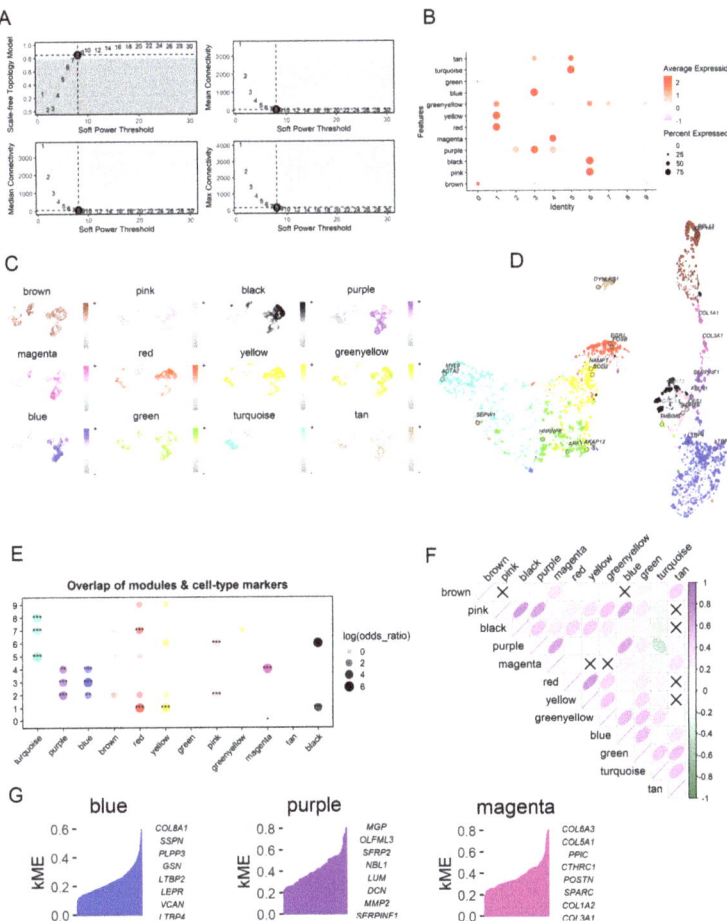

Figure 5. Identification of potential IPF-fibroblast-related genes associated with IPF by high-dimensional weighted gene co-expression network analysis (hdWGCNA). (**A**) The selection of soft-thresholding powers; left panels: the impact of soft-threshold power on the scale-free topology fit index; right panels: average network connectivity under different weighting coefficients. (**B**) Module activities in different fibroblast clusters. The hdWGCNA algorithm was used to estimate the module score. (**C**) UMAP plots as in Figure 2A, colored by MEs for the 12 co-expression gene modules. (**D**) An UMAP diagram Illustrating the co-expression network in fibroblasts. The edges show co-expression connections between genes and module hub genes, while each node represents a single gene. Point size is scaled by kME. Nodes are colored by co-expression module assignment. The top two hub genes per module are labeled. Network edges were downsampled for visual clarity. (**E**) Gene overlap within different modules (* $p < 0.05$, ** $p < 0.01$, *** $p < 0.001$). (**F**) The matrix plot visually represents the inter-module relationships by depicting the correlation between module eigengenes. (**G**) Three IPF-fibroblast-related gene modules were obtained, and the top hub gene was presented according to the hdWGCNA pipeline.

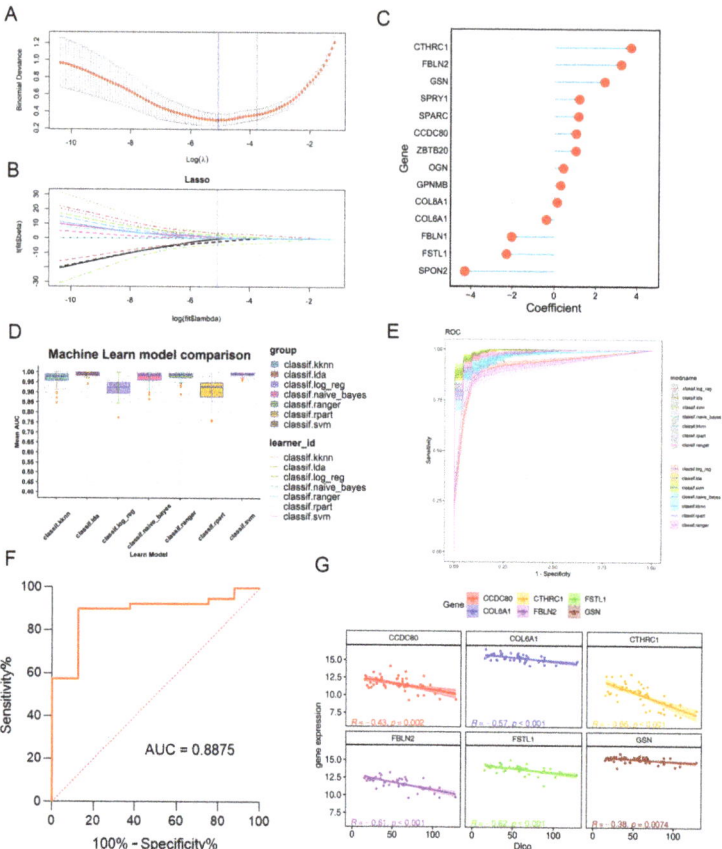

Figure 6. Construction of a machine learning model using bulk transcriptomic data. (**A,B**). LASSO regression was used to narrow down the IPF-fibroblast-related genes associated with IPF. (**A**) LASSO algorithm for selection features for IPF-fibroblasts. (**B**) Selection of genes with a non-zero coefficient for the construction of a model. (**C**) Coefficients of the identified genes within the prediction model. (**D**) Construction of the IPF-related fibroblast model through seven machine learning algorithms. (**E**) ROC values of all seven algorithms were shown in the training cohort. (**F**) ROC curve of the "svm" machine learning algorithm model in the validation cohort. (**G**) Correlation of IPF-fibroblast-related genes and diffusion lung capacity for CO (% DLCO) in control and IPF patients.

2.6. External Validation of an IPF-Fibroblast-Related Prognostic Signature

To gain a deeper understanding of the correlation between the gene signature of IPF-fibroblasts and the process of fibrogenesis, we conducted an analysis of the IPF-fibroblast-related gene (IFRG) scores in both normal and fibrotic lung tissue. The results revealed a noteworthy increase in the IFRG scores within IPF lung tissue, indicating a potential association between the IFRGs and fibrogenesis (Figures 7A and S5). We employed NMF clustering to classify IPF patients into two subtypes, namely, subtype 1 and subtype 2, using the clustering criteria derived from the differentially expressed IFRGs (Figure 7B). After comparing the two subtypes, we observed that patients with type 2 had a higher DLCO index, indicating better lung function in this group (Figure 7C). Given that fibroblasts associated with IPF may play a role in initiating and progressing the disease by releasing secretory proteins, we conducted a screening process using the IFRGs to identify these IPF-fibroblast-associated secretory proteins. Then, we evaluated the risk ratio for each secretory protein and identified five stable essential prognostic genes through multivariate Cox regression (Figure 7D). We displayed the distribution of risk scores based on pseudogenes, overall survival of IPF patients, and corresponding pseudogene expression profiles in another GEO dataset (GSE70866) to intuitively understand the prognostic effect of identified secretory protein-encoding genes. The results indicated that SPON2, FSTL1, CCDC80, COL8A1, and FBLN2 demonstrated high expressions in the high-risk group (Figure 7E). The patients in the high-risk group with the five-gene signature had a poor prognosis (Figure 7F). Furthermore, patients with IPF who exhibited elevated levels of these five gene expressions had shorter survival times (Figure 7J,K).

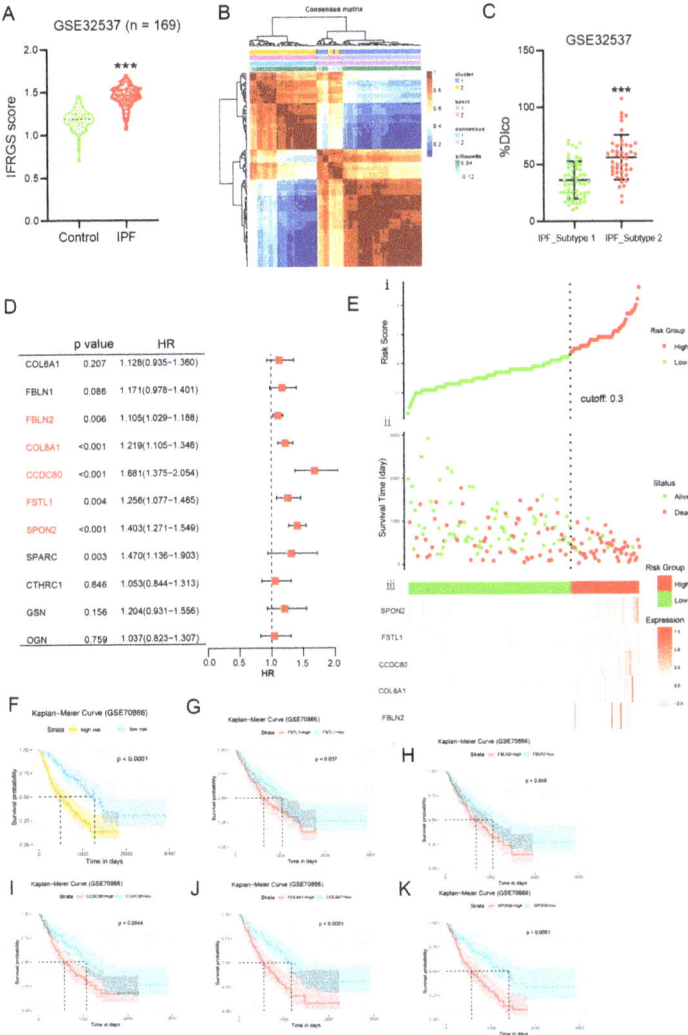

Figure 7. External validation of an IPF-fibroblast-related prognostic signature. (**A**) Boxplots depict the IPF-fibroblast-related genes (IFRGs) score levels in lung tissue of patients with IPF (n = 131) and normal controls (n = 39). Results are expressed as means ± SD (*** $p < 0.001$ vs. control). (**B**) A The heatmap of gene expression clusters for 131 IPF samples by unsupervised NMF illustrates two distinct expression patterns. (**C**) Box plots show the differences in diffusion lung capacity for CO (% DLCO)

between two clusters (means ± SD; *** $p < 0.001$ vs. IPF_subtype 1). (**D**) Univariate Cox analysis of 11 IFRGs encoding secretory proteins associated with overall survival. (**E**) The risk score distribution, patient status, and mRNA expression heatmaps of the prognostic five-gene risk signature. (**F**) Kaplan–Meier curves for patients with high- or low-risk scores. (**G–K**) Kaplan–Meier survival analyses of IPF patients based on the expression of the identified genes.

3. Discussion

Despite extensive research on human idiopathic pulmonary fibrosis (IPF), the underlying mechanisms responsible for the development of these diseases remain poorly understood. In addition, the available treatments for preventing or treating IPF are limited and often ineffective [27,28]. Fibrotic lung tissues comprise various cell subpopulations exhibiting diverse genetic and phenotypic traits. However, the precise mechanisms underlying the emergence of this heterogeneity during fibrosis development remain unclear. Herein, we conducted a thorough analysis of the fibroblast landscape in human idiopathic pulmonary fibrosis (IPF) and successfully identified two predominant subpopulations primarily found within fibrotic lung tissues, which have been designated as IPF-fibroblasts. Subsequently, we extensively investigated the distinctive characteristics and key regulatory pathways of distinct fibroblast subtypes. This in-depth exploration not only enhances our understanding of the pathogenesis of pulmonary fibrosis but also identifies potential targets for clinical therapies aimed at treating these conditions.

Previous single-cell RNA sequencing (scRNA-seq) studies that examined the heterogeneity of various lung-resident cell populations in pulmonary fibrosis have provided valuable insights but have only provided a limited snapshot of the overall landscape [29,30]. Through our extensive characterization of fibroblast heterogeneity, we have successfully identified a distinct fibroblast subtype, known as IPF-fibroblast, which exhibits a predominant presence in fibrotic lung tissue. We characterize the molecular features and identify novel markers of these fibroblast subpopulations related to IPF. We identify LTBP1, DPT, INHBA, and CTHRC1 as core enriched genes for IPF-fibroblasts in fibrotic lung tissues; however, their role in fibrogenesis remains largely elusive and requires further exploration. In addition, our findings revealed that IPF-fibroblasts exhibit elevated expression levels of genes associated with profibrotic processes, such as cellular response to transforming growth factor beta and collagen fibril organization [31,32]. Remarkably, our observations unveiled intriguing immunosuppressive characteristics in IPF-fibroblasts, implying their potential role in immunoregulation within the microenvironment of fibrotic pulmonary tissue. Consistent with this, previous studies have demonstrated that lung-resident fibroblasts play a crucial role in reshaping the local immune landscape, thereby facilitating the progression of the disease [14]. These scRNA-seq data reveal the dynamic molecular features of IPF-fibroblasts during development and provide a valuable resource for further studies.

The interactions between fibroblasts and other cell types in the lung are dynamic and multifaceted, playing a vital role in maintaining lung homeostasis and facilitating tissue repair processes. scRNA-seq analysis enables the identification of cell-surface receptors and their ligands that mediate the communication between fibroblasts and other cell types in the lung tissue. Through cell–cell communication analysis, it has been observed that IPF-fibroblasts have the potential to promote the differentiation of normal fibroblasts into fibrotic fibroblasts through CTGF/ITGA5 interaction [33]. In addition, we predicted that the PARs and GDF signaling pathways, which are mediated by NK cells and epithelial cells, respectively, would support the maintenance of the IPF-fibroblast phenotype. Furthermore, the trajectory study of IPF-fibroblasts changing from state 1 to state 2 further supports earlier findings and suggests that the majority of IPF-fibroblasts are probably derived from the activation of resident normal fibroblasts [7]. These results suggest that the identified IPF-fibroblast subpopulation might have an important role in fibrogenesis and may serve as target cells for a fibrosis treatment.

In this work, high-dimensional weighted gene co-expression network analysis (hdWGCNA), which is an advanced bioinformatics strategy for cell-associated gene module detection [34], and LASSO analysis combined with univariate analysis were further applied to develop a prognostic IPF-fibroblast-related gene (IFRG) signature. One of the characteristics of the IPF-fibroblasts in fibrotic lung tissue is their high expression of secretory proteins such as CCDC80, CTHRC1, COL6A1, FBLN2, FSTL1, and GSN. Previous studies suggested that some of these proteins, such as FSTL1, increased and could promote epithelial–mesenchymal transition in the lung [35,36]. Gelsolin (GSN), a protein that severs and caps actin filaments and plays a pivotal role in regulating actin assembly, has been reported to be involved in fibroblast activation during the development of myocardial fibrosis [37]. While there is currently a lack of reported evidence linking GSN to IPF, our research indicates that it plays a significant role in the activation process of fibroblasts associated with IPF. Furthermore, in line with our research findings, studies have revealed that the specific subset of cells with elevated CTHCR1 expression within fibrotic lung tissue demonstrates the highest level of collagen expression [38]. Notably, our findings showed that the aforementioned proteins were inversely linked with the percent anticipated diffusing capacity for lung carbon monoxide (% DLCO), emphasizing the clinical importance of IFRGs in pulmonary fibrosis. Based on the above findings, we first employed seven machine learning algorithms to construct an IPF-fibroblast-related predictive model for IPF. Then, we applied cross-validation and ROC curve analysis to assess the model's performance. After comparing the algorithms' performance on the validation set, we finally selected the algorithm with the best performance as our final model. Recently, a study utilized machine learning to develop an IPF prediction model focused on the midkine gene [39]. Similar to our research, this study employed three different algorithmic models (SVM, Adaboost, and random forest) and determined, through ROC curve analysis, that the SVM algorithm exhibited the highest accuracy. These findings align with our own research, indicating that the SVM algorithm possesses a considerable advantage and accuracy in constructing predictive models.

Overall, the model can effectively predict the prognosis of IPF patients, providing valuable assistance for disease prediction and intervention.

Despite the importance of these data for advancing knowledge, we should acknowledge some limitations. First, despite our best efforts to ensure the robustness of our clustering analysis of fibroblasts in fibrotic lung tissue, larger datasets could further improve and refine our clustering results. Second, being computational and omic in nature, our work requires experimental validation of the IPF-fibroblast markers derived from our findings for identifying and characterizing fibroblast subtypes in fibrotic lung tissues.

4. Materials and Methods

4.1. Data Acquisition

In total, four independent public datasets were downloaded from the NCBI GEO databases (http://www.ncbi.nlm.nih.gov/geo/, accessed on 2 September 2023). Specifically, the single-cell RNA-seq dataset GSE128033 was utilized to analyze the heterogeneity of fibroblasts in normal and fibrotic lung tissues [40]. We selected the lower lobe samples from the dataset, including normal and fibrotic lung tissues, as pulmonary fibrosis often originates in the lower lobe in clinical practice [41]. Three bulk RNA-seq datasets (GSE32537, GSE110147, and GSE70866) were employed for the construction and validation of our predictive model. Essential information about the samples of the four given datasets is displayed in Table 1. For analyses of data from a public database, patient consent and ethics committee approval were not required.

4.2. scRNA-seq Data Processing

ScRNA-seq data processing was performed using the R 'Seurat' package (version: 4.3.0) as previously described [42]. Briefly, cells with gene expression levels below 300 genes or above 6500 genes, as well as those with mitochondrial gene expression exceeding 10%, were excluded, ensuring the inclusion of the majority of cells in the utilized datasets. The SCTransform function was then applied to normalize and scale raw counts, followed by principal component analysis (PCA). To mitigate batch effects across dissociated scRNA-seq raw data, the R 'Harmony' package (version: 0.1.1) was employed. Clustering analysis was performed based on the edge weights between any two cells, and a shared nearest-neighbor graph was generated using the Louvain algorithm, implemented in the FindNeighbors and FindClusters functions. The resulting cells were visualized using the uniform manifold approximation and projection (UMAP) algorithm. A similar procedure was applied for subclustering analysis. The Seurat "FindMarkers" function was used to identify preferentially expressed genes within clusters as well as differentially expressed genes (DEGs) between fibrotic- and normal-derived cells. Each cell cluster was subsequently annotated using known cell-type marker genes. The specific expression patterns of the identified genes at the single-cell level were visualized using the "scRNAtoolVis" package (version 0.0.5).

Table 1. Overview of the information of analyzed datasets.

Dataset	Year	Area	Species	Platform	Data Type	Number of Samples	
						Normal	IPF
GSE128033	2019	United States	Homo	GPL18573	scRNA-seq	10	8
GSE32537	2011	United States	Homo	GPL6244	Bulk RNA-seq	39	131
GSE110147	2018	Canada	Homo	GPL6244	Bulk RNA-seq	11	22
GSE70866	2015	Germany	Homo	GPL14550	Bulk RNA-seq	20	212

4.3. Trajectory and Cell–Cell Communication Analysis

Unsupervised pseudotemporal analysis was conducted using the "Monocle" package (version 2.26.0) with the DDR-Tree algorithm and default parameters to investigate the trajectory of fibroblasts. Subsequently, the 'plot_pseudotime_heatmap' function was utilized to generate a heatmap, visually representing the dynamic expression of module genes along the pseudotime trajectories of fibroblasts in fibrotic lung tissues. To identify potential interactions between and within fibroblasts and other cell populations, the "CellChat" package (version 1.6.1) was employed with the default settings of the recommended pipelines [43].

4.4. Enrichment Analysis

The Seurat "FindMarkers" function was used to identify the DEGs of each cell subcluster. A fold change (|FC|) greater than 2 and an adjusted p-value less than 0.05 were used as the cut-off criteria. Based on the DEGs, the gene set enrichment analysis (ssGSEA) and gene ontology (GO) enrichment analyses between the cell subgroups were performed using the "clusterProfiler" package (version 4.7.1003). The functional enrichment result was shown using the "GseaVis" package (version 0.0.8).

4.5. High Dimensional Weighted Gene Co-Expression Network Analysis (hdWGCNA)

To identify potential IPF-fibroblast-related genes associated with IPF, we performed hdWGCNA by using the "hdWGCNA" package (version 0.1.1.9010) [34]. Briefly, metacells were constructed separately for each sample and each cell cluster with the hdWGCNA function MetacellsByGroups, aggregating 50 cells per metacell. For each cell population, we first subset the Seurat object for the cell population of interest and then performed the standard hdWGCNA pipeline by sequentially running the following functions with default parameters: TestSoftPowers, ConstructNetwork, ModuleEigengenes, ModuleConnectivity, and RunModuleUMAP.

4.6. Machine Learning-Based Construction of an IPF-Fibroblast-Related Predictive Model

In order to assess the predictive potential of IPF-fibroblast-related genes (IFRGs) identified by hdWGCNA in the development of IPF, we collected two transcriptome sequencing datasets, namely, GSE32537 and GSE110147. These datasets were systematically collected and utilized as the training and testing cohorts, respectively. The identified IFRG signature was utilized to build a prediction model in the training set. This was accomplished by employing seven machine learning algorithms through the "mlr3" package (version 0.16.0). The machine learning algorithms included in our analysis are as follows: log_reg (logistic regression), Ida (iterative dichotomizer 3), ranger (random forest), SVM (support vector machines), nave_bayes (naive Bayes classifier), rpart (recursive partitioning and regression trees), and kknn (k nearest neighbors). The model generation procedure was as follows: (a) Prognostic IFRGs (immune-related functional genes) were identified in the training cohort using least absolute shrinkage and selection operator (LASSO) and univariate Cox regression analyses [44]. (b) Seven machine learning algorithms were then applied to the prognostic IFRGs to create prediction models. Each algorithm was validated using 5-times-repeated tenfold cross-validation. (c) For each model, the "timeROC" R-package (version: 0.4) was utilized to generate ROC curves and evaluate the predictive capacity of the model. The model with the highest accuracy was selected as the final predictive model. (d) The predictive ability of the final model was evaluated using independent testing sets.

4.7. Non–Negative Matrix Factorization (NMF) of IPF by IFRGs and Survival Analysis

To delve deeper into the subtypes of IPF, we employed the non-negative matrix factorization (NMF) algorithm from the "NMF" package (version 0.20.6) to gain additional insights. Firstly, we chose 30 IFRGs (identified by hdWGCNA) from the GSE32537 dataset. Following that, we employed the expression matrix of these selected genes in the NMF analysis to identify unique subtypes of IPF. Survival analysis of the identified secretory protein-encoding genes was conducted using the "survival" package (version 2.1.2). Survival status and risk scores of IPF patients in the high- and low-risk groups were analyzed through the "ggrisk" package (version 1.3). Statistical significance was set at p value < 0.05.

4.8. Statistical Analysis

All statistical analyses and data visualizations were performed using the R software (version 4.2.1). Pearson's correlation coefficients were used to assess the correlations between two continuous variables. For quantitative data, either a two-tailed, unpaired Student t-test or a one-way analysis of variance (ANOVA) with Tukey's multiple comparisons test was used to compare values between subgroups. $p < 0.05$ was considered to be statistically significant.

5. Conclusions

Overall, our study contributes to a better understanding of the heterogeneity within the fibroblast population in fibrotic lung tissue. The identification of distinct molecular and biological characteristics of fibroblast subtypes, along with the prognostic utility of signature genes, provides valuable insights into the pathogenesis of IPF. These findings have the potential to improve disease prediction and facilitate targeted interventions for patients with IPF.

Supplementary Materials: The following supporting information can be downloaded at: https://www.mdpi.com/article/10.3390/ijms25010094/s1, Figure S1: Fibrotic and normal lung fibroblasts subcluster into distinct cell populations; Figure S2: Enrichment analyses of the gene clusters; Figure S3: Cell communication analysis in fibroblast subpopulations; Figure S4: High-dimensional weighted gene co-expression network analysis (hdWGCNA) reveals module-specific hub genes in IPF-related fibroblasts.

Author Contributions: Conceptualization, J.H. and X.H.; methodology, J.H.; software, J.H.; validation, J.H., Y.Y. and X.H.; writing—original draft preparation, J.H.; writing—review and editing, J.H.; visualization, J.H.; supervision, X.H.; funding acquisition, J.H. and X.H. All authors have read and agreed to the published version of the manuscript.

Funding: This research was funded by the National Natural Science Foundation of China (NSFC No. 82300085), the Basic Science (Natural Science) Research Projects in Higher Education Institutions in Jiangsu Province (23KJB310014), the Jiangsu Provincial Double-Innovation Doctor Program (JSSCBS20220453), and the Jiangsu Key Discipline Fund for the 14th Five-Year Plan (Biology).

Institutional Review Board Statement: Not applicable.

Informed Consent Statement: Not applicable.

Data Availability Statement: The data that support this study are available within the article and its supplementary data files or available from the authors upon request.

Conflicts of Interest: The authors declare no conflict of interest.

References

1. Spagnolo, P.; Kropski, J.A.; Jones, M.G.; Lee, J.S.; Rossi, G.; Karampitsakos, T.; Maher, T.M.; Tzouvelekis, A.; Ryerson, C.J. Idiopathic pulmonary fibrosis: Disease mechanisms and drug development. *Pharmacol. Ther.* **2021**, *222*, 107798. [CrossRef] [PubMed]
2. Moss, B.; Ryter, S.; Rosas, I. Pathogenic Mechanisms Underlying Idiopathic Pulmonary Fibrosis. *Annu. Rev. Pathol.* **2022**, *17*, 515–546. [CrossRef] [PubMed]
3. Sgalla, G.; Biffi, A.; Richeldi, L. Idiopathic pulmonary fibrosis: Diagnosis, epidemiology and natural history. *Respirology* **2016**, *21*, 427–437. [CrossRef] [PubMed]
4. King, T.E., Jr.; Tooze, J.A.; Schwarz, M.I.; Brown, K.R.; Cherniack, R.M. Predicting survival in idiopathic pulmonary fibrosis: Scoring system and survival model. *Am. J. Respir. Crit. Care Med.* **2001**, *164*, 1171–1181. [CrossRef] [PubMed]
5. Mei, Q.; Liu, Z.; Zuo, H.; Yang, Z.; Qu, J. Idiopathic Pulmonary Fibrosis: An Update on Pathogenesis. *Front. Pharmacol.* **2021**, *12*, 797292. [CrossRef] [PubMed]
6. Dean, C.H.; Lloyd, C.M. Lung Alveolar Repair: Not All Cells Are Equal. *Trends Mol. Med.* **2017**, *23*, 871–873. [CrossRef] [PubMed]
7. Hung, C.; Linn, G.; Chow, Y.H.; Kobayashi, A.; Mittelsteadt, K.; Altemeier, W.A.; Gharib, S.A.; Schnapp, L.M.; Duffield, J.S. Role of lung pericytes and resident fibroblasts in the pathogenesis of pulmonary fibrosis. *Am. J. Respir. Crit. Care Med.* **2013**, *188*, 820–830. [CrossRef]
8. Bamberg, A.; Redente, E.F.; Groshong, S.D.; Tuder, R.M.; Cool, C.D.; Keith, R.C.; Edelman, B.L.; Black, B.P.; Cosgrove, G.P.; Wynes, M.W.; et al. Protein Tyrosine Phosphatase-N13 Promotes Myofibroblast Resistance to Apoptosis in Idiopathic Pulmonary Fibrosis. *Am. J. Respir. Crit. Care Med.* **2018**, *198*, 914–927. [CrossRef]
9. Lederer, D.J.; Martinez, F.J. Idiopathic Pulmonary Fibrosis. *N. Engl. J. Med.* **2018**, *378*, 1811–1823. [CrossRef]
10. Günther, A.; Korfei, M.; Mahavadi, P.; von der Beck, D.; Ruppert, C.; Markart, P. Unravelling the progressive pathophysiology of idiopathic pulmonary fibrosis. *Eur. Respir. Rev.* **2012**, *21*, 152–160. [CrossRef]
11. Liu, G.; Philp, A.M.; Corte, T.; Travis, M.A.; Schilter, H.; Hansbro, N.G.; Burns, C.J.; Eapen, M.S.; Sohal, S.S.; Burgess, J.K.; et al. Therapeutic targets in lung tissue remodelling and fibrosis. *Pharmacol. Ther.* **2021**, *225*, 107839. [CrossRef] [PubMed]
12. Lynch, M.D.; Watt, F.M. Fibroblast heterogeneity: Implications for human disease. *J. Clin. Investig.* **2018**, *128*, 26–35. [CrossRef] [PubMed]
13. Valenzi, E.; Bulik, M.; Tabib, T.; Morse, C.; Sembrat, J.; Trejo Bittar, H.; Rojas, M.; Lafyatis, R. Single-cell analysis reveals fibroblast heterogeneity and myofibroblasts in systemic sclerosis-associated interstitial lung disease. *Ann. Rheum. Dis.* **2019**, *78*, 1379–1387. [CrossRef] [PubMed]
14. Gong, Z.; Li, Q.; Shi, J.; Wei, J.; Li, P.; Chang, C.H.; Shultz, L.D.; Ren, G. Lung fibroblasts facilitate pre-metastatic niche formation by remodeling the local immune microenvironment. *Immunity* **2022**, *55*, 1483–1500.e9. [CrossRef] [PubMed]
15. Hanley, C.J.; Waise, S.; Ellis, M.J.; Lopez, M.A.; Pun, W.Y.; Taylor, J.; Parker, R.; Kimbley, L.M.; Chee, S.J.; Shaw, E.C.; et al. Single-cell analysis reveals prognostic fibroblast subpopulations linked to molecular and immunological subtypes of lung cancer. *Nat. Commun.* **2023**, *14*, 387. [CrossRef] [PubMed]
16. Ramachandran, P.; Dobie, R.; Wilson-Kanamori, J.R.; Dora, E.F.; Henderson, B.E.P.; Luu, N.T.; Portman, J.R.; Matchett, K.P.; Brice, M.; Marwick, J.A.; et al. Resolving the fibrotic niche of human liver cirrhosis at single-cell level. *Nature* **2019**, *575*, 512–518. [CrossRef] [PubMed]
17. Kuppe, C.; Ibrahim, M.M.; Kranz, J.; Zhang, X.; Ziegler, S.; Perales-Patón, J.; Jansen, J.; Reimer, K.C.; Smith, J.R.; Dobie, R.; et al. Decoding myofibroblast origins in human kidney fibrosis. *Nature* **2021**, *589*, 281–286. [CrossRef]
18. Reyfman, P.A.; Walter, J.M.; Joshi, N.; Anekalla, K.R.; McQuattie-Pimentel, A.C.; Chiu, S.; Fernandez, R.; Akbarpour, M.; Chen, C.I.; Ren, Z.; et al. Single-Cell Transcriptomic Analysis of Human Lung Provides Insights into the Pathobiology of Pulmonary Fibrosis. *Am. J. Respir. Crit. Care Med.* **2019**, *199*, 1517–1536. [CrossRef]
19. Rajkomar, A.; Dean, J.; Kohane, I. Machine Learning in Medicine. *N. Engl. J. Med.* **2019**, *380*, 1347–1358. [CrossRef]
20. Gupta, R.; Srivastava, D.; Sahu, M.; Tiwari, S.; Ambasta, R.K.; Kumar, P. Artificial intelligence to deep learning: Machine intelligence approach for drug discovery. *Mol. Divers.* **2021**, *25*, 1315–1360. [CrossRef]
21. Maddali, M.V.; Kalra, A.; Muelly, M.; Reicher, J.J. Development and validation of a CT-based deep learning algorithm to augment non-invasive diagnosis of idiopathic pulmonary fibrosis. *Respir. Med.* **2023**, *219*, 107428. [CrossRef] [PubMed]

22. El-Hasnony, I.M.; Elzeki, O.M.; Alshehri, A.; Salem, H. Multi-Label Active Learning-Based Machine Learning Model for Heart Disease Prediction. *Sensors* **2022**, *22*, 1184. [CrossRef] [PubMed]
23. Wu, Z.; Chen, H.; Ke, S.; Mo, L.; Qiu, M.; Zhu, G.; Zhu, W.; Liu, L. Identifying potential biomarkers of idiopathic pulmonary fibrosis through machine learning analysis. *Sci. Rep.* **2023**, *13*, 16559. [CrossRef] [PubMed]
24. Swanson, K.; Wu, E.; Zhang, A.; Alizadeh, A.A.; Zou, J. From patterns to patients: Advances in clinical machine learning for cancer diagnosis, prognosis, and treatment. *Cell* **2023**, *186*, 1772–1791. [CrossRef] [PubMed]
25. Pan, J.; Hofmanninger, J.; Nenning, K.H.; Prayer, F.; Röhrich, S.; Sverzellati, N.; Poletti, V.; Tomassetti, S.; Weber, M.; Prosch, H.; et al. Unsupervised machine learning identifies predictive progression markers of IPF. *Eur. Radiol.* **2023**, *33*, 925–935. [CrossRef] [PubMed]
26. Wu, L.D.; Li, F.; Chen, J.Y.; Zhang, J.; Qian, L.L.; Wang, R.X. Analysis of potential genetic biomarkers using machine learning methods and immune infiltration regulatory mechanisms underlying atrial fibrillation. *BMC Med. Genom.* **2022**, *15*, 64. [CrossRef] [PubMed]
27. Lancaster, L.H.; de Andrade, J.A.; Zibrak, J.D.; Padilla, M.L.; Albera, C.; Nathan, S.D.; Wijsenbeek, M.S.; Stauffer, J.L.; Kirchgaessler, K.U.; Costabel, U. Pirfenidone safety and adverse event management in idiopathic pulmonary fibrosis. *Eur. Respir. Rev.* **2017**, *26*, 170057. [CrossRef]
28. Glass, D.S.; Grossfeld, D.; Renna, H.A.; Agarwala, P.; Spiegler, P.; DeLeon, J.; Reiss, A.B. Idiopathic pulmonary fibrosis: Current and future treatment. *Clin. Respir. J.* **2022**, *16*, 84–96. [CrossRef]
29. Habermann, A.C.; Gutierrez, A.J.; Bui, L.T.; Yahn, S.L.; Winters, N.I.; Calvi, C.L.; Peter, L.; Chung, M.I.; Taylor, C.J.; Jetter, C.; et al. Single-cell RNA sequencing reveals profibrotic roles of distinct epithelial and mesenchymal lineages in pulmonary fibrosis. *Sci. Adv.* **2020**, *6*, eaba1972. [CrossRef]
30. Adams, T.S.; Schupp, J.C.; Poli, S.; Ayaub, E.A.; Neumark, N.; Ahangari, F.; Chu, S.G.; Raby, B.A.; DeIuliis, G.; Januszyk, M.; et al. Single-cell RNA-seq reveals ectopic and aberrant lung-resident cell populations in idiopathic pulmonary fibrosis. *Sci. Adv.* **2020**, *6*, eaba1983. [CrossRef]
31. Budi, E.H.; Schaub, J.R.; Decaris, M.; Turner, S.; Derynck, R. TGF-β as a driver of fibrosis: Physiological roles and therapeutic opportunities. *J. Pathol.* **2021**, *254*, 358–373. [CrossRef] [PubMed]
32. Kong, W.; Lyu, C.; Liao, H.; Du, Y. Collagen crosslinking: Effect on structure, mechanics and fibrosis progression. *Biomed. Mater.* **2021**, *16*, 062005. [CrossRef] [PubMed]
33. Leask, A. Potential therapeutic targets for cardiac fibrosis: TGFbeta, angiotensin, endothelin, CCN2, and PDGF, partners in fibroblast activation. *Circ. Res.* **2010**, *106*, 1675–1680. [CrossRef] [PubMed]
34. Morabito, S.; Reese, F.; Rahimzadeh, N.; Miyoshi, E.; Swarup, V. hdWGCNA identifies co-expression networks in high-dimensional transcriptomics data. *Cell. Rep. Methods* **2023**, *3*, 100498. [CrossRef] [PubMed]
35. Gervasi, M.; Bianchi-Smiraglia, A.; Cummings, M.; Zheng, Q.; Wang, D.; Liu, S.; Bakin, A.V. JunB contributes to Id2 repression and the epithelial-mesenchymal transition in response to transforming growth factor-β. *J. Cell Biol.* **2012**, *196*, 589–603. [CrossRef] [PubMed]
36. Liu, T.; Liu, Y.; Miller, M.; Cao, L.; Zhao, J.; Wu, J.; Wang, J.; Liu, L.; Li, S.; Zou, M.; et al. Autophagy plays a role in FSTL1-induced epithelial mesenchymal transition and airway remodeling in asthma. *Am. J. Physiol. Lung Cell Mol. Physiol.* **2017**, *313*, L27–L40. [CrossRef]
37. Jana, S.; Aujla, P.; Hu, M.; Kilic, T.; Zhabyeyev, P.; McCulloch, C.A.; Oudit, G.Y.; Kassiri, Z. Gelsolin is an important mediator of Angiotensin II-induced activation of cardiac fibroblasts and fibrosis. *FASEB J.* **2021**, *35*, e21932. [CrossRef]
38. Tsukui, T.; Sun, K.H.; Wetter, J.B.; Wilson-Kanamori, J.R.; Hazelwood, L.A.; Henderson, N.C.; Adams, T.S.; Schupp, J.C.; Poli, S.D.; Rosas, I.O.; et al. Collagen-producing lung cell atlas identifies multiple subsets with distinct localization and relevance to fibrosis. *Nat. Commun.* **2020**, *11*, 1920. [CrossRef]
39. Zhang, S.; Zhang, T.; Wang, L.; Wang, H.; Wu, J.; Cai, H.; Mo, C.; Yang, J. Machine learning identified MDK score has prognostic value for idiopathic pulmonary fibrosis based on integrated bulk and single cell expression data. *Front. Genet.* **2023**, *14*, 1246983. [CrossRef]
40. Morse, C.; Tabib, T.; Sembrat, J.; Buschur, K.L.; Bittar, H.T.; Valenzi, E.; Jiang, Y.; Kass, D.J.; Gibson, K.; Chen, W.; et al. Proliferating SPP1/MERTK-expressing macrophages in idiopathic pulmonary fibrosis. *Eur. Respir. J.* **2019**, *54*, 1802441. [CrossRef]
41. Martinez, F.J.; Collard, H.R.; Pardo, A.; Raghu, G.; Richeldi, L.; Selman, M.; Swigris, J.J.; Taniguchi, H.; Wells, A.U. Idiopathic pulmonary fibrosis. *Nat. Rev. Dis. Primers* **2017**, *3*, 17074. [CrossRef] [PubMed]
42. Zhang, C.; Shen, H.; Yang, T.; Li, T.; Liu, X.; Wang, J.; Liao, Z.; Wei, J.; Lu, J.; Liu, H.; et al. A single-cell analysis reveals tumor heterogeneity and immune environment of acral melanoma. *Nat. Commun.* **2022**, *13*, 7250. [CrossRef] [PubMed]
43. Jin, S.; Guerrero-Juarez, C.F.; Zhang, L.; Chang, I.; Ramos, R.; Kuan, C.H.; Myung, P.; Plikus, M.V.; Nie, Q. Inference and analysis of cell-cell communication using CellChat. *Nat. Commun.* **2021**, *12*, 1088. [CrossRef] [PubMed]
44. Tibshirani, R. The lasso method for variable selection in the Cox model. *Stat. Med.* **1997**, *16*, 385–395. [CrossRef]

Disclaimer/Publisher's Note: The statements, opinions and data contained in all publications are solely those of the individual author(s) and contributor(s) and not of MDPI and/or the editor(s). MDPI and/or the editor(s) disclaim responsibility for any injury to people or property resulting from any ideas, methods, instructions or products referred to in the content.

Article

Regulation of STAT1 and STAT4 Expression by Growth Factor and Interferon Supplementation in Sjögren's Syndrome Cell Culture Models

Jean-Luc C. Mougeot [1,2,*], Thomas E. Thornburg [1,2], Braxton D. Noll [1,2], Michael T. Brennan [1,2] and Farah Bahrani Mougeot [1,2,*]

[1] Translational Research Laboratories, Oral Medicine, Oral & Maxillofacial Surgery, Atrium Health-Carolinas Medical Center, Charlotte, NC 28203, USA; thomas.thornburg@atriumhealth.org (T.E.T.); braxton.n@icloud.com (B.D.N.); mike.brennan@atriumhealth.org (M.T.B.)
[2] Department of Otolaryngology, Head & Neck Surgery, Wake Forest University School of Medicine, Winston-Salem, NC 27157, USA
* Correspondence: jean-luc.mougeot@atriumhealth.org (J.-L.C.M.); farah.mougeot@atriumhealth.org (F.B.M.)

Abstract: Our goal was to investigate the effects of epidermal growth factor (EGF) and interferons (IFNs) on signal transducer and activator of transcription STAT1 and STAT4 mRNA and active phosphorylated protein expression in Sjögren's syndrome cell culture models. iSGECs (immortalized salivary gland epithelial cells) and A253 cells were treated with EGF, IFN-alpha, -beta, -gamma, or mitogen-activated protein kinase p38 alpha (p38-MAPK) inhibitor for 0–24–48–72 h. STAT1 and STAT4 mRNA expression was quantified by qRT-PCR. Untreated and treated cells were compared using the delta-delta-CT method based on glyceraldehyde-3-phosphate dehydrogenase (GAPDH) normalized relative fold changes. phospho-tyrosine-701-STAT1 and phospho-serine-721-STAT4 were detected by Western blot analysis. STAT4 mRNA expression decreased 48 h after EGF treatment in A253 cells, immortalized salivary gland epithelial cells iSGECs nSS2 (sicca patient origin), and iSGECs pSS1 (anti-SSA negative Sjögren's Syndrome patient origin). EGF and p38-MAPK inhibitor decreased A253 STAT4 mRNA levels. EGF combined with IFN-gamma increased phospho-STAT4 and phospho-STAT1 after 72 h in all cell lines, suggesting additive effects for phospho-STAT4 and a major effect from IFN-gamma for phospho-STAT1. pSS1 and nSS2 cells responded differently to type I and type II interferons, confirming unique functional characteristics between iSGEC cell lines. EGF/Interferon related pathways might be targeted to regulate STAT1 and STAT4 expression in salivary gland epithelial cells. Further investigation is required learn how to better target the Janus kinases/signal transducer and activator of transcription proteins (JAK/STAT) pathway-mediated inflammatory response in Sjögren's syndrome.

Keywords: STAT4; STAT1; cytokine regulation; interferons; EGF

1. Introduction

Sjögren's syndrome (SS) is a chronic autoimmune disease impacting over 200,000 people in united states annually, targeting mucous membranes and glands of the eyes and mouth which can lead to other problems, such as dental cavities, oral thrush, and vision problems [1]. Salivary glands affected by SS commonly exhibit an epithelitis implying independent regulation of the immune response beyond reactivity to immune cell infiltration [2]. Epithelitis is marked by increased cytokine expression, limited Fas cell surface death receptor-mediated apoptosis, and p38-MAPK pathway activation promoting cluster of differentiation 40-mediated epithelial cell death [3–5]. In previous studies, the p38-MAPK pathway was demonstrated to be an upstream activator for phosphorylation and full transcriptional activity of STAT4 [6].

Determining the interactions between these pathways may allow for improved drug intervention treatment for SS patients. STAT4, involved in the JAK/STAT signaling pathway, is a confirmed SS susceptibility gene with active roles in cytokine expression and secretion, such as interferons (IFNs) and tumor necrosis factor-*alpha* in peripheral blood mononuclear cells (PBMCs) [7–10]. It is, however, unclear, the extent to which STAT4 and associated polymorphisms relate to immune cells and/or salivary gland epithelial cells during SS development.

The JAK/STAT pathway mediates a variety of immune responses in both epithelial and immune related cell types [9]. Among the STAT proteins, there is little information describing how STAT4 specifically interacts with other STAT proteins within this pathway [11]. Moreover, this family of proteins interacts after phosphorylation, in which the proteins often dimerize and translocate to the nucleus where they regulate gene expression [12]. The JAK/STAT pathway, involving phospho-STAT4 activity, plays an important role in autoimmune mediated pathways and may facilitate aberrant inflammatory responses [13]. Additionally, cytokines such as IFNs bind and activate JAK-1/JAK-2, activating STAT proteins through phosphorylation, possibly propagating aberrant inflammatory responses in disease [14]. Targeting these factors or downstream targets in an appropriate manner within the salivary gland epithelia might serve to reduce the local inflammation and infiltration. Inhibition of JAK-1 has been studied for its potential as a possible treatment for SS [3,15]. JAK-1 inhibition may suppress both IFN type I and type II activities which can lead to downregulation of factors that promote infiltration of immune cells in salivary glands of SS patients [15].

Based on cross talk between pathways and potential effects on EGF, Wnt, and transforming growth factor-*beta* signaling pathways, it is inferred that small molecule inhibitors targeting JAK/STAT4 pathway may tamper with salivary gland epithelial cell homeostasis [16]. Therefore, this pathway needs to be better targeted to resolve inflammation and to preserve salivary gland function in SS [16]. EGF has been associated with the severity of intraoral manifestations in SS [17]. In addition, abnormal neovascularization in SS was shown to involve vascular endothelial growth factor receptor 2 (VEGFR2)/nuclear factor-*kappa* B (NF-κB) pathway leading to exacerbation of autoimmunity [18,19]. Also, there is an increased risk of cerebrovascular events and myocardial infarction in SS patients [20]. Although EGF and VEGF levels in saliva are not specific to the disease, SS patients exhibit a dysregulation of response to growth factors in addition to interferons [21].

The relationship between STAT4 and STAT1 and the response to growth factors and interferons are unclear. Therefore, since modulation of STAT signaling may provide new therapeutic avenues in general, our objective was to understand how this dysregulation affects expression of STAT4 and STAT1 at the mRNA level and phosphorylated protein levels using iSGECs and A253 cells line [22]. Additionally, since p38-MAPK and JAK/STAT pathways can act in concert in inflammation in multiple disease models, we tested the effects of p38 inhibitor on STAT1 and STAT4 expression in these cell lines [6,23,24].

2. Results

The overall experimental design is presented in Figure 1.

2.1. Regulation of STAT4 Expression by EGF and p38 Inhibitor

The mRNA expression of STAT4 was initially determined in A253 and nSS2 cells (Figure 2). The cells were cultured and treated with p38 inhibitor (20 μM) and EGF (10 ng/mL), and a combination of both. STAT4 expression was reduced in both cell lines in response to all three treatments of p38 inhibitor or EGF alone and the combination of both. In A253 cells the largest reduction resulted from EGF supplementation with an approximately two-fold reduction (Figure 2A). STAT4 expression in nSS2 ISGECs was similarly downregulated by EGF supplementation, but the combination of EGF and p38 inhibitor yielded an enhanced effect resulting in an approximately three-fold reduction (Figure 2B). The overall results indicate p38 inhibitor effects might be more pronounced on nSS2 cells.

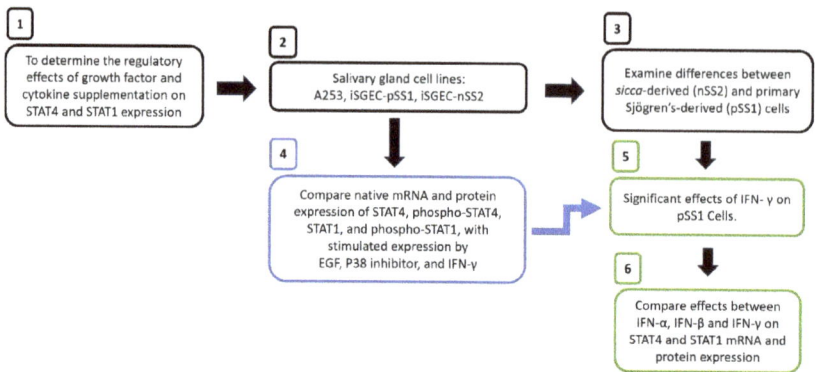

Figure 1. Determination of STAT1 and STAT4 expression and their phosphorylated forms in growth factors and interferon stimulated salivary gland cell line models for Sjögren's Syndrome. Legend. (1) Experimental strategy used to determine regulatory effects of interferon proteins and growth factor supplementation on the expression of STAT4 and STAT1 is shown. (2) In this study, the A253 mucoepidermoid carcinoma of the salivary gland cell line was compared as a control to the iSGEC-pSS1 (primary Sjögren's syndrome) and iSGEC-nSS2 (*sicca*) cell lines derived from primary salivary gland epithelial cells (SGECs) previously established in our laboratory. Previous studies have shown that interferons activate the JAK/STAT pathway resulting in the phosphorylation of STAT proteins and activating them as transcription factors. In Sjögren's syndrome, STAT4 is known to contain susceptibility polymorphisms and to mediate cytokines (IFN, TNF-α) production. This led us to directly compare three cell lines: pSS1, an immortalized salivary gland cell line derived from a Sjögren's patient (anti-SSA negative but focus score = 1.8); nSS2, an immortalized salivary gland cell line derived from a *sicca* patient (focus score = 0.16); A253, a mucoepidermoid salivary gland control cell line. (3) The pSS1 and nSS2 cell lines were directly compared to better understand their commonalities and differences (4) We investigated the effects of EGF, p38 inhibitor, and IFN-γ on STAT1 and STAT4 mRNA and active protein expressions, in all three cell lines. (5) Since the effects of IFN-γ on the pSS1 cell line were prominent for phospho-STAT1, we determined whether this effect was unique to IFN-γ, or if it was a more general response to interferons. (6) Protein and gene expressions following supplementation with IFN-α, IFN-β and IFN-γ were determined in each cell line.

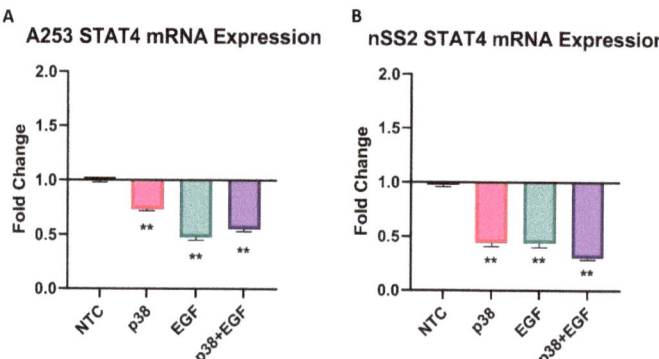

Figure 2. Effects of EGF and p38-inhibitor on STAT4 expression in salivary gland cell lines. Legend. (**A**) Expression of STAT4 mRNA following treatment of A253 cells with EGF and p38-I alone or in combination, is shown. (**B**) Results from supplementation with EGF and p38-I alone or in combination for nSS2 cells are shown. EFG is epidermal growth factor; p38-I is p38 inhibitor (SB203580); NTC is non-treated control. Error bars represent mean (+/−) standard error (SE) based on 3 independent replicates. Unpaired *t*-test was used to determine significance (** $p < 0.01$).

2.2. Effects of EGF and IFN-γ on Protein Expression of Phospho-STAT4 and Phospho-STAT1

To better understand the effects of EGF and IFN-γ supplementations in A253 and iSGECs on phospho-STAT4 and phospho-STAT1 protein production, the presence of both proteins was determined by Western blot. In all three cell lines, phospho-STAT4 presence was slightly increased by EGF ($p = 0.11262$)) and further increased by the combination of EGF and IFN-γ ($p = 0.003906$) (Figure 3A). Notably, A253 and pSS1 cells demonstrated an approximate average of 8- and 4-fold increase in phospho-STAT4 levels, respectively, after supplementation with both EGF and IFN-γ (Figure 3B). These effects are in the opposite direction compared to our described mRNA expression results for A253 and nSS2 cells in which STAT4 mRNA was downregulated in response to the addition of EGF (Figure 2). Phospho-STAT1 was expressed at lower levels in both A253 and nSS2 cells compared to pSS iSGECs ($p < 0.05$, paired t-test) (Figure 3C). All three cell lines displayed a large increase in phospho-STAT1 relative expression with supplementation of both EGF and IFN-γ combined, this effect being solely attributed to IFN-γ (Figure 3D). Of these, pSS1 iSGECs showed the largest presence of phospho-STAT1 compared to A253 cells and nSS2 iSGECs (Figure 3D).

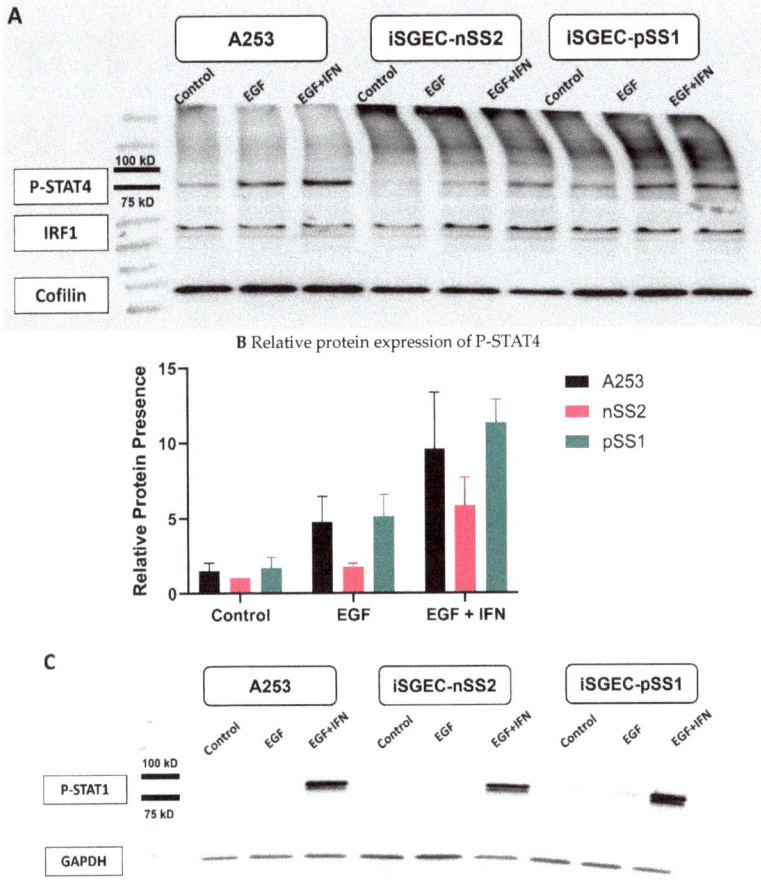

Figure 3. Cont.

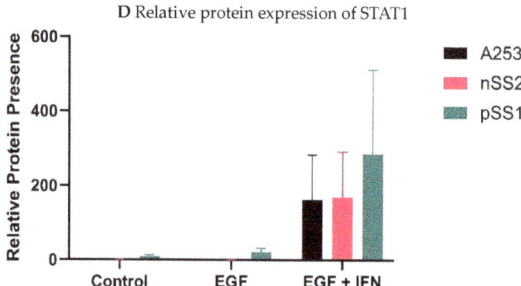

Figure 3. Western blot analysis of salivary gland cell lines treated with EGF and gamma-interferon. Legend. Representative Western blots and semi-quantitative Western blot analysis (from 3–4 independent experimental replicates) of cells treated with EGF and EGF combined with IFN-γ are shown. (**A**) Phospho-STAT4 (P-STAT4) and (**C**) Phospho-STAT1 (P-STAT1) protein levels were determined 72 h post-dosing with EGF and combined EGF and IFN-γ relative to untreated control, in A253, pSS1 and nSS2 whole cell lysates. (**B,D**) P-STAT4 and P-STAT1 presence was determined by semiquantitative analysis based on relative band intensity. Equal protein amounts were loaded in each lane and target bands were normalized to protein expression of cofilin or GAPDH. Loading controls and target proteins of the same blot were individually optimized for exposure requirements and reconstituted for imaging While incremental differences were observed for P-STAT4 across the three treatments, no changes were observed for interferon regulatory factor 1 (IRF1) (A). Effects from combined treatment with EGF and IGF were overall significant across the three cell lines compared to control and EGF treatment alone for both P-STAT1 and P-STAT4 ($p < 0.01$, Wilcoxon signed-rank test with Bonferroni correction).

2.3. Regulation of STAT1 and STAT4 mRNA and Active Protein Expression by Interferons α, β and γ

To further investigate the large effect of IFN-γ on STAT1 and STAT4 mRNA and protein expression, the effects of IFN-α and IFN-β protein supplementation were also assessed. The effects of IFN-α, IFN-β and IFN-γ on STAT4 and STAT1 mRNA expression in A253 cells, pSS1 and nSS2 ISGECs were determined. STAT4 expression was upregulated in all cell lines by IFN-α and IFN-β but was not changed by IFN-γ in pSS1 iSGECs (Figure 4A).

While comparisons of STAT4 levels between cell lines did not show significant differences, there was a noticeable difference in expression between pSS1 cells and the other two cell lines when treated with IFN-γ (Figure 4B). In contrast, STAT1 expression was found to be upregulated in all three cell lines with supplementation of type I and type II IFNs (Figure 4C). There was a significant difference between pSS1 and nSS2 cells in STAT1 expression with type I IFN treatment (IFN-α and IFN-β), where nSS2 cells displayed higher expression (Figure 4D). This trend was also observed in the IFN-γ treatment group but was not found to be statistically significant. These data show that nSS2 cells respond to IFN proteins in a unique manner.

Following observation of an increase in STAT1 mRNA expression with type I and type II IFNs, expression changes of phospho-STAT1 and phospho-STAT4 were determined by Western blot analysis. Each cell line was grown with supplementation of IFN-α, IFN-β and IFN-γ and was harvested for protein after 72 h. Type I interferon IFN-β supplementation showed low relative expression in nSS2 cells and pSS1 cells for phospho-STAT1, whereas IFN-γ supplementation yielded a statistically significant increase in phospho-STAT1 in both pSS1 and nSS2 cells (Figure 5A). A253 cells were not significantly responsive to both type I and type II IFNs ($p < 0.05$). The relative expression of phospho-STAT1 based on the normalized control Cyclophilin A, is shown is Figure 5C. Western blots of phospho-STAT4 protein expression for the three cell lines are shown in Figure 5D,E. The relative protein expression for phospho-STAT4 is shown in Figure 5F. The data show a slight marginal effect of IFN-γ on pSS1 cells.

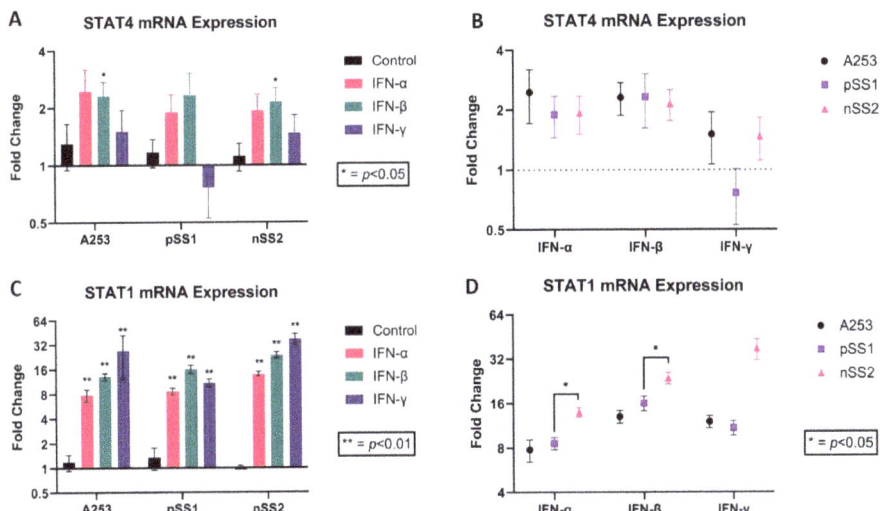

Figure 4. mRNA expression analysis of salivary gland cell lines A253, nSS2, and pSS1 treated with interferons α, β and γ. Legend. mRNA expression of STAT4 and STAT1 determined by RT-PCR was normalized to GAPDH. (**A**) Bar graph representing STAT4 mRNA relative expression in all cell lines treated with type I (IFN-alpha and -beta) and type II (IFN-gamma) interferons. (**B**) Plot showing differences in effects from type I and type II IFNs. (**C,D**) Bar graph representing STAT4 and STAT1 mRNA relative expression in all cell lines treated with type I (IFN-alpha and -beta) and type II (IFN-gamma) interferons and plot showing differences in effects of type I and type II IFNs. Error bars represent mean (+/−) standard error (SE) based on 9 independent experimental replicates. Mann–Whitney U-test was used to determine significance (** $p < 0.01$; * $p < 0.05$).

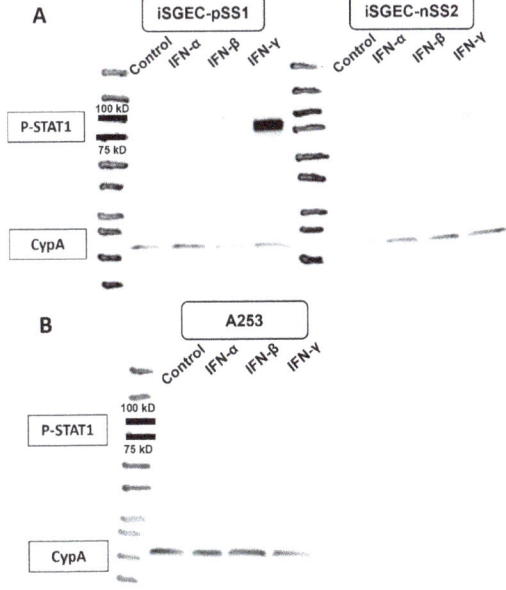

Figure 5. Cont.

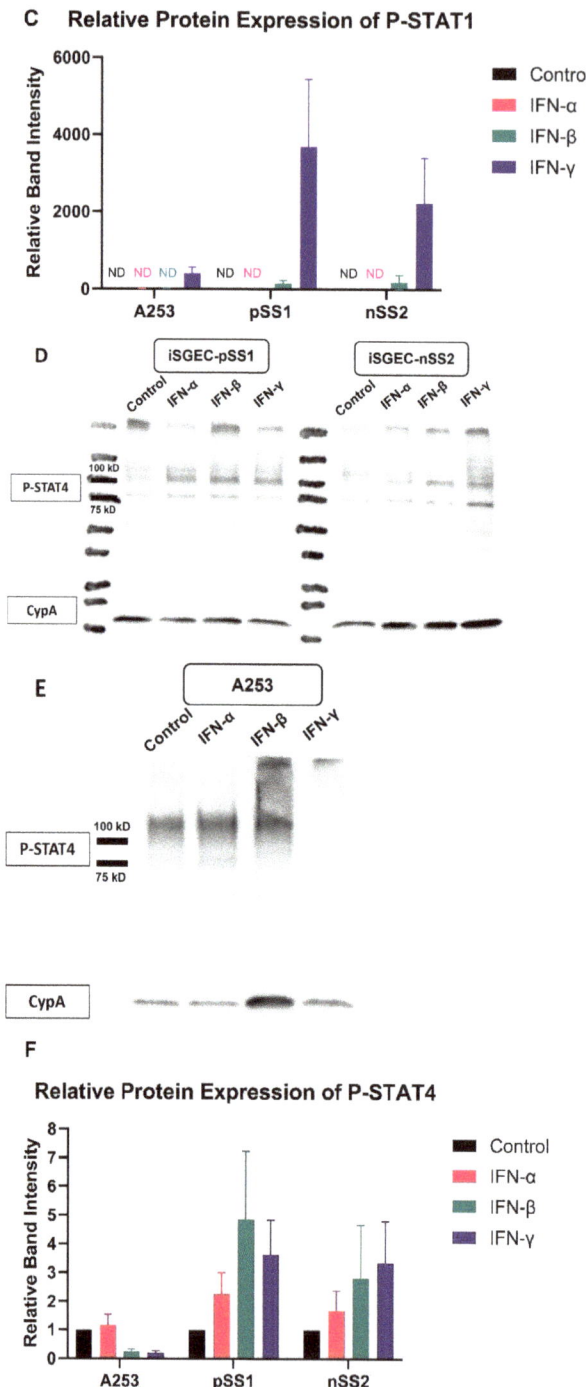

Figure 5. Active protein expression analysis of salivary gland cell lines treated with interferons α, β and γ. Legend. Representative Western blots and semi-quantitative Western blot analysis (from 3–4 independent experimental replicates) of IFN treated cells are shown. (**A,B**) Phospho-STAT1 protein

levels were determined 72 h post dosing with IFN-α, IFN-β, and IFN-γ in A253, pSS1 and nSS2 whole cell lysates. (**C**) phospho-STAT1 (P-STAT1) presence was measured by semiquantitative analysis based on relative band intensity. (**D,E**) phospho-STAT4 (P-STAT4) protein levels were determined 72 h post dosing with IFN-α, IFN-β, and IFN-γ in A253, pSS1 and nSS2 whole cell lysates. (**F**) P-STAT4 expression was measured by semiquantitative analysis measuring relative band intensity. Equal protein amounts were loaded in each lane and target bands were normalized to protein expression of cyclophilin A. Error bars represent the mean (+/−) standard error (SE). However, the Control value was set to 1 by default without error bar for better representation of the differences between cell lines. Loading controls and target proteins of the same blot were individually optimized for exposure requirements and reconstituted for imaging. IFN-γ had a significant effect on P-STAT1 expression for pSS1 and nSS2 cell lines ($p < 0.05$, paired *t*-test with Bonferroni correction). There was a marginal effect of IFN-γ on P-STAT4 expression in pSS1 cells ($p = 0.049$). ND is not detected.

3. Discussion

In this study, for the first time, we have shown differences in responses to EGF and interferons regarding the expression of STAT1 and STAT4 between *sicca* and SS patients' derived iSGECs at both mRNA and active protein levels. Multiple genetic polymorphisms in the STAT1-STAT4 risk locus located on chromosome 2 have been confirmed for their association with Sjögren's syndrome [25]. While both STAT1 and STAT4 have been associated with SS, it is unclear how polymorphisms affect these adjacent genes, whether in PBMCs or in salivary gland epithelial cells that are no longer considered as passive bystanders [26–29].

While we demonstrated differential reactivity to interferons and EGF between pSS1 and nSS2 iSGEC cell lines (more prominent for pSS1), we recognize that the SS-related biology of the cells will need to be confirmed in in vivo and ex vivo models for SS. In addition, Theander et al. 2015 [30] showed that SS autoantibodies may be present 20 years before SS diagnosis, which raises the question as to whether genetic polymorphisms, somatic mutations, or epigenetic changes may have occurred in these patients' salivary glands that would promote autoimmunity. Ideally, single cell analysis would help the field to develop iSGECs that are representative of defined categories of *sicca* and SS patient categories, carrying recently confirmed/identified single nucleotide polymorphisms (SNPs) [25], or would have been engineered to carry such SNPs by clustered regularly interspaced short palindromic repeats (CRISPR) gene editing. Thus, an appropriate panel of iSGECs carrying SS-associated SNPs might be suitable for initial drug screening assays or assays which will contribute to the understanding of onset and progression of the disease. Moreover, the role of interferons in SS pathogenesis and clinical trials testing inhibitors of IFN-related pathways have been extensively reviewed by Del Papa et al. 2021 [31]. Whether polymorphisms impact immune cells or SGECs in patients, testing IFN reactivity of iSGECs from different sources might contribute to the understanding of inefficacy of drugs tested so far in clinical trials.

3.1. STAT4 mRNA Expression Downregulation by EGF and p38 Inhibitor

Previous studies in mice models for Sjögren's syndrome have shown that treatment with p38 inhibitors have positive effects on the pathology of the disease [32–34]. In addition, there is an inverse correlation between EGF salivary levels and SS progression. Our results show downregulation of STAT4 at the mRNA level whether A253 or nSS2 cells were treated with EGF or p38 inhibitor alone or in combination, and to a greater extent for nSS2 cells. The significance of this result is unclear but might reflect an imbalance in JAK/STAT pathway [26]. Indeed, salivary gland epithelium homeostasis is maintained by numerous signaling pathways. EGF receptors are known to be involved in signaling pathways necessary for the TDL-4-mediated activation of downstream NF-κB pathway [26].

3.2. Effects of EGF and IFN-γ on Protein Expression of Phospho-STAT4 and Phospho-STAT1

The role of type I and type II interferons in the development of Sjögren's syndrome may not be understated [31]. However, their effects on the production of active forms of STAT1 and STAT have been mainly investigated in PBMCs and infiltrating immune cells within the salivary gland [31]. In this study, EGF increased phospho-STAT4 in all three cell lines A253, nSS2, and pSS1, and to a larger extent when EGF was combined with IFN-γ, suggests an additive effect. Since, EGF was shown to reduce STAT4 at the mRNA levels (Figure 2), there is a possibility of regulation of the production of phosphorylated STAT4 protein. Such regulation might involve microRNA stabilization process or increased phosphorylation due to an unknown mechanism. It remains also unclear how the p38-MAPK pathway is tied to this mechanism. In addition, phospho-STAT1 was remarkably overexpressed in all three cell lines when EGF and IFN-γ were used in combination, although such effect was most likely due to IFN-γ. In this case, pSS1 showed the highest induction which was higher than in nSS2. The results suggest more prominent deregulation of the JAK/STAT pathway in pSS cells compared to nSS2 cells. This result, overall, warrants further investigation of the multi-protein transcriptional complex involved on chromosome 2 in absence or presence of polymorphisms. The effect of polymorphisms in STAT1 and STAT4 associated with Sjögren's syndrome in cell culture models is currently under investigation in our laboratory.

3.3. Regulation of STAT1 and STAT4 mRNA and Active Protein Expression by Interferons α, β and γ

We further investigated the interferon type I and type II responses in the three cell lines A253, nSS2, and pSS1 to determine whether there were significant differences in STAT1 and STAT4 mRNA expression. As shown in results, there were two noticeable differences. IFN-γ induced an opposite trend in STAT4 mRNAS expression in pSS1 cells compared to A253 and nSS2 cells, suggesting a fundamental difference in reactivity. Additionally, the results of 9 replicates showed variability in response to IFN-γ in pSS1 cells. Assuming minimal technical error, this result might reflect an intrinsic feature in pSS1 possibly associated with epigenetic changes upon cell passaging, as processing of outliers did not improve statistical significance. Furthermore, type I and type II interferons induced similar responses in all three cell lines regarding STAT1 mRNA expression. However, nSS2 and pSS1 cells behaved differently, with lower STAT1 mRNA expression in pSS1 compared to nSS2. The results suggest unique properties of pSS1 cells compared to nSS2, possibly reflecting Sjögren's syndrome specific pathological features.

At the active protein level, IFN-γ had the largest effect on the expression of phospho-STAT1 in pSS1 cells, confirming intrinsic differences between pSS1 and nSS2 cells. In contrast, there was a small effect by type I and II IFNs on the expression of phospho-STAT4, for which a trend of greater effect in pSS1 was observed. Overall, results show consistency among the differences observed at both mRNA and active protein expression levels for the two genes.

3.4. Limitations

This study used two immortalized cell lines derived from one sicca and one anti-SSA negative Sjögren patients' salivary gland epithelial cells. Also, the study could have further benefitted from iSGECs derived from an anti-SSA positive Sjögren patient. Efforts in our lab are underway to circumvent these limitations. Interpretation of our data is in part limited due to insufficient data in animal models relevant to interferon and JAK/STAT related pathways in salivary gland epithelial cells. Additional experimentation using for instance siRNA and CRISPR knockdown technologies will be required to better understand the interaction between EGF, interferons, and the p38-MAPK pathway in regulating STAT1 and STAT4 and to characterize the impact of such interaction on JAK/STAT pathway in vitro and in vivo.

4. Materials and Methods

4.1. Reagents

Reagents used included: Epilife media with HKGS (Human Keratinocyte Growth Serum; medium K) (ThermoFisher Scientific, Waltham, MA, USA), McCoys 5A Medium; medium 5A) (Cytiva, Marlborough, MA, USA), FBS (VWR International, Radnor, PA, USA), LB Agar with AMP, Human immortalized salivary gland epithelial cells (iSGECs) (established in our laboratory), salivary gland cell line A253 (ATCC—American Type Culture Collection, Manassas, VA, USA), EGF (ThermoFisher Scientific, MA, USA), p38 Inhibitor (SB203580) (Cell Signaling Technology, Danvers, MA, USA), Quick-RNA Miniprep Kit (Zymo Research, Irvine, CA, USA), HEX random primers (IDT—Integrated DNA Technologies, Coralville, IA, USA), DTT, SmartScribe reverse transcriptase kit (Takara Bio, San Jose, CA, USA), dNTPs (NEB—New England Biolabs, Ipswich, MA, USA), SYBR green master mix (Qiagen, Germantown, MD, USA), primer (IDT—Integrated DNA Technologies, IA, USA), Western blot transfer membrane (Cytiva, MA, USA), and antibodies (Supplementary Table S1).

4.2. Cell Culture

Human iSGECs, and A253 (ATCC) were inoculated into media K and 5A (10% FBS), respectively, and treated with IFN-gamma (10 ng/mL) (Cell Signaling Technology), IFN-alpha (10 ng/mL) (Cell Signaling Technology), IFN-beta (10 ng/mL) (Proteintech, Rosemont, IL, USA), EGF (10 ng/mL) (ThermoFisher), and/or p38 inhibitor (20 µM) (Cell Signaling Technology) for 24 and 48 h. Cells were cultured in CO_2 incubator (37 °C; 5% CO_2). After 24 and 48 h incubation, the cells adhered to the well surface indicating successful recovery. Cells were then used for RNA extraction, conversion to cDNA for further qRT-PCR.

4.3. Quantitative Real-Time RT-PCR

Total RNA was isolated from cells using the Quick-RNA Miniprep kit (Zymo) following the manufacturer protocol. RNA (500 ng) was reverse transcribed from each sample using SmartScribe reverse transcriptase kit (Takara) following the provided instructions. Random hexamers (IDT) and dNTP mix (NEB) were used. Expression levels of STAT4 and STAT1 were determined relative to GAPDH based on the $\Delta\Delta CT$ method using SYBR Green mix (Qiagen). qRT-PCR primers are listed in Supplementary Table S2.

4.4. Western Blot

Cells were grown in 6-well plates and were serum starved for 24 h. Treated and untreated cells were grown for 72 h before harvesting whole cell lysates for nuclear protein using MPER (Mammalian Protein Extraction Reagent) (ThermoFisher). Levels of target proteins, phospho-tyrosine-701-STAT1 and phospho-serine-721-STAT4 in whole cell lysates of treated and untreated iSGECs and A253 cells, were determined by Western blot. Primary antibodies were used at appropriate dilutions (Supplementary Table S1). Secondary antibodies anti-mouse IgG-HRP and anti-rabbit IgG-HRP were used according to primary antibodies. Supersignal West Pico and Femto solutions were used for signal detection (ThermoFisher). ImageQuant LAS4000 (GE) was used for imaging and Image Studio™ Lite v2.5 (Li-COR Biosciences – U.S.) for data processing.

4.5. Statistical Analysis

Differences of expression in the cell lines of qRT-PCR results relative to the control or between cell lines were analyzed by Mann–Whitney U-test (9 replicates for interferons treatments) or t-test (3 replicates for p38 and EGF treatments) (alpha = 0.05). Wilcoxon signed-rank test with Bonferroni correction was used on Western blot data normalized to the loading control to compare treatments with EGF alone and EGF and IFN-γ combined. Paired t-test with Bonferroni correction was used on Western blot data normalized to the loading control to compare treatments with IFN-α, IFN-β and IFN-γ relative to the experimental control.

5. Conclusions

Immortalized salivary gland epithelial cells derived from *sicca* and Sjögren syndrome patients may constitute useful tools to pinpoint and investigate specific disease mechanisms and genetic abnormalities. Indeed, *sicca* and Sjögren syndrome patients represent a broad spectrum of disease etiologies. Thus, future genomic single cell analysis may provide further clues as to how better target the JAK/STAT pathway to modulate the immune response to improve clinical outcomes in Sjögren's syndrome patients.

Supplementary Materials: The following supporting information can be downloaded at: https://www.mdpi.com/article/10.3390/ijms25063166/s1.

Author Contributions: Conceptualization: J.-L.C.M., F.B.M. and B.D.N.; Experiments: B.D.N. and T.E.T.; Writing: J.-L.C.M., F.B.M., B.D.N. and T.E.T.; Review and Revisions: J.-L.C.M., F.B.M., M.T.B., B.D.N. and T.E.T. All authors have read and agreed to the published version of the manuscript.

Funding: This study was funded by Atrium Health research fund.

Institutional Review Board Statement: Not applicable.

Informed Consent Statement: Not applicable.

Data Availability Statement: The data presented in this study are available on request from the corresponding author. The data are not publicly available as no genome-wide data have been generated/analysed in this study.

Acknowledgments: We thank our project manager Micaela Beckman for her excellent editorial skills.

Conflicts of Interest: The authors declare no conflicts of interest.

References

1. Parisis, D.; Chivasso, C.; Perret, J.; Soyfoo, M.S.; Delporte, C. Current State of Knowledge on Primary Sjögren's Syndrome, an Autoimmune Exocrinopathy. *J. Clin. Med.* **2020**, *9*, 2299. [CrossRef]
2. Ogawa, Y.; Takeuchi, T.; Tsubota, K. Autoimmune Epithelitis and Chronic Inflammation in Sjögren's Syndrome-Related Dry Eye Disease. *Int. J. Mol. Sci.* **2021**, *22*, 11820. [CrossRef] [PubMed]
3. Selmi, C.; Gershwin, M.E. Chronic Autoimmune Epithelitis in Sjögren's Syndrome and Primary Biliary Cholangitis: A Comprehensive Review. *Rheumatol. Ther.* **2017**, *4*, 263–279. [CrossRef] [PubMed]
4. Manganelli, P.; Fietta, P. Apoptosis and Sjögren syndrome. *Semin. Arthritis Rheum.* **2003**, *33*, 49–65. [CrossRef] [PubMed]
5. Ping, L.; Ogawa, N.; Zhang, Y.; Sugai, S.; Masaki, Y.; Weiguo, X. p38 mitogen-activated protein kinase and nuclear factor-κB facilitate CD40-mediated salivary epithelial cell death. *J. Rheumatol.* **2012**, *39*, 1256–1264. [CrossRef]
6. Visconti, R.; Gadina, M.; Chiariello, M.; Chen, E.H.; Stancato, L.F.; Gutkind, J.S.; O'Shea, J.J. Importance of the MKK6/p38 pathway for interleukin-12–induced STAT4 serine phosphorylation and transcriptional activity. *Blood* **2000**, *96*, 1844–1852. [CrossRef]
7. Korman, B.D.; Kastner, D.L.; Gregersen, P.K.; Remmers, E.F. STAT4: Genetics, mechanisms, and implications for autoimmunity. *Curr. Allergy Asthma Rep.* **2008**, *8*, 398–403. [CrossRef]
8. Deng, Y.; Tsao, B.P. Genetic susceptibility to systemic lupus erythematosus in the genomic era. *Nat. Rev. Rheumatol.* **2010**, *6*, 683–692. [CrossRef]
9. Hu, Q.; Bian, Q.; Rong, D.; Wang, L.; Song, J.; Huang, H.S.; Zeng, J.; Mei, J.; Wang, P.Y. JAK/STAT pathway: Extracellular signals, diseases, immunity, and therapeutic regimens. *Front. Bioeng. Biotechnol.* **2023**, *11*, 1110765. [CrossRef]
10. Gestermann, N.; Mekinian, A.; Comets, E.; Loiseau, P.; Puechal, X.; Hachulla, E.; Gottenberg, J.E.; Mariette, X.; Miceli-Richard, C. STAT4 is a confirmed genetic risk factor for Sjögren's syndrome and could be involved in type 1 interferon pathway signaling. *Genes Immun.* **2010**, *11*, 432–438. [CrossRef]
11. Awasthi, N.; Liongue, C.; Ward, A.C. STAT proteins: A kaleidoscope of canonical and non-canonical functions in immunity and cancer. *J. Hematol. Oncol.* **2021**, *14*, 198. [CrossRef]
12. O'Shea, J.J.; Holland, S.M.; Staudt, L.M. JAKs and STATs in immunity, immunodeficiency, and cancer. *N. Engl. J. Med.* **2013**, *368*, 161–170. [CrossRef]
13. Banerjee, S.; Biehl, A.; Gadina, M.; Hasni, S.; Schwartz, D.M. JAK-STAT Signaling as a Target for Inflammatory and Autoimmune Diseases: Current and Future Prospects. *Drugs* **2017**, *77*, 521–546. [CrossRef] [PubMed]
14. Hu, X.; Li, J.; Fu, M.; Zhao, X.; Wang, W. The JAK/STAT signaling pathway: From bench to clinic. *Signal Transduct. Target. Ther.* **2021**, *6*, 402. [CrossRef]
15. Schwartz, D.M.; Kanno, Y.; Villarino, A.; Ward, M.; Gadina, M.; O'Shea, J.J. Erratum: JAK inhibition as a therapeutic strategy for immune and inflammatory diseases. *Nat. Rev. Drug Discov.* **2018**, *17*, 78. [CrossRef]

16. Luo, K. Signaling Cross Talk between TGF-β/Smad and Other Signaling Pathways. *Cold Spring Harb. Perspect. Biol.* **2017**, *9*, a022137. [CrossRef]
17. Azuma, N.; Katada, Y.; Kitano, S.; Nishioka, A.; Sekiguchi, M.; Kitano, M.; Hashimoto, N.; Matsui, K.; Iwasaki, T.; Sano, H. Salivary epidermal growth factor (EGF) in Sjögren's syndrome: Association between salivary EGF levels and the severity of intraoral manifestations. *Nihon Rinsho Meneki Gakkai Kaishi* **2016**, *39*, 42–50. [CrossRef] [PubMed]
18. Sisto, M.; Lisi, S.; Lofrumento, D.D.; D'amore, M.; Frassanito, M.A.; Ribatti, D. Sjögren's syndrome pathological neovascularization is regulated by VEGF-A-stimulated TACE-dependent crosstalk between VEGFR2 and NF-κB. *Genes Immun.* **2012**, *13*, 411–420. [CrossRef] [PubMed]
19. Huang, Z.Y.; Yuan, Y.R.; Kong, Y.L.; Zhang, T.; Liang, Y. Vascular Endothelial Growth Factor-A Is Associated with Platelets and Complement 4 in Patients with Primary Sjögren's Syndrome. *Ann. Clin. Lab. Sci.* **2020**, *50*, 790–796. [PubMed]
20. Bartoloni, E.; Baldini, C.; Schillaci, G.; Quartuccio, L.; Priori, R.; Carubbi, F.; Bini, V.; Alunno, A.; Bombardieri, S.; De Vita, S.; et al. Cardiovascular disease risk burden in primary Sjögren's syndrome: Results of a population-based multicentre cohort study. *J. Intern. Med.* **2015**, *278*, 185–192. [CrossRef]
21. Błochowiak, K.J.; Trzybulska, D.; Olewicz-Gawlik, A.; Sikora, J.J.; Nowak-Gabryel, M.; Kocięcki, J.; Witmanowski, H.; Sokalski, J. Levels of EGF and VEGF in patients with primary and secondary Sjögren's syndrome. *Adv. Clin. Exp. Med.* **2018**, *27*, 455–461. [CrossRef] [PubMed]
22. Miklossy, G.; Hilliard, T.S.; Turkson, J. Therapeutic modulators of STAT signalling for human diseases. *Nat. Reviews. Drug Discov.* **2013**, *12*, 611–629. [CrossRef]
23. George, G.; Shyni, G.L.; Abraham, B.; Nisha, P.; Raghu, K.G. Downregulation of TLR4/MyD88/p38MAPK and JAK/STAT pathway in RAW 264.7 cells by Alpinia galanga reveals its beneficial effects in inflammation. *J. Ethnopharmacol.* **2021**, *275*, 114132. [CrossRef]
24. Thomas, C.; Couch, D.; Wang, B. p38-MAPK and JAK/STAT Pathway Inhibition Reduces Indoxyl Sulfate-Inducted Impairment of Human Edothelial Cells. *Heart Lung Circ.* **2022**, *31*, S313–S314. [CrossRef]
25. Khatri, B.; Tessneer, K.L.; Rasmussen, A.; Aghakhanian, F.; Reksten, T.R.; Adler, A.; Alevizos, I.; Anaya, J.M.; Aqrawi, L.A.; Baecklund, E.; et al. Genome-wide association study identifies Sjögren's risk loci with functional implications in immune and glandular cells. *Nat. Commun.* **2022**, *13*, 4287. [CrossRef] [PubMed]
26. Gandolfo, S.; Ciccia, F. JAK/STAT Pathway Targeting in Primary Sjögren Syndrome. *Rheumatol. Immunol. Res.* **2022**, *3*, 95–102. [CrossRef] [PubMed]
27. Pertovaara, M.; Silvennoinen, O.; Isomäki, P. Cytokine-induced STAT1 activation is increased in patients with primary Sjögren's syndrome. *Clin. Immunol.* **2016**, *165*, 60–67. [CrossRef]
28. Colafrancesco, S.; Ciccacci, C.; Priori, R.; Latini, A.; Picarelli, F.; Arienzo, F.; Novelli, G.; Valesini, G.; Perricone, C.; Borgiani, P. STAT4, TRAF3IP2, IL10, and HCP5 Polymorphisms in Sjögren's Syndrome: Association with Disease Susceptibility and Clinical Aspects. *J. Immunol. Res.* **2019**, *2019*, 7682827. [CrossRef]
29. Rivière, E.; Pascaud, J.; Tchitchek, N.; Boudaoud, S.; Paoletti, A.; Ly, B.; Dupré, A.; Chen, H.; Thai, A.; Allaire, N.; et al. Salivary gland epithelial cells from patients with Sjögren's syndrome induce B-lymphocyte survival and activation. *Ann. Rheum. Dis.* **2020**, *79*, 1468–1477. [CrossRef]
30. Theander, E.; Jonsson, R.; Sjöström, B.; Brokstad, K.; Olsson, P.; Henriksson, G. Prediction of Sjögren's Syndrome Years Before Diagnosis and Identification of Patients With Early Onset and Severe Disease Course by Autoantibody Profiling. *Arthritis Rheumatol.* **2015**, *67*, 2427–2436. [CrossRef]
31. Del Papa, N.; Minniti, A.; Lorini, M.; Carbonelli, V.; Maglione, W.; Pignataro, F.; Montano, N.; Caporali, R.; Vitali, C. The Role of Interferons in the Pathogenesis of Sjögren's Syndrome and Future Therapeutic Perspectives. *Biomolecules* **2021**, *11*, 251. [CrossRef] [PubMed]
32. Nakamura, H.; Kawakami, A.; Yamasaki, S.; Kawabe, Y.; Nakamura, T.; Eguchi, K. Expression of mitogen activated protein kinases in labial salivary glands of patients with Sjögren's syndrome. *Ann. Rheum. Dis.* **1999**, *58*, 382–385. [CrossRef] [PubMed]
33. Cao, N.; Shi, H.; Chen, C.; Zheng, L.; Yu, C. Inhibition of the TLR9-dependent p38 MAPK signaling pathway improves the pathogenesis of primary Sjögren's syndrome in the NOD/Ltj mouse. *J. Biol. Regul. Homeost. Agents* **2021**, *35*, 1103–1108. [CrossRef] [PubMed]
34. Ma, X.; Zou, J.; He, L.; Zhang, Y. Dry eye management in a Sjögren's syndrome mouse model by inhibition of p38-MAPK pathway. *Diagn. Pathol.* **2014**, *9*, 5. [CrossRef]

Disclaimer/Publisher's Note: The statements, opinions and data contained in all publications are solely those of the individual author(s) and contributor(s) and not of MDPI and/or the editor(s). MDPI and/or the editor(s) disclaim responsibility for any injury to people or property resulting from any ideas, methods, instructions or products referred to in the content.

Article

Corylin Attenuates CCl₄-Induced Liver Fibrosis in Mice by Regulating the GAS6/AXL Signaling Pathway in Hepatic Stellate Cells

Chin-Chuan Chen [1,2,†], Chi-Yuan Chen [1,3,†], Chau-Ting Yeh [4], Yi-Tsen Liu [1], Yann-Lii Leu [1,2], Wen-Yu Chuang [5,6], Yin-Hwa Shih [7], Li-Fang Chou [8], Tzong-Ming Shieh [9,*] and Tong-Hong Wang [1,2,3,4,*]

1. Biobank, Chang Gung Memorial Hospital, Tao-Yuan 33305, Taiwan; chinchuan@mail.cgu.edu.tw (C.-C.C.); d49417002@gmail.com (C.-Y.C.); crea456m@gmail.com (Y.-T.L.); ylleu@mail.cgu.edu.tw (Y.-L.L.)
2. Graduate Institute of Natural Products, Chang Gung University, Tao-Yuan 33303, Taiwan
3. Graduate Institute of Health Industry and Technology, Research Center for Chinese Herbal Medicine and Research Center for Food and Cosmetic Safety, Chang Gung University of Science and Technology, Tao-Yuan 33303, Taiwan
4. Liver Research Center, Department of Hepato-Gastroenterology, Chang Gung Memorial Hospital, Tao-Yuan 33305, Taiwan; chautingy@gmail.com
5. Department of Anatomic Pathology, Chang Gung Memorial Hospital, Tao-Yuan 33305, Taiwan; s12126@cgmh.org.tw
6. College of Medicine, Chang Gung University, Tao-Yuan 33303, Taiwan
7. Department of Healthcare Administration, Asia University, Taichung 41354, Taiwan; evashih@gm.asia.edu.tw
8. Kidney Research Center, Chang Gung Memorial Hospital, Tao-Yuan 33305, Taiwan; d928209@gmail.com
9. School of Dentistry, China Medical University, Taichung 40402, Taiwan
* Correspondence: tmshieh@mail.cmu.edu.tw (T.-M.S.); cellww@gmail.com (T.-H.W.); Tel.: +886-4-2205-3366-6 (ext. 2316) (T.-M.S.); +886-3-3281200 (ext. 7713) (T.-H.W.)
† These authors contributed equally to the work.

Citation: Chen, C.-C.; Chen, C.-Y.; Yeh, C.-T.; Liu, Y.-T.; Leu, Y.-L.; Chuang, W.-Y.; Shih, Y.-H.; Chou, L.-F.; Shieh, T.-M.; Wang, T.-H. Corylin Attenuates CCl₄-Induced Liver Fibrosis in Mice by Regulating the GAS6/AXL Signaling Pathway in Hepatic Stellate Cells. *Int. J. Mol. Sci.* **2023**, *24*, 16936. https://doi.org/10.3390/ijms242316936

Academic Editors: Sabrina Lisi and Margherita Sisto

Received: 3 November 2023
Revised: 27 November 2023
Accepted: 28 November 2023
Published: 29 November 2023

Copyright: © 2023 by the authors. Licensee MDPI, Basel, Switzerland. This article is an open access article distributed under the terms and conditions of the Creative Commons Attribution (CC BY) license (https://creativecommons.org/licenses/by/4.0/).

Abstract: Liver fibrosis is reversible when treated in its early stages and when liver inflammatory factors are inhibited. Limited studies have investigated the therapeutic effects of corylin, a flavonoid extracted from *Psoralea corylifolia* L. (Fabaceae), on liver fibrosis. Therefore, we evaluated the anti-inflammatory activity of corylin and investigated its efficacy and mechanism of action in ameliorating liver fibrosis. Corylin significantly inhibited inflammatory responses by inhibiting the activation of mitogen-activated protein kinase signaling pathways and the expression of interleukin (IL)-1β, IL-6, and tumor necrosis factor-alpha in human THP-1 and mouse RAW264.7 macrophages. Furthermore, corylin inhibited the expression of growth arrest-specific gene 6 in human hepatic stellate cells (HSCs) and the activation of the downstream phosphoinositide 3-kinase/protein kinase B pathway. This inhibited the activation of HSCs and the expression of extracellular matrix proteins, including α-smooth muscle actin and type I collagen. Additionally, corylin induced caspase 9 and caspase 3 activation, which promoted apoptosis in HSCs. Moreover, in vivo experiments confirmed the regulatory effects of corylin on these proteins, and corylin alleviated the symptoms of carbon tetrachloride-induced liver fibrosis in mice. These findings revealed that corylin has anti-inflammatory activity and inhibits HSC activation; thus, it presents as a potential adjuvant in the treatment of liver fibrosis.

Keywords: corylin; anti-inflammation; liver fibrosis; hepatic stellate cell; growth arrest-specific gene 6/AXL signaling pathway

1. Introduction

Liver fibrosis, caused by viral or metabolic chronic liver diseases, is a major challenge of global health [1,2]. In Taiwan, approximately 60% of patients with liver cancer have previously suffered from viral hepatitis B and C infections [3–5]. Chronic hepatitis causes repeated liver inflammation and activates hepatic stellate cells (HSCs) to secrete collagen for tissue repair. Consequently, the extracellular matrix (ECM) accumulates during repeated inflammation and repair, leading to liver fibrosis and liver cirrhosis. Patients with cirrhosis are 60–250 times more likely to develop liver cancer than those without liver disease [6].

Liver fibrosis can be reversed by administering treatment at an early stage and inhibiting the factors that cause liver inflammation [7]. Current treatment options for liver fibrosis can be classified based on three strategies: the inhibition of liver inflammation, the inhibition of HSC activation, and the acceleration of ECM breakdown. In the clinical setting, the administration of antiviral drugs alone, such as entecavir or lamivudine, or interferon (IFN) alone, may help ameliorate liver fibrosis caused by viral hepatitis; however, these treatments show low efficacy [8,9]. The treatment outcomes of liver fibrosis can be effectively improved if an appropriate adjuvant is administered to suppress inflammation or inhibit HSC activation. However, no clinically effective drugs are currently available for treating liver fibrosis with low side effects; thus, continuous research and development are required.

HSCs play a key role in the progression of liver fibrosis. When liver tissues are injured or stimulated by oxidative stress or inflammatory cytokines, HSCs are activated followed by proliferation and transformation into fibrogenic cells, thereby synthesizing large amounts of ECM. HSC activation is regulated by various pathways, with the growth arrest-specific 6 (GAS6)/AXL receptor tyrosine kinase (AXL) being a key regulatory pathway [10–12]. AXL, a member of the TYRO-AXL-MER (TAM) receptor tyrosine kinase (RTK) family, is mainly expressed in neural, vascular, immune, and stellate cells and is involved in the regulation of cellular physiological processes, such as growth, survival, differentiation, adhesion, and migration [13]. The binding of AXL to its ligand protein, GAS6, initiates autophosphorylation, which further activates the downstream phosphoinositide 3-kinase/protein kinase B (PI3K/AKT), rat sarcoma/rapidly accelerated fibrosarcoma kinase/mitogen-activated protein kinase (MAPK) kinase/extracellular signal-regulated kinase (RAS/RAF/MEK/ERK), and wingless-related integration site (Wnt) signaling pathways, thereby promoting cell growth, migration, and angiogenesis and inhibiting apoptosis [14]. The GAS6/AXL signaling pathway regulates HSC proliferation and activation, which play an important role in liver fibrosis development. The treatment of carbon tetrachloride (CCl_4)-exposed mice with an AXL inhibitor effectively alleviates fibrosis symptoms [10]. Therefore, at present, TAM receptors, including AXL, are considered as key targets for treating liver fibrosis [15,16].

Natural products contain diverse pharmacophores and highly complex stereochemistry, and most of them have low physiological toxicity. Therefore, natural products have always represented important sources for new drug development [17,18]. Compounds, such as paclitaxel, curcumin, camptothecin, and their derivatives, have been used to treat various cancers, such as breast cancer, lung cancer, colorectal cancer, and melanoma, as they significantly prolong patient survival time [19–23]. Other natural compounds, such as resveratrol, metformin, magnolol, sulforaphane, and diallyl disulfide, exhibit anti-inflammatory activity and have the potential to be used in the treatment of inflammatory diseases [24–28].

Psoralea corylifolia L. (cullen corylifolium; Fabaceae) is an herb widely used for treating bacterial infections, inflammation, and cancers in many Asian countries [29–31]. Its polyphenolic extracts, such as psoralen, isopsoralen, and psoralidin; flavonoid extracts, such as bavachin, isobavachalcone, and neobavaisoflavone; and the phenolic extract backuchiol have all been identified as biologically active with different therapeutic effects [32]. Corylin, a flavonoid isolated from the fruits of *P. corylifolia* L., exerts an anti-inflammatory effect by inhibiting the expression of inducible nitric oxide synthase and cyclooxygenase which is increased during bacterial infections [33,34]. In addition, the antioxidant, anti-aging, and anti-tumor activities of corylin have also been reported recently [35–38], and have also shown the therapeutic potential of corylin in hyperlipidemia, insulin resistance, atherosclerosis, hepatocellular carcinoma, and neurological diseases [39–41]. We previously showed that corylin ameliorates obesity by activating adipocyte browning and reduces hepatic steatosis and hepatic fibrosis in high-fat diet (HFD)-fed mice [38]. However, the molecular mechanism of corylin's anti-inflammation and anti-hepatic fibrosis effects have not yet been fully clarified. Therefore, in this study, we investigated the anti-inflammatory and therapeutic effects of corylin on liver fibrosis and further clarified its downstream regulatory mechanisms. Our findings showed that corylin has anti-inflammatory activity and inhibits HSC activation; thus, it can be used as a potential adjuvant in the treatment of liver fibrosis.

2. Results

2.1. Corylin Treatment Suppressed Lipopolysaccharide-Induced Pro-Inflammatory Cytokine Production in THP-1 and RAW264.7 Cells

To determine whether corylin exhibits anti-inflammatory activity, we treated human monocyte THP-1 cells and RAW 264.7 mouse macrophage cells with different concentrations of corylin for 2 h followed by lipopolysaccharide (LPS) treatment for 24 h to induce an inflammatory response. The culture media were collected to perform an enzyme-linked immunosorbent assay (ELISA) to analyze the expression of pro-inflammatory cytokines. LPS treatment significantly increased the expression of cytokines, such as interleukin (IL)-1β, IL-6, and tumor necrosis factor alpha (TNF-α), in THP-1 and RAW264.7 cells. The expression of these pro-inflammatory cytokines was significantly reduced in corylin-treated cells compared to that in the control group (dimethyl sulfoxide (DMSO)-treated), indicating that corylin exhibited anti-inflammatory activity and inhibited the expression of pro-inflammatory cytokines (Figure 1).

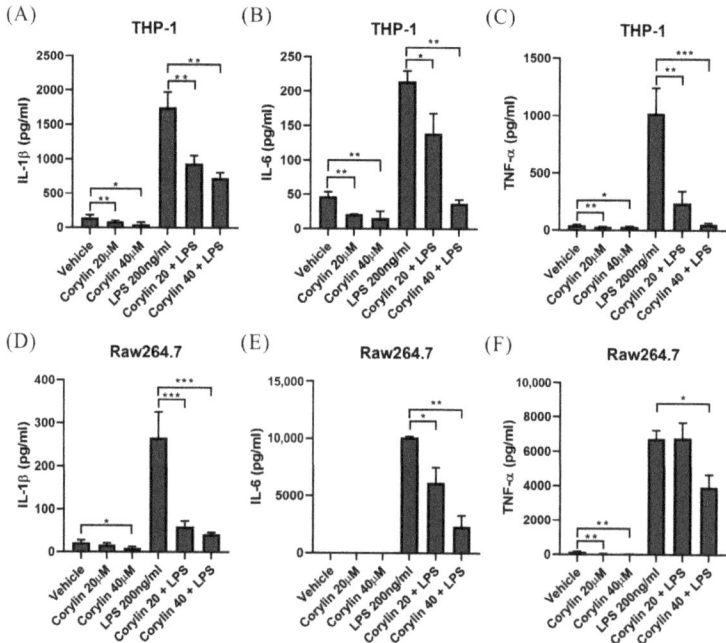

Figure 1. Corylin significantly inhibits IL-1β, IL-6, and TNF-α expression in THP-1 and RAW264.7 cells. Expression of pro-inflammatory cytokines, according to enzyme-linked immunosorbent assays, in (**A–C**) THP-1 and (**D–F**) RAW264.7 cells that were treated with different concentrations of corylin or vehicle for 2 h and then treated with 200 ng/mL LPS to induce an inflammatory response for 24 h. All data are expressed as the mean ± standard deviations of three independent experiments. $p < 0.05$ (*), $p < 0.01$ (**), $p < 0.001$ (***). LPS, lipopolysaccharide; IL-1β, interleukin 1 beta; TNF-α, tumor necrosis factor alpha; TGF-β, transforming growth factor beta.

2.2. Corylin Treatment Inhibited the Activation of MAPK Signaling Pathways in LPS-Stimulated THP1 and RAW264.7 Cells

To further determine whether the anti-inflammatory effect of corylin was associated with MAPK signaling pathways, THP-1 and RAW264.7 cells were pre-treated with corylin and stimulated with LPS. Subsequently, the phosphorylation levels of c-Jun N-terminal kinase, ERK, and p38 proteins were analyzed using Western blotting. LPS treatment activated the aforementioned MAPKs, which subsequently upregulated pro-inflammatory cytokines. However, in corylin-treated cells, the activation of these kinases was significantly inhibited (Figure 2), leading to the decreased expression of pro-inflammatory cytokines. Therefore, the anti-inflammatory activity of corylin was mediated by blocking the activation of MAPK signaling pathways.

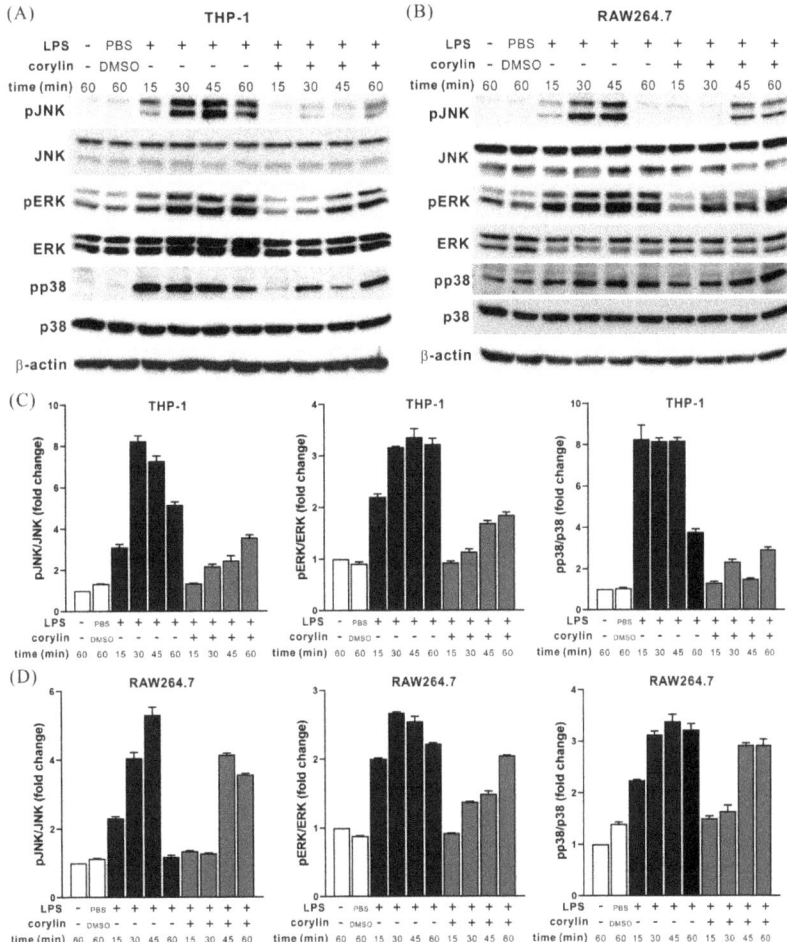

Figure 2. Corylin inhibits mitogen-activated protein kinase activation in THP-1 and RAW264.7 cells. (**A**) THP-1 cells and (**B**) RAW264.7 cells stimulated with 200 ng/mL LPS and 40 mM corylin for the indicated time period, and the activities of JNK, ERK, and p38 examined using Western blot analysis with phosphospecific antibodies are shown. (**C,D**) The total protein levels of JNK, ERK, and p38 were measured, and quantitative results are shown. LPS, lipopolysaccharide; JNK, c-Jun N-terminal kinase; ERK, extracellular signal-regulated kinase; p-JNK, phosphorylated JNK; IL-1β, interleukin 1 beta; TNF-α, tumor necrosis factor alpha; TGF-β, transforming growth factor beta.

2.3. Corylin Treatment Alleviated the Symptoms of CCl$_4$-Induced Liver Fibrosis in Mice

To confirm the anti-inflammatory activity of corylin and its efficacy in treating liver fibrosis in vivo, BALB/c mice were intraperitoneally injected with CCl$_4$ (0.5 µL/g body weight) twice a week for six weeks to induce liver fibrosis. Further, the mice were intraperitoneally injected with/without corylin (30 mg/kg of body weight). After six weeks of CCl$_4$ treatment, the mice exhibited significant fibrosis of the liver tissue, whereas liver fibrosis in corylin-treated mice was significantly alleviated compared with that in mice without corylin treatment (Figure 3A–C). Serological analysis also revealed that liver function indicator levels, including aspartate aminotransferase (AST) and alanine transaminase (ALT), in CCl$_4$-treated mice were 8–10-fold higher than those in untreated mice, indicating that their liver tissues were in an inflammatory and injured state. In contrast, liver function indicator levels in corylin-treated mice were significantly reduced, indicating that corylin effectively inhibited liver tissue inflammation and injury caused by CCl$_4$ (Figure 3D).

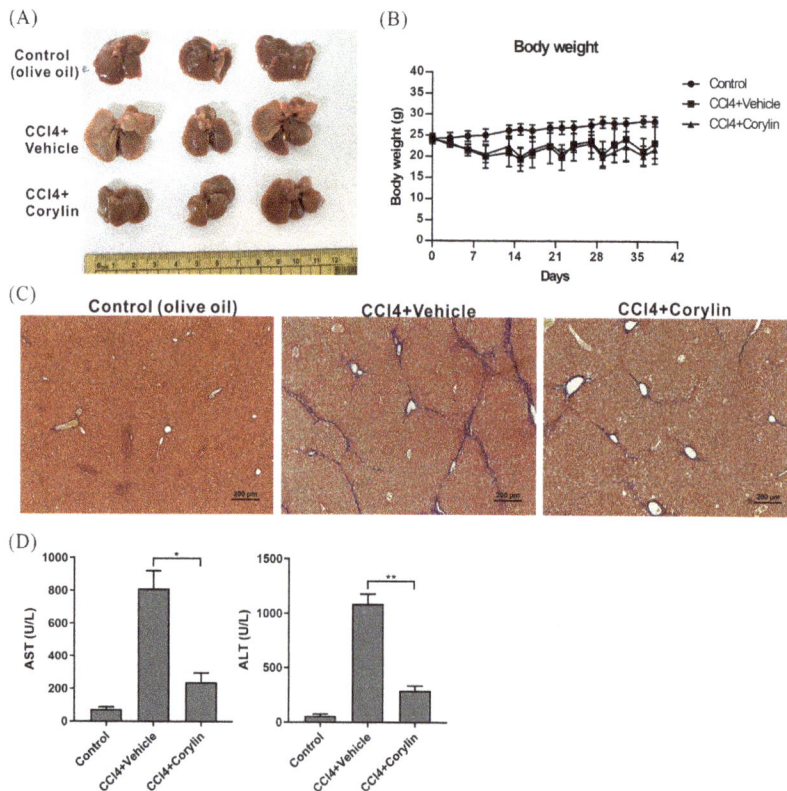

Figure 3. Effect of corylin on liver fibrosis in mice induced by CCl$_4$ treatment. (**A**) Mice livers treated with CCl$_4$, CCl$_4$ + corylin, or olive oil (NC), as described in Section 4, are shown. Corylin treatment reduces liver fibrosis symptoms in CCl$_4$-exposed mice six weeks after drug administration. (**B**) Body weights measured every three days after CCl$_4$ injection are shown. (**C**) Masson's trichrome staining reveals the effects of corylin on CCl$_4$-induced liver fibrosis. (**D**) Effect of corylin on serum AST and ALT levels in mice. $p < 0.05$ (*), $p < 0.01$ (**). CCl$_4$, carbon tetrachloride; AST, aspartate aminotransferase; ALT, alanine transaminase.

2.4. Corylin Treatment Inhibited HSC Activation

Liver fibrosis is caused by the excessive accumulation of ECM proteins, such as collagen, which are secreted by activated HSCs. To determine whether corylin inhibits

HSC activation, HHSteC cells were treated with corylin or a vehicle for 2 h, followed by transforming growth factor-β (TGF-β) treatment for 24 h to stimulate cell activation. Western blotting was performed to analyze the expression of alpha-smooth muscle actin (α-SMA) and collagen 1A to determine the effects of corylin on HSC activation. The expression of α-SMA and collagen 1A decreased significantly in HHSteC cells treated with corylin compared with that in the control group, indicating that corylin inhibited HSC activation (Figure 4A,B). In addition, immunohistochemical staining also showed that the expression of α-SMA and collagen 1A was significantly reduced in the tissues of corylin-treated mice compared to that in mice in the control group (Figure 4C,D). Therefore, corylin retards the progression of liver fibrosis in mice by inhibiting HSC activation.

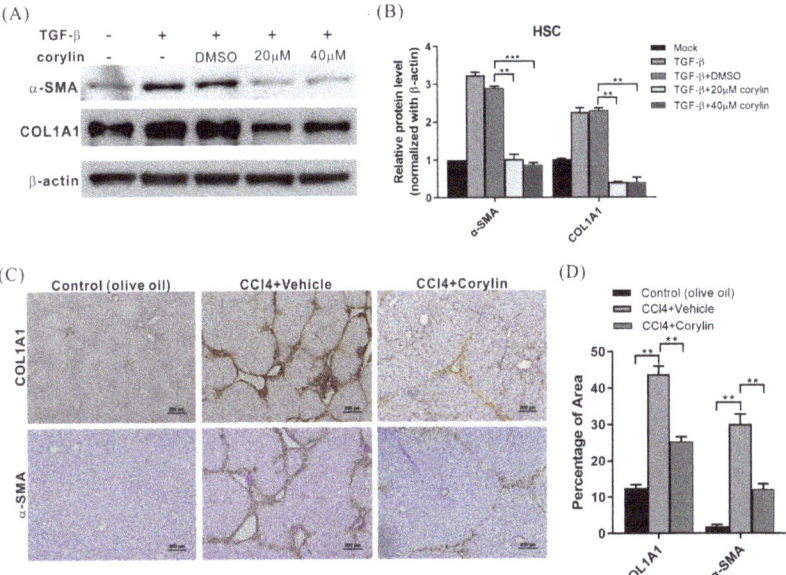

Figure 4. Corylin inhibits HSC activation. (**A**) Western blot analysis of human HSCs HHSteC cells treated with corylin or vehicle for 2 h, followed by TGF-β (4 ng/mL) treatment for 24 h to stimulate cell activation, to analyze α-SMA and COL1A1 protein expression, and to determine the effects of corylin on HSC activation. Quantitative results are shown in (**B**). All data are expressed as the mean ± standard deviations of three independent experiments. $p < 0.01$ (**), $p < 0.001$ (***). (**C**) Immunohistochemical staining representing the effects of corylin on COL1A1 and α-SMA expression in mouse livers. Quantitative results are shown in (**D**). $p < 0.01$ (**). HSCs, hepatic stellate cells; COL1A1, collagen 1A; α-SMA, smooth muscle-actin; TGF-β, transforming growth factor beta; DMSO, dimethyl sulfoxide.

2.5. Corylin Treatment Inhibited HSC Activation by Suppressing GAS6 Expression and Downstream PI3K/AKT Pathway Activation

The GAS6/AXL signaling pathway is important in regulating HSC activation [10]. To determine the effects of corylin on the expression of GAS6 and AXL and their downstream regulatory pathways, HHSteC cells were treated with corylin or vehicle for 2 h, followed by TGF-β treatment for 24 h to stimulate cell activation. Western blotting was performed to analyze the effects of corylin on the GAS6/AXL signaling pathway. GAS6 expression was significantly reduced in corylin-treated cells compared to that in the control group (DMSO-treated), and the activation of the downstream PI3K/AKT signaling pathway was also significantly inhibited. This indicated that corylin inhibits GAS6 expression and downstream signaling pathway activation in HSCs, which subsequently inhibits the expression of ECM proteins, such as α-SMA and collagen (Figure 5).

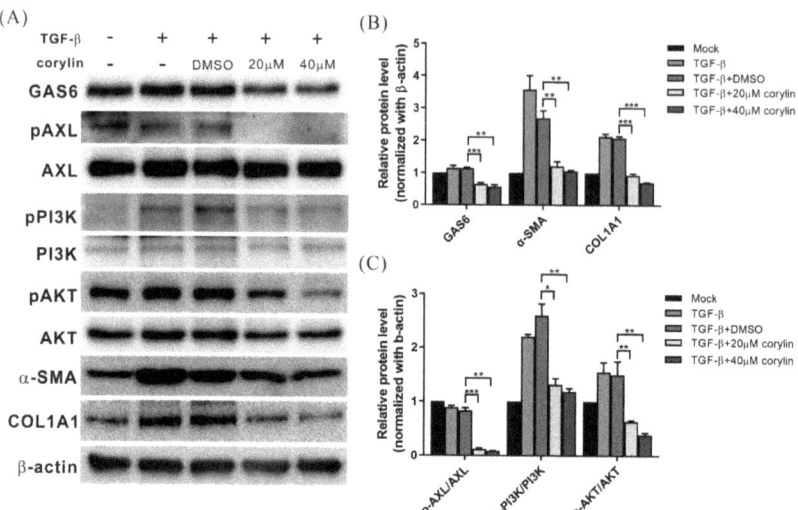

Figure 5. Corylin inhibits HSC activation by suppressing GAS6 expression and downstream PI3K/AKT signaling. (**A**) Western blotting results of cell lysates of HHSteC cells treated with corylin or DMSO for 2 h, followed by TGF-β (4 ng/mL) treatment for 24 h to stimulate cell activation and to determine the effects of corylin on the GAS6/AXL signaling pathway. β-Actin is the internal control. Quantitative results are shown in (**B**,**C**). All data are expressed as the mean ± standard deviations of three independent experiments. $p < 0.05$ (*), $p < 0.01$ (**), $p < 0.001$ (***). HSCs, hepatic stellate cells; GAS6, growth arrest-specific gene 6; PI3K/AKT, phosphoinositide 3-kinase/protein kinase B; p-PI3K, phosphorylated PI3K; DMSO, dimethyl sulfoxide; TGF-β, transforming growth factor beta; COL1A1, collagen 1A; α-SMA, smooth muscle-actin.

2.6. Corylin Inhibited the Expression of MMP Inhibitors, TIMP-1 and TIMP-2, in HSCs

In addition to ECM, HSCs express matrix metalloproteinase (MMP)-2 and MMP-9 along with their inhibitors, tissue inhibitor of metalloproteinase (TIMP)-1 and TIMP-2, to regulate ECM breakdown. To evaluate the effects of corylin on the expression of these proteins, HHSteC cells were treated with corylin or vehicle for 2 h, followed by TGF-β treatment to stimulate cell activation. Cell lysates were collected after 24 h for Western blotting to analyze the expression of MMP-2, TIMP-1, and TIMP-2. The results showed that TIMP-1 and TIMP-2 expression in the corylin-treated group was significantly lower than that in the control group. In contrast, there was a minor decrease in the expression of MMP-2 (Figure 6), indicating that corylin may upregulate the activity of MMP2 by inhibiting the expression of TIMP1 and TIMP2, thereby accelerating ECM breakdown.

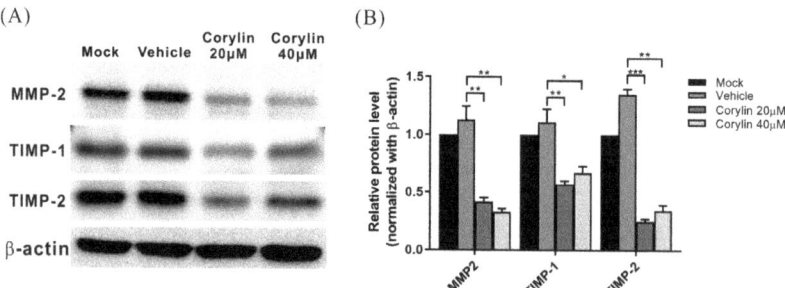

Figure 6. Corylin inhibits TGF-β-induced TIMP-1 and TIMP-2 expression in HHSteC cells. (**A**) Western blotting results of cell lysates of HHSteC cells treated with different concentrations of corylin or

vehicle for 2 h and then treated with 4 ng/mL TGF-β to induce an inflammatory response for 24 h; to analyze the expression of MMP-2, TIMP-1, and TIMP-2. Quantitative results are shown in (**B**). All data are expressed as the mean ± standard deviations of three independent experiments. $p < 0.05$ (*), $p < 0.01$ (**), $p < 0.001$ (***). TIMP-1, tissue inhibitor of metalloproteinase 1; MMP-2, matrix metalloproteinase 2.

2.7. Corylin Treatment Promoted HSC Apoptosis

To further evaluate the effects of corylin on HSC physiology, HHSteC cells were treated with different concentrations of corylin for 48 h. The cells were harvested and subjected to flow cytometry and terminal deoxynucleotidyl transferase dUTP nick-end labeling (TUNEL) assay analysis to assess the apoptosis and cell cycle statuses. Compared to those in the control group, corylin-treated cells were mostly arrested in the S phase, and the number of cells in the sub-G1 phase were significantly increased, indicating that corylin inhibits cell cycle progression and induces apoptosis (Figure 7A,B). TUNEL assay analysis also showed that the number of apoptotic cells in the corylin-treated group increased compared to that in the control group (Figure 7C). Furthermore, Western blotting showed that the levels of cleaved caspase 3 and caspase 9 were significantly increased in corylin-treated HHSteC cells, indicating that corylin promotes HSC apoptosis (Figure 7D).

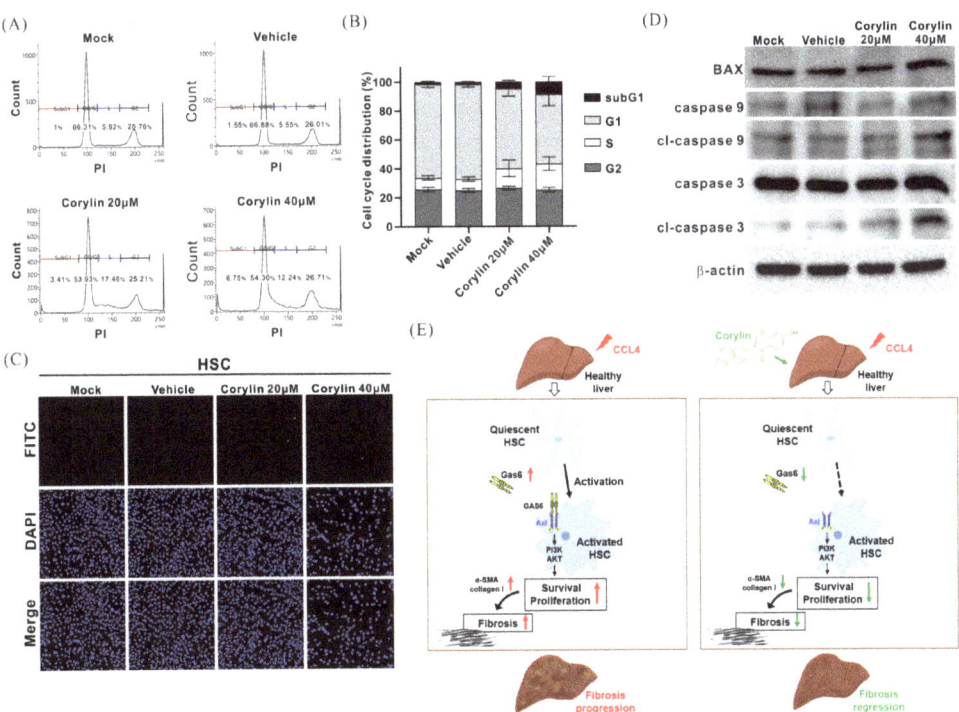

Figure 7. Effects of corylin on apoptosis in HHSteC cells. The apoptotic cell rate and cell cycle status according to (**A**) flow cytometry and (**C**) TUNEL assays of cells incubated with a vehicle (DMSO) and different concentrations of corylin (20 and 40 µM) for 48 h. (**B**) Quantitative results of flow cytometry. Error bars present the mean ± standard deviation from three independent experiments. (**D**) Effects of corylin on apoptosis-related protein expression according to Western blotting analysis. b-actin is the internal control. (**E**) Schematic representation summarizing the anti-liver fibrosis mechanisms of corylin. TUNEL, terminal deoxynucleotidyl transferase dUTP nick-end labeling. Red up arrow indicates up-regulation, and green down arrow means down-regulation.

3. Discussion

Chronic hepatitis leads to liver fibrosis and cirrhosis, which are risk factors for liver cancer. However, the early suppression of factors that cause liver injury and inflammation along with the administration of anti-inflammatory drugs can help reverse the progression of liver fibrosis. In this study, we found that corylin, a flavonoid extracted from the fruits of *P. corylifolia*, exhibited anti-inflammatory activity and inhibited the macrophage-mediated secretion of pro-inflammatory cytokines, such as IL-1β, IL-6, and TNF-α. Corylin also inhibited the expression of GAS6 and the downstream activation of the PI3K/AKT signaling pathway in HSCs, thereby inhibiting HSC activation and the expression of ECM proteins, including α-SMA and collagen. Moreover, corylin treatment alleviated the symptoms of CCl_4-induced liver fibrosis in a mouse model. These findings suggest that corylin has the potential to be used in the treatment of hepatitis and liver fibrosis. To the best of our knowledge, this is the first study to demonstrate that corylin inhibits GAS6 expression and subsequently inhibits HSC activation to alleviate the symptoms of liver fibrosis in mice (Figure 7E).

RTK AXL is expressed in most tissues and is involved in the regulation of diverse cellular physiological processes, including growth, survival, differentiation, adhesion, and migration. GAS6/AXL signaling is also involved in the regulation of macrophage polarization and the inflammatory response [42,43]. The downstream signaling pathway of AXL activation induces macrophage polarization into the M2 type, which subsequently downregulates pro-inflammatory cytokines, such as IL-6, TNF, type-I IFNs, and IL-12. Most tumor cells activate the downstream signaling pathway of TAM receptors by secreting GAS6, which subsequently inhibits macrophage activation and pro-inflammatory cytokine expression, creating an immune-tolerant environment around the tumor, which helps cancer cells survive during the immune response [44,45]. In the present study, we also analyzed the effect of corylin on the expression of GAS6 and AXL in THP1 cells. However, there was no significant change in their expression, indicating that corylin does not inhibit macrophage activation and pro-inflammatory cytokine expression by regulating the AXL signaling pathway. This also suggests that corylin is cell-specific in its regulation of physiological processes.

GAS6 supports hematopoietic stem cell growth and promotes fibroblast and endothelial cell survival [46,47]. In addition, GAS6/AXL signaling induces the accumulation of mesangial cells in kidney fibrosis [48], vascular smooth muscle cells in response to intimal vascular injury [49], and cardiac fibroblasts during the wound-healing process [50], thereby suggesting that GAS6 plays an important role in tissue fibrosis. Furthermore, GAS6 modulates HSC and HSC/myofibroblastoma survival during liver repair after acute injury [51]. The results of our study showed that corylin promoted HSC apoptosis, and part of this effect may have been achieved by inhibiting GAS6 expression.

Activated HSCs not only regulate the expression of ECM proteins, but also regulate the expression and secretion of MMP-2 and MMP-9 along with their inhibitors, TIMP-1 and TIMP-2, to regulate ECM breakdown [52,53]. During fibrogenesis, this equilibrium is disturbed, and the expression of TIMPs and MMPs is increased leading to an excess of TIMPs and subsequent matrix degradation. In the present study, we found that corylin inhibited the expression of ECM proteins, including α-SMA and collagen, in HSCs and also inhibited the expression of MMP inhibitors, TIMP-1 and TIMP-2, which may upregulate MMP-2 activity to accelerate ECM breakdown. In addition, the increased expression of MMP-2, MMP-9, and TIMP-1 has been regarded as an indicator of HSC activation [54]. The corylin-mediated inhibition of these proteins also indicates the inhibitory effect of corylin on HSC activation. Furthermore, corylin promoted apoptosis in HSCs. These findings showed that corylin simultaneously regulated multiple pathways to inhibit the progression of liver fibrosis. In addition to HSCs, MMP-2 and MMP-9 are expressed in most inflammatory cells, such as lymphocytes, neutrophils, macrophages, and Kupffer cells [55]. Thus, the effects of corylin on the expression of TIMP-1 and MMP-9 in macrophages and

Kupffer cells should be further studied to clarify the mechanism by which corylin inhibits liver fibrosis.

In this study, we demonstrated the anti-inflammatory activity of corylin, an extract of *P. corylifolia*, and its efficacy in treating liver fibrosis. Corylin has no obvious physiological toxicity and thus has great potential to be used as an adjuvant in clinical treatment. We have clarified the anti-inflammatory molecular mechanism of corylin and its potential for clinical application.

4. Materials and Methods

4.1. Cell Lines

The human monocyte cell line THP1, and mouse macrophage cell line RAW 264.7 were purchased from the American Type Culture Collection (Manassas, VA, USA). The aforementioned cells were cultured in Dulbecco's modified Eagle's medium containing 10% fetal bovine serum at 37 °C in a 5% CO_2 atmosphere. The human HSC cell line HHSteC was purchased from the ScienCell Research Laboratories (Carlsbad, CA, USA) and cultured using Stellate Cell Medium.

4.2. Materials and Reagents

Primary antibodies against AXL, phosphorylated (phospho)-AXL, GAS6, phospho-PI3K, PI3K, phospho-AKT, AKT, IL-1β, IL-6, MMP-2, MMP-9, TIMP-1, TIMP2, cleaved caspase-3, caspase-3, cleaved caspase-9, and caspase-9 were purchased from Genetex (Irvine, CA, USA), ABclonal (Woburn, MA, USA), and Cell Signaling Technology (Beverly, MA, USA). Secondary antibodies were purchased from Santa Cruz Biotechnology (Santa Cruz, CA, USA). Pre-stained protein marker and TOOLSmart RNA extractor were purchased from BIOTOOLS (Taipei, Taiwan). Corylin powder (purity above 98% as measured by high-performance liquid chromatography) was purchased from Shanghai BS Bio-Tech (Shanghai, China).

4.3. Western Blot Analysis

Cells treated with different concentrations of corylin for 24 and 48 h were harvested and washed twice with phosphate-buffered saline (PBS) and then lysed in 200 μL of radioimmunoprecipitation assay lysis buffer (BIOTOOLS) containing a protease inhibitor. Protein (30 μg) from the supernatant was loaded onto a sodium dodecyl sulfate polyacrylamide gel, followed by Western blot analysis to detect the levels of target proteins. Detailed information of antibodies used in the experiments is shown Table S1. The immuno-reactive bands were revealed using an enhanced chemiluminescence system (NEN Life Science Products, Boston, MA, USA) and detected using UVP ChemStudio Imaging Systems (Analytik Jena, Upland, CA, USA). The intensity of each band was quantified using ImageQuant 5.2 (GE Healthcare, Piscataway, NJ, USA).

4.4. Enzyme-Linked Immunosorbent Assay

THP-1 and RAW 264.7 cells were treated with different concentrations of corylin for 2 h followed by treatment with lipopolysaccharide (LPS) for 24 h to induce an inflammatory response. The protein levels of IL-1β, IL-6, and TNF-α in the culture medium were measured using ELISA kits (BioLegend, San Diego, CA, USA) according to the manufacturer's instructions.

4.5. Flow Cytometry

HHSteC cells were treated with DMSO, 20 μM corylin, and 40 μM corylin for 24 h, followed by trypsinization and then washed twice and incubated in PBS containing 0.12% Triton X-100, 0.12 mM ethylenediaminetetraacetic acid, and 100 mg/mL ribonuclease A. Propidium iodide (50 μg/mL) was then added to each sample for 20 min at 4 °C. Cell cycle distribution was analyzed using flow cytometry (Beckman Coulter Epics Elite, Beckman Coulter, Brea, CA, USA).

4.6. Terminal Deoxynucleotidyl Transferase dUTP Nick-End Labeling Assay

The apoptosis status of HHSteC cells was determined using a DeadEnd™ Fluorometric TUNEL Assay Kit (Promega, Madison, WI, USA) according to the manufacturer's protocol. Briefly, HHSteC cells were treated with DMSO, 20 μM corylin, or 40 μM corylin for 24 h. The cells were then subjected to a TUNEL assay. The cells were counted using a microscope (magnification, ×100). Cells in five different microscopic fields/dishes were analyzed for each experiment.

4.7. Mice

Male BALB/c mice (age, 6–8 weeks; National Laboratory animal center, Taipei, Taiwan) were housed under pathogen-free conditions with a 12 h light/12 h dark schedule and fed autoclaved standard chow and water. All animal experiments were approved by the Institutional Animal Care and Use Committee (IACUC) at Chang-Gung memorial Hospital (IACUC approval no.: 2019032009, approval date: 2019/6/11).

4.8. CCl_4-Induced Liver Fibrosis Mouse Model

A total of 20 mice were randomly assigned to three groups: negative control (control, $n = 6$), CCl_4 treatment + DMSO (vehicle, $n = 7$), and CCl_4 treatment + corylin (30 mg/kg, $n = 7$). CCl_4 was liquefied in olive oil to obtain a 10% CCl_4 solution that was injected intraperitoneally into mice (0.5 μL/g body weight) twice a week for six weeks. At the beginning of the second week, mice were intraperitoneally injected with 100 μL of corylin (at a dose of 30 mg/kg of body weight) or an equal volume of DMSO as a control for 3 d per week. Negative control mice were treated with olive oil alone. At the endpoint, blood samples were collected to measure the levels of serum AST and ALT. Liver tissues were collected for further assays such as histology and Western blotting.

4.9. Masson's Trichrome Staining

Masson's trichrome staining was performed at the Chang Gung Memorial Hospital Department of Anatomic Pathology as follows: First, 5 μm thick formalin-fixed paraffin-embedded (FFPE) sections were deparaffinized and hydrated in distilled water. Subsequently, Bouin's fixative was used as a mordant for 1 h at 56 °C. The FFPE sections were cooled and washed in running water until the yellow color disappeared. The samples were stained in Weigert's hematoxylin stain for 10 min, thoroughly washed in tap water for 10 min, stained in an acid fuchsin solution for 15 min, and then rinsed in distilled water for 3 min. After rinsing, the slides were treated with phosphomolybdic acid solution for 10 min and rinsed in distilled water for 10 min. Finally, the slides were stained with a light-green solution for 2 min and rinsed in distilled water. After thorough dehydration using alcohol, the slides were mounted, and coverslips were placed onto them.

4.10. Immunohistochemistry

Mouse liver tissues were fixed in formalin and embedded in paraffin, and 2 μm thick consecutive sections were cut and subjected to immunohistochemical staining using a BOND III autostainer (Leica Biosystems, Wetzlar, Germany) as described previously [35].

4.11. Data Analysis

The quantitative real-time polymerase chain reaction and Western blot data were recorded as continuous variants and analyzed using the Student's *t*-test. All statistical analyses were performed using SPSS 16.0 (IBM, Armonk, NY, USA) and Excel 2007. All statistical tests were two-sided, and *p* values < 0.05 (*), <0.01 (**), and <0.001 (***) were considered significant.

Supplementary Materials: The following supporting information can be downloaded at: https://www.mdpi.com/article/10.3390/ijms242316936/s1.

Author Contributions: Conceptualization, C.-C.C. and T.-H.W.; methodology and investigation, C.-C.C., C.-Y.C., Y.-L.L., Y.-T.L., C.-T.Y., Y.-H.S. and L.-F.C.; analysis and interpretation, Y.-T.L., T.-H.W., C.-C.C., C.-Y.C. and W.-Y.C.; ImageJ analysis, Y.-T.L., Y.-H.S. and W.-Y.C.; writing—original draft preparation, T.-H.W. and C.-C.C.; review and editing, T.-M.S. and T.-H.W.; project administration—funding acquisition, T.-M.S. and T.-H.W.; Primary responsibility for final content, C.-C.C. and T.-H.W. All authors have read and agreed to the published version of the manuscript.

Funding: This research was partially supported by the National Science and Technology Council of Taiwan (grants 111-2320-B-182A-011 and 112-2320-B-182A-009-MY3) and the Chang Gung Medical Research Program, Taiwan (CMRPG3J0863 and CMRPG3M0182).

Institutional Review Board Statement: All animal experiments were approved by the Institutional Animal Care and Use Committee (IACUC) at Chang-Gung memorial Hospital (IACUC approval no.: 2019032009, approval date: 11 June 2019).

Informed Consent Statement: Not applicable.

Data Availability Statement: All data analyzed during this study are included in this article. Further inquiries can be directed towards the corresponding author.

Acknowledgments: We thank the Biobank at the Chang Gung Memorial Hospital, Lin-Kou, Taiwan for tissue processing and Yi-Ting Tsai, Yu-Wen Jiang, Fang-Ching Chuan, Li-Shan Wei, Hsin-Wei Lin, and Kai-Yin Chen for their technical assistance in this project.

Conflicts of Interest: The authors declare no conflict of interest.

References

1. Roehlen, N.; Crouchet, E.; Baumert, T.F. Liver Fibrosis: Mechanistic Concepts and Therapeutic Perspectives. *Cells* **2020**, *9*, 875. [CrossRef] [PubMed]
2. Asrani, S.K.; Devarbhavi, H.; Eaton, J.; Kamath, P.S. Burden of liver diseases in the world. *J. Hepatol.* **2019**, *70*, 151–171. [CrossRef] [PubMed]
3. Liao, S.H.; Chen, C.L.; Hsu, C.Y.; Chien, K.L.; Kao, J.H.; Chen, P.J.; Chen, T.H.; Chen, C.H. Long-term effectiveness of population-wide multifaceted interventions for hepatocellular carcinoma in Taiwan. *J. Hepatol.* **2021**, *75*, 132–141. [CrossRef] [PubMed]
4. Trepo, C.; Chan, H.L.; Lok, A. Hepatitis B virus infection. *Lancet* **2014**, *384*, 2053–2063. [CrossRef] [PubMed]
5. Kee, K.M.; Chen, C.H.; Hu, J.T.; Huang, Y.H.; Wang, T.E.; Chau, G.Y.; Chen, K.H.; Chen, Y.L.; Lin, C.C.; Hung, C.F.; et al. Secular Trends of Clinical Characteristics and Survival of Hepatocellular Carcinoma in Taiwan from 2011 to 2019. *Viruses* **2022**, *15*, 126. [CrossRef] [PubMed]
6. Bengtsson, B.; Widman, L.; Wahlin, S.; Stal, P.; Bjorkstrom, N.K.; Hagstrom, H. The risk of hepatocellular carcinoma in cirrhosis differs by etiology, age and sex: A Swedish nationwide population-based cohort study. *United Eur. Gastroenterol. J.* **2022**, *10*, 465–476. [CrossRef] [PubMed]
7. Popov, Y.; Schuppan, D. Targeting liver fibrosis: Strategies for development and validation of antifibrotic therapies. *Hepatology* **2009**, *50*, 1294–1306. [CrossRef] [PubMed]
8. Tang, L.S.Y.; Covert, E.; Wilson, E.; Kottilil, S. Chronic Hepatitis B Infection: A Review. *JAMA* **2018**, *319*, 1802–1813. [CrossRef]
9. Grossi, G.; Vigano, M.; Loglio, A.; Lampertico, P. Hepatitis B virus long-term impact of antiviral therapy nucleot(s)ide analogues (NUCs). *Liver Int.* **2017**, *37* (Suppl. S1), 45–51. [CrossRef]
10. Barcena, C.; Stefanovic, M.; Tutusaus, A.; Joannas, L.; Menendez, A.; Garcia-Ruiz, C.; Sancho-Bru, P.; Mari, M.; Caballeria, J.; Rothlin, C.V.; et al. Gas6/Axl pathway is activated in chronic liver disease and its targeting reduces fibrosis via hepatic stellate cell inactivation. *J. Hepatol.* **2015**, *63*, 670–678. [CrossRef]
11. Holstein, E.; Binder, M.; Mikulits, W. Dynamics of Axl Receptor Shedding in Hepatocellular Carcinoma and Its Implication for Theranostics. *Int. J. Mol. Sci.* **2018**, *19*, 4111. [CrossRef] [PubMed]
12. Smirne, C.; Rigamonti, C.; De Benedittis, C.; Sainaghi, P.P.; Bellan, M.; Burlone, M.E.; Castello, L.M.; Avanzi, G.C. Gas6/TAM Signaling Components as Novel Biomarkers of Liver Fibrosis. *Dis. Markers* **2019**, *2019*, 2304931. [CrossRef]
13. Graham, D.K.; DeRyckere, D.; Davies, K.D.; Earp, H.S. The TAM family: Phosphatidylserine sensing receptor tyrosine kinases gone awry in cancer. *Nat. Rev. Cancer* **2014**, *14*, 769–785. [CrossRef] [PubMed]
14. Wu, G.; Ma, Z.; Hu, W.; Wang, D.; Gong, B.; Fan, C.; Jiang, S.; Li, T.; Gao, J.; Yang, Y. Molecular insights of Gas6/TAM in cancer development and therapy. *Cell Death Dis.* **2017**, *8*, e2700. [CrossRef] [PubMed]
15. Zagorska, A.; Traves, P.G.; Jimenez-Garcia, L.; Strickland, J.D.; Oh, J.; Tapia, F.J.; Mayoral, R.; Burrola, P.; Copple, B.L.; Lemke, G. Differential regulation of hepatic physiology and injury by the TAM receptors Axl and Mer. *Life Sci. Alliance* **2020**, *3*, e202000694. [CrossRef] [PubMed]

16. Tutusaus, A.; de Gregorio, E.; Cucarull, B.; Cristobal, H.; Areste, C.; Graupera, I.; Coll, M.; Colell, A.; Gausdal, G.; Lorens, J.B.; et al. A Functional Role of GAS6/TAM in Nonalcoholic Steatohepatitis Progression Implicates AXL as Therapeutic Target. *Cell. Mol. Gastroenterol. Hepatol.* **2020**, *9*, 349–368. [CrossRef] [PubMed]
17. Wang, Z.; Li, J.; Ji, Y.; An, P.; Zhang, S.; Li, Z. Traditional herbal medicine: A review of potential of inhibitory hepatocellular carcinoma in basic research and clinical trial. *Evid. Based Complement. Altern. Med.* **2013**, *2013*, 268963. [CrossRef] [PubMed]
18. Li-Weber, M. Targeting apoptosis pathways in cancer by Chinese medicine. *Cancer Lett.* **2013**, *332*, 304–312. [CrossRef]
19. Ye, M.X.; Li, Y.; Yin, H.; Zhang, J. Curcumin: Updated molecular mechanisms and intervention targets in human lung cancer. *Int. J. Mol. Sci.* **2012**, *13*, 3959–3978. [CrossRef]
20. Hong, M.; Tan, H.Y.; Li, S.; Cheung, F.; Wang, N.; Nagamatsu, T.; Feng, Y. Cancer Stem Cells: The Potential Targets of Chinese Medicines and Their Active Compounds. *Int. J. Mol. Sci.* **2016**, *17*, 893. [CrossRef]
21. Terlikowska, K.M.; Witkowska, A.M.; Zujko, M.E.; Dobrzycka, B.; Terlikowski, S.J. Potential application of curcumin and its analogues in the treatment strategy of patients with primary epithelial ovarian cancer. *Int. J. Mol. Sci.* **2014**, *15*, 21703–21722. [CrossRef] [PubMed]
22. Shi, M.; Gu, A.; Tu, H.; Huang, C.; Wang, H.; Yu, Z.; Wang, X.; Cao, L.; Shu, Y.; Wang, H.; et al. Comparing nanoparticle polymeric micellar paclitaxel and solvent-based paclitaxel as first-line treatment of advanced non-small-cell lung cancer: An open-label, randomized, multicenter, phase III trial. *Ann. Oncol.* **2021**, *32*, 85–96. [CrossRef] [PubMed]
23. Gokduman, K. Strategies Targeting DNA Topoisomerase I in Cancer Chemotherapy: Camptothecins, Nanocarriers for Camptothecins, Organic Non-Camptothecin Compounds and Metal Complexes. *Curr. Drug Targets* **2016**, *17*, 1928–1939. [CrossRef] [PubMed]
24. Hsu, Y.A.; Chen, C.S.; Wang, Y.C.; Lin, E.S.; Chang, C.Y.; Chen, J.J.; Wu, M.Y.; Lin, H.J.; Wan, L. Anti-Inflammatory Effects of Resveratrol on Human Retinal Pigment Cells and a Myopia Animal Model. *Curr. Issues Mol. Biol.* **2021**, *43*, 716–727. [CrossRef] [PubMed]
25. Kristofi, R.; Eriksson, J.W. Metformin as an anti-inflammatory agent: A short review. *J. Endocrinol.* **2021**, *251*, R11–R22. [CrossRef] [PubMed]
26. Cicalau, G.I.P.; Babes, P.A.; Calniceanu, H.; Popa, A.; Ciavoi, G.; Iova, G.M.; Ganea, M.; Scrobota, I. Anti-Inflammatory and Antioxidant Properties of Carvacrol and Magnolol, in Periodontal Disease and Diabetes Mellitus. *Molecules* **2021**, *26*, 6899. [CrossRef] [PubMed]
27. Mazarakis, N.; Snibson, K.; Licciardi, P.V.; Karagiannis, T.C. The potential use of l-sulforaphane for the treatment of chronic inflammatory diseases: A review of the clinical evidence. *Clin. Nutr.* **2020**, *39*, 664–675. [CrossRef]
28. He, H.; Ma, Y.; Huang, H.; Huang, C.; Chen, Z.; Chen, D.; Gu, Y.; Wang, X.; Chen, J. A comprehensive understanding about the pharmacological effect of diallyl disulfide other than its anti-carcinogenic activities. *Eur. J. Pharmacol.* **2021**, *893*, 173803. [CrossRef]
29. Huang, M.Y.; Tu, C.E.; Wang, S.C.; Hung, Y.L.; Su, C.C.; Fang, S.H.; Chen, C.S.; Liu, P.L.; Cheng, W.C.; Huang, Y.W.; et al. Corylin inhibits LPS-induced inflammatory response and attenuates the activation of NLRP3 inflammasome in microglia. *BMC Complement. Altern. Med.* **2018**, *18*, 221. [CrossRef]
30. Hung, Y.L.; Fang, S.H.; Wang, S.C.; Cheng, W.C.; Liu, P.L.; Su, C.C.; Chen, C.S.; Huang, M.Y.; Hua, K.F.; Shen, K.H.; et al. Corylin protects LPS-induced sepsis and attenuates LPS-induced inflammatory response. *Sci. Rep.* **2017**, *7*, 46299. [CrossRef]
31. Chang, Z.Y.; Liu, H.M.; Leu, Y.L.; Hsu, C.H.; Lee, T.Y. Modulation of Gut Microbiota Combined with Upregulation of Intestinal Tight Junction Explains Anti-Inflammatory Effect of Corylin on Colitis-Associated Cancer in Mice. *Int. J. Mol. Sci.* **2022**, *23*, 2667. [CrossRef]
32. Chopra, B.; Dhingra, A.K.; Dhar, K.L. *Psoralea corylifolia* L. (Buguchi)—Folklore to modern evidence: Review. *Fitoterapia* **2013**, *90*, 44–56. [CrossRef] [PubMed]
33. Xiong, Z.; Wang, D.; Xu, Y.; Li, F. Osteoblastic differentiation bioassay and its application to investigating the activity of fractions and compounds from *Psoralea corylifolia* L. *Pharmazie* **2003**, *58*, 925–928. [PubMed]
34. Wang, D.; Li, F.; Jiang, Z. Osteoblastic proliferation stimulating activity of Psoralea corylifolia extracts and two of its flavonoids. *Planta Med.* **2001**, *67*, 748–749. [CrossRef] [PubMed]
35. Chen, C.C.; Chen, C.Y.; Ueng, S.H.; Hsueh, C.; Yeh, C.T.; Ho, J.Y.; Chou, L.F.; Wang, T.H. Corylin increases the sensitivity of hepatocellular carcinoma cells to chemotherapy through long noncoding RNA RAD51-AS1-mediated inhibition of DNA repair. *Cell Death Dis.* **2018**, *9*, 543. [CrossRef] [PubMed]
36. Chen, C.Y.; Chen, C.C.; Shieh, T.M.; Hsueh, C.; Wang, S.H.; Leu, Y.L.; Lian, J.H.; Wang, T.H. Corylin Suppresses Hepatocellular Carcinoma Progression via the Inhibition of Epithelial-Mesenchymal Transition, Mediated by Long Noncoding RNA GAS5. *Int. J. Mol. Sci.* **2018**, *19*, 380. [CrossRef]
37. Li, N.; Liu, T.; Zhu, S.; Yang, Y.; Wang, Z.; Zhao, Z.; Liu, T.; Wang, X.; Qin, W.; Yan, Y.; et al. Corylin from Psoralea fructus (*Psoralea corylifolia* L.) protects against UV-induced skin aging by activating Nrf2 defense mechanisms. *Phytother. Res.* **2022**, *36*, 3276–3294. [CrossRef] [PubMed]
38. Wang, T.H.; Tseng, W.C.; Leu, Y.L.; Chen, C.Y.; Lee, W.C.; Chi, Y.C.; Cheng, S.F.; Lai, C.Y.; Kuo, C.H.; Yang, S.L.; et al. The flavonoid corylin exhibits lifespan extension properties in mouse. *Nat. Commun.* **2022**, *13*, 1238. [CrossRef] [PubMed]
39. Patel, D.K. Biological Importance, Therapeutic Benefits, and Analytical Aspects of Active Flavonoidal Compounds 'Corylin' from Psoralea corylifolia in the Field of Medicine. *Infect. Disord. Drug Targets* **2023**, *23*, e250822208005. [CrossRef]

40. Zheng, Z.G.; Zhang, X.; Liu, X.X.; Jin, X.X.; Dai, L.; Cheng, H.M.; Jing, D.; Thu, P.M.; Zhang, M.; Li, H.; et al. Inhibition of HSP90beta Improves Lipid Disorders by Promoting Mature SREBPs Degradation via the Ubiquitin-proteasome System. *Theranostics* **2019**, *9*, 5769–5783. [CrossRef]
41. Chen, C.C.; Li, H.Y.; Leu, Y.L.; Chen, Y.J.; Wang, C.J.; Wang, S.H. Corylin Inhibits Vascular Cell Inflammation, Proliferation and Migration and Reduces Atherosclerosis in ApoE-Deficient Mice. *Antioxidants* **2020**, *9*, 275. [CrossRef]
42. Zhou, L.; Matsushima, G.K. Tyro3, Axl, Mertk receptor-mediated efferocytosis and immune regulation in the tumor environment. *Int. Rev. Cell Mol. Biol.* **2021**, *361*, 165–210. [PubMed]
43. Huang, H.; Jiang, J.; Chen, R.; Lin, Y.; Chen, H.; Ling, Q. The role of macrophage TAM receptor family in the acute-to-chronic progression of liver disease: From friend to foe? *Liver Int.* **2022**, *42*, 2620–2631. [CrossRef] [PubMed]
44. Lee, C.H.; Chun, T. Anti-Inflammatory Role of TAM Family of Receptor Tyrosine Kinases Via Modulating Macrophage Function. *Mol. Cells* **2019**, *42*, 1–7.
45. Wu, G.; Ma, Z.; Cheng, Y.; Hu, W.; Deng, C.; Jiang, S.; Li, T.; Chen, F.; Yang, Y. Targeting Gas6/TAM in cancer cells and tumor microenvironment. *Mol. Cancer* **2018**, *17*, 20. [CrossRef] [PubMed]
46. Dormady, S.P.; Zhang, X.M.; Basch, R.S. Hematopoietic progenitor cells grow on 3T3 fibroblast monolayers that overexpress growth arrest-specific gene-6 (GAS6). *Proc. Natl. Acad. Sci. USA* **2000**, *97*, 12260–12265. [CrossRef] [PubMed]
47. Zuo, P.Y.; Chen, X.L.; Lei, Y.H.; Liu, C.Y.; Liu, Y.W. Growth arrest-specific gene 6 protein promotes the proliferation and migration of endothelial progenitor cells through the PI3K/AKT signaling pathway. *Int. J. Mol. Med.* **2014**, *34*, 299–306. [CrossRef] [PubMed]
48. Yanagita, M.; Arai, H.; Ishii, K.; Nakano, T.; Ohashi, K.; Mizuno, K.; Varnum, B.; Fukatsu, A.; Doi, T.; Kita, T. Gas6 regulates mesangial cell proliferation through Axl in experimental glomerulonephritis. *Am. J. Pathol.* **2001**, *158*, 1423–1432. [CrossRef]
49. Melaragno, M.G.; Wuthrich, D.A.; Poppa, V.; Gill, D.; Lindner, V.; Berk, B.C.; Corson, M.A. Increased expression of Axl tyrosine kinase after vascular injury and regulation by G protein-coupled receptor agonists in rats. *Circ. Res.* **1998**, *83*, 697–704. [CrossRef]
50. Stenhoff, J.; Dahlback, B.; Hafizi, S. Vitamin K-dependent Gas6 activates ERK kinase and stimulates growth of cardiac fibroblasts. *Biochem. Biophys. Res. Commun.* **2004**, *319*, 871–878. [CrossRef]
51. Bellan, M.; Cittone, M.G.; Tonello, S.; Rigamonti, C.; Castello, L.M.; Gavelli, F.; Pirisi, M.; Sainaghi, P.P. Gas6/TAM System: A Key Modulator of the Interplay between Inflammation and Fibrosis. *Int. J. Mol. Sci.* **2019**, *20*, 5070. [CrossRef] [PubMed]
52. Lachowski, D.; Cortes, E.; Rice, A.; Pinato, D.; Rombouts, K.; Del Rio Hernandez, A. Matrix stiffness modulates the activity of MMP-9 and TIMP-1 in hepatic stellate cells to perpetuate fibrosis. *Sci. Rep.* **2019**, *9*, 7299. [CrossRef] [PubMed]
53. Robert, S.; Gicquel, T.; Bodin, A.; Lagente, V.; Boichot, E. Characterization of the MMP/TIMP Imbalance and Collagen Production Induced by IL-1beta or TNF-alpha Release from Human Hepatic Stellate Cells. *PLoS ONE* **2016**, *11*, e0153118. [CrossRef] [PubMed]
54. Roeb, E. Matrix metalloproteinases and liver fibrosis (translational aspects). *Matrix Biol.* **2018**, *68–69*, 463–473. [CrossRef] [PubMed]
55. Geervliet, E.; Bansal, R. Matrix Metalloproteinases as Potential Biomarkers and Therapeutic Targets in Liver Diseases. *Cells* **2020**, *9*, 1212. [CrossRef]

Disclaimer/Publisher's Note: The statements, opinions and data contained in all publications are solely those of the individual author(s) and contributor(s) and not of MDPI and/or the editor(s). MDPI and/or the editor(s) disclaim responsibility for any injury to people or property resulting from any ideas, methods, instructions or products referred to in the content.

Article

Cathelicidin Treatment Silences Epithelial–Mesenchymal Transition Involved in Pulmonary Fibrosis in a Murine Model of Hypersensitivity Pneumonitis

Marta Kinga Lemieszek [1,*], Marcin Golec [2], Jacek Zwoliński [3], Jacek Dutkiewicz [3] and Janusz Milanowski [4]

1. Department of Medical Biology, Institute of Rural Health, 20-090 Lublin, Poland
2. Heidelberg Institute of Global Health (HIGH), Faculty of Medicine and University Hospital, Heidelberg University, 69117 Heidelberg, Germany
3. Department of Biological Health Hazards and Parasitology, Institute of Rural Health, 20-090 Lublin, Poland
4. Department of Pneumonology, Oncology and Allergology, Medical University of Lublin, 20-059 Lublin, Poland
* Correspondence: lemieszek.marta@imw.lublin.pl

Abstract: Pulmonary fibrosis is becoming an increasingly common pathology worldwide. Unfortunately, this disorder is characterized by a bad prognosis: no treatment is known, and the survival rate is dramatically low. One of the most frequent reasons for pulmonary fibrosis is hypersensitivity pneumonitis (HP). As the main mechanism of pulmonary fibrosis is a pathology of the repair of wounded pulmonary epithelium with a pivotal role in epithelial–mesenchymal transition (EMT), we assumed that EMT silencing could prevent disease development. Because of several biological features including wound healing promotion, an ideal candidate for use in the treatment of pulmonary fibrosis seems to be cathelicidin. The aim of the studies was to understand the influence of cathelicidin on the EMT process occurring during lung fibrosis development in the course of HP. Cathelicidin's impact on EMT was examined in a murine model of HP, wherein lung fibrosis was induced by chronic exposure to extract of *Pantoea agglomerans* (SE-PA) by real-time PCR and Western blotting. Studies revealed that mouse exposure to cathelicidin did not cause any side changes in the expression of investigated genes/proteins. Simultaneously, cathelicidin administered together or after SE-PA decreased the elevated level of myofibroblast markers (*Acta2*/α-smooth muscle actin, *Cdh1*/N-cadherin, *Fn1*/Fibronectin, *Vim*/vimentin) and increased the lowered level of epithelial markers (*Cdh1*/E-cadherin, *Ocln*/occludin). Cathelicidin provided with SE-PA or after cessation of SE-PA inhalations reduced the expression of EMT-associated factors (*Ctnnd1*/β-catenin, *Nfkb1*/NFκB, *Snail1*/Snail, *Tgfb1*/TGFβ1 *Zeb1*/ZEB1, *Zeb2*/ZEB2) elevated by *P. agglomerans*. Cathelicidin's beneficial impact on the expression of genes/proteins involved in EMT was observed during and after the HP development; however, cathelicidin was not able to completely neutralize the negative changes. Nevertheless, significant EMT silencing in response to cathelicidin suggested the possibility of its use in the prevention/treatment of pulmonary fibrosis.

Keywords: defense peptides; immune peptides; pulmonary fibrosis; hypersensitivity pneumonitis; extrinsic allergic alveolitis

1. Introduction

Hypersensitivity pneumonitis (HP) or extrinsic allergic alveolitis (EAA) is a heterogenic group of interstitial lung diseases in which the chronic inhalation of a wide variety of organic dust provokes in susceptible subjects a hypersensitivity reaction with inflammation in the terminal bronchioles, the pulmonary interstitium, and the alveolar tree. This inflammation often organizes into granulomas and may progress to pulmonary fibrosis, which leads to the elimination of pathologically changed areas of lung tissue from the gas exchange, causing hypoxia and, in advanced cases, death [1–3]. HP can be provoked by a diverse range of antigens, including bacteria, fungi, mycobacteria, plant and animal proteins, chemicals, and metals [1]. Depending on the source and the type of antigens, several varieties of HP have been distinguished. One of the most common is farmer's lung, induced by the inhalation of organic dust coming from agricultural products, mostly hay, straw, grain, and moldy plants, which are the source of a range of antigens, e.g., *Pantoea agglomerans, Saccharopolyspora rectivirgula, Streptomyces thermohygroscopicus, Streptomyces albus, Thermoactinomyces vulgaris, Thermoactinomyces viridis, Aspergillus fumigatus, Aspergillus niger, Absidia corymbifera,* and *Micropolyspora faeni*. Other very common types of HP are bird fancier's lung (antigens: avian feathers, droppings, and serum); lung of mushroom growers (antigens: *Thermoactinomyces vulgaris, Micropolyspora faeni*); grain fever (antigens: *Pantoea agglomerans, Sitophilus granaries*); cheese disease (antigens: *Penicillium casei, Acarus siro*); and humidifier lung (antigens: *Alternaria alternata, Aureobasidium* spp., *Aspergillus* spp., *Bacillus* spp., *Cephalosporium* spp., *Fusarium* spp., *Trochoderma viridae*) [4,5]. Although the above-mentioned examples of different varieties of HP may suggest the presence of HP-associated environments, it has to be noted that extrinsic allergic alveolitis is caused by similar antigens in distinct environments, e.g., in differs ranges of agricultural environments in eastern Poland, one of the most important causes of HP is *Pantoea agglomerans* [5]. Nevertheless, because of the great variety and distribution of HP-induced antigens, millions of individuals are exposed to them as part of their occupational, home, or recreational environments. Thus, HP is estimated to be one of the most frequent reasons for pulmonary fibrosis with the known etiology; nevertheless, its global prevalence is relatively rare, especially when compared with idiopathic pulmonary fibrosis [6,7].

Emerging evidence suggests that pulmonary fibrosis is the pathology of respiratory repair following chronic lung epithelial injury [3,8–11]. In the case of hypersensitivity pneumonitis (HP), repeated injuries of the respiratory epithelium caused by chronic organic dust exposure leads to disorders of tissue repair. Successful wound repair requires close coordination of epithelial cell proliferation, migration, and differentiation with mesenchymal cell recruitment, proliferation, differentiation, and subsequent extracellular matrix remodeling. Deregulation of wound repair response leads to pathological scar formation and excessive deposition of extracellular matrix components, which rebuild and destroy normal tissue architecture [12–14]. Extracellular matrix deposition under physiological and pathological conditions is regulated primarily by myofibroblasts, which combine the features of fibroblasts and smooth muscle cells. These spindle-shaped cells produce a diverse range of cytokines, growth factors, and extracellular matrix components [12,13,15]. Several studies revealed that the most important source of myofibroblasts is an epithelial–mesenchymal transition (EMT). EMT describes the global process during which epithelial cells undergo local conversion, including loss of cell–cell adhesion and apical–basal polarity, and gain a mesenchymal phenotype including elongated shape, enhanced motility and invasiveness, and increased production of extracellular matrix [12,16–18]. Among three different types of EMT, a pivotal role in tissue regeneration and organ fibrosis is played by EMT type 2 [12,17]. EMT type 2 begins as a part of a normal repair-associated event that generates fibroblasts, myofibroblasts, and other related cells in order to reconstruct tissues following injury. EMT type 2 is linked to inflammation and, in the case of "physiological" repair, this process ceases once inflammation is attenuated. This is the signal that the wound is closed. In the case of pulmonary fibrosis, EMT type 2 and inflammation are ongoing, coupled in a vicious circle, until the fibrotic process reaches the point where it cannot be attenuated by calming down the inflammation [3,12,17]. The EMT initiation is triggered by cellular signaling mechanisms including Wnt/β-catenin and TGFβ pathways [12,17–20]. N-cadherin, α-SMA, vimentin, and fibronectin have been shown to be reliable biomarkers characterizing mesenchymal products generated by the EMT process occurring during fibrosis development. Additionally, E-cadherin and occludins were proven to be useful in the identification of epithelial cells undergoing an EMT associated with chronic inflammation [12,17]. Other studies demonstrated that EMT in lungs was characterized by downregulation of E-cadherin, occludins, cytokeratin, and aquaporins, while α-SMA, vimentin, collagens, N-cadherin, fibronectin, and desmin were upregulated [21–24]. Snail, a direct transcriptional repressor for the E-cadherin gene, was demonstrated as a target for EMT-promoting signaling pathways [25]. Recently, additional transcription factors ZEB1 and ZEB2 were identified as E-cadherin repressors and mediators of the EMT [26]. Our earlier studies [27], conducted in mice strain C57BL/6J chronically exposed to the antigen of *Pantoea agglomerant* (well-documented etiological factor of HP) [5], have also shown downregulation of epithelial markers (*Cdh1, Cldn1, Jup, Ocln*) and upregulation of myofibroblasts markers (*Acta2, Cdh2, Fn1, Vim*) in lung tissue in response to bacterial antigen treatment. Furthermore, the mentioned alterations in gene expression typical for EMT correlated with an increase in fibrosis markers (hydroxyproline, collagens) as well as histological changes characteristic for lung fibrosis development [27]. Considering the role of EMT in pulmonary fibrosis [24,28–30], including the above-mentioned studies conducted by the authors that indicated EMT significance for the HP development in a murine model [27], it seemed reasonable to base HP therapy on inhibiting/counteracting EMT. Because of pleiotropic activities, an ideal candidate for the proposed therapeutic strategy seemed to be cathelicidin.

Cathelicidins belong to a large, conserved group of antimicrobial peptides and represent an important part of innate immunity [31]. These host defense peptides directly kill bacteria as well as some enveloped viruses, parasites, and fungi by perturbing their cell membranes [32,33]. Furthermore, they can also neutralize the biological activities of endotoxin [34,35]. Cathelicidins increase the natural abilities of immune cells to fight infection in several different ways, including attraction and recognition of pathogens, enhancement of phagocytosis, and stimulation of production and release of pro-inflammatory compounds [32,36–40]. Cathelicidins also accelerate epithelial cell proliferation, migration, and promotion of wound closure which, all together, play an important role in the maintenance of tissue homeostasis by supervising regenerative processes [41–43]. It needs to be highlighted that recent studies by the authors revealed the beneficial impact of cathelicidin on the development of lung fibrosis in the course of HP, which was associated with restoring the balance in quantity of immune cells (NK cells, macrophages, lymphocytes: Tc, Th, Treg, B), cytokine production (IFNγ, TNFα, TGFβ1, IL1β, IL4, IL5, IL10, IL12α, IL13), and synthesis of extracellular matrix components (hydroxyproline, collagens) [44]. Cathelicidin treatment also effectively protected lung tissue structure from pathological changes induced by antigen of *P. agglomerans* (HP trigger) [44]. Despite the fact that many advantageous biological activities of cathelicidin have been discovered, the presented study is the first to focus on understanding the influence of cathelicidin on the EMT process occurring during lung fibrosis development in the course of HP. Thus, the aim of the presented study was the identification of the molecular mechanism responsible for the antifibrotic properties of cathelicidin previously described by our research team [44].

2. Results

2.1. Cathelicidin Restored the Balance in the Expression of Genes Responsible for EMT

Changes in the expression of genes involved in epithelial–mesenchymal transition in lung tissue homogenates obtained from mice chronically exposed to cathelicidin and/or saline extract of *P. agglomerans* were examined by real-time PCR (Figure 1, Table 1). The study revealed that chronic exposure of mice to cathelicidin did not cause any changes in the expression of all investigated genes. On the contrary *P. agglomerans* treatment induced alterations in the gene expression characteristic for epithelial–mesenchymal transition: downregulation of epithelial markers (*Cdh1, Ocln*), as well as significant upregulation of mesenchymal markers (*Cdh2, Acta2, Fn1, Vim*). The decrease in epithelial markers, on average, was 21.6% after 14 days and 40.6% after 28 days of SE-PA exposure, while the increase in mesenchymal markers, on average, was 26.7% and 39.1% at the mentioned time points, respectively. The level of mRNA coding transcription factors responsible for EMT (*Snail1, Zeb1, Zeb2*) was also upregulated by SE-PA exposure; however, statistically significant changes were noted in the case of *Snail1* at both investigated time points (increase by 35.4% and 82.1%, respectively), and just after 28 days of inhalation with SE-PA in the case of *Zeb1* and *Zeb2* (average increase by 48.5%). Expression of representatives of signaling pathways leading to mesenchymal differentiation (*Ctnnd1, Nfkb1, Tgfb1*) in response to *P. agglomerans* reached the following levels: 1.236, 1.249, and 1.698 (2 weeks of inhalations) and 1.415, 1.565, and 2.255 (4 weeks of inhalations). The levels of almost all the investigated mRNA observed in mice 14 days after cessation of SE-PA chronic exposure were quite similar to the data obtained from mice only after 28 days of inhalation with *P. agglomerans*; the exception was *Ocln*, whose expression increased by 27.3% during 2 weeks without exposure. Cathelicidin administered together with SE-PA and after SE-PA treatment significantly increased the expression level of *Cdh1* and *Ocln* lowered by *P. agglomerans*. Expression of other investigated mRNAs recorded in the homogenates of lungs collected from mice treated with both CRAMP and SE-PA was lower compared with samples obtained from animals exposed to bacterial extract; nevertheless, differences observed in the case of *Zeb1* and *Zeb2* in both tested time points, as well as changes recorded in the case of *Vim, Nfkb1*, and *Tgfb1* after 14 days of exposure were not statistically significant. Statistically significant differences in the expression of *Cdh2, Acta2, Fn1, Snail*, and *Ctnnd1* recorded during comparison "SE-PA + CRAMP 14d." vs. "SE-PA 14d.", on average, were 13.1%, while in the expression of *Cdh2, Acta2, Fn1, Vim, Snail, Ctnnd1, Nfkb1*, and *Tgfb1* recorded during comparison of "SE-PA + CRAMP 28d." vs. "SE-PA 28d.", on average, were 21.1%. The cathelicidin administration after 28 days of mice exposure to SE-PA also decreased the elevated by *P. agglomerans* treatment level of all tested mesenchymal markers (on average by 11.7%), as well as factors involved in EMT (on average by 22.0%). It has to be noted that cathelicidin administered after cessation of SE-PA treatment better restored the balance in the mRNA levels of *Snail, Zeb1*, and *Zeb2*; on the contrary, alterations induced by *P. agglomerans* in the expression of other investigated genes better neutralized CRAMP together with bacterial extract. However, significant differences (at least 5%) were noted in the cases of *Cdh2, Ctnnd1, Fn1*, and *Tgfb1*.

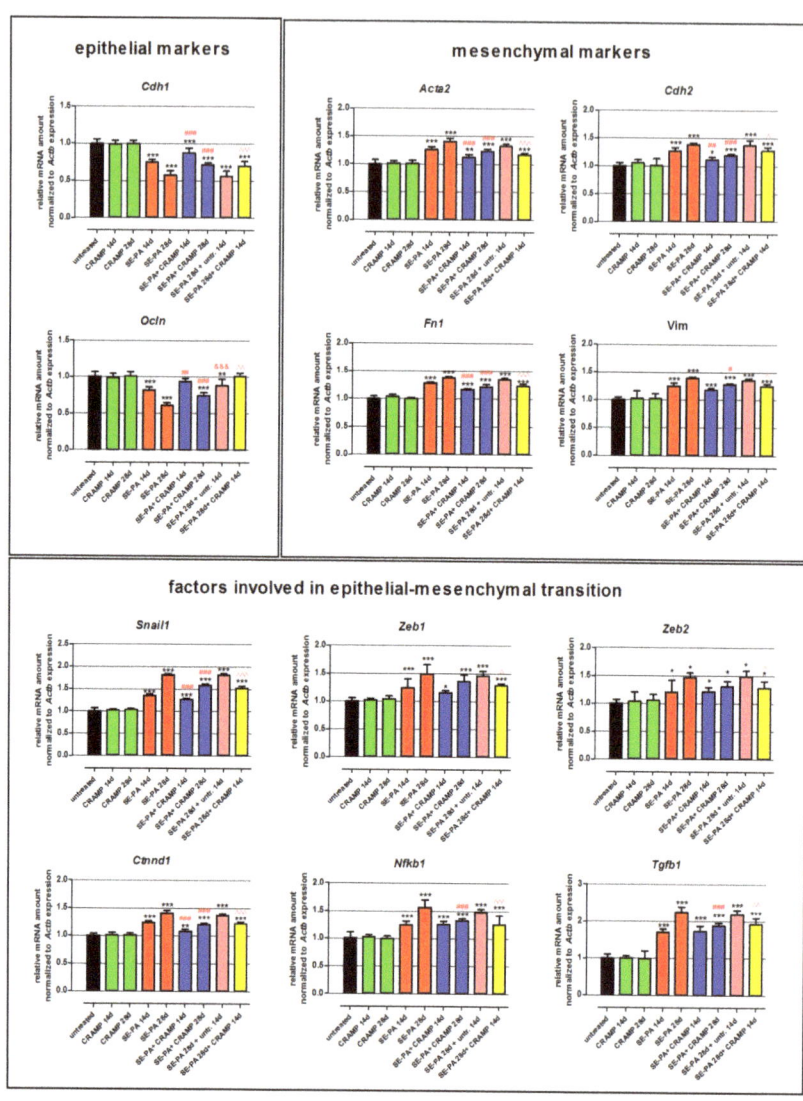

Figure 1. Alterations in the expression of genes involved in epithelial–mesenchymal transition (EMT) in response to cathelicidin (CRAMP) and/or saline extract of *Pantoea agglomerans* (SE-PA) treatment. Expression of gene coding epithelial markers, mesenchymal markers, and factors involved in epithelial–mesenchymal transition. Gene expression was investigated in homogenates of lungs collected from untreated mice (control) and animals exposed to investigated compounds for 14, 28, or 42 days using the real-time PCR method. Results are presented as the mean of relative mRNA amount ± SD. Each research group consisted of 8 mice: 6 treated and 2 untreated animals. Samples were collected from all animals and analyzed in 3 replications. Statistical significance from a one-way ANOVA test followed by Tukey's post hoc test: compared to untreated mice at $p < 0.05$ (*), $p < 0.01$ (**), $p < 0.001$ (***); SE-PA + CRAMP 14 d/28 d vs. SE-PA 14 d/28 d (comparison within corresponding time points) at $p < 0.05$ (#), $p < 0.01$ (##), $p < 0.001$ (###); SE-PA 28 d + CRAMP 14 d vs. SE-PA 28 d + untreated 14 d at $p < 0.05$ (ˆ), $p < 0.01$ (ˆˆ), $p < 0.001$ (ˆˆˆ); SE-PA 28 d + untreated 14 d vs. SE-PA 28 d at $p < 0.001$ (&&&).

Table 1. Expression of genes involved in EMT in response to cathelicidin (CRAMP) and/or saline extract of *Pantoea agglomerans* (SE-PA) treatment. Results are presented as the mean of relative mRNA amount ± SD.

	Untreated	CRAMP 14 d	CRAMP 28 d	SE-PA 14 d	SE-PA 28 d	SE-PA + CRAMP 14 d	SE-PA + CRAMP 28 d	SE-PA 28 d + Untreated 14 d	SE-PA 28 d + CRAMP 14 d
Cdh1	1.000 ± 0.058	0.988 ± 0.051	0.996 ± 0.042	0.748 ± 0.036	0.579 ± 0.059	0.873 ± 0.069	0.715 ± 0.024	0.566 ± 0.066	0.701 ± 0.061
Ocln	1.003 ± 0.064	0.986 ± 0.057	1.008 ± 0.062	0.820 ± 0.041	0.610 ± 0.035	0.936 ± 0.040	0.744 ± 0.042	0.883 ± 0.087	1.006 ± 0.051
Acta2	1.003 ± 0.075	1.010 ± 0.051	1.006 ± 0.065	1.261 ± 0.055	1.408 ± 0.058	1.121 ± 0.059	1.231 ± 0.046	1.329 ± 0.039	1.175 ± 0.032
Cdh2	1.003 ± 0.056	1.053 ± 0.065	1.005 ± 0.013	1.270 ± 0.065	1.389 ± 0.030	1.119 ± 0.051	1.200 ± 0.023	1.375 ± 0.092	1.280 ± 0.062
Fn1	1.001 ± 0.046	1.043 ± 0.034	0.995 ± 0.027	1.285 ± 0.014	1.376 ± 0.018	1.168 ± 0.018	1.216 ± 0.046	1.346 ± 0.029	1.233 ± 0.038
Vim	1.001 ± 0.043	1.020 ± 0.014	1.025 ± 0.094	1.253 ± 0.054	1.390 ± 0.029	1.181 ± 0.027	1.275 ± 0.018	1.355 ± 0.030	1.248 ± 0.037
Snail1	1.003 ± 0.069	1.023 ± 0.024	1.033 ± 0.028	1.354 ± 0.030	1.821 ± 0.027	1.264 ± 0.023	1.580 ± 0.035	1.813 ± 0.041	1.525 ± 0.047
Zeb1	1.003 ± 0.058	1.021 ± 0.034	1.046 ± 0.058	1.241 ± 0.165	1.490 ± 0.176	1.158 ± 0.044	1.355 ± 0.131	1.466 ± 0.088	1.288 ± 0.034
Zeb2	1.001 ± 0.073	1.043 ± 0.163	1.058 ± 0.106	1.210 ± 0.212	1.479 ± 0.081	1.214 ± 0.080	1.310 ± 0.098	1.495 ± 0.100	1.283 ± 0.125
Ctnnd1	1.003 ± 0.036	1.005 ± 0.052	1.004 ± 0.042	1.236 ± 0.030	1.415 ± 0.046	1.079 ± 0.031	1.201 ± 0.029	1.376 ± 0.029	1.218 ± 0.034
Nfkb1	1.005 ± 0.106	1.024 ± 0.037	0.996 ± 0.049	1.249 ± 0.066	1.565 ± 0.014	1.256 ± 0.056	1.324 ± 0.040	1.483 ± 0.055	1.258 ± 0.016
Tgfb1	1.005 ± 0.101	1.008 ± 0.053	0.983 ± 0.211	1.698 ± 0.090	2.255 ± 0.132	1.725 ± 0.153	1.886 ± 0.075	2.188 ± 0.112	1.928 ± 0.153

2.2. Cathelicidin Eliminated Negative Changes in the Expression of Proteins Associated with EMT

Changes in the expression of proteins involved in epithelial–mesenchymal transition were investigated by Western blotting in homogenates of lung tissue collected from mice treated with cathelicidin and/or saline extract of *Pantoea agglomerans* (Figure 2, Table 2). Western blots revealed that cathelicidin used alone did not impact the expression of proteins associated with EMT: E-cadherin, N-cadherin, β-catenin, fibronectin, NFκB, vimentin, occludin, Snail, α-smooth muscle actin (α-SMA), TGFβ, ZEB1, ZEB2. On the contrary, chronic exposure of mice to saline extract of *P. agglomerans* significantly decreased the expression of epithelial markers (E-cadherin, occludin) and distinctly increased the level of mesenchymal markers (N-cadherin, fibronectin, vimentin, α-SMA). Expression of epithelial markers were lowered, on average, 48.8% and 56.4% after 14 and 28 days of SE-PA exposure, respectively. Simultaneously, the expression of mesenchymal markers grew, in response to 14 and 28 days of mice treatment with *P. agglomerans*, on average by 53.7% and 101.4%, respectively. Furthermore, a distinct increase in transcription factors responsible for EMT (Snail, ZEB1, ZEB2) and key members of signaling pathways involved in EMT (β-catenin, NFκB, TGFβ) was also observed in SE-PA-treated mice. The most significant changes were recorded in the case of TGFβ, in which expression on the 14th and 28th days of the experiment reached 257.3% and 376.0% of the control, respectively. Equally important changes were observed in the case of Snail and β-catenin, whose expressions increased to 224.2% and 247.6% of the control at the above-mentioned time points, respectively. It has to be noted that the expression of all investigated proteins recorded at 28 days of SE-PA treatment and 14 days after cessation of SE-PA chronic exposure was maintained at a similar level. Cathelicidin administered together with SE-PA and after cessation of SE-PA treatment significantly increased the expression of epithelial markers lowered by inhalations with *P. agglomerans*, and the most significant improvement was observed in the case of occludin, the expression of which after 14 days of CRAPM + SE-PA exposure approached the level of the control. The expression of almost all mesenchymal markers (except vimentin in time point of 14th days) elevated by the bacterial extract was significantly lower in the lung samples collected from mice treated simultaneously with SE-PA and CRAMP as well as animals treated with defense peptide for the next 14 days after cessation of antigen administration. The most spectacular restoration of the level of mesenchymal markers altered by SE-PA was observed in the case of α-SMA, the expression of which after 14 days of cathelicidin administration together with SE-PA dropped to the level registered in untreated mice; the differences between the compared research groups (SE-PA + CRAMP 14d vs. SE-PA 14d) was 38.5%. The beneficial effect of cathelicidin given together with or after *P. agglomerans* treatment was also observed in the expression of factors involved of EMT, the level of which significantly decreased, compared to the data recorded in lung samples collected from mice treated with SE-PA. Among the investigated factors, the most significant improvement was observed in the expression of TGFβ, the level of which decreased by 115.8% (SE-PA + CRAMP 14d vs. SE-PA 14d), 187.4% (SE-PA + CRAMP 28d vs. SE-PA 28d), and 185.0% (SE-PA 28d + CRAMP 14d vs. SE-PA 28d + untreated 14d). The second factor with the highest amplitude of changes was ZEB1, the level of which decreased by 124.3% (SE-PA + CRAMP 28d vs. SE-PA 28d) and 103.7% (SE-PA 28d + CRAMP 14d vs. SE-PA 28d + untreated 14d); however, the difference of 16.7% in the expression of ZEB1 observed between "SE-PA + CRAMP 14d" and "SE-PA 14d" was not statistically significant. The beneficial impact of cathelicidin administered together or after *P. agglomerans* exposure on the expression of investigated proteins altered by SE-PA was quite similar in most of the performed experiments. Nevertheless, in the case of Snail and β-catenin, cathelicidin worked much better if given together with SE-PA; on the contrary, the defense peptide administered after cessation of SE-PA exposure better restored the balance in the expression of ZEB2, NFκB, and N-cadherin.

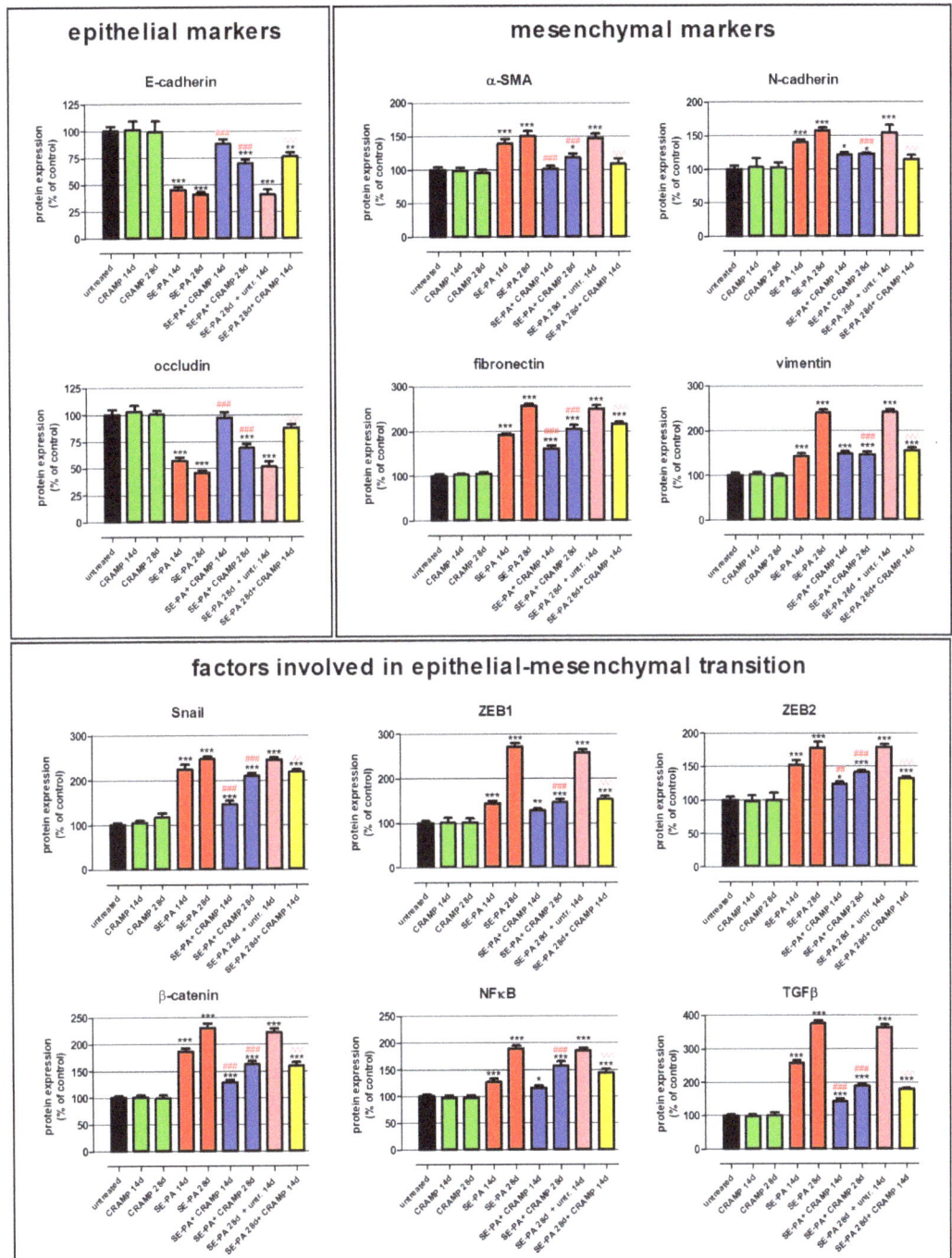

Figure 2. Cont.

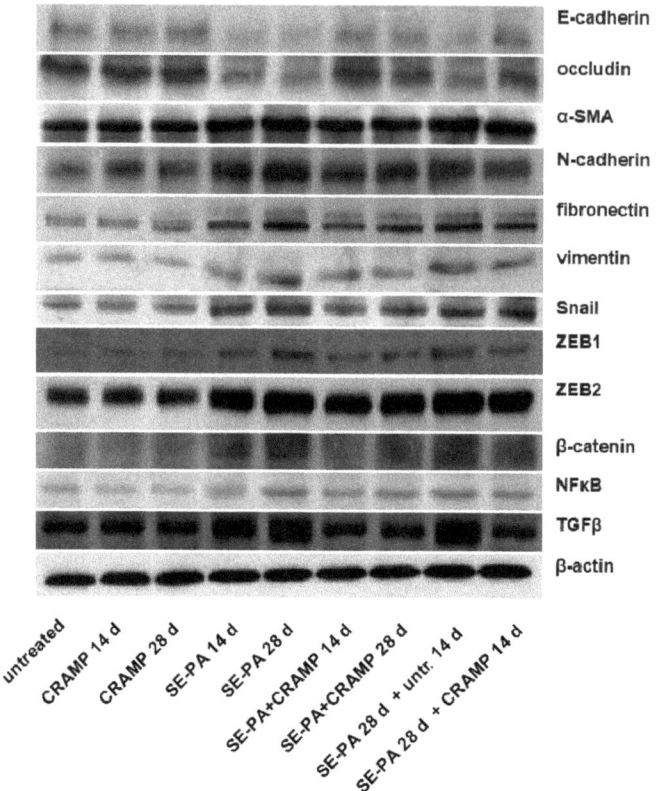

Figure 2. Alterations in the expression of proteins involved in epithelial–mesenchymal transition (EMT) in response to cathelicidin (CRAMP) and/or saline extract of *Pantoea agglomerans* (SE-PA) treatment. Protein expression was investigated in homogenates of lungs collected from untreated mice (control) and animals exposed to investigated compounds for 14, 28, or 42 days using the Western blotting method. Examination of ß-actin expression level was used as the internal control. Representative Western blots with densitometric analyses of epithelial markers, mesenchymal markers, and factors involved in epithelial–mesenchymal transition. Results of the densitometric analysis are presented as the mean of protein expression ± SD. Each research group consisted of 8 mice: 6 treated and 2 untreated animals. Samples collected from animals belonging to the common research group were mixed in equal volumes and then analyzed in 3 replications. Statistical significance from one-way ANOVA test followed by Tukey's post hoc test: compared to untreated mice at $p < 0.05$ (*), $p < 0.01$ (**), $p < 0.001$ (***); SE-PA + CRAMP 14 d/28 d vs. SE-PA 14 d/28 d (comparison within corresponding time points) at $p < 0.01$ (##), $p < 0.001$ (###); SE-PA 28 d + CRAMP 14 d vs. SE-PA 28 d + untreated 14 d at $p < 0.01$ (⁀), $p < 0.001$ (⁀⁀⁀).

Table 2. Expression of proteins involved in EMT in response to cathelicidin (CRAMP) and/or saline extract of *Pantoea agglomerans* (SE-PA) treatment. Results are presented as the mean of protein expression ± SD.

	Untreated	CRAMP 14 d	CRAMP 28 d	SE-PA 14 d	SE-PA 28 d	SE-PA + CRAMP 14 d	SE-PA + CRAMP 28 d	SE-PA 28 d + Untreated 14 d	SE-PA 28 d + CRAMP 14 d
E-cadherin	100.0 ± 4.5	101.1 ± 8.4	99.4 ± 10.1	45.4 ± 2.7	41.2 ± 2.1	88.3 ± 4.4	70.2 ± 3.5	40.9 ± 4.8	76.3 ± 3.8
Occludin	100.0 ± 5.2	102.8 ± 6.0	100.4 ± 3.7	56.9 ± 3.1	46.0 ± 2.5	97.2 ± 5.4	69.4 ± 3.8	51.5 ± 5.2	87.9 ± 3.5
α-SMA	100.0 ± 5.0	98.9 ± 4.9	96.1 ± 4.8	139.5 ± 7.0	150.8 ± 7.5	101.3 ± 5.1	118.4 ± 6.3	147.3 ± 7.4	109.3 ± 7.9
N-cadherin	100.0 ± 5.6	103.1 ± 14.0	102.7 ± 7.7	140.2 ± 4.1	157.6 ± 4.2	121.6 ± 3.6	122.5 ± 2.3	154.5 ± 11.1	114.9 ± 6.3
Fibronectin	100.0 ± 4.5	102.8 ± 2.8	104.8 ± 4.6	192.8 ± 2.8	257.2 ± 4.2	161.4 ± 7.3	205.6 ± 9.3	251.0 ± 7.8	216.9 ± 4.8
Vimentin	100.0 ± 5.0	102.2 ± 5.1	98.1 ± 4.9	142.4 ± 7.1	240.0 ± 7.2	148.4 ± 5.2	145.6 ± 7.3	241.6 ± 5.9	155.9 ± 6.5
Snail	100.0 ± 5.0	105.0 ± 5.3	117.7 ± 9.2	224.2 ± 11.2	247.6 ± 5.8	146.6 ± 7.3	209.9 ± 6.2	246.1 ± 4.2	217.9 ± 5.7
ZEB1	100.0 ± 5.1	100.9 ± 11.4	101.5 ± 9.1	144.3 ± 5.7	270.8 ± 8.5	127.6 ± 4.6	146.5 ± 7.4	257.1 ± 8.1	153.3 ± 7.8
ZEB2	100.0 ± 5.0	98.2 ± 9.1	99.9 ± 11.1	152.0 ± 7.6	177.6 ± 9.0	123.6 ± 3.6	141.0 ± 3.0	178.6 ± 5.1	132.3 ± 2.8
β-catenin	100.0 ± 3.6	100.5 ± 4.8	98.9 ± 5.9	186.8 ± 6.7	231.4 ± 8.3	129.6 ± 4.6	163.2 ± 5.8	222.0 ± 7.9	160.1 ± 6.3
NFκB	100.0 ± 3.5	97.3 ± 4.9	97.4 ± 4.9	127.2 ± 6.4	189.6 ± 5.6	117.5 ± 5.9	157.7 ± 7.9	185.3 ± 5.8	144.2 ± 7.2
TGFβ	100.0 ± 4.0	97.9 ± 5.3	100.3 ± 9.7	257.3 ± 7.9	376.0 ± 8.6	141.5 ± 8.8	188.6 ± 7.6	364.7 ± 7.6	179.7 ± 3.1

3. Discussion

Despite the well-described EMT role in pulmonary fibrosis [24,27–30,45], as well as several pieces of data indicating cathelicidin involvement in the EMT process [41–43,46], there is no evidence regarding the cathelicidin influence of EMT in the course of HP development. The current study fills this gap of knowledge, describing the cathelicidin impact on the expression of genes and proteins involved in EMT under developing lung fibrosis in the course of HP. The study was conducted in created and validated by our research group the murine model of HP, wherein pulmonary fibrosis was induced by an extract of *Pantoea agglomerans* administered daily for 28 days to fibrosis mice strain C57BL/6J [47–51]. Pathological changes observed in mice in response to chronic exposure to the antigen of *P. agglomerans* were similar to the clinical picture of HP, which was confirmed on the genome and proteome level as well as in the microscopic image of lung tissue and in the characteristics of the immune response. Furthermore, earlier studies by our team also revealed that depending on the time of exposure, successive stages of disease development can be obtained: acute with a strong inflammatory response (7–14 days of exposure) and chronic with significant signs of fibrosis (28 days of exposure) [47–51]. It should be emphasized that according to our best knowledge, the mentioned model is the only research model of HP that, under laboratory conditions, reproduces the environmental exposure to organic dust causing lung fibrosis. For this reason, the use of this model in the presented study increases the chances of effective translation of the obtained results into clinical practice, thus accelerating the introduction of a new HP therapeutic strategy. An additional reason for selecting the mentioned model for the current study was the previously mentioned data [27], which demonstrated the relationship between pulmonary fibrosis induced by mice chronic exposure to *P. agglomerans* with the antigen ability to provoke epithelial–mesenchymal transition type 2, which is the focus of this study. First of all, the study revealed that the chronic exposure of mice to cathelicidin did not cause any changes in the expression of all investigated molecules, neither the genes nor their proteins products: *Acta2*/α-smooth muscle actin, *Cdh1*/E-cadherin, *Cdh2*/N-cadherin, *Ctnnd1*/β-catenin, *Fn1*/fibronectin, *Nfkb1*/NFκB, *Vim*/vimentin, *Ocln*/occludin, *Snail1*/Snail, *Tgfb1*/TGFβ, *Zeb1*/ZEB1, *Zeb2*/ZEB2. On the contrary, SE-PA treatment induced alterations characteristic for EMT: downregulation of epithelial markers (*Cdh1*/E-cadherin, *Ocln*/occludin) and significant upregulation of mesenchymal markers (*Cdh2*/N-cadherin, *Fn1*/fibronectin, *Vim*/vimentin, *Acta2*/α-smooth muscle actin). Chronic exposure of mice to *P. agglomerans* also increased the level of transcription factors responsible for the EMT process (*Snail1*/Snail, *Zeb1*/ZEB1, *Zeb2*/ZEB2) and selected members of signaling pathways, leading to mesenchymal differentiation (*Ctnnd1*/β-catenin, *Nfkb1*/NFκB, *Tgfb1*/TGFβ). The most significant changes were noted in the expressions of *Tgfb1*/TGFβ and *Snail1*/Snail. The observed changes highlight the key roles of these two molecules in the execution of the EMT program in response to SE-PA exposure. TGFβ has been shown to play a pivotal role in pulmonary fibrosis, not only through induction of EMT in alveolar epithelial cells and its ability to attract and stimulate proliferation of fibroblasts and myofibroblasts but also as potent inducers of ECM production, including collagen and other matrix proteins [28,29,52,53]. Since *Snail* is an immediate-early response gene for TGFβ, the observed significant increase in its expression seems to be logical. Indeed, most EMT studies revealed Snail induction in response to TGFβ and furthermore demonstrated a correlation between the elevated level of this transcription factor with the repression of genes coding epithelial markers, as well as concomitant activation of mesenchymal gene expression [54–56]. While the amplitudes of changes in the level of expression of other investigated genes in response to SE-PA were similar, analysis of protein expression indicated two additional molecules (ZEB1 and β-catenin) that, in addition to the TGFβ and Snail, were proven to be very sensitive to the action of the tested bacterial extract. As in the case of the Snails, the expression of transcription factor ZEB1 is activated by TGFβ and directly represses the expression of epithelial marker genes, increases expression of the mesenchymal markers, and additionally promotes cell migration [29,54,57–59]. β-catenin is physiologically part of a major cell-surface adhesion complex, but if released from there becomes an important part of Wnt signaling, the involvement of which in EMT has been proven in many studies [29,30,57]. Released β-catenin interacts with transcription factors LEF1 or TCF and as a complex translocates into the nucleus and regulates the transcription of several genes associated with EMT, e.g., decreasing the expression of E-cadherin gene, increasing the expression of genes coding vimentin and fibronectin, and stimulating the production of matrix metalloproteinases [30,57,60].

The conducted studies revealed that cathelicidin administered with SE-PA or after 28 days of *P. agglomerans* exposure restored the balance in the expression of genes and proteins disturbed by mice exposure to the mentioned extract. CRAMP increased the lowered by SE-PA level of *Cdh1*/E-cadherin and *Ocln*/occludin. Cathelicidin particularly effectively neutralized the negative effect of *P. agglomerans* on the expression of *Ocln*/occludin—14 days of CRAMP treatment together with SE-PA or after cessation of SE-PA inhalations increased the expression of the investigated molecule to the level observed in untreated mice. Additionally, CRAMP administered with bacterial extract or after SE-PA exposure decreased the elevated by SE-PA level of *Cdh2*/N-cadherin, *Acta2*/α-SMA, *Fn1*/fibronectin, and *Vim*/vimentin. Nevertheless, complete neutralization of the alteration induced by SE-PA was observed in the case of α-SMA after 14 days of mice exposure to both CRAMP and SE-PA. The similar beneficial impact of the tested peptide on α-SMA expression was noted in mice after 14 days of CRAMP treatment preceded by 28 days of SE-PA inhalations. Alterations in the expression of factors associated with EMT observed in mice chronically exposed to SE-PA were also leveled by cathelicidin treatment. Nevertheless, the effectiveness of cathelicidin treatment depends on the way of cathelicidin administration. The investigation of gene expression revealed three different patterns of CRAMP response: (1) *Snail1* and *Ctnnd1*: significant differences in the expression after 14 and 28 days of animals exposure to SE-PA used alone and together with CRAMP; (2) *Nfkb1* and *Tgfb*: significant differences in the expression after 28 days of animals exposure to SE-PA used alone and together with CRAMP; (3) differences in the expression of *Zeb1* and *Zeb2* after 14 and 28 days of animals exposure to SE-PA used alone and together with CRAMP were not significant. Nevertheless, alteration in the expression of all investigated factors was restored when CRAMP was provided after cessation of SE-PA inhalations. Results obtained from Western blotting revealed that mice inhalation with CRAMP after 28 days of SE-PA exposure as well as CRAMP administration for 4 weeks together with SE-PA significantly lowered the pathological level of all investigated factors. Furthermore, the expression of Snail, ZEB2, β-catenin, and TGFβ was also improved in mice exposed to both CRAMP and *P. agglomerans* antigene for 14 days.

It should be emphasized that changes recorded on genome and proteome levels correspond with alterations in lung tissue morphology previously described by our team [44] (see the Supplementary Materials Figure S1). Similar to the presented data, histological examination of murine lungs stained with hematoxylin and eosin (H&E) or Masson trichrome (TRI) revealed the lack of any changes in tissue morphology after 14 and 28 days of animal exposure to cathelicidin. On the contrary, mice inhalations with SE-PA induced changes typical for HP; in particular, significant inflammatory response with interstitial infiltrations of lymphocyte and macrophage (mean score for inflammation in time points 14 and 28: 1.6 and 2.0) as well as signs of fibrosis (mean score for fibrosis in time points 14 and 28: 0.8 and 1.6) associated with abnormal collagen deposition, leading to thickening of alveolar walls, which intensified during the time of exposure. The mentioned alterations perfectly corresponded with the herein presented increase in expression of mesenchymal markers and transcription factors involved in EMT. Moreover, the histological examination also did not show statistically significant changes between lungs collected 14 days after cessation of mice chronic exposure to SE-PA to lungs obtained directly after 28 days of inhalation with *P. agglomerans* (SE-PA 28d + untreated 14d vs. SE-PA 28d). Furthermore, the evaluation of lung sections also revealed the beneficial impact of cathelicidin treatment on fibrosis development. Comparison data obtained from the following research groups: SE-PA + CRAMP 28d vs. SE-PA 28d as well as SE-PA 28d + CRAMP 14d vs. SE-PA 28d + untreated 14d, demonstrated a significant decrease in fibrosis scores by 36.8% and 38.9%, respectively. The above-mentioned antifibrotic properties of cathelicidin correspond with data obtained in presented studies, that revealed the beneficial impact of CRAMP treatment in the inhibition of the expression of both mesenchymal markers and factors involved in EMT, which have been pathologically elevated by *P. agglomerans*. It needs to be highlighted that the concordance of changes observed at the level of genes and proteins as well as in the histological evaluation indicated that the previously described beneficial properties of cathelicidin should be associated with its ability to modulate the course of EMT [44].

The presented results indicate CRAMP as an effective inhibitor of EMT associated with pulmonary fibrosis in HP, which is the first such evidence according to the best of our knowledge. Nevertheless, the ability of cathelicidin to inhibit pathological EMT was reported previously by Cheng et al., who showed that LL-37 inhibited TGFβ-induced gene expression of *ACTA2* and *VIM* as well as protein expression of E-cadherin, Twist1, and Slug in human colon cancer HT-29 cells. Furthermore, Cheng et al. demonstrated in vivo that administration of cathelicidin expression viral vector or synthetic mCRAMP inhibited vimentin and E-cadherin expression and collagen deposition, leading to suppression of EMT and consequently to colon cancer development [46]. Additionally, Zheng et al. revealed the inhibition of cardiac fibrosis in diabetic mice heart treated with CRAMP, which was a consequence of an increase in endothelial markers (CD31, cadherin), a decrease in fibroblast markers (collagen I, collagen III, vimentin), and a reduction of transcription factors (Snial1, Snial2, Twist1, Twist2). Zheng et al. connected the anti-fibrotic CRAMP abilities with the silencing of signal transduction in the TGFβ/Smad pathway [61]. Despite the fact that some alterations in the expression of genes and proteins in response to CRAMP observed by Cheng et al. and Zheng et al. correspond with the currently presented results, it has to be stressed that there are differences. Cheng et al. investigated the impact of cathelicidin on EMT type 3 [46], while Zheng et al. discovered the beneficial impact of CRAMP on endothelial–mesenchymal transition [61]. Consequently, the presented study is the first report showing the influence of cathelicidin on EMT type 2. Furthermore, obtained data indicated that maintaining the physiological level of cathelicidin in the respiratory tract plays an important role in the supervision of regenerative processes and consequently prevention of pathological wound healing leading to fibrosis.

4. Materials and Methods

4.1. Reagents

Unless otherwise indicated, the chemicals used in the study were purchased from Sigma-Aldrich Co. (St. Louis, MO, USA) LLC. Murine cathelicidin (ISRLAGLLRKGGEKIGEKLKKI GQKIKNFFQKLVPQPE) was purchased from Novozym Polska s.c. Poznań Science and Technology Park, Poznań, Poland. The preparation of Pantoea agglomerans extract, as well as its main composition, has been described previously [27].

4.2. Animal Inhalations

Three-month-old female C57BL/6J mice were purchased from Mossakowski Medical Research Centre of the Polish Academy of Sciences in Warsaw, Poland. The conditions in which the animals were kept, as well as the procedure for their preparation for the study, has been described previously [44]. Mice were exposed to a finely dispersed aerosol of the saline extract of *P. agglomerans* (SE-PA; 10 mg/mL; 5 mL/single inhalation), cathelicidin (CRAMP; 1.44 µg/mL; 5 mL/single inhalation), or phosphate-buffered saline (PBS; 5 mL/single inhalation), administered separately or in combination (Table 3). The mice were treated to each investigated factor for one hour daily for 14, 28, or 42 days. Inhalations were carried out using the Buxco Inhalation Tower (Data Sciences International, St. Paul, MN, USA) under the following conditions: airflow 1.5 L/min; pressure 0.5 cm H_2O; room temperature; nebulization rate 84 µL/min. Before and after the indicated time of treatment, the animals were sacrificed by cervical dislocation with spinal cord rupture, after which lung samples were collected, frozen in liquid nitrogen, and stored at $-80\ °C$ until evaluation.

Table 3. The research group description.

Name of Research Group	PBS (Time of Exposure)	SE-PA (Time of Exposure)	CRAMP (Time of Exposure)	Factors Administration Sequence
untreated ($n = 16$)	-	-	-	-
CRAMP 14 d ($n = 6$)	1 h a day for 14 days	-	1 h a day for 14 days	one by one on the same day
CRAMP 28 d ($n = 6$)	1 h a day for 28 days	-	1 h a day for 28 days	one by one on the same day
SE-PA 14 d ($n = 6$)	1 h a day for 14 days	1 h a day for 14 days	-	one by one on the same day
SE-PA 28 d ($n = 6$)	1 h a day for 28 days	1 h a day for 28 days	-	one by one on the same day
SE-PA + CRAMP 14 d ($n = 6$)	-	1 h a day for 14 days	1 h a day for 14 days	one by one on the same day
SE-PA + CRAMP 28 d ($n = 6$)	-	1 h a day for 28 days	1 h a day for 28 days	one by one on the same day
SE-PA 28 d + untreated 14 d ($n = 6$)	1 h a day for 28 days	1 h a day for 28 days	-	one by one on the same day
	-	-	-	After 28 days of treatment, mice stayed an additional 14 days in the experiment without exposure
SE-PA 28 d + CRAMP 14 d ($n = 6$)	1 h a day for 28 days	1 h a day for 28 days	-	one by one on the same day
	-	-	1 h a day for 14 days	after 28 days of mice exposure to (PBS + SE-PA), CRAMP was applied for an additional 14 days

PBS—phosphate-buffered saline; SE-PA—saline extract of Pantoea agglomerans; CRAMP—cathelicidin; n—number of animals in research group; d—days.

4.3. Evaluation of Gene Expression

Gene expression evaluation has been described previously [62]; nevertheless, what should be mentioned are the TaqMan Gene Expression Assays (sets of probes and primers) used in the research: Mm00725412_s1 for *Acta2*; Mm02619580_g1 for *Actb*; Mm01247357_m1 for *Cdh1*; Mm01162497_m1 for *Cdh2*; Mm01334599_m1 for *Ctnnd1*; Mm01256744_m1 for *Fn1*; Mm00476361_m1 for *Nfkb1*; Mm00500912_m1 for *Ocln*; Mm01178820_m1 for *Tgfb1*; Mm00441533_g1 for *Snail1*; Mm01333430_m1 for *Vim*; Mm00495564_m1 for *Zeb1*; Mm00497196_m1 for *Zeb2*. Relative expression was calculated using the efficiency method (relative advanced quantification) and normalized to the expression of *Actb*.

4.4. Evaluation of Proteins Expression

Examination of protein expression has been described previously [62]. The investigation has been conducted using primary antibodies directed against β-actin, β-catenin, E-cadherin, N-cadherin, Snail, vimentin (Cell Signaling Technology, Danvers, MA, USA), α-smooth muscle actin, fibronectin, occludin, NFκB, TGFβ, ZEB1, and ZEB2 (ThermoFisher Scientific, Waltham, MA, USA). The amount of protein was densitometrically determined using ImageJ software.

4.5. Statistical Analysis

The data were presented as the mean value and standard deviation (SD). Statistical analysis was performed using linear regression analysis, as well as the one way-ANOVA with Tukey's post hoc test, and column statistics were used for comparisons. Significance was accepted at $p < 0.05$.

5. Conclusions

In summary, the beneficial impact of cathelicidin treatment on the expression of genes/proteins involved in EMT was observed both during and after the development of hypersensitivity pneumonitis. Despite the fact that cathelicidin was not able to completely neutralize the negative changes induced by *P. agglomerans*, EMT silencing in response to

CRAMP was significant. Furthermore, cathelicidin used alone did not cause any side-changes in the expression of the investigated genes and proteins. It needs to be highlighted that the presented study is the first report showing the influence of cathelicidin on EMT type 2. Due to the importance of EMT for pulmonary fibrosis, the presented results suggest the possibility of using cathelicidin in the prevention and treatment of this pathological process. Unfortunately, exogenous cathelicidin was not able to eliminate the negative changes completely, perhaps because of the fact that CRAMP inhalations did not restore the physiological level of cathelicidin disturbed by the disease development. Nevertheless, in the face of the absence of a safe and successful strategy for the prevention and treatment of pulmonary fibrosis in the course of HP, even a slight beneficial effect of cathelicidin, especially in terms of the lack of side effects, deserves attention and presentation in order to create the basis for the development an effective therapeutic strategy in the future.

Supplementary Materials: The supporting information can be downloaded at https://www.mdpi.com/article/10.3390/ijms232113039/s1.

Author Contributions: Conceptualization, M.K.L., M.G. and J.M.; methodology, M.K.L. and J.D.; formal analysis, M.K.L.; investigation, M.K.L. and J.Z.; writing—original draft preparation, M.K.L.; writing—review and editing, M.K.L., M.G., J.Z., J.D. and J.M.; visualization, M.K.L.; supervision, J.D. and J.M.; project administration, M.K.L.; funding acquisition, M.K.L. All authors have read and agreed to the published version of the manuscript.

Funding: The studies were supported by the National Science Centre, Poland (grant no. 2015/19/D/NZ7/02952, 2016).

Institutional Review Board Statement: The experimental protocols and procedures listed below conform to the Guide for the Care and Use of Laboratory Animals and were approved by the Local Ethics Committee for Animal Experimentation in Lublin, Poland (Resolution nos. 39/2016 and 2/2017).

Informed Consent Statement: Not applicable.

Data Availability Statement: All data analyzed during this study are included in this article. Further inquiries can be directed towards the corresponding author.

Conflicts of Interest: M.G. and J.D. have no conflict of interest to declare. M.K.L., J.Z., and J.M. received payment for performing research in project no. 2015/19/D/NZ7/02952 from the National Science Centre, Poland. The funders had no role in the design of the study; in the collection, analyses, or interpretation of data; in the writing of the manuscript; or in the decision to publish the results.

References

1. Spagnolo, P.; Rossi, G.; Cavazza, A.; Bonifazi, M.; Paladini, I.; Bonella, F.; Sverzellati, N.; Costabel, U. Hypersensitivity pneumonitis: A comprehensive review. *J. Investig. Allergy Clin. Immunol.* **2015**, *25*, 237–250.
2. Ho, Y.Y.; Lagares, D.; Tager, A.M.; Kapoor, M. Fibrosis—A lethal component of systemic sclerosis. *Nat. Rev. Rheumatol.* **2014**, *10*, 390–402. [CrossRef]
3. Ziesche, R.; Golec, M.; Samaha, E. The RESOLVE concept: Approaching pathophysiology of fibroproliferative disease in aged individuals. *Biogerontology* **2013**, *14*, 679–685. [CrossRef] [PubMed]
4. Nogueira, R.; Melo, N.; e Bastos, H.N.; Martins, N.; Delgado, L.; Morais, A.; Mota, P.C. Hypersensitivity pneumonitis: Antigen diversity and disease implications. *Pulmonology* **2019**, *25*, 97–108. [CrossRef]
5. Dutkiewicz, J.; Mackiewicz, B.; Lemieszek, M.K.; Golec, M.; Skórska, C.; Góra-Florek, A.; Milanowski, J. Pantoea agglomerans: A mysterious bacterium of evil and good. Part II–deleterious effects: Dustborne endotoxins and allergens–focus on grain dust, other agricultural dusts and wood dust. *Ann. Agric. Environ. Med.* **2016**, *23*, 6–29. [CrossRef] [PubMed]
6. Kaul, B.; Cottin, V.; Collard, H.R.; Valenzuela, C. Variability in global prevalence of interstitial lung disease. *Front. Med.* **2021**, *8*, 751181. [CrossRef]
7. Hyldgaard, C.; Hilberg, O.; Muller, A.; Bendstrup, E. A cohort study of interstitial lung diseases in central Denmark. *Respir. Med.* **2014**, *108*, 793–799. [CrossRef]
8. Egea-Zorrilla, A.; Vera, L.; Saez, B.; Pardo-Saganta, A. Promises and challenges of cell-based therapies to promote lung regeneration in idiopathic pulmonary fibrosis. *Cells* **2022**, *11*, 2595. [CrossRef]
9. Salton, F.; Ruaro, B.; Confalonieri, P.; Confalonieri, M. Epithelial-mesenchymal transition: A major pathogenic driver in idiopathic pulmonary fibrosis? *Medicina* **2020**, *56*, 608. [CrossRef]

10. Sakai, N.; Tager, A.M. Fibrosis of two: Epithelial cell-fibroblast interactions in pulmonary fibrosis. *Biochim. Biophys. Acta (BBA)-Mol. Basis Dis.* **2013**, *1832*, 911–921. [CrossRef]
11. Vaughan, A.E.; Chapman, H.A. Regenerative activity of the lung after epithelial injury. *Biochim. et Biophys. Acta (BBA)-Mol. Basis Dis.* **2013**, *1832*, 922–930. [CrossRef]
12. Wynn, T.A.; Ramalingam, T.R. Mechanisms of fibrosis: Therapeutic translation for fibrotic disease. *Nat. Med.* **2012**, *18*, 1028–1040. [CrossRef] [PubMed]
13. Wynn, T.A. Common and unique mechanisms regulate fibrosis in various fibroproliferative diseases. *J. Clin. Investig.* **2007**, *117*, 524–529. [CrossRef] [PubMed]
14. Horowitz, J.; Thannickal, V. Epithelial-mesenchymal interactions in pulmonary fibrosis. *Semin. Respir. Crit. Care Med.* **2006**, *27*, 600–612. [CrossRef] [PubMed]
15. Guarino, M.; Tosoni, A.; Nebuloni, M. Direct contribution of epithelium to organ fibrosis: Epithelial-mesenchymal transition. *Hum. Pathol.* **2009**, *40*, 1365–1376. [CrossRef] [PubMed]
16. Confalonieri, P.; Volpe, M.C.; Jacob, J.; Maiocchi, S.; Salton, F.; Ruaro, B.; Confalonieri, M.; Braga, L. Regeneration or repair? The role of alveolar epithelial cells in the pathogenesis of idiopathic pulmonary fibrosis (IPF). *Cells* **2022**, *11*, 2095. [CrossRef]
17. Kalluri, R.; Weinberg, R.A. The basics of epithelial-mesenchymal transition. *J. Clin. Investig.* **2009**, *119*, 1420–1428. [CrossRef]
18. Thiery, J.P.; Sleeman, J.P. Complex networks orchestrate epithelial–mesenchymal transitions. *Nat. Rev. Mol. Cell Biol.* **2006**, *7*, 131–142. [CrossRef]
19. Voulgari, A.; Pintzas, A. Epithelial-mesenchymal transition in cancer metastasis: Mechanisms, markers and strategies to overcome drug resistance in the clinic. *Biochim. Biophys. Acta (BBA) Rev. Cancer* **2009**, *1796*, 75–90. [CrossRef]
20. Moustakas, A.; Heldin, C.-H. Signaling networks guiding epithelial?mesenchymal transitions during embryogenesis and cancer progression. *Cancer Sci.* **2007**, *98*, 1512–1520. [CrossRef]
21. Hackett, T.-L.; Warner, S.M.; Stefanowicz, D.; Shaheen, F.; Pechkovsky, D.V.; Murray, L.A.; Argentieri, R.; Kicic, A.; Stick, S.M.; Bai, T.R.; et al. Induction of epithelial–mesenchymal transition in primary airway epithelial cells from patients with asthma by transforming growth factor-β1. *Am. J. Respir. Crit. Care Med.* **2009**, *180*, 122–133. [CrossRef] [PubMed]
22. Kim, J.H.; Jang, Y.S.; Eom, K.-S.; Hwang, Y.I.; Kang, H.R.; Jang, S.H.; Kim, C.H.; Park, Y.B.; Lee, M.G.; Hyun, I.G.; et al. Transforming Growth factor β1 induces epithelial-to-mesenchymal transition of A549 cells. *J. Korean Med. Sci.* **2007**, *22*, 898–904. [CrossRef] [PubMed]
23. Kasai, H.; Allen, J.T.; Mason, R.M.; Kamimura, T.; Zhang, Z. TGF-β1 induces human alveolar epithelial to mesenchymal cell transition (EMT). *Respir. Res.* **2005**, *6*, 56. [CrossRef] [PubMed]
24. Willis, B.C.; Liebler, J.M.; Luby-Phelps, K.; Nicholson, A.G.; Crandall, E.D.; du Bois, R.M.; Borok, Z. Induction of epithelial-mesenchymal transition in alveolar epithelial cells by transforming growth factor-β1: Potential role in idiopathic pulmonary fibrosis. *Am. J. Pathol.* **2005**, *166*, 1321–1332. [CrossRef]
25. Cano, A.; Pérez-Moreno, M.A.; Rodrigo, I.; Locascio, A.; Blanco, M.J.; Del Barrio, M.G.; Portillo, F.; Nieto, M.A. The transcription factor Snail controls epithelial–mesenchymal transitions by repressing E-cadherin expression. *Nat. Cell Biol.* **2000**, *2*, 76–83. [CrossRef]
26. Comijn, J.; Berx, G.; Vermassen, P.; Verschueren, K.; van Grunsven, L.; Bruyneel, E.; Mareel, M.; Huylebroeck, D.; van Roy, F. The two-handed e box binding zinc finger protein sip1 downregulates e-cadherin and induces invasion. *Mol. Cell* **2001**, *7*, 1267–1278. [CrossRef]
27. Lemieszek, M.K.; Rzeski, W.; Golec, M.; Mackiewicz, B.; Zwoliński, J.; Dutkiewicz, J.; Milanowski, J. Pantoea agglomerans chronic exposure induces epithelial-mesenchymal transition in human lung epithelial cells and mice lungs. *Ecotoxicol. Environ. Saf.* **2020**, *194*, 110416. [CrossRef]
28. Tanjore, H.; Xu, X.C.; Polosukhin, V.V.; Degryse, A.L.; Li, B.; Han, W.; Sherrill, T.P.; Plieth, D.; Neilson, E.G.; Blackwell, T.S.; et al. Contribution of epithelial-derived fibroblasts to bleomycin-induced lung fibrosis. *Am. J. Respir. Crit. Care Med.* **2009**, *180*, 657–665. [CrossRef]
29. Willis, B.C.; Borok, Z. TGF-β-induced EMT: Mechanisms and implications for fibrotic lung disease. *Am. J. Physiol. Cell. Mol. Physiol.* **2007**, *293*, L525–L534. [CrossRef]
30. Chilosi, M.; Poletti, V.; Zamò, A.; Lestani, M.; Montagna, L.; Piccoli, P.; Pedron, S.; Bertaso, M.; Scarpa, A.; Murer, B.; et al. Aberrant Wnt/β-catenin pathway activation in idiopathic pulmonary fibrosis. *Am. J. Pathol.* **2003**, *162*, 1495–1502. [CrossRef]
31. Bals, R.; Wilson, J.M. Cathelicidins—A family of multifunctional antimicrobial peptides. *Cell. Mol. Life. Sci.* **2003**, *60*, 711–720. [CrossRef] [PubMed]
32. Van Harten, R.M.; van Woudenbergh, E.; van Dijk, A.; Haagsman, H.P. Cathelicidins: Immunomodulatory antimicrobials. *Vaccines* **2018**, *6*, 63. [CrossRef] [PubMed]
33. Kościuczuk, E.M.; Lisowski, P.; Jarczak, J.; Strzałkowska, N.; Jóźwik, A.; Horbańczuk, J.; Krzyżewski, J.; Zwierzchowski, L.; Bagnicka, E. Cathelicidins: Family of antimicrobial peptides. A review. *Mol. Biol. Rep.* **2012**, *39*, 10957–10970. [CrossRef] [PubMed]
34. Mookherjee, N.; Brown, K.; Bowdish, D.; Doria, S.; Falsafi, R.; Hokamp, K.; Roche, F.M.; Mu, R.; Doho, G.H.; Pistolic, J.; et al. Modulation of the TLR-mediated inflammatory response by the endogenous human host defense peptide LL-37. *J. Immunol.* **2006**, *176*, 2455–2464. [CrossRef]

35. Bowdish, D.M.E.; Davidson, D.J.; Scott, M.G.; Hancock, R.E.W. Immunomodulatory activities of small host defense peptides. *Antimicrob. Agents Chemother.* **2005**, *49*, 1727–1732. [CrossRef] [PubMed]
36. Kress, E.; Merres, J.; Albrecht, L.-J.; Hammerschmidt, S.; Pufe, T.; Tauber, S.C.; Brandenburg, L.-O. CRAMP deficiency leads to a pro-inflammatory phenotype and impaired phagocytosis after exposure to bacterial meningitis pathogens. *Cell Commun. Signal.* **2017**, *15*, 32. [CrossRef]
37. Gupta, K.; Subramanian, H.; Ali, H. Modulation of host defense peptide-mediated human mast cell activation by LPS. *Innate Immun.* **2016**, *22*, 21–30. [CrossRef]
38. Ugarova, T.P.; Moreno, B.; Podolnikova, N.P.; Lishko, V.K. Identification of human cathelicidin peptide LL-37 as a ligand for macrophage integrin $\alpha_M\beta_2$ (Mac-1, CD11b/CD18) that promotes phagocytosis by opsonizing bacteria. *Res. Rep. Biochem.* **2016**, *6*, 39–55. [CrossRef]
39. Neumann, A.; Berends, E.T.M.; Nerlich, A.; Molhoek, E.M.; Gallo, R.; Meerloo, T.; Nizet, V.; Naim, H.Y.; Von Köckritz-Blickwede, M. The antimicrobial peptide LL-37 facilitates the formation of neutrophil extracellular traps. *Biochem. J.* **2014**, *464*, 3–11. [CrossRef]
40. Wan, M.; van der Does, A.; Tang, X.; Lindbom, L.; Agerberth, B.; Haeggström, J.Z. Antimicrobial peptide LL-37 promotes bacterial phagocytosis by human macrophages. *J. Leukoc. Biol.* **2014**, *95*, 971–981. [CrossRef]
41. Ramos, R.; Silva, J.P.; Rodrigues, A.C.; Costa, R.; Guardão, L.; Schmitt, F.; Soares, R.; Vilanova, M.; Domingues, L.; Gama, M. Wound healing activity of the human antimicrobial peptide LL37. *Peptides* **2011**, *32*, 1469–1476. [CrossRef] [PubMed]
42. Carretero, M.; Escámez, M.J.; García, M.; Duarte, B.; Holguín, A.; Retamosa, L.; Jorcano, J.L.; del Río, M.; Larcher, F. In vitro and in vivo wound healing-promoting activities of human cathelicidin LL-37. *J. Investig. Dermatol.* **2008**, *128*, 223–236. [CrossRef]
43. Shaykhiev, R.; Beisswenger, C.; Kändler, K.; Senske, J.; Püchner, A.; Damm, T.; Behr, J.; Bals, R. Human endogenous antibiotic LL-37 stimulates airway epithelial cell proliferation and wound closure. *Am. J. Physiol. Cell. Mol. Physiol.* **2005**, *289*, L842–L848. [CrossRef] [PubMed]
44. Lemieszek, M.K.; Sawa-Wejksza, K.; Golec, M.; Dutkiewicz, J.; Zwoliński, J.; Milanowski, J. Beneficial impact of cathelicidin on hypersensitivity pneumonitis treatment—In vivo studies. *PLoS ONE* **2021**, *16*, e0251237. [CrossRef]
45. Kim, K.K.; Kugler, M.C.; Wolters, P.J.; Robillard, L.; Galvez, M.G.; Brumwell, A.N.; Sheppard, D.; Chapman, H.A. Alveolar epithelial cell mesenchymal transition develops *in vivo* during pulmonary fibrosis and is regulated by the extracellular matrix. *Proc. Natl. Acad. Sci. USA* **2006**, *103*, 13180–13185. [CrossRef] [PubMed]
46. Cheng, M.; Ho, S.; Yoo, J.H.; Tran, D.H.; Bakirtzi, K.; Su, B.; Tran, D.H.; Kubota, Y.; Ichikawa, R.; Koon, H.W. Cathelicidin suppresses colon cancer development by inhibition of cancer associated fibroblasts. *Clin. Exp. Gastroenterol.* **2015**, *8*, 13–29. [CrossRef]
47. Golec, M.; Wielscher, M.; Lemieszek, M.K.; Vierlinger, K.; Skórska, C.; Huetter, S.; Sitkowska, J.; Mackiewicz, B.; Góra-Florek, A.; Ziesche, R.; et al. Middle age enhances expression of innate immunity genes in a female mouse model of pulmonary fibrosis. *Biogerontology* **2017**, *18*, 253–262. [CrossRef] [PubMed]
48. Golec, M.; Lemieszek, M.; Skórska, C.; Sitkowska, J.; Zwoliński, J.; Mackiewicz, B.; Góra-Florek, A.; Milanowski, J.; Dutkiewicz, J. Cathelicidin related antimicrobial peptide, laminin, Toll-like receptors and chemokines levels in experimental hypersensitivity pneumonitis in mice. *Pathol. Biol.* **2015**, *63*, 130–135. [CrossRef]
49. Lemieszek, M.K.; Chilosi, M.; Golec, M.; Skórska, C.; Dinnyes, A.; Mashayekhi, K.; Vierlinger, K.; Huaux, F.; Wielscher, M.; Hofner, M.; et al. Age influence on hypersensitivity pneumonitis induced in mice by exposure to *Pantoea agglomerans*. *Inhal. Toxicol.* **2013**, *25*, 640–650. [CrossRef]
50. Lemieszek, M.; Chilosi, M.; Golec, M.; Skórska, C.; Huaux, F.; Yakoub, Y.; Pastena, C.; Daniele, I.; Cholewa, G.; Sitkowska, J.; et al. Mouse model of hypersensitivity pneumonitis after inhalation exposure to different microbial antigens associated with organic dusts. *Ann. Agric. Environ. Med.* **2011**, *18*, 71–80.
51. Golec, M.; Skórska, C.; Lemieszek, M.; Dutkiewicz, J. A novel inhalation challenge to study animal model of allergic alveolitis. *Ann. Agric. Environ. Med.* **2009**, *16*, 173–175. [PubMed]
52. Tatler, A.L.; Jenkins, G. TGF-β activation and lung fibrosis. *Proc. Am. Thorac. Soc.* **2012**, *9*, 130–136. [CrossRef]
53. Yue, X.; Shan, B.; Lasky, J.A. TGF-β: Titan of lung fibrogenesis. *Curr. Enzym. Inhib.* **2010**, *6*, 67–77. [CrossRef]
54. Xu, J.; Lamouille, S.; Derynck, R. TGF-beta-induced epithelial to mesenchymal transition. *Cell Res.* **2009**, *19*, 156–172. [CrossRef]
55. Peinado, H.; Quintanilla, M.; Cano, A. Transforming growth factor β-1 induces Snail transcription factor in epithelial cell lines: Mechanisms for epithelial mesenchymal transitions. *J. Biol. Chem.* **2003**, *278*, 21113–21123. [CrossRef]
56. Carver, E.A.; Jiang, R.; Lan, Y.; Oram, K.F.; Gridley, T. The mouse snail gene encodes a key regulator of the epithelial-mesenchymal transition. *Mol. Cell. Biol.* **2001**, *21*, 8184–8188. [CrossRef] [PubMed]
57. Chilosi, M.; Caliò, A.; Rossi, A.; Gilioli, E.; Pedica, F.; Montagna, L.; Pedron, S.; Confalonieri, M.; Doglioni, C.; Ziesche, R.; et al. Epithelial to mesenchymal transition-related proteins ZEB1, β-catenin, and β-tubulin-III in idiopathic pulmonary fibrosis. *Mod. Pathol.* **2016**, *30*, 26–38. [CrossRef]
58. Spaderna, S.; Schmalhofer, O.; Wahlbuhl, M.; Dimmler, A.; Bauer, K.; Sultan, A.; Hlubek, F.; Jung, A.; Strand, D.; Eger, A.; et al. The transcriptional repressor ZEB1 promotes metastasis and loss of cell polarity in cancer. *Cancer Res.* **2008**, *68*, 537–544. [CrossRef]
59. Shirakihara, T.; Saitoh, M.; Miyazono, K. Differential regulation of epithelial and mesenchymal markers by δEF1 proteins in epithelial–mesenchymal transition induced by TGF-β. *Mol. Biol. Cell* **2007**, *18*, 3533–3544. [CrossRef]
60. Nelson, W.J.; Nusse, R. Convergence of Wnt, ß-catenin, and cadherin pathways. *Science* **2004**, *303*, 1483–1487. [CrossRef]

61. Zheng, X.; Peng, M.; Li, Y.; Wang, X.; Lu, W.; Wang, X.; Shan, Y.; Li, R.; Gao, L.; Qiu, C. Cathelicidin-related antimicrobial peptide protects against cardiac fibrosis in diabetic mice heart by regulating endothelial-mesenchymal transition. *Int. J. Biol. Sci.* **2019**, *15*, 2393–2407. [CrossRef] [PubMed]
62. Lemieszek, M.K.; Golec, M.; Zwoliński, J.; Dutkiewicz, J.; Milanowski, J. Cathelicidin influence on pathological activation of Wnt pathway in murine model of hypersensitivity pneumonitis. *Ann. Agric. Environ. Med.* **2022**, *29*, 358–364. [CrossRef]

Article

New Insights via RNA Profiling of Formalin-Fixed Paraffin-Embedded Lung Tissue of Pulmonary Fibrosis Patients

Dymph Klay [1], Karin M. Kazemier [2,3], Joanne J. van der Vis [1,4], Hidde M. Smits [2], Jan C. Grutters [1,3] and Coline H. M. van Moorsel [1,*]

[1] Interstitial Lung Diseases Center of Excellence, Department of Pulmonology, St. Antonius Hospital, 3435 CM Nieuwegein, The Netherlands
[2] Center of Translational Immunology, University Medical Center Utrecht, 3584 CX Utrecht, The Netherlands
[3] Division of Heart and Lungs, University Medical Center Utrecht, 3584 CX Utrecht, The Netherlands
[4] Department of Clinical Chemistry, ILD Center of Excellence, St. Antonius Hospital, 3435 CM Nieuwegein, The Netherlands
* Correspondence: c.van.moorsel@antoniusziekenhuis.nl

Citation: Klay, D.; Kazemier, K.M.; van der Vis, J.J.; Smits, H.M.; Grutters, J.C.; van Moorsel, C.H.M. New Insights via RNA Profiling of Formalin-Fixed Paraffin-Embedded Lung Tissue of Pulmonary Fibrosis Patients. *Int. J. Mol. Sci.* **2023**, *24*, 16748. https://doi.org/10.3390/ijms242316748

Academic Editors: Sabrina Lisi and Margherita Sisto

Received: 5 October 2023
Revised: 16 November 2023
Accepted: 17 November 2023
Published: 25 November 2023

Copyright: © 2023 by the authors. Licensee MDPI, Basel, Switzerland. This article is an open access article distributed under the terms and conditions of the Creative Commons Attribution (CC BY) license (https://creativecommons.org/licenses/by/4.0/).

Abstract: In sporadic idiopathic pulmonary fibrosis (sIPF) and pulmonary fibrosis caused by a mutation in telomere (TRG-PF) or surfactant related genes (SRG-PF), there are a number of aberrant cellular processes known that can lead to fibrogenesis. We investigated whether RNA expression of genes involved in these processes differed between sIPF, TRG-PF, and SRG-PF and whether expression levels were associated with survival. RNA expression of 28 genes was measured in lung biopsies of 26 sIPF, 17 TRG-PF, and 6 SRG-PF patients. Significant differences in RNA expression of *TGFBR2* ($p = 0.02$) and *SFTPA2* ($p = 0.02$) were found between sIPF, TRG-PF, and SRG-PF. Patients with low (<median) expression of *HSPA5* ($p = 0.04$), *COL1A1* ($p = 0.03$), and *ATF4* (0.005) had significantly longer survival rates than patients with high ($\geq$median) expression of these genes. In addition, we scored for low (0) or high (1) expression of six endoplasmic reticulum (ER) stress genes (*HSP90B1*, *DDIT3*, *EDEM1*, *HSPA5*, *ATF4*, and *XBP1*) and found that patients with high expression in a low number of ER stress genes (total score 0–1) had longer survival rates than patients with high expression in a high number of ER stress genes (total score 2–6) ($p = 0.03$). In conclusion, there are minor differences between sIPF, TRG-PF, and SRG-PF and high expression in a high number of ER stress genes significantly associated with shorter survival time, suggesting that ER stress may be a target for therapy for PF.

Keywords: idiopathic pulmonary fibrosis; surfactant related gene mutation; telomere related gene mutation; SFTPC; SFTPA2; RTEL1; TERT; lung tissue; RNA expression

1. Introduction

Idiopathic pulmonary fibrosis (IPF) is characterized by damage to the alveolar epithelium and accumulation of extracellular matrix in the interstitium. IPF has a poor prognosis with a median survival of approximately 3–4 years [1,2]. The disease is highly heterogeneous, while in the majority of patients etiology is unknown, some of the patients have genetic pulmonary fibrosis. Pathogenic mutations causing pulmonary fibrosis have been identified in two different groups of genes, surfactant related genes (SRG) such as *SFTPC* and *SFTPA2*, or telomere related genes (TRG) such as *TERT* and *RTEL1* [3]. In a study by Snetselaar et al. [4], telomere length was measured in white blood cells of patients with sporadic IPF (sIPF) and SRG-PF and TRG-PF. It was shown that some sIPF patients had short telomeres comparable with TRG-PF, whereas others had a telomere length comparable with SRG-PF. This is in congruence with a recent report showing the wide range in telomere length in patients with pulmonary fibrosis who do not carry a likely pathogenic mutation [5]. This suggests that disease drivers in sIPF may overlap with those in specific genetic groups.

In previous studies, several molecular processes involved in IPF have been identified such as extracellular matrix deposition [6,7], endoplasmic reticulum (ER) stress [8,9], senescence [10,11], and hypoxia [12]. In addition, different cell types of the bronchoalveolar compartment, and particularly the alveolar type 2 cell (AT2), were found to be involved in the pathogenesis of IPF [13]. IPF expression studies showed an increase in expression of fibrogenesis related genes such as *TGFB1*, *ACTA2* [14–16], and hypoxia genes including *HIF1A* and *EPAS1* [12,17,18], but also involvement of cellular processes such as autophagy [19,20]. Mutations in surfactant and telomere genes were shown to each lead to different processes in the lung, and it is unknown to what extent these processes overlap. Heterozygous mutations in SRG have a toxic gain of function effect on surfactant processing [21–23] and may cause ER stress with upregulation of ER stress associated genes, such as *HSPA5* and *XBP1* [24–26]. On the other hand, heterozygous TRG mutations cause haploinsufficiency leading to excessive shortening of telomeres and were shown to increase DNA damage related processes including upregulation of *TP53BP1* and *TP53* [27] and cause senescence with altered expression of *CDKN2A* and *CDKN1A* [28]. Because disease cause and outcome are so heterogeneous in PF, a better understanding of the involvement of the different processes and their relation to patient survival is warranted and may aid development of therapies targeting specific processes in patients. To determine whether TRG-PF and SRG-PF have distinct expressions of genes involved in disease pathogenesis, we measured RNA expression in diagnostic lung biopsies of sIPF, TRG-PF, and SRG-PF. In addition, we investigated whether different levels of RNA expression are associated with survival.

2. Results

RNA expression of 28 genes involved in IPF pathogenesis was measured in FFPE surgical lung biopsies of three groups of patients. (26 sIPF, 17 TRG-PF, and 6 SRG-PF). Clinical characteristics of the three patient groups are displayed in Table 1. Telomere length in tissue and blood was significantly different between the three groups ($p = 0.001$ and $p < 0.001$, respectively). Telomere length was lower in TRG-PF patients than the other two patient groups. In addition, male predominance was present in sIPF (92.3%) and TRG-PF patients (82.4%), but not in SRG-PF patients (50%). The 28 genes are associated with different processes, such as senescence, DNA damage, endoplasmic reticulum (ER) stress, surfactant homeostasis, extracellular matrix (ECM), autophagy, hypoxia, and protein degradation. Statistical analysis showed no difference in RNA expression of 25 genes (expression of *TP53*, *CDKN2A*, and *CDKN1A* was not measured in SRG-PF patients) between TRG-PF and SRG-PF patients. Comparison of RNA expression of the 28 genes between sIPF, TRG-PF, and SRG-PF showed a significant difference in RNA expression of *TGFBR2* ($p = 0.02$) and *SFTPA2* ($p = 0.02$). Post-hoc analysis showed significantly lower expression of *TGFBR2* in TRG-PF compared to sIPF patients ($p = 0.001$) and lower expression of *SFTPA2* in SRG-PF compared to sIPF patients ($p = 0.002$, Figure 1). Two out of three patients with an *SFTPC* mutation had the lowest levels of *SFTPC* expression. The median *SFTPC* expression in the *SFTPC* mutation carriers was just 0.03. This was not significantly different from the *SFPTC* level of 0.05 in the patients carrying an *SFTPA2* mutation. Similarly, two out of three patients carrying an *SFTPA2* mutation had the lowest SFTPA2 expression and the median *SFTPA2* expression in the entire *SFTPA2* group was just 0.008. However, this was not significantly lower than the *SFTPA2* expression in the patients carrying an *SFTPC* mutation (0.004).

Table 1. Demographics and clinical characteristics of patient groups at time of biopsy.

Characteristic	sIPF	TRG-PF	SRG-PF	p
N	26	17	6	
Age biopsy, years, median (IQR)	62.8 (12.5)	58.5 (9.1)	41.9 (30.6)	0.01
Male, n (%)	24 (92.3)	14 (82.4)	3 (50)	0.04

Table 1. Cont.

Characteristic	sIPF	TRG-PF	SRG-PF	p
Ever smoker, n (%)	19 (82.6)	14 (82.3)	2 (40.0)	0.06
FVC % predicted, median (IQR)	81.4 (25.9)	77.3 (24.7)	54.3 (31.6)	0.06
DLCO % predicted, median (IQR)	49.9 (15.2)	41.6 (9.6)	30.7 (20.1)	0.009
T/S ratio tissue, median (IQR)	0.839 (0.127)	0.772 (0.074)	0.830 (0.087)	0.001
T/S ratio blood, median (IQR)	0.795 (0.128)	0.690 (0.180)	0.901 (0.274)	<0.001
T/S ratio blood observed, expected, median (IQR)	−0.147 (0.149)	−0.271 (0.151)	−0.049 (0.229)	<0.001
Transplant event, n (%)	7 (26.9)	2 (11.7)	3 (50)	0.16
Deaths, n (%)	15 (57.7)	14 (82.4)	2 (33)	0.07
Overall survival time, mo, median (SE)	42.8 (19.1)	21.5 (9.3)	NA	0.05

FVC % predicted and DLCO % predicted values are ±12 months from time of biopsy. The T/S ratio is a measure for telomere length. IQR: interquartile range; mo: months; sIPF sporadic idiopathic pulmonary fibrosis; TRG-PF: telomere related gene mutation pulmonary fibrosis; SRG-PF: surfactant related gene mutation pulmonary fibrosis; FVC: forced vital capacity; DLCO: diffusing capacity of the lung for carbon monoxide. Smoking status: n is 23 sIPF, 16 TRG-PF, 5 SRG-PF. FVC %predicted n is 19 sIPF, 14 TRG-PF, 4 SRG-PF. DLCO % predicted: n is 17 sIPF, 13 TRG-PF, 5 SRG-PF.

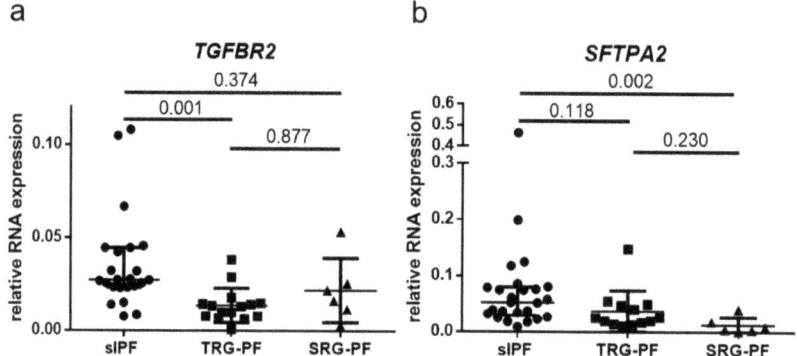

Figure 1. Relative RNA expression in lung tissue of sporadic IPF (sIPF) patients and pulmonary fibrosis patients with a telomere related gene (TRG-PF) or surfactant related gene (SRG-PF) mutation (**a**). Relative RNA expression of *TGFBR2*. A significant difference in RNA expression was found between the three groups (Kruskal–Wallis test $p = 0.02$). Post hoc analysis showed significantly higher expression in sIPF compared to TRG-PF ($p = 0.001$). There were no significant differences in expression between sIPF and SRG-PF (0.374) or between TRG-PF and SRG-PF (0.877). (**b**). Relative RNA expression of *SFTPA2*. A significant difference in RNA expression was found between the three groups (Kruskal–Wallis test $p = 0.02$). Post hoc analysis showed significantly higher expression in sIPF compared to SRG-PF patients ($p = 0.002$). There were no significant differences in expression between sIPF and TRG-PF (0.118) or between TRG-PF and SRG-PF (0.23).

2.1. Clustering of Genes and Patients

Unsupervised two-way hierarchical clustering of RNA expression of the 28 genes in 49 patients demonstrated five major vertical clusters of genes and two major horizontal clusters of patients (Figure 2). The upper cluster of patients contains 10 sIPF, 12 TRG-PF, and 3 SRG-PF patients, whereas the lower cluster contains 16 sIPF, 5 TRG-PF, and 3 SRG-PF patients. The distribution of patient groups was not significantly different between the two clusters ($p = 0.12$). Analysis of the clinical characteristics of the two clusters showed no differences except for a significantly higher number of deaths in the lower cluster compared to the upper cluster ($p = 0.04$, Table 2).

The 28 measured genes can be categorized in gene process groups: surfactant homeostasis (*SFTPC*, *SFTPA2*, and *SFTPB*), ER-stress (*HSP90B1*, *EDEM1*, *DDIT3*, *ATF4*, *XBP1*, and *HSPA5*), DNA-damage (*TP53BP1*, *H2AX*, and *TP53*), extracellular matrix (*ACTA2*,

VIM, COL1A1, COL1A2, COL3A1, SMAD4, and TGFBR2), hypoxia (HIF1A and EPAS1), senescence (CDKN2A and CDKN1A), protein degradation (PSMD11) and autophagy (MAP1LC3B), and AEC1 involvement (HOPX and CAV1) and bronchiolar involvement (SCGB1A1). HOPX and CAV1 encode proteins regulating the repair of AEC1 after injury. SCGB1A1 encodes a secretoglobin that exerts an anti-inflammatory function in the small airways. Genes were not randomly distributed over the five major vertical clusters (Figure 2). Both senescence genes were part of one cluster, and all other three surfactant homeostasis genes were part of another cluster. Furthermore, the three collagen encoding genes formed a separate cluster.

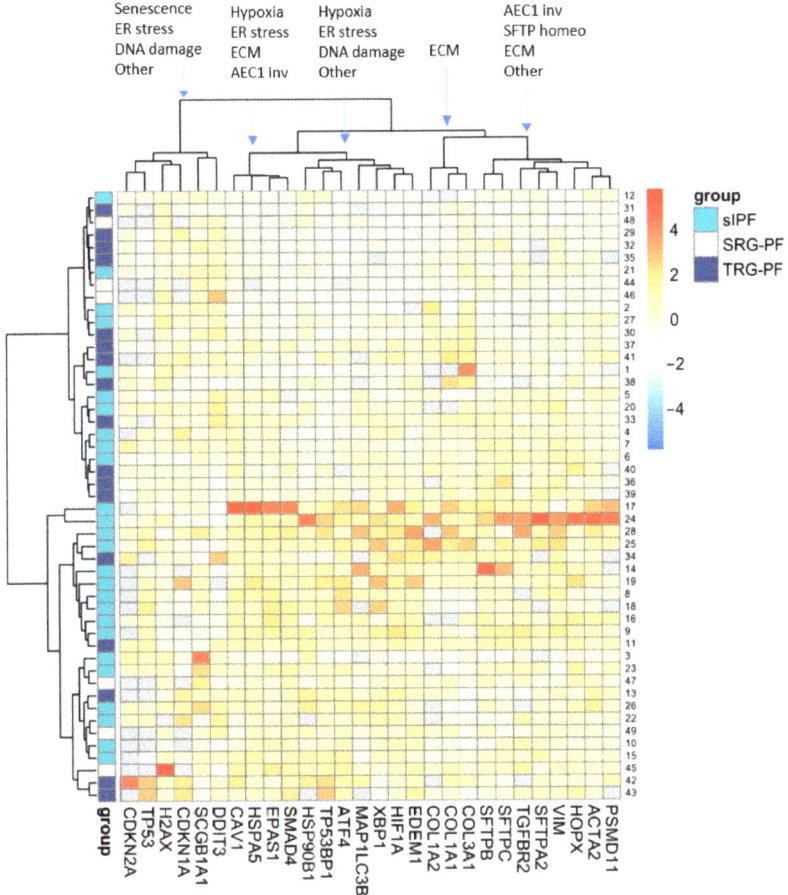

Figure 2. Unsupervised two-way hierarchical clustering of RNA expression of 28 genes in 49 patients. Rows represent individual patients and columns represent genes. Grey blocks indicate that RNA expression of a certain gene was not measured due to shortage of tissue or undetectable expression. Clustering of process related genes in each of the five main gene clusters is indicated at the top with arrows pointing to the cluster. AEC1 inv: alveolar epithelial type 1 involvement, SFTP homeo: surfactant homeostasis; ECM: extracellular matrix; other: this includes bronchiolar involvement, autophagy, or protein degradation.

Table 2. Demographics and clinical characteristics of two patient clusters.

Characteristic	Upper Cluster	Lower Cluster	p
N	25	24	
Group, sIPF/TRG-PF/SRG-PF	10/12/3	16/5/3	0.12
Age biopsy, years, median (IQR)	58.0 (13.5)	61.0 (13.8)	0.34
Male, n (%)	20 (80)	21 (87.5)	0.70
Ever smoker, n (%)	19 (79.2)	16 (80.0)	1.00
FVC % predicted, median (IQR)	82.6 (24.9)	71.7 (26.2)	0.40
DLCO % predicted, median (IQR)	45.2 (14.4)	46.7 (25.3)	0.91
T/S ratio tissue, median (IQR)	0.806 (0.081)	0.831 (0.127)	0.05
T/S ratio blood, median (IQR)	0.784 (0.189)	0.797 (0.141)	0.17
T/S ratio blood, observed, expected, median (IQR)	−0.161 (0.237)	−0.131 (0.134)	0.10
Transplant event, n (%)	8 (32.0)	4 (16.7)	0.32
Deaths, n (%)	12 (48.0)	19 (79.2)	0.04
Overall survival time, median (SE), mo	43.0 (17.8)	22.6 (6.8)	0.08

FVC % predicted and DLCO % predicted values are ±12 months of biopsy date; IQR: interquartile range; mo: months; n: number; sIPF sporadic idiopathic pulmonary fibrosis; TRG-PF: telomere related gene mutation pulmonary fibrosis; SRG-PF: surfactant related gene mutation pulmonary fibrosis; FVC: forced vital capacity; DLCO: diffusing capacity of the lung for carbon monoxide. Smoking status: n is 24 in upper cluster, 20 in lower cluster. FVC % predicted n is 21 in upper cluster, 16 in lower cluster. DLCO % predicted: n is 21 in upper cluster, 14 in lower cluster. T/S ratio is a measure for telomere length.

2.2. Survival

We investigated whether overall survival was different between patients with high and low RNA expression of a gene. Comparison showed significantly longer survival in patients with low (<median) RNA expression of ER stress genes *ATF4* (64 months vs. 22 months; $p = 0.005$), *HSPA5* (64 months vs. 22 months; $p = 0.04$), and for ECM gene *COL1A1* (64 months vs. 22 months; $p = 0.03$) compared to patients with high (≥median) RNA expression of these genes (Figure 3a–c).

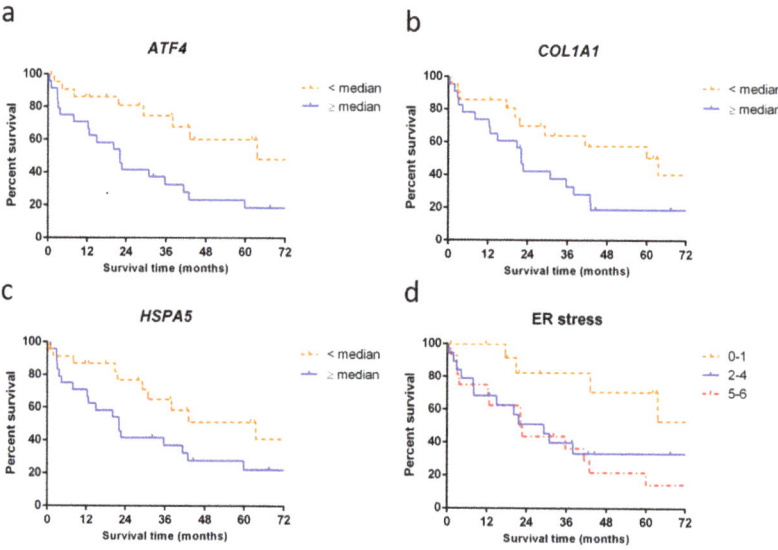

Figure 3. Survival of patients with pulmonary fibrosis stratified by RNA expression levels. (**a**) Survival curve showing significantly longer survival in patients with low expression of ATF4 (64 months, $n = 23$, dashed line) compared to patients with high expression of ATF4 (22 months, $n = 24$, solid line,

p = 0.005); (**b**) Survival curve showing a significantly longer survival in patients with low expression of COL1A1 (64 months, n = 22, dashed line) compared to patients with high expression of COL1A1 (22 months, n = 23, solid line, p = 0.03). (**c**) Survival curve showing a significantly longer survival in patients with low RNA expression of HSPA5 (64 months, n= 24, dashed line) compared to patients with high expression of HSPA5 (22 months, n = 24, solid line, p = 0.04). (**d**) Accumulative score in patients for high (+1 point) or low (0 points) expression in each of six ER stress genes (*HSP90B1, DDIT3, EDEM1, HSPA5, ATF4,* and *XBP1*). Survival curve showing significantly longer median survival in patients with high RNA expression of 0–1 ER stress genes (95 months, n = 14, dashed line) compared to patients with high expression of 2–4 ER stress genes (29 months, n = 19, solid line) and high expression of 5–6 ER stress genes (22 months, n = 16, dash-dotted line, p = 0.03).

Next, we grouped three or more genes that play a role in the same processes into process expression groups: surfactant homeostasis (*SFTPB, SFTPC,* and *SFTPA2*), DNA damage (*TP53BP1, H2AX,* and *TP53*), ECM (*ACTA2, VIM, TGFBR2, COL1A1,* and *SMAD4*), and ER stress (*HSP90B1, DDIT3, EDEM1, HSPA5, ATF4,* and *XBP1*). For each patient, a process expression score was calculated based on the sum of above (+1) or below (0) median expression for each gene in the process. Comparison of overall survival between high, average, and low scoring patients showed a significant difference for ER stress: patients with low expression of ER stress genes (score 0–1) showed a longer survival time than patients with average or high expression of these genes (score 2–4 or 5–6). Median survival in low expressing ER stress patients was 95 months versus 29 and 22 months in the average and high expressing patients, respectively (p = 0.03, Figure 3d). Survival analysis results of surfactant homeostasis, DNA damage, and extracellular matrix showed no significant differences (see Supplementary Materials). Results for processes represented by only two genes were also included in the Supplementary Materials.

3. Discussion

Several processes have been associated with the development of PF in general and with surfactant or telomere related PF specifically. In this study, we quantified RNA expression of proteins that have previously been shown to be involved in pulmonary fibrosis in lung biopsies of patients with sporadic and genetic IPF and analyzed the effect on survival. There was no significant clustering of patient groups and genes and no significant differences in RNA expression between the groups with TRG-PF and SRG-PF. This is the first study that showed significantly worse survival in PF patients with high expression of two or more ER stress genes compared to PF patients with low expression of five or all six ER stress genes.

Several gene expression studies have been performed on lung tissue from patients with pulmonary fibrosis. First, it was shown that expression in IPF patients differed significantly from control subjects. In a study by Wang et al. [29], microarray data from lung tissue of 131 IPF patients with UIP on pathology and 12 controls were used. Comparison of IPF patients and controls resulted in 988 differentially expressed genes. In addition, by ward clustering and principal component analysis of gene expression profiles of 131 IPF patients, six patient clusters were found. The six clusters differed in disease severity and differential expression compared to controls in genes that play a role in processes such as extracellular matrix organization, regulation of cell migration, collagen catabolic process, cilium, cilium assembly, angiogenesis, and lung alveolar morphology. Furthermore, in a whole genome oligonucleotide microarray study by Yang et al. [30], RNA expression of 41,000 genes and transcripts was measured in lung tissue of 16 sporadic idiopathic interstitial pneumonia patients, 10 familial idiopathic interstitial pneumonia patients, and 9 normal control subjects. In total, 135 transcripts were found to be upregulated or downregulated more than 1.8-fold in pulmonary fibrosis patients compared to normal control subjects. After hierarchical clustering, four patient clusters were identified; all controls except for two samples clustered together and all familial idiopathic interstitial pneumonia except for three samples clustered together. Comparison between sporadic idiopathic interstitial pneumonia (IIP) and familial IIP patients resulted in 142 transcripts from 62 genes with known functions that were more than 1.8-fold upregulated or downregulated. Interestingly, they found that genes from the same functional categories, such as calcium/potassium binding, cell adhesion, cell proliferation and death, ECM degradation, and cytokines/chemokines, are differentially expressed between sporadic and familial IIP as well as between IIP and normal control subjects, but then to a larger extent in the familial IIP patients. But because no genetic analysis was performed, it remains unclear whether patients carrying a TRG or SRG mutation had been included.

In our study, we included patients with familial disease of known cause and divided them over two groups: patients carrying an SRG and patients carrying a TRG mutation and studied expression of genes involved in mutation driven aberrant processes in PF. To our surprise, we detected no differences in expression of any of the studied genes between the TRG and SRG-PF groups. However, when we compared sIPF, TRG-PF, and SRG-PF, we found significant differences in RNA expression of *TGFBR2* and *SFTPA2* between the three groups. The cluster analysis did not result in clustering of patients in their respective sIPF, TRG-PF, or SRG-PF patient group. On the other hand, there was clustering of genes belonging to the same processes. As surfactant homeostasis genes clustered together and senescence genes also clustered together, this indicates that the expression levels may correctly inform about the activity of the process.

Lawson et al. studied protein expression in AECs in lung tissue of three patients with pulmonary fibrosis and an *SFTPC* mutation, ten patients with familial interstitial pneumonia without an *SFTPC* mutation, and ten patients with sIPF. They observed expression of the ER stress proteins BiP, EDEM, and XBP1 in lung tissues of all patients [8]. In addition, Carleo et al. investigated protein patterns in bronchoalveolar lavage (BAL) fluid of 10 familial and 17 sporadic IPF patients. In total, 22 proteins were found to be differentially expressed between familial and sporadic IPF. The upregulated proteins in sIPF played a role in oxidative stress response and the upregulated proteins in familial IPF (including SP-A2 encoded by *SFTPA2*) played a role in immune response, coagulation system, wounding response, and ion homeostasis [31]. The increase of SP-A2 in BAL fluid of familial IPF patients in Carleo et al. is in contrast with the lower RNA expression of *SFTPA2* in lung tissue of the SRG-PF patients in our study. Although their small familial cohort was not analyzed for carriage of genetic mutations, it is unlikely that their familial cohort included any patients with an SRG mutation. SRG mutations occur in 3–8% of patients with familial disease, while TRG mutations are present in approximately 35% of familial patients [3]. However, we did not observe a difference in *SFTPA2* RNA expression between our sIPF and familial TRG-PF patients. The level of RNA expression may therefore not reflect protein expression, while increased SP-A2 in BAL fluid may aid identification of familial patients, and the lower RNA expression of *SFTPA2* in lung tissue may aid identification of patients with surfactant-related pathology.

Furthermore, the SP-A2 protein level in blood was shown to be a predictive marker for progressive disease [32,33]. Although it remained unclear what the source is of SP-A2 in blood, and how increased levels in blood mechanistically related to progression of disease in the lung. When analyzing the survival of patients, TRG-PF showed the shortest survival (Table 1), but the survival of patients with SRG-PF could not be calculated due to the low number. However, in a recent study, survival of SRG-PF patients was shown to be comparable with that of sIPF [34]. Our study suggests that in patients with SRG-PF, levels may be influenced by the presence of surfactant related mutations, and thus future investigations into biomarkers for PF may benefit from stratification of patients by genetic cause of disease.

When we studied the impact of RNA expression of each gene on survival in patients, we found that patients with low expression of *HSPA5*, *COL1A1*, or *ATF4* showed a significantly longer survival rate than patients with high expression of one of these genes. In a study by Tsitoura et al. [35], they observed increased *COL1A1* RNA expression in BAL cells of patients with 53 IPF and 62 non-IPF ILD compared to 19 controls. In addition, they also found that high levels of *COL1A1* expression were associated with worse survival.

However, this was only found in the non-IPF ILD patients and not in the IPF cohort. For ER stress gene *ATF4*, as far as we know, no association with survival was investigated before in other studies, although it was found that ATF4 expression was increased in IPF lung tissue and colocalized with apoptosis markers CHOP and cleaved caspase 3, and encoded by *DDIT3* and *CASP3*, respectively, in alveolar epithelial cells overlying fibroblast foci [9]. For ER stress marker GRP78/BiP encoded by *HSPA5*, conflicting results have been found; in one study elevated GRP94 and CHOP, encoded by *HSP90B1* and *DDIT3*, respectively, and reduced expression of GRP78 were observed in alveolar epithelial cells type 2 (AEC2s) from IPF lung tissue compared to AEC2s from normal donors [36], whereas in another study increased GRP78/BiP expression was observed in IPF lung tissue compared to control lung tissue. In the latter study, strong GRP78/BiP staining was found in alveolar epithelial cells in regions of fibroblasts foci [37]. Fibroblasts foci are aggregates of proliferating fibroblasts and myofibroblasts thought to represent areas of active fibrosis. Therefore, the proximity of strong BiP staining (encoded by *HSPA5*) to fibroblast foci suggests a role in fibrogenesis. Together with our finding that high expression of ER stress genes, including *ATF4* and *HSPA5*, results in low survival, the evidence suggests that ER stress in IPF may be an important factor for progression of fibrogenesis. It is known that the cause of ER stress may be different in each patient [38]. However, the absence of differences in expression between our groups of patients and the presence of a correlation with survival suggest that ER stress is an important factor in the progression of pulmonary fibrosis independent of underlying genetic cause. Two recent studies show the feasibility of targeting ER stress in pulmonary fibrosis. Kropski et al. [39] showed that ER stress may be targeted by pharmacological chaperones such as sodium phenylbutyrate, which resulted in reduced ER stress in preclinical disease models or by drugs that selectively inhibit oligomerized IRE1α, and which have been investigated in animal models. Chen et al. [40] showed that inhibition of IRE1α by an endoribonuclease inhibitor alleviates CS-induced pulmonary inflammation and fibrogenesis in a mouse model.

A limitation of our study is that only bulk RNA expression was measured. Differences in RNA expression between cell types, such as in single cell RNA sequencing, were not investigated. In addition, no spatial transcriptomics and immunohistochemistry was used to localize which cell populations were involved in the regulation of RNA expression in PF. Furthermore, the results were not verified at the protein level and therefore effects of posttranscriptional regulation of gene expression during fibrogenesis was not investigated. The number of patients, especially those with SRG-PF, was quite low. However, the low number of SRG-PF cases was limited by the rarity of these patients compared to those with sIPF. Inclusions of more cases might have shown additional significant differences in the clinical characteristics between the two clusters. In addition, the number of genes measured may be too small to create distinct clusters. The low number of differences in RNA expression between sIPF, TRG-PF, and SRG-PF may also explain why based on most recent finding these patients appear to benefit from the same treatment [32,41,42]. In addition, although all samples were diagnostic lung biopsies, a considerably low DLCO % predicted in especially TRG-PF and SRG-PF patients was present, which suggests already advanced lung disease. It remains possible that more differences, particularly mutation associated differences, would be present if early disease samples had been analyzed. While screening family members at risk may aid early disease detection, genetic analysis is likely to reduce the need for biopsies. In a recent survey, 72% of pulmonologists reported that they might modify their diagnostic work-up according to the results of genetic testing of whom 78% would postpone or exclude surgical lung biopsy [43].

In conclusion, analysis of expression of fibrosis related genes in sIPF and TRG-PF and SRG-PF patients showed almost similar gene expression between the groups. The finding that high expression of ER stress genes leads to worse survival time supports development of therapies targeting ER stress that may be beneficial to all PF patients, including those with genetic pulmonary fibrosis. Further studies are needed to investigate this association in more detail.

4. Materials and Methods

4.1. Patients and Tissue Selection

Diagnostic surgical lung biopsies and blood obtained between 1997 and 2016 of 26 sIPF, 17 TRG-PF (*RTEL* $n = 3$, *TERT* $n = 14$), and 6 SRG-PF (*SFTPC* $n = 3$, *SFTPA2* $n = 3$) patients were included in this study. Diagnoses were based on the ATS/ERS/JRS/ALAT guidelines [44]. All subjects signed written consent for the study approved by the Medical Research Ethics Committees United (MEC-U) of the St. Antonius hospital (R05-08A).

4.2. DNA/RNA Isolation from Lung Tissue and Blood

After removal of paraffin with paraffin dissolver (Macherey-Nagel, Düren, Germany), RNA and DNA was isolated from formalin-fixed paraffin embedded tissue using All-Prep DNA/RNA FFPE kit (Qiagen Benelux BV, Venlo, The Netherlands). DNA and RNA concentration and purity were measured using a NanoDrop spectrophotometer. For isolation of genomic DNA from peripheral white blood cells, a magnetic beads-based method (chemagic DNA blood 10k kit; Perkin Elmer Inc., Waltham, MA, USA) was used.

4.3. Telomere Length Measurements in Lung Tissue and Blood

T/S ratio was measured in isolated DNA from tissue and blood by monochrome multiplex quantitative polymerase chain reaction (MMqPCR) as described before in [45–47]. The T/S ratio is a measure for telomere length. Telomere length adjusted for age was calculated by the difference between the observed T/S ratio and the age-adjusted normal value (T/S expected).

4.4. Real-Time PCR

cDNA (6ng per reaction), prepared from RNA using i-script (Bio-Rad GmbH laboratories B.V), was amplified using iQ SYBR Green Supermix (Bio-Rad GmbH laboratories B.V, Lunteren, The Netherlands) and gene specific primers (see Supplementary Materials) in a CFX96 Bio-Rad qPCR machine with the following run conditions: 3 min at 95 °C followed by 45 cycles of 10 s at 95 °C, 20 s at 61 °C, and 25 s at 72 °C. RNA expression was calculated by delta Ct method using the mean of three reference genes: *ACTB*, *RPL13A*, and *EEF1A1*. RNA expression was investigated for genes involved in the following processes: bronchiolar involvement (*SCGB1A1*), alveolar epithelial cell type 1 (AEC1) involvement (*HOPX* and *CAV1*), surfactant homeostasis (*SFTPC*, *SFTPA2*, *SFTPB*), hypoxia (*HIF1A*, *EPAS1*), protein degradation (*PSMD11*), autophagy (*MAP1LC3B*), senescence (*CDKN2A*, *CDKN1A*), ER-stress (*HSP90B1*, *EDEM1*, *DDIT3*, *ATF4*, *XBP1*, *HSPA5*), DNA-damage (*TP53BP1*, *H2AX*, *TP53*) and extracellular matrix (*ACTA2*, *VIM*, *COL1A1*, *COL1A2*, *COL3A1*, *SMAD4*, *TGFBR2*). List of used primers can be found in Table S1.

4.5. Statistical Analysis

For analysis of differences in demographics and clinical characteristics between sIPF, TRG-PF, and SRG-PF patients, the chi-square test was used for discrete variables and the Kruskal–Wallis test was used for continuous variables, as appropriate with a small sample size. The Mann–Whitney U test was used to compare RNA expression between TRG-PF and SRG-PF followed by false discovery rate (FDR) to correct for multiple comparisons. The Kruskal–Wallis test was used to analyze differences in RNA expression between sIPF, TRG-PF, and SRG-PF followed by FDR to correct for multiple comparisons. Unsupervised two-way hierarchical clustering of RNA expression using Ward's clustering method with Euclidean distance metric was used to create a heatmap. Chi-square and Mann–Whitney U test were used for discrete and continuous variables, respectively, to compare demographics and clinical characteristics between the two horizontal clusters in the heatmap. For survival analyses, median RNA expression was calculated for each gene and patients were divided into two groups containing patients with <median expression of that gene and patients with ≥median expression of that gene. Additionally, when three or more genes play a role in the same process, these were also analyzed together as one process-

group, surfactant homeostasis (*SFTPB*, *SFTPC*, *SFTPA2*), DNA damage (*TP53BP1*, *TP53*, *H2AX*), ECM (*ACTA2*, *VIM*, *TGFBR2*, *COL1A1*, *SMAD4*), and ER stress (*HSP90B1*, *DDIT3*, *EDEM1*, *HSPA5*, *ATF4*, *XBP1*). For ECM, only one, *COL1A1*, of the three collagen genes was included in the survival analysis to prevent overrepresentation of collagen genes. For each process-group of genes, we determined the score for each patient. Per gene, RNA expression $\geq$median is +1 point and <median is 0 points. The sum of these scores resulted in an accumulative score per process per patient. For each process, we performed survival analysis dividing the patients into groups. Survival was compared using the Kaplan–Meier method with log-rank tests. Survival time was determined from time of lung biopsy until death. Patients were censored when lost to follow-up or when they underwent lung transplantation or at the last contact date. Statistical analyses were performed in SPSS v.26 and R v.4.2.2 (including the following packages: tidyverse version 1.3.2, ggsci v. 2.9, readxl v. 1.4.1, pheatmap v. 1.0.12, ggpubr v. 0.5.0).

Supplementary Materials: The following supporting information can be downloaded at https://www.mdpi.com/article/10.3390/ijms242316748/s1.

Author Contributions: Conceptualization, D.K., K.M.K., J.J.v.d.V. and C.H.M.v.M.; Data curation, D.K. and K.M.K.; Formal analysis, D.K., K.M.K. and H.M.S.; Funding acquisition, D.K., J.J.v.d.V., J.C.G. and C.H.M.v.M.; Investigation, D.K., K.M.K. and J.J.v.d.V.; Methodology, D.K., K.M.K., J.J.v.d.V. and C.H.M.v.M.; Supervision, C.H.M.v.M.; Validation, K.M.K.; Visualization, D.K. and H.M.S.; Writing—original draft, D.K.; Writing—review and editing, D.K., K.M.K., J.J.v.d.V., H.M.S., J.C.G. and C.H.M.v.M. All authors have read and agreed to the published version of the manuscript.

Funding: This research was funded by ZonMW–TopZorg St. Antonius Science Corner grant (Topzorg grant number 842002001), ZonMW TZO vision grant (grant number 10070012010004), and the Prof. Dr. Jaap Swierenga foundation. The funders had no role in study design, data collection and analysis, decision to publish, or preparation of the manuscript.

Institutional Review Board Statement: This study was conducted according to the guidelines of the Declaration of Helsinki and approved by the Medical Research Ethics Committees United (MEC-U) of the St. Antonius Hospital (R05-08A).

Informed Consent Statement: Informed consent was obtained from all subjects involved in this study.

Data Availability Statement: Data can be requested from the corresponding author.

Acknowledgments: We would like to thank the patients who participated in this study.

Conflicts of Interest: The authors declare no conflict of interest.

References

1. Strongman, H.; Kausar, I.; Maher, T.M. Incidence, Prevalence, and Survival of Patients with Idiopathic Pulmonary Fibrosis in the UK. *Adv. Ther.* **2018**, *35*, 724. [CrossRef]
2. Kaunisto, J.; Salomaa, E.-R.; Hodgson, U.; Kaarteenaho, R.; Kankaanranta, H.; Koli, K.; Vahlberg, T.; Myllärniemi, M. Demographics and survival of patients with idiopathic pulmonary fibrosis in the FinnishIPF registry. *ERJ Open Res.* **2019**, *5*, 00170–2018. [CrossRef] [PubMed]
3. van Moorsel, C.H.M.; van der Vis, J.J.; Grutters, J.C. Genetic disorders of the surfactant system: Focus on adult disease. *Eur. Respir. Rev.* **2021**, *30*, 200085. [CrossRef] [PubMed]
4. Snetselaar, R.; van Moorsel, C.H.M.; Kazemier, K.M.; van der Vis, J.J.; Zanen, P.; van Oosterhout, M.F.M.; Grutters, J.C. Telomere length in interstitial lung diseases. *Chest* **2015**, *148*, 1011–1018. [CrossRef]
5. Zhang, D.; Newton, C.A.; Wang, B.; Povysil, G.; Noth, I.; Martinez, F.J.; Raghu, G.; Goldstein, D.; Garcia, C.K. Utility of whole genome sequencing in assessing risk and clinically-relevant outcomes for pulmonary fibrosis. *Eur. Respir. J.* **2022**, *1*, 2200577. [CrossRef] [PubMed]
6. Chien, J.W.; Richards, T.J.; Gibson, K.F.; Zhang, Y.; Lindell, K.O.; Shao, L.; Lyman, S.K.; Adamkewicz, J.I.; Smith, V.; Kaminski, N.; et al. Serum lysyl oxidase-like 2 levels and idiopathic pulmonary fibrosis disease progression. *Eur. Respir. J.* **2014**, *43*, 1430–1438. [CrossRef] [PubMed]
7. Kuhn, C.; Boldt, J.; King, T.E.; Crouch, E.; Vartio, T.; McDonald, J.A. An immunohistochemical study of architectural remodeling and connective tissue synthesis in pulmonary fibrosis. *Am. Rev. Respir. Dis.* **1989**, *140*, 1693–1703. [CrossRef]

8. Lawson, W.E.; Crossno, P.F.; Polosukhin, V.V.; Roldan, J.; Cheng, D.-S.; Lane, K.B.; Blackwell, T.R.; Xu, C.; Markin, C.; Ware, L.B.; et al. Endoplasmic reticulum stress in alveolar epithelial cells is prominent in IPF: Association with altered surfactant protein processing and herpesvirus infection. *Am. J. Physiol. Lung Cell Mol. Physiol.* **2008**, *294*, L1119–L1126. [CrossRef]
9. Korfei, M.; Ruppert, C.; Mahavadi, P.; Henneke, I.; Markart, P.; Koch, M.; Lang, G.; Fink, L.; Bohle, R.-M.; Seeger, W.; et al. Epithelial endoplasmic reticulum stress and apoptosis in sporadic idiopathic pulmonary fibrosis. *Am. J. Respir. Crit. Care Med.* **2008**, *178*, 838–846. [CrossRef]
10. Minagawa, S.; Araya, J.; Numata, T.; Nojiri, S.; Hara, H.; Yumino, Y.; Kawaishi, M.; Odaka, M.; Morikawa, T.; Nishimura, S.L.; et al. Accelerated epithelial cell senescence in IPF and the inhibitory role of SIRT6 in TGF-β-induced senescence of human bronchial epithelial cells. *Am. J. Physiol. Lung Cell Mol. Physiol.* **2011**, *300*, L391–L401. [CrossRef]
11. Disayabutr, S.; Kim, E.K.; Cha, S.-I.; Green, G.; Naikawadi, R.P.; Jones, K.D.; Golden, J.A.; Schroeder, A.; Matthay, M.A.; Kukreja, J.; et al. miR-34 miRNAs Regulate Cellular Senescence in Type II Alveolar Epithelial Cells of Patients with Idiopathic Pulmonary Fibrosis. *PLoS ONE* **2016**, *11*, e0158367. [CrossRef]
12. Aquino-Gálvez, A.; González-Ávila, G.; Jiménez-Sánchez, L.L.; Maldonado-Martínez, H.A.; Cisneros, J.; Toscano-Marquez, F.; Castillejos-López, M.; Torres-Espíndola, L.M.; Velázquez-Cruz, R.; Rodríguez, V.H.O.; et al. Dysregulated expression of hypoxia-inducible factors augments myofibroblasts differentiation in idiopathic pulmonary fibrosis. *Respir. Res.* **2019**, *20*, 130. [CrossRef]
13. Xu, Y.; Mizuno, T.; Sridharan, A.; Du, Y.; Guo, M.; Tang, J.; Wikenheiser-Brokamp, K.A.; Perl, A.-K.T.; Funari, V.A.; Gokey, J.J.; et al. Single-cell RNA sequencing identifies diverse roles of epithelial cells in idiopathic pulmonary fibrosis. *J. Clin. Investig. Insight.* **2017**, *1*, e90558. [CrossRef]
14. Guiot, J.; Henket, M.; Corhay, J.L.; Moermans, C.; Louis, R. Sputum biomarkers in IPF: Evidence for raised gene expression and protein level of IGFBP-2, IL-8 and MMP-7. *PLoS ONE* **2017**, *12*, e0171344. [CrossRef]
15. Roach, K.M.; Wulff, H.; Feghali-Bostwick, C.; Amrani, Y.; Bradding, P. Increased constitutive αSMA and Smad2/3 expression in idiopathic pulmonary fibrosis myofibroblasts is KCa3.1-dependent. *Respir. Res.* **2014**, *15*, 155. [CrossRef]
16. Li, Y.; Jiang, D.; Liang, J.; Meltzer, E.B.; Gray, A.; Miura, R.; Wogensen, L.; Yamaguchi, Y.; Noble, P.W. Severe lung fibrosis requires an invasive fibroblast phenotype regulated by hyaluronan and CD44. *J. Exp. Med.* **2011**, *208*, 1459. [CrossRef]
17. Shochet, G.E.; Bardenstein-Wald, B.; McElroy, M.; Kukuy, A.; Surber, M.; Edelstein, E.; Pertzov, B.; Kramer, M.R.; Shitrit, D. Hypoxia Inducible Factor 1A Supports a Pro-Fibrotic Phenotype Loop in Idiopathic Pulmonary Fibrosis. *Int. J. Mol. Sci.* **2021**, *22*, 3331. [CrossRef]
18. Delbrel, E.; Soumare, A.; Naguez, A.; Label, R.; Bernard, O.; Bruhat, A.; Fafournoux, P.; Tremblais, G.; Marchant, D.; Gille, T.; et al. HIF-1α triggers ER stress and CHOP-mediated apoptosis in alveolar epithelial cells, a key event in pulmonary fibrosis. *Sci. Rep.* **2018**, *8*, 17939. [CrossRef]
19. Patel, A.S.; Lin, L.; Geyer, A.; Haspel, J.A.; An, C.H.; Cao, J.; Rosas, I.O.; Morse, D. Autophagy in Idiopathic Pulmonary Fibrosis. *PLoS ONE* **2012**, *7*, e41394. [CrossRef]
20. Im, J.; Hergert, P.; Nho, R.S. Translational Research in Acute Lung Injury and Pulmonary Fibrosis: Reduced FoxO3a expression causes low autophagy in idiopathic pulmonary fibrosis fibroblasts on collagen matrices. *Am. J. Physiol.-Lung Cell Mol. Physiol.* **2015**, *309*, L552. [CrossRef]
21. Legendre, M.; Butt, A.; Borie, R.; Debray, M.-P.; Bouvry, D.; Filhol-Blin, E.; Desroziers, T.; Nau, V.; Copin, B.; Moal, F.D.-L.; et al. Functional assessment and phenotypic heterogeneity of SFTPA1 and SFTPA2 mutations in interstitial lung diseases and lung cancer. *Eur. Respir. J.* **2020**, *56*, 2002806. [CrossRef]
22. Nogee, L.M.; Dunbar, A.E.; Wert, S.E.; Askin, F.; Hamvas, A.; Whitsett, J.A. A mutation in the surfactant protein C gene associated with familial interstitial lung disease. *N. Engl. J. Med.* **2001**, *344*, 573–579. [CrossRef]
23. Klay, D.; Hoffman, T.W.; Harmsze, A.M.; Grutters, J.C.; van Moorsel, C.H.M. Systematic review of drug effects in humans and models with surfactant-processing disease. *Eur. Respir. Rev.* **2018**, *27*, 170135. [CrossRef]
24. Maguire, J.A.; Mulugeta, S.; Beers, M.F. Endoplasmic reticulum stress induced by surfactant protein C BRICHOS mutants promotes proinflammatory signaling by epithelial cells. *Am. J. Respir. Cell Mol. Biol.* **2011**, *44*, 404–414. [CrossRef]
25. Maitra, M.; Wang, Y.; Gerard, R.D.; Mendelson, C.R.; Garcia, C.K. Surfactant protein A2 mutations associated with pulmonary fibrosis lead to protein instability and endoplasmic reticulum stress. *J. Biol. Chem.* **2010**, *285*, 22103–22113. [CrossRef]
26. Nguyen, H.; Uhal, B.D. The unfolded protein response controls ER stress-induced apoptosis of lung epithelial cells through angiotensin generation. *Am. J. Physiol.-Lung Cell Mol. Physiol.* **2016**, *311*, L846–L854. [CrossRef]
27. Alder, J.K.; Barkauskas, C.E.; Limjunyawong, N.; Stanley, S.E.; Kembou, F.; Tuder, R.M.; Hogan, B.L.M.; Mitzner, W.; Armanios, M. Telomere dysfunction causes alveolar stem cell failure. *Proc. Natl. Acad. Sci. USA* **2015**, *112*, 5099–5104. [CrossRef]
28. Liu, T.; De Los Santos, F.G.; Zhao, Y.; Wu, Z.; Rinke, A.E.; Kim, K.K.; Phan, S.H. Telomerase reverse transcriptase ameliorates lung fibrosis by protecting alveolar epithelial cells against senescence. *J. Biol. Chem.* **2019**, *294*, 8861–8871. [CrossRef]
29. Wang, Y.; Yella, J.; Chen, J.; McCormack, F.X.; Madala, S.K.; Jegga, A.G. Unsupervised gene expression analyses identify IPF-severity correlated signatures, associated genes and biomarkers. *BMC Pulm. Med.* **2017**, *17*, 133. [CrossRef]
30. Yang, I.V.; Burch, L.H.; Steele, M.P.; Savov, J.D.; Hollingsworth, J.W.; McElvania-Tekippe, E.; Berman, K.G.; Speer, M.C.; Sporn, T.A.; Brown, K.K.; et al. Gene Expression Profiling of Familial and Sporadic Interstitial Pneumonia. *Am. J. Respir. Crit. Care Med.* **2007**, *175*, 45. [CrossRef]

31. Carleo, A.; Bargagli, E.; Landi, C.; Bennett, D.; Bianchi, L.; Gagliardi, A.; Carnemolla, C.; Perari, M.G.; Cillis, G.; Armini, A.; et al. Comparative proteomic analysis of bronchoalveolar lavage of familial and sporadic cases of idiopathic pulmonary fibrosis. *J. Breath Res.* **2016**, *10*, 26007. [CrossRef]
32. Takahashi, H.; Fujishima, T.; Koba, H.; Murakami, S.; Kurokawa, K.; Shibuya, Y.; Shiratori, M.; Kuroki, Y.; Abe, S. Serum surfactant proteins A and D as prognostic factors in idiopathic pulmonary fibrosis and their relationship to disease extent. *Am. J. Respir. Crit. Care Med.* **2000**, *162*, 1109–1114. [CrossRef]
33. Greene, K.E.; King, T.E.; Kuroki, Y.; Bucher-Bartelson, B.; Hunninghake, G.W.; Newman, L.S.; Nagae, H.; Mason, R.J. Serum surfactant proteins-A and -D as biomarkers in idiopathic pulmonary fibrosis. *Eur. Respir. J.* **2002**, *19*, 439–446. [CrossRef]
34. Klay, D.; Grutters, J.C.; van der Vis, J.J.; Platenburg, M.G.; Kelder, J.C.; Tromp, E.; van Moorsel, C.H. Progressive Disease With Low Survival in Adult Patients With Pulmonary Fibrosis Carrying Surfactant-Related Gene Mutations: An Observational Study. *Chest* **2023**, *163*, 870–880. [CrossRef]
35. Tsitoura, E.; Trachalaki, A.; Vasarmidi, E.; Mastrodemou, S.; Margaritopoulos, G.A.; Kokosi, M.; Fanidis, D.; Galaris, A.; Aidinis, V.; Renzoni, E.; et al. Collagen 1a1 Expression by Airway Macrophages Increases In Fibrotic ILDs and Is Associated With FVC Decline and Increased Mortality. *Front. Immunol.* **2021**, *12*, 645548. [CrossRef]
36. Borok, Z.; Horie, M.; Flodby, P.; Wang, H.; Liu, Y.; Ganesh, S.; Firth, A.L.; Minoo, P.; Li, C.; Beers, M.F.; et al. Grp78 loss in epithelial progenitors reveals an age-linked role for endoplasmic reticulum stress in pulmonary fibrosis. *Am. J. Respir. Crit. Care Med.* **2020**, *201*, 198–211. [CrossRef]
37. Baek, H.A.; Kim, D.S.; Park, H.S.; Jang, K.Y.; Kang, M.J.; Lee, D.G.; Moon, W.S.; Chae, H.J.; Chung, M.J. Involvement of endoplasmic reticulum stress in myofibroblastic differentiation of lung fibroblasts. *Am. J. Respir. Cell Mol. Biol.* **2012**, *46*, 731–739. [CrossRef]
38. Tanjore, H.; Blackwell, T.S.; Lawson, W.E. Emerging evidence for endoplasmic reticulum stress in the pathogenesis of idiopathic pulmonary fibrosis. *Am. J. Physiol.-Lung Cell Mol. Physiol.* **2012**, *302*, 721–729. [CrossRef]
39. Kropski, J.A.; Blackwell, T.S. Endoplasmic reticulum stress in the pathogenesis of fibrotic disease. *J. Clin. Invest.* **2018**, *128*, 64–73. [CrossRef]
40. Chen, X.; Li, C.; Liu, J.; He, Y.; Wei, Y.; Chen, J. Inhibition of ER stress by targeting the IRE1α-TXNDC5 pathway alleviates crystalline silica-induced pulmonary fibrosis. *Int. Immunopharmacol.* **2021**, *95*, 107519. [CrossRef]
41. Justet, A.; Klay, D.; Porcher, R.; Cottin, V.; Ahmad, K.; Molina, M.M.; Nunes, H.; Reynaud-Gaubert, M.; Naccache, J.M.; Manali, E.; et al. Safety and efficacy of pirfenidone and nintedanib in patients with idiopathic pulmonary fibrosis and carrying a telomere-related gene mutation. *Eur. Respir. J.* **2021**, *57*, 2003198. [CrossRef]
42. Borie, R.; Kannengiesser, C.; Antoniou, K.; Bonella, F.; Crestani, B.; Fabre, A.; Froidure, A.; Galvin, L.; Griese, M.; Grutters, J.C.; et al. European Respiratory Society Statement on Familial Pulmonary Fibrosis. *Eur. Respir. J.* **2022**, *61*, 2201383. [CrossRef]
43. Terwiel, M.; Borie, R.; Crestani, B.; Galvin, L.; Bonella, F.; Fabre, A.; Froidure, A.; Griese, M.; Grutters, J.C.; Johannson, K.; et al. Genetic testing in interstitial lung disease: An international survey. *Respirology* **2022**, *27*, 747–757. [CrossRef]
44. Raghu, G.; Collard, H.R.; Egan, J.J.; Martinez, F.J.; Behr, J.; Brown, K.K.; Colby, T.V.; Cordier, J.-F.; Flaherty, K.R.; Lasky, J.A.; et al. An Official ATS/ERS/JRS/ALAT Statement: Idiopathic pulmonary fibrosis: Evidence-based guidelines for diagnosis and management. *Am. J. Respir. Crit. Care Med.* **2011**, *183*, 788–824. [CrossRef]
45. Cawthon, R.M. Telomere length measurement by a novel monochrome multiplex quantitative PCR method. *Nucleic Acids Res.* **2009**, *37*, e21. [CrossRef]
46. Snetselaar, R.; Van Batenburg, A.A.; Van Oosterhout, M.F.M.; Kazemier, K.M.; Roothaan, S.M.; Peeters, T.; Van Der Vis, J.J.; Goldschmeding, R.; Grutters, J.C.; Van Moorsel, C.H.M. Short telomere length in IPF lung associates with fibrotic lesions and predicts survival. *PLoS ONE* **2017**, *12*, e0189467. [CrossRef]
47. van Batenburg, A.A.; Kazemier, K.M.; van Oosterhout, M.F.M.; van der Vis, J.J.; van Es, H.W.; Grutters, J.C.; Goldschmeding, R.; van Moorsel, C.H.M. From organ to cell: Multi-level telomere length assessment in patients with idiopathic pulmonary fibrosis. *PLoS ONE* **2020**, *15*, e0226785. [CrossRef]

Disclaimer/Publisher's Note: The statements, opinions and data contained in all publications are solely those of the individual author(s) and contributor(s) and not of MDPI and/or the editor(s). MDPI and/or the editor(s) disclaim responsibility for any injury to people or property resulting from any ideas, methods, instructions or products referred to in the content.

Novel Peritoneal Sclerosis Rat Model Developed by Administration of Bleomycin and Lansoprazole

Kosei Kunitatsu [1], Yuta Yamamoto [2,*], Shota Nasu [2], Akira Taniji [2], Shuji Kawashima [1], Naoko Yamagishi [2], Takao Ito [2], Shigeaki Inoue [1] and Yoshimitsu Kanai [2]

[1] Department of Emergency and Critical Care Medicine, Wakayama Medical University, 811-1 Kimiidera, Wakayama 641-8509, Japan
[2] Department of Anatomy and Cell Biology, Wakayama Medical University, 811-1 Kimiidera, Wakayama 641-8509, Japan
* Correspondence: yuta-y@wakayama-med.ac.jp; Tel.: +81-73-441-0616; Fax: +81-73-441-0860

Abstract: In our preliminary experiment, peritoneal sclerosis likely induced by peritoneal dialysis was unexpectedly observed in the livers of rats given bleomycin and lansoprazole. We examined whether this peritoneal thickening around the liver was time-dependently induced by administration of both drugs. Male Wistar rats were injected with bleomycin and/or lansoprazole for 2 or 4 weeks. The 3YB-1 cell line derived from rat fibroblasts was treated by bleomycin and/or lansoprazole for 24 h. The administration of both drugs together, but not individually, thickened the peritoneal tissue around the liver. There was accumulation of collagen fibers, macrophages, and eosinophils under mesothelial cells. Expressions of *Col1a1*, *Mcp1* and *Mcp3* genes were increased in the peritoneal tissue around the liver and in 3YB-1 cells by the administration of both drugs together, and *Opn* genes had increased expressions in this tissue and 3YB-1 cells. Mesothelial cells indicated immunoreactivity against both cytokeratin, a mesothelial cell marker, and αSMA, a fibroblast marker, around the livers of rats given both drugs. Administration of both drugs induced the migration of macrophages and eosinophils and induced fibrosis associated with the possible activation of fibroblasts and the possible promotion of the mesothelial–mesenchymal transition. This might become a novel model of peritoneal sclerosis for peritoneal dialysis.

Keywords: fibrosis; collagen 1a1; bleomycin; lansoprazole; *Mcp1*; *Mcp3*; *Opn*

1. Introduction

The peritoneum is the largest serous membrane; it partially or fully covers the intra-abdominal organs [1]. It is contained within an epithelial monolayer of mesothelial cells and loose connective tissues including fibroblasts [2]. Peritoneal dialysis is a renal replacement therapy for renal failure; it slowly induces inflammation and fibrosis in the peritoneum [3,4]. Sub-mesothelial connective tissue is increased in thickness by peritoneal dialysis in a time-dependent manner [5]. In rat models of peritoneal sclerosis using microbicides and alcohol, there was reportedly an increased expression of monocyte chemoattractant protein (Mcp)-1 and transforming growth factor (Tgf)-β1. This increase was inhibited by a specific compound which ameliorated peritoneal thickening [6]. Peritoneal fibrosis induced by peritoneal dialysis in the long term may therefore be associated with the migration of macrophages and the production of the extracellular matrix, including type I collagen.

Lansoprazole is a proton pump inhibitor used for gastric and esophageal ulcers [7]. The expression of anti-oxidative stress proteins is increased in response to reactive oxygen species (ROS) via the nuclear factor erythroid 2-related factor 2 (Nrf2) pathway, and the upregulation of these genes ameliorates the cell damage induced by ROS. Lansoprazole induction was also reported to increase the expression of antioxidative stress protein genes through the activation of Nrf2 [8]. Lansoprazole ameliorated inflammation in the intestines, liver, and kidneys via the upregulation of anti-oxidative stress protein genes [8–10]. Lansoprazole inhibited the increased upregulation of interleukin 6 in rat hearts injured by cisplatin and fibrosis in non-alcoholic steatohepatitis model rats [11,12]. Bleomycin is an anticancer drug for squamous cell carcinoma, and it is used as an ROS inducer or for making an animal model of pulmonary inflammation [13]. We therefore examined whether lansoprazole ameliorates the pulmonary inflammation induced by bleomycin. Our preliminary experiment incidentally found that the peritoneal tissue was remarkably thick, a gross abnormality, around the livers of rats given both bleomycin and lansoprazole (BLM + LAP) subcutaneously for 4 weeks.

This study examined whether the peritoneal thickening around the liver was time-dependently induced by administration of BLM + LAP. To understand the mechanism, we analyzed histological changes in the peritoneal tissue around rat livers. To explore the genes associated with these histological changes, changes in gene expression induced by BLM + LAP treatment were analyzed in the livers of rats and 3Y1-B cells derived from rat fibroblast cells.

2. Results
2.1. Peritoneal Changes around the Livers of Rats Given BLM + LAP
2.1.1. Histological Changes in the Peritoneal Tissue around the Liver

We examined whether tissue injury induced by bleomycin was ameliorated by lansoprazole, which has been shown to have anti-inflammatory effects in several organs [8–10,12]. The peritoneal tissue at the inferior border of the liver thickened in a time-dependent manner (Figure 1A). We analyzed the liver histologically with hematoxylin eosin and Masson–Goldner staining. Reactive mesothelial cells were observed in rats given BLM + LAP for 2 weeks (Figure 1B), and collagen fiber under the mesothelial cells was observed in two out of the six rats given BLM + LAP for 2 weeks and in all rats given BLM + LAP for 4 weeks. Collagen fiber under the mesothelial cells of peritoneal tissue around the liver was not observed in rats given either bleomycin or lansoprazole separately (Figure 1E). The thickness of peritoneal tissue around the liver increased in rats given BLM + LAP in a time-dependent manner (Figure 1C). A two-way analysis of variance (ANOVA) indicated that the effects of lansoprazole (2 weeks: $F(1, 17) = 0.00$, $p = 0.96$, 4 weeks: $F(1, 17) = 0.06$, $p = 0.81$) and bleomycin (2 weeks: $F(1, 17) = 0.00$, $p = 0.98$, 4 weeks: $F(1, 17) = 0.01$, $p = 0.91$) were not significant, but the effect of their interaction was significant (2 weeks: $F(1, 17) = 5.83$, $p < 0.05$, 4 weeks: $F(1, 17) = 28.13$, $p < 0.05$). The migration of eosinophils (Figure 1B: indicated by arrowheads) and macrophages (Figure 1D) was detected in the thickened peritoneal tissue of rats given BLM + LAP for 4 weeks.

2.1.2. Effect of BLM + LAP on Rat Liver Injuries

Serum AST and ALT concentrations were measured in rats administered these drugs for 4 weeks to evaluate their effect on liver injuries (Figure 2A). A two-way ANOVA indicated no significant effects of lansoprazole ($F(1, 17) = 0.05$, $p = 0.82$) or bleomycin ($F(1, 17) = 2.42$, $p = 0.14$) and their interaction ($F(1, 17) = 3.30$, $p = 0.09$) in AST. There were no significant effects of lansoprazole ($F(1, 17) = 4.09$, $p = 0.06$) or bleomycin ($F(1, 17) = 0.02$, $p = 0.89$), but a significant difference in their interaction ($F(1, 17) = 6.62$, $p < 0.05$) was revealed in ALT. Dunnet's test indicated that there were no significant differences in the concentrations of AST and ALT compared with the control group. The expressions of *HO-1, catalase (Cat), glutathione S-transferase alpha 2 (Gsta2), NAD(P)H quinone dehydrogenase 1 (Nqo1),* and *glutathione peroxidase 1 (Gpx1)* genes were also measured via quantitative RT-PCR. The expression of the *Nqo1* gene was significantly increased in the lansoprazole, bleomycin and BLM + LAP groups (Supplemental Figure S1). To evaluate the effect of the secretion of bile due to peritoneal thickening, we measured the total bilirubin (Figure 2B). A two-way ANOVA indicated that there were no significant effects of lansoprazole ($F(1, 17) = 1.73$, $p = 0.21$), bleomycin ($F(1, 17) = 0.14$, $p = 0.71$), or their interaction ($F(1, 17) = 2.17$, $p = 0.16$) in total bilirubin. Dunnet's test indicated that there were no significant differences in the concentrations compared with the control group. There was no detection of γGTP in the serum of any of the rats.

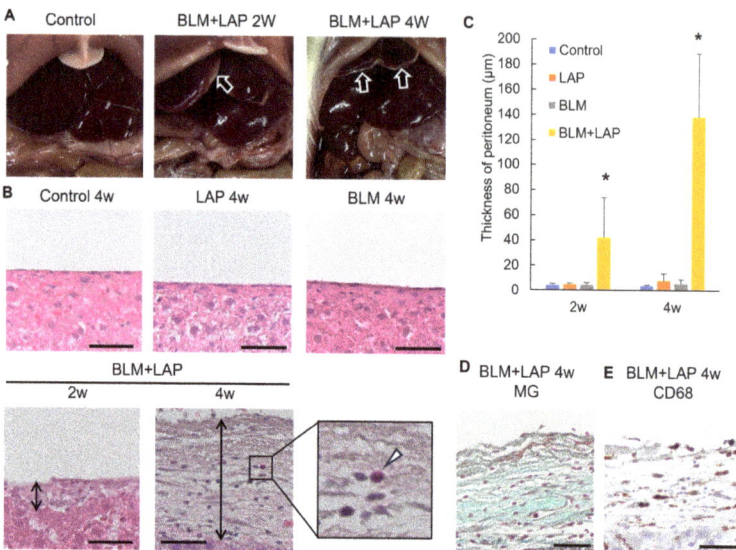

Figure 1. Peritoneal thickening induced by administration of BLM + LAP. We observed peritoneal thickening at the edge of the liver. The phenotype is indicated by arrows (**A**). Peritoneal thickening was induced only by administration of BLM + LAP. Double arrows indicate the thickness of peritoneum. Arrowhead indicates eosinophils in the sub-mesothelial connective tissue of the liver thickened by administration of BLM + LAP (**B**). This increased in a time-dependent manner (**C**). Hematoxylin and eosin stains indicated eosinophils in the thickened peritoneal tissue (**B**). Masson–Goldner stains indicated collagen fibers in the thickened peritoneal tissue (**D**). Immunohistochemistry using the anti-CD68 antibody indicated macrophages in the thickened peritoneal tissue (**E**). Scale bars indicate 50 μm. Asterisks indicate significant difference compared with the control ($p < 0.05$). Five rats each were allocated to the control, lansoprazole, and bleomycin groups, and six rats were allocated to the BLM + LAP group. p values were calculated by Dunnet's test. LAP: lansoprazole, BLM: bleomycin.

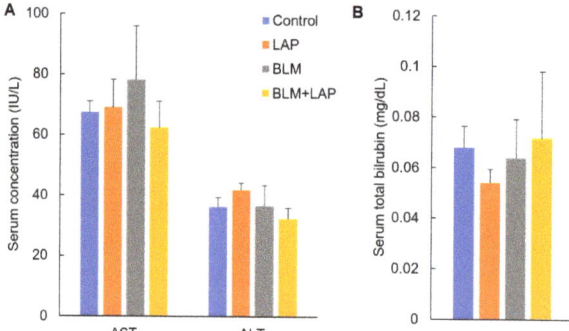

Figure 2. Administration of BLM + LAP for four weeks did not induce liver injury or inhibit the secretion of bile. The concentrations of ASL and ALT, markers of damaged hepatic cells, were unchanged in serum (**A**), and there was no change in the concentration of total bilirubin, the marker of bile stasis, in serum (**B**). Five rats each were allocated to the control, lansoprazole, and bleomycin groups, and six rats were allocated to the BLM + LAP group. p values were calculated by Dunnet's test.

2.1.3. Expression Changes of Genes Associated with Fibrosis in the Peritoneal Tissue around the Liver

To explore the mechanism of peritoneal tissue thickening, we measured the expression of *collagen type 1 alpha 1 chain (Col1a1)* and *transforming growth factor beta 1 (Tgfb1)* genes. A two-way ANOVA indicated significant differences in the effect of drug interactions on the expression of *Col1a1* genes in the liver of rats administered drugs for 4 weeks (lansoprazole (2 weeks: $F(1, 16) = 2.91$, $p = 0.11$, 4 weeks: $F(1, 17) = 0.05$, $p = 0.83$), bleomycin (2 weeks: $F(1, 16) = 3.58$, $p = 0.08$, 4 weeks: $F(1, 17) = 0.45$, $p = 0.51$) and their interaction (2 weeks: $F(1, 16) = 2.10$, $p = 0.17$, 4 weeks: $F(1, 17) = 8.01$, $p < 0.05$)). Dunnett's test indicated that the expression of *Col1a1* was significantly increased only in the livers of rats given BLM + LAP for 4 weeks (Figure 3A). A two-way ANOVA indicated significant differences in the effect of lansoprazole on the expression of *Tgfb1* genes in the liver of rats administered drugs for 2 weeks (lansoprazole (2 weeks: $F(1, 16) = 5.91$, $p < 0.05$, 4 weeks: $F(1, 17) = 1.84$, $p = 0.19$), bleomycin (2 weeks: $F(1, 16) = 0.20$, $p = 0.66$, 4 weeks: $F(1, 17) = 2.57$, $p = 0.13$) and their interaction (2 weeks: $F(1, 16) = 4.01$, $p = 0.06$, 4 weeks: $F(1, 17) = 0.65$, $p = 0.43$)). Dunnett's test indicated that there was no change in expression of *Tgfb1* in any of the groups (Figure 3B). To examine whether these changes in expression occurred in the liver ubiquitously, we measured the expression of these genes in the outer and inner parts of liver (Figure 3C). A two-way ANOVA indicated significant differences in the effects of drug interaction on the expression of *Col1a1* genes in the outer parts of liver in rats administered drugs for 4 weeks (lansoprazole ($F(1, 17) = 0.01$, $p = 0.94$), bleomycin ($F(1, 17) = 0.26$, $p = 0.62$) and their interaction ($F(1, 17) = 7.65$, $p < 0.05$)). However, there were no significant differences in any effects of the drug or their interaction on the expression of *Col1a1* genes in the inner parts of livers from rats treated for 4 weeks (lansoprazole ($F(1, 17) = 2.27$, $p = 0.15$), bleomycin ($F(1, 17) = 0.41$, $p = 0.53$) and their interaction ($F(1, 17) = 0.00$, $p = 0.96$)). Dunnett's test indicated that an increase in *Col1a1* gene expression was observed in the outer parts of the livers of rats that were given BLM + LAP for 4 weeks, but it was not observed in the inner parts of the livers (Figure 3D).

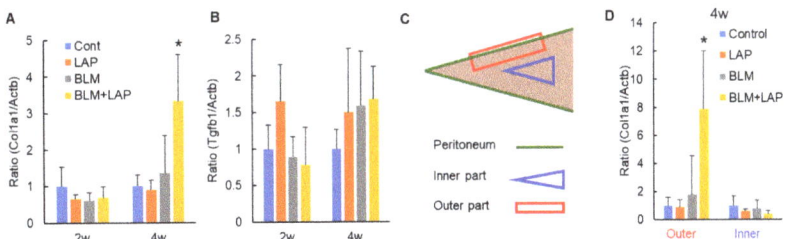

Figure 3. Expression changes of genes associated with fibrosis in the livers of rats given BLM + LAP. The expression of Col1a1 changed in a time-dependent manner (**A**). The expression of Tgfb1 was not changed by bleomycin or lansoprazole separately (**B**). To examine whether the peritoneal thickening was associated with Col1a1 or Tgfb1 in detail, the liver samples were divided into two groups: samples from the inner parts and samples from the outer parts (**C**). Col1a1 was only upregulated by BLM + LAP in the outer part of liver (**D**). Five rats each were allocated to the control, lansoprazole, or bleomycin groups, and six rats were allocated to the BLM + LAP group. One rat in the BLM + LAP group was excluded due to a missing sample. p values were calculated by Dunnet's test. Asterisks indicate a significant difference from the control group ($p < 0.05$).

2.1.4. Expression Changes of Genes Associated with Migration of Macrophages and Eosinophils in the Peritoneal Tissue around the Liver

Migration of macrophages and eosinophils was observed in the peritoneal tissue thickened by BLM + LAP. Ip10, Mip1a1, Mcp1, Mcp3 and Rantes promoted the migration of macrophages, and Mcp3 and Rantes also promoted the migration of eosinophils. We measured the expression of chemokine genes expressed in fibroblasts. Focusing on the outer part, a two-way ANOVA indicated that a synergy between lansoprazole and bleomycin was observed in the upregulation of *Mcp1* genes (lansoprazole (F(1, 16) = 0.00, p = 0.98), bleomycin (F(1, 16) = 0.63, p = 0.44) and their interaction (F(1, 16) = 9.47, p < 0.05)) but not *Mcp3* genes (lansoprazole (F(1, 16) = 0.00, p = 0.98), bleomycin (F(1, 16) = 1.41, p = 0.25) and their interaction (F(1, 12) = 2.52, p = 0.13)) in the outer part of the livers from rats treated for 4 weeks (Supplemental Table S1). Dunnett's test indicated that the expressions of *Mcp1* and *Mcp3* genes tended to be increased by BLM + LAP after 2 weeks (Figure 4A), and they were significantly increased after 4 weeks (Figure 4B). However, upregulation of *Mcp1* and *Mcp3* genes was not observed in the inner part (Figure 4C,D). Expression of *Mcp1* and *Mcp3* genes was also significantly increased in 3Y1-B cells derived from rat fibroblast cells treated by BLM + LAP (Figure 4E), although synergy between lansoprazole and bleomycin was observed in the upregulation of *Mcp1* genes (lansoprazole (F(1, 12) = 3.38, p = 0.09), bleomycin (F(1, 12) = 9.23, p < 0.05) and their interaction (F(1, 12) = 6.42, p < 0.05)) but not *Mcp3* genes (lansoprazole (F(1, 12) = 2.50, p = 0.14), bleomycin (F(1, 12) = 1.85, p = 0.20) and their interaction (F(1, 12) = 2.17, p = 0.17)).

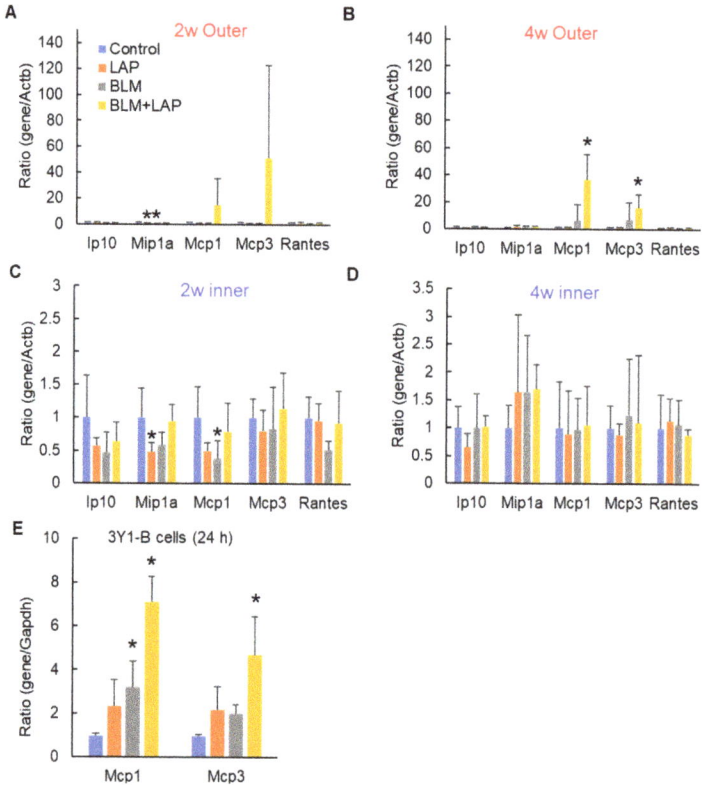

Figure 4. Expressions of chemokines inducing the migration of macrophages or eosinophils in the liver. The expression of chemokines in fibroblast cells was measured in the outer (**A**,**B**) and inner parts (**C**,**D**). Expressions of *Mcp1* and *Mcp3* were measured in 3Y1-B cells (**E**). These chemokines promote the migration of macrophages, and Mcp3 and Rantes also promote the migration of eosinophils. Asterisks indicate a significant difference from the control group (p < 0.05).

2.2. BLM + LAP Increased the Expression of Genes Associated with the Migration and Proliferation of Fibroblasts

Activation of fibroblasts might be promoted in the peritoneal tissues of the liver in the BLM + LAP group because collagen fiber was accumulated in the peritoneal membrane (Figure 1D). The osteopontin (*Opn*) gene is expressed in the activated fibroblasts. A two-way ANOVA indicated significant differences in the effect of drug interactions on the expression of the *Opn* gene in the outer parts of the livers of rats treated for 4 weeks (lansoprazole (F(1, 16) = 0.36, p = 0.55), bleomycin (F(1, 16) = 0.23, p = 0.64) and their interaction (F(1, 16) = 7.27, p < 0.05)). Meanwhile, there were significant differences in the effect of lansoprazole on the expression of *Opn* in 3YB-1 cells treated for 24 h (lansoprazole (F(1, 12) = 6.82, p < 0.05), bleomycin (F(1, 12) = 2.35, p = 0.15) and their interaction (F(1, 12) = 0.34, p = 0.57)). Expression of the *Opn* gene was significantly increased in the outer part of the livers of rats that were given BLM + LAP for 4 weeks (Figure 5A). This increase was also detected in 3Y1-B cells (Figure 5B).

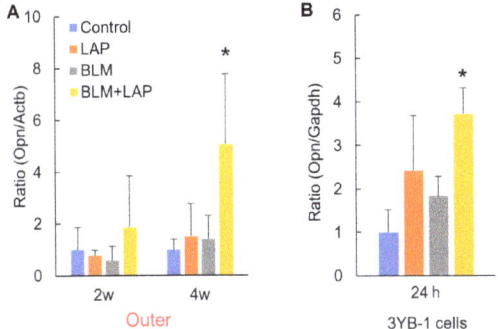

Figure 5. BLM + LAP increased the expression of the *Opn* gene. The expression of the *Opn* gene in activated fibroblasts was increased by BLM + LAP in the outer part of liver (**A**) and in 3Y1-B cells (**B**). Five rats each were allocated to control, lansoprazole, and bleomycin groups, and six rats were allocated to the BLM + LAP group. Cell culture experiments were performed four times. p values were calculated by Dunnet's test. Asterisks indicate a significant difference from the control group (p < 0.05).

2.3. BLM + LAP Increased the Number of Mesothelial Cells Indicating Immunoreactivity against Cytokeratin and αSMA.

Reactive mesothelial cells were detected in the peritoneal tissue of the livers in the BLM + LAP group after 2 weeks (Figure 1B). To examine whether the reactive mesothelial cells were differentiated to fibroblasts, immunohistochemistry was performed in the liver tissue of rats administered BLM + LAP for 2 weeks with anti-cytokeratin, which is expressed in mesothelial cells and anti-α smooth muscle actin (αSMA), which is expressed in fibroblasts. The form of mesothelium cells, excluding one case in the bleomycin group, was not changed with lansoprazole or bleomycin, and reactive mesothelial cells indicated immunoreactivity against both cytokeratin and αSMA in the BLM + LAP group (Figure 6) and one of the bleomycin groups (Supplemental Figure S2). The immunoreactivity against cytokeratin was weak in one case in which the peritoneal tissue was thickened (Supplemental Figure S2).

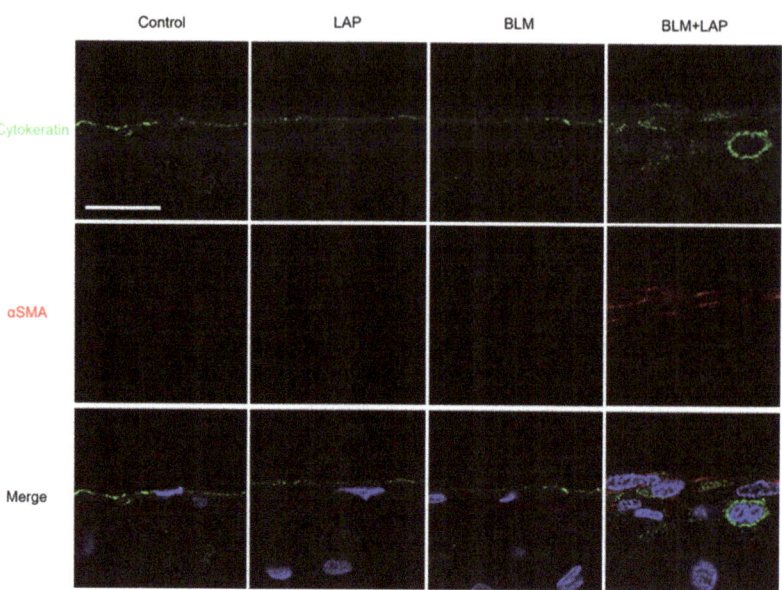

Figure 6. Administration of BLM + LAP was associated with mesothelial–mesenchymal transition. The peritoneal tissue around the liver was double-stained for cytokeratin (green) and αSMA (red) followed by DAPI (blue). Scale bar indicates 60 μm.

3. Discussion

This study demonstrated that the administration of BLM + LAP time-dependently thickened the peritoneal tissue around rat livers. Thickening of the peritoneal tissue was induced by the infiltration of macrophages and eosinophils and the accumulation of collagen fibers (Figure 1). This phenotype might be associated with the upregulation of *Col1a1*, *Mcp1* and *Mcp3* genes in the peritoneal tissue of the liver (Figures 3 and 4). The expression of the *Opn* gene, which is expressed in activated fibroblasts, was increased by BLM + LAP in 3YB-1 cells derived from rat fibroblast cells, as well as in peritoneal liver tissues (Figure 5). Immunohistochemistry indicated the possibility that the reactive mesothelial cells might also induce the mesothelial–mesenchymal transition (MMT) by BLM + LAP (Figure 6). Administration of BLM + LAP could therefore affect the fibroblasts in the peritoneal tissue regarding gene expression changes associated with peritoneal thickening, and it might be associated with a MMT in mesothelial cells (Figure 7).

Bleomycin, the agonist of Toll-like receptor (TLR) 2, induced inflammation and fibrosis, as shown by the expression of proinflammatory cytokines and Tgfb1 in mouse lungs [13,14]. This fibrosis resulted in the activation of fibroblasts [15,16]. However, the reported bleomycin-induced fibrosis was pulmonary fibrosis, not pleural fibrosis. In this study, peritoneal fibrosis in the liver, but not liver fibrosis, was revealed to be time dependent in some of the rats given BLM + LAP (Figure 1). The phenotype was observed in the rats given either bleomycin or lansoprazole for 28 days (Figure 1). This phenotype may therefore be different from the previously reported bleomycin-induced fibrosis. Peritoneal fibrosis induced by the administration of BLM + LAP for 28 days might induce bile statis, so we measured the concentrations of γGTP and bilirubin in serum. The concentration of γGTP was lower than the detection limit in all rats, and the concentration of bilirubin was not increased by the administration of BLM + LAP (Figure 2). Peritoneal fibrosis may therefore be unnoticeable if the fibrosis occurs in patients undergoing chemotherapy with bleomycin and taking lansoprazole for gastric ulcers.

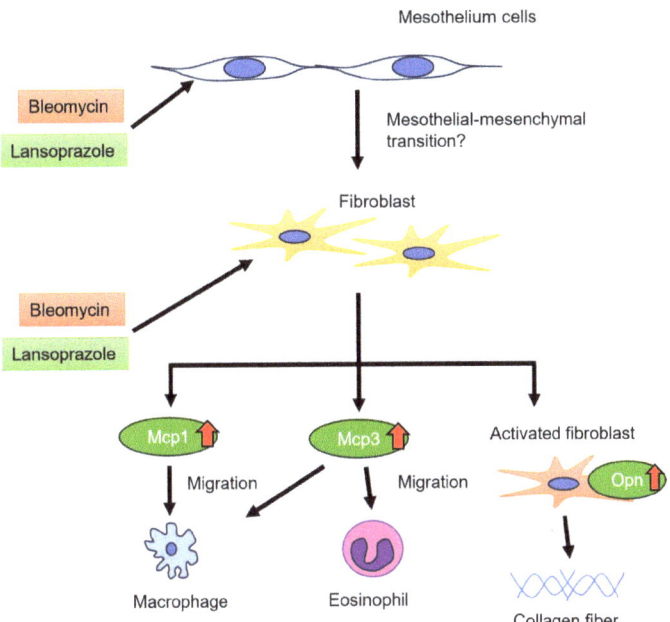

Figure 7. Scheme of this study.

Administration of BLM + LAP increased the expression of chemokine genes (*Mcp1* and *Mcp3*), which promote the migration of macrophages and eosinophils and promote the activation of fibroblasts, which expressed the *Opn* gene in sub-mesothelial connective tissues. The mesothelial–mesenchymal transition might be progressed by the administration of BLM + LAP.

Bleomycin as an ROS inducer might injure the hepatocytes in the liver, but no effect of bleomycin and/or lansoprazole on the toxicity of hepatocytes was observed in serum AST and ALT levels (Figure 2A). The expression of the *Nqo1* gene was increased by lansoprazole and bleomycin (Supplemental Figure S1). The upregulation of the *Nqo1* gene might be induced by lansoprazole via activation of Nrf2 or ROS induced by bleomycin. Thus, there might be no heavy cellular damage in the livers of rats given bleomycin due to the cytoprotection effect of Nqo1. The peritoneal sclerosis in this study induced by BLM + LAP might not be associated with the remarkable exposure to ROS.

Peritoneal dialysis over a long time induced encapsulating peritoneal sclerosis (EPS), the accumulation of collagen fiber, and the migration of macrophages and eosinophils [17,18]. The expression of Col1a1 was increased in patients that underwent peritoneal dialysis [19]. The peritoneal sclerosis induced by the administration of BLM + LAP had histological changes similar to that of EPS (Figure 1). The expressions of *Col1a1*, *Mcp1* and *Mcp3* genes in peritoneal tissues of the liver were increased by the administration of BLM + LAP (Figures 3 and 4). The migration of macrophages is associated with several chemokines. The chemokines which were expressed in fibroblasts and induced the migration of macrophages were IP-10 [20,21], Mcp1 [22,23], Mcp3 [24,25] and Rantes [24]. Eosinophils are also attracted by Mcp3 and Rantes [26]. The accumulation of collagen fiber might therefore be associated with the upregulation of the *Col1a1* gene, and the migration of macrophages might be associated with the upregulation of the *Mcp1* and *Mcp3* genes. The migration of eosinophils might also be associated with the upregulation of the *Mcp3* gene. Expression changes of these genes were observed in 3YB-1 cells treated with BLM + LAP. BLM + LAP might therefore increase the expression of Mcp1 and Mcp3 genes in the fibroblast cells of the peritoneum following the migration of macrophages and eosinophils.

Opn is a cytokine which activates fibroblasts, producing extracellular matrix including collagen 1 [27,28]. Bleomycin induces the maturation of Tgfβ from latent Tgfβ in mouse lungs [29], and maturated Tgfβ induces the activation of fibroblasts which secrete Opn protein [30]. Extracellular matrix, including collagen type 1, is produced by activated fibroblasts [31]. Thus, the production of extracellular matrix is associated with the Tgfβ/SMAD and Tgfβ/non-SMAD pathways, including the Tgfβ/p38 pathway [32]. Bleomycin is also an agonist of TLR2 and promotes the production of inflammatory cytokines and chemokines [14,33]. Tgfβ/non-SMAD pathways including the Tgfβ/p38 pathway and the TLR2 pathway also include the TNF receptor-associated factor 6 (TRAF6). Lansoprazole enhances polyubiquitination of TRAF6 through binding lansoprazole to the deubiquitinate enzyme and cylindromatosis, and activates the Tgf-β-activated kinase-1 (TAK1)–p38 pathway [34]. Bleomycin might have promoted the production of Col1a1 via the non-SMAD pathway and the production of Mcp1 and Mcp3 via TLR2 signaling in the activated fibroblasts expressing Opn, but this phenomenon was not observed with the dosage and during the period we applied in this study (Figures 3–5). Lansoprazole, as the activator of the TAK1-p38 pathway, might enhance the Tgfβ/p38 pathway in fibroblasts, and this might increase the expression of *Col1a1*, *Mcp1* and *Mcp3* in the liver around the peritoneum and fibroblast cells. An effect of drug interactions between lansoprazole and bleomycin on the upregulation of the *Mcp1* gene was observed in the outer part of the liver and fibroblast cells, but the effect of drug interactions between lansoprazole and bleomycin on the upregulation of the *Mcp3* gene was not observed. Thus, the upregulation of the *Mcp1* gene might be associated with the synergistic effect of BLM + LAP, while the upregulation of the *Mcp3* gene might be associated with the additive effect of BLM + LAP.

The MMT results from peritoneal dialysis and is induced by cytokines, including Tgfβ [35,36]. The expression of mesothelial cell marker proteins including cytokeratin is decreased and the expression of fibroblast marker proteins including αSMA is increased in MMT [35]. The mesothelial cells of hepatic peritoneal tissues indicated immunoreactivity against both cytokeratin and αSMA in rats given BLM + LAP for 14 days (Figure 6). Cells which were immunoreactive against both cytokeratin and αSMA were also detected in the hepatic peritoneal tissue in one out of five rats in the bleomycin group (Supplemental Figure S1). Bleomycin induced the epithelial–mesenchymal transition via Tgfβ in mouse lungs [37]. Bleomycin might therefore slightly induce the MMT in the hepatic peritoneal tissue, but the MMT might be enhanced by lansoprazole via the Tgfβ/p38 pathway.

This study did not, however, examine whether administration of BLM + LAP promotes the MMT in mesothelial cells from peritoneal tissue, because mesothelial cell lines derived from peritoneal tissue are not deposited in cell banks. Future studies are required to demonstrate that the MMT is induced by bleomycin and enhanced by lansoprazole in mesothelial cells from peritoneal tissue. The peritoneal thickness might be enhanced by lansoprazole rather than other proton pump inhibitors because this enhancement is the result of the binding of lansoprazole to cylindromatosis [34]. Further study is required to examine whether the peritoneal sclerosis in the BLM + LAP group also occurs with the combination of bleomycin and other proton pump inhibitors. Peritoneal sclerosis might be associated with the activation and enhancement of the Tgfβ pathway, but there was no significant change in expression of the *Tgfb1* gene. One reason for this is that bleomycin might mature Tgfβ but not increase the expression of Tgfb1. Another reason could be that there was a comparatively small number of rats used to examine whether the expression of Tgfb1 was truly increased by the administration of BLM + LAP.

The animal model of peritoneal dialysis was prepared using peritoneal dialysates with high glucose and/or chlorhexidine gluconate levels [24,38]. These solutions indicate that the cytotoxicity, as well as bleomycin and high glucose solutions, may affect the blood glucose levels in animals. The administration of BLM + LAP does not directly affect the mesothelial cells in peritoneal tissue. This novel animal model of peritoneal sclerosis may aid in understanding the mechanism of the development of peritoneal fibrosis induced by peritoneal dialysis.

4. Materials and Methods

4.1. Chemicals

The following chemicals were purchased and used in this study: lansoprazole (129-05863, Fujifilm Wako, Osaka, Japan), bleomycin (21800AMX10210, Nippon Kayaku, Tokyo, Japan), and carboxymethyl cellulose (CMC) (039-01335, Fujifilm Wako).

4.2. Animals

Five-week-old male Wistar rats were purchased from Kiwa Laboratory Animals (Wakayama, Japan). Two or three rats were housed in a plastic rat cage (24.7 cm × 40.9 cm × 19.7 cm) with free access to tap water and laboratory animal feed (Oriental Yeast Co., Ltd., Tokyo, Japan) under a 12 h light/dark cycle (lights on/off at 8:00 a.m./p.m.) at 25 °C ± 1 °C and 50–60% humidity. All animals were used for experiments after an acclimation period of 1 week.

4.3. Drug Administration

Twenty-one six-week-old male Wistar rats were divided into four groups (control (n = 5), lansoprazole (n = 5), bleomycin (n = 5), BLM + LAP (n = 6) groups) in every time course (2-week or 4-week experiment). Saline was subcutaneously injected into the right side of the back in control and lansoprazole groups, and bleomycin (1 mg/mL) was subcutaneously injected into the right side of the back in bleomycin and BLM + LAP groups. A 0.5% CMC solution was subcutaneously injected into the left side of the back in control and bleomycin groups, and a lansoprazole suspension with 0.5% CMC (30 mg/mL) was subcutaneously injected into the right side of the back in lansoprazole and BLM + LAP groups. Drug administration was performed over 14 or 28 days. One day after the last drug administration, blood samples were collected under anesthesia with isoflurane, and liver tissue samples were obtained following perfusion with 4% formalin neutral buffer solution.

4.4. Cell Culture

3Y1-B clone 1-6 (3Y1-B) cells were derived from a non-oncogenic rat fibroblast cell line (JCRB0734, Japanese Collection of Research Bioresources, Osaka, Japan) (PMID: 166944). Cells were maintained in Dulbecco's modified Eagle medium (DMEM) without phenol red (Fujifilm Wako) and supplemented with 10% fetal bovine serum (Sigma-Aldrich, St. Louis, MO, USA) at 37 °C with 5% CO_2. Lansoprazole was dissolved in dimethyl sulfoxide (DMSO) (Fujifilm Wako). Cells were treated for 24 h in the culture media with drugs (control media: 0.5% DMSO, media: 0.5% DMSO and 50 µM lansoprazole, media: 0.5% DMSO and 10 µg/mL bleomycin, and BLM + LAP media: 0.5% DMSO, 50 µM LAP and 10 µg/mL BLM). Cell culture experiments were performed four times.

4.5. Histology

Obtained liver tissues were fixed by 4% formalin neutral buffer solution for a day and embedded in paraffin. Specimens in paraffin blocks were cut at an interval of 5 µm and stained with hematoxylin and eosin solutions (Muto Pure Chemicals, Tokyo, Japan). Masson–Goldner staining was performed with an MG staining kit (Melck Millipore, Burlington, MA, USA).

For immunohistochemistry, the sections were incubated in HistoVT One (06380-05 Nacalai Tesque, Kyoto, Japan) for 20 min at 90 °C. The blocking process was performed with Blocking One (03953-66, Nacalai Tesque) after incubation in 0.3% H_2O_2/methanol for 30 min. Mouse anti-CD68 antibody (1:100, MCA341R, Bio-Rad Laboratories, Hercules, CA, USA) was used as the primary antibody, and anti-mouse IgG antibody (1:100, BA-2001, Vector Laboratories, Burlingame, CA, USA) was used as the secondary antibody. To detect the antibodies, we used an elite ABC kit (PK-6100, Vector Laboratories) and diaminobenzidine solutions (0.1 mg/mL, Dojindo Laboratories, Kumamoto, Japan).

For fluorescent immunohistochemistry, the sections were incubated in citric acid buffer (pH 6.0) for 15 min in a microwave. The blocking process was performed with 1%

normal goat serum (S-1000, Vector Laboratories) in phosphate-buffered saline. Mouse anti-cytokeratin antibody (1:100, GTX75521, Gene Tex, Irvine, CA, USA) and rabbit anti-αSMA (1:400, GTX100034, Gene Tex) were used as the primary antibodies, and anti-mouse IgG antibody conjugated with Alexa Fluor (1:100, A-11029, Invitrogen, Waltham, MA, USA) and anti-rabbit IgG antibody conjugated with Alexa Fluor 578 (1:100, A-11011, Invitrogen) were used as the secondary antibody. To stain the nucleus, we used a 4′,6-diamidino-2-phenylindole (DAPI) solution (1:2000, Dojindo Laboratories).

4.6. Quantitative PCR

To extract RNA samples from tissue samples (approximately 20 mg), Sepasol RNA 1 Super G (Nacalai Tesque) was used. ReverTra Ace (FSQ-301, TOYOBO, Osaka, Japan) was used to synthesize first-strand cDNA samples from RNA samples (500 ng). Quantitative RT-PCR analyses were performed with the Brilliant III Ultra-Fast SYBR Green QPCR Master Mix (Agilent Technologies, Tokyo, Japan) and an AriaMX Real-time PCR system (Agilent Technologies). The primer sets used are indicated in Table 1. The relative expression ratio of each gene (gene/Actb) was normalized to that of the control group.

Table 1. Primer sequences for quantitative PCR.

Symbol	Forward	Reverse
Col1a1	AGGCTGGTGTGATGGGATT	AGGGCCTTGTTCACCTCTCT
Tgfb1	GACCGCAACAACGCAAT	GGCACTGCTTCCCGAAT
HO-1	ACAGGGTGACAGAAGAGGCTAA	CTGTGAGGGACTCTGGTCTTTG
Cat	GCCTGTGTGAGAACATTGC	CCTGTACGTAGGTGTGAATTG
Gsta2	CTTCTCCTCTATGTTGAAGAGTTTG	TTTTGCATCCACGGGAA
Nqo1	CAGCGGCTCCATGTACT	GACCTGGAAGCCACAGAAG
Gpx1	GGACTACACCGAAATGAATGAT	CTCGCACTTCTCAAACAATG
Mcp1	TTG TCA CCA AGC TCA AGA GA	CAC ATT CAA AGG TGC TGA AG
Mcp3	GCATGGAAGTCTGTGCTGAA	CGTTCCTACCCCTTAGGAC
Ip-10	TCCTGCAAGTCTATCCTGTC	TGGCTTCTCTCTAGTTACGG
Mip-1a	GCGAGTACCAGTCCCTTCT	GGTGCTGAGCAGGTAACAGA
Rantes	TCGTCTTTGTCACTCGAAGG	GAGCAAGCAATGACAGGAAA
Opn	AGTGGTTTGCCTTTGCCTGTT	TCAGCCAAGTGGCTACAGCAT
Actb	GGAGATTACTGCCCTGGCTCCTA	GACTCATCGTACTCCTGCTTGCT
Gapdh	AGGTTGTCTCCTGTGACTTC	CTGTTGCTGTAGCCATATTC

To extract RNA samples from cell samples, Sepasol RNA 1 Super G (Nacalai Tesque) was used. ReverTra Ace (FSQ-301, TOYOBO, Osaka, Japan) was used to synthesize first-strand cDNA samples from RNA samples (500 ng). Quantitative RT-PCR analyses were performed with a KAPA SYBR FAST qPCR Kit (KAPA Biosystems, Wilmington, MA, USA) and a CFX96 real-time PCR system (Bio-Rad Laboratories). The primer sets used are indicated in Table 1. The relative expression ratio of each gene (gene/Gapdh) was normalized to that of the control group.

4.7. Statistical Analysis

All statistical analyses were performed using JMP statistical software, version 14.3 (SAS Institute Inc., Cary, NC, USA). Statistical analyses were performed using a two-way ANOVA and Dunnett's test. Values of $p < 0.05$ were considered to be statistically significant. Results are expressed as means ± standard deviation.

5. Conclusions

Administration of BLM + LAP induced the thickening of hepatic peritoneal tissue via the synergistic or additive effect of lansoprazole and bleomycin. This mechanism might be associated with the migration of macrophages and eosinophils via Mcp1 and Mcp3 secretion in fibroblasts treated by both drugs and the accumulation of collagen via the fibroblasts activated by both drugs. In addition, the slight induction of the MMT by bleomycin might be enhanced by lansoprazole. The peritoneal thickening seen here is different to the animal model for peritoneal sclerosis resulting from peritoneal dialysis, and it could become a novel peritoneal sclerosis model.

Supplementary Materials: The following supporting information can be downloaded at: https://www.mdpi.com/article/10.3390/ijms242216108/s1.

Author Contributions: Conceptualization, Y.Y.; methodology, K.K., Y.Y. and N.Y.; formal analysis, Y.Y.; investigation, K.K., Y.Y., S.N., A.T., S.K. and T.I.; writing—original draft preparation, K.K. and Y.Y.; supervision, S.I. and Y.K.; project administration, Y.Y.; funding acquisition, Y.Y. and S.K. All authors have read and agreed to the published version of the manuscript.

Funding: This study was supported by JSPS KAKENHI Grant Number (17K17066, 22K09187) and by Medical Research Support from Wakayama Medical University.

Institutional Review Board Statement: The experimental protocol was approved by the Wakayama Medical University Animal Experiment Committee (No. 989).

Data Availability Statement: Data are contained within the article and supplementary materials.

Acknowledgments: We acknowledge proofreading and editing by Benjamin Phillis at the Clinical Study Support Center at Wakayama Medical University and the statistical consultation by Kensuke Tanioka at Department of Biomedical Sciences and Informatics, Doshisha University.

Conflicts of Interest: The authors declare no conflict of interest.

References

1. Bermo, M.S.; Koppula, B.; Kumar, M.; Leblond, A.; Matesan, M.C. The Peritoneum: What Nuclear Radiologists Need to Know. *Semin. Nucl. Med.* **2020**, *50*, 405–418. [CrossRef]
2. do Amaral, R.; Arcanjo, K.D.; El-Cheikh, M.C.; de Oliveira, F.L. The Peritoneum: Health, Disease, and Perspectives regarding Tissue Engineering and Cell Therapies. *Cells Tissues Organs* **2017**, *204*, 211–217. [CrossRef]
3. Jagirdar, R.M.; Bozikas, A.; Zarogiannis, S.G.; Bartosova, M.; Schmitt, C.P.; Liakopoulos, V. Encapsulating Peritoneal Sclerosis: Pathophysiology and Current Treatment Options. *Int. J. Mol. Sci.* **2019**, *20*, 5765. [CrossRef]
4. Masola, V.; Bonomini, M.; Borrelli, S.; Di Liberato, L.; Vecchi, L.; Onisto, M.; Gambaro, G.; Palumbo, R.; Arduini, A. Fibrosis of Peritoneal Membrane as Target of New Therapies in Peritoneal Dialysis. *Int. J. Mol. Sci.* **2022**, *23*, 4831. [CrossRef]
5. Williams, J.D.; Craig, K.J.; Topley, N.; Von Ruhland, C.; Fallon, M.; Newman, G.R.; Mackenzie, R.K.; Williams, G.T. Morphologic changes in the peritoneal membrane of patients with renal disease. *J. Am. Soc. Nephrol.* **2002**, *13*, 470–479. [CrossRef]
6. Li, Z.; Zhang, L.; He, W.; Zhu, C.; Yang, J.; Sheng, M. Astragalus membranaceus inhibits peritoneal fibrosis via monocyte chemoattractant protein (MCP)-1 and the transforming growth factor-beta1 (TGF-beta1) pathway in rats submitted to peritoneal dialysis. *Int. J. Mol. Sci.* **2014**, *15*, 12959–12971. [CrossRef]
7. Sachs, G.; Shin, J.M.; Briving, C.; Wallmark, B.; Hersey, S. The pharmacology of the gastric acid pump: The H+,K+ ATPase. *Annu. Rev. Pharmacol. Toxicol.* **1995**, *35*, 277–305. [CrossRef]
8. Yamashita, Y.; Ueyama, T.; Nishi, T.; Yamamoto, Y.; Kawakoshi, A.; Sunami, S.; Iguchi, M.; Tamai, H.; Ueda, K.; Ito, T.; et al. Nrf2-inducing anti-oxidation stress response in the rat liver--new beneficial effect of lansoprazole. *PLoS ONE* **2014**, *9*, e97419. [CrossRef]
9. Khaleel, S.A.; Alzokaky, A.A.; Raslan, N.A.; Alwakeel, A.I.; Abd El-Aziz, H.G.; Abd-Allah, A.R. Lansoprazole halts contrast induced nephropathy through activation of Nrf2 pathway in rats. *Chem. Biol. Interact.* **2017**, *270*, 33–40. [CrossRef]
10. Yoda, Y.; Amagase, K.; Kato, S.; Tokioka, S.; Murano, M.; Kakimoto, K.; Nishio, H.; Umegaki, E.; Takeuchi, K.; Higuchi, K. Prevention by lansoprazole, a proton pump inhibitor, of indomethacin -induced small intestinal ulceration in rats through induction of heme oxygenase-1. *J. Physiol. Pharmacol.* **2010**, *61*, 287–294.
11. Hassanein, E.H.M.; Ali, F.E.M.; Mohammedsaleh, Z.M.; Atwa, A.M.; Elfiky, M. The involvement of Nrf2/HO-1/cytoglobin and Ang-II/NF-kappaB signals in the cardioprotective mechanism of lansoprazole against cisplatin-induced heart injury. *Toxicol. Mech. Methods* **2023**, *33*, 316–326. [CrossRef]
12. Nishi, T.; Yamamoto, Y.; Yamagishi, N.; Iguchi, M.; Tamai, H.; Ito, T.; Tsuruo, Y.; Ichinose, M.; Kitano, M.; Ueyama, T. Lansoprazole prevents the progression of liver fibrosis in non-alcoholic steatohepatitis model rats. *J. Pharm. Pharmacol.* **2018**, *70*, 383–392. [CrossRef] [PubMed]
13. Kim, S.N.; Lee, J.; Yang, H.S.; Cho, J.W.; Kwon, S.; Kim, Y.B.; Her, J.D.; Cho, K.H.; Song, C.W.; Lee, K. Dose-response Effects of Bleomycin on Inflammation and Pulmonary Fibrosis in Mice. *Toxicol. Res.* **2010**, *26*, 217–222. [CrossRef]
14. Yang, H.Z.; Cui, B.; Liu, H.Z.; Chen, Z.R.; Yan, H.M.; Hua, F.; Hu, Z.W. Targeting TLR2 attenuates pulmonary inflammation and fibrosis by reversion of suppressive immune microenvironment. *J. Immunol.* **2009**, *182*, 692–702. [CrossRef] [PubMed]

15. Chen, L.J.; Ye, H.; Zhang, Q.; Li, F.Z.; Song, L.J.; Yang, J.; Mu, Q.; Rao, S.S.; Cai, P.C.; Xiang, F.; et al. Bleomycin induced epithelial-mesenchymal transition (EMT) in pleural mesothelial cells. *Toxicol. Appl. Pharmacol.* **2015**, *283*, 75–82. [CrossRef] [PubMed]
16. Ju, N.; Hayashi, H.; Shimamura, M.; Baba, S.; Yoshida, S.; Morishita, R.; Rakugi, H.; Nakagami, H. Prevention of bleomycin-induced pulmonary fibrosis by a RANKL peptide in mice. *Sci. Rep.* **2022**, *12*, 12474. [CrossRef]
17. Braun, N.; Fritz, P.; Ulmer, C.; Latus, J.; Kimmel, M.; Biegger, D.; Ott, G.; Reimold, F.; Thon, K.P.; Dippon, J.; et al. Histological criteria for encapsulating peritoneal sclerosis—A standardized approach. *PLoS ONE* **2012**, *7*, e48647. [CrossRef]
18. Marchant, V.; Tejera-Munoz, A.; Marquez-Exposito, L.; Rayego-Mateos, S.; Rodrigues-Diez, R.R.; Tejedor, L.; Santos-Sanchez, L.; Egido, J.; Ortiz, A.; Valdivielso, J.M.; et al. IL-17A as a Potential Therapeutic Target for Patients on Peritoneal Dialysis. *Biomolecules* **2020**, *10*, 1361. [CrossRef]
19. Si, M.; Wang, Q.; Li, Y.; Lin, H.; Luo, D.; Zhao, W.; Dou, X.; Liu, J.; Zhang, H.; Huang, Y.; et al. Inhibition of hyperglycolysis in mesothelial cells prevents peritoneal fibrosis. *Sci. Transl. Med.* **2019**, *11*, eaav5341. [CrossRef]
20. Pein, M.; Insua-Rodriguez, J.; Hongu, T.; Riedel, A.; Meier, J.; Wiedmann, L.; Decker, K.; Essers, M.A.G.; Sinn, H.P.; Spaich, S.; et al. Metastasis-initiating cells induce and exploit a fibroblast niche to fuel malignant colonization of the lungs. *Nat. Commun.* **2020**, *11*, 1494. [CrossRef]
21. Szentes, V.; Gazdag, M.; Szokodi, I.; Dezsi, C.A. The Role of CXCR3 and Associated Chemokines in the Development of Atherosclerosis and during Myocardial Infarction. *Front. Immunol.* **2018**, *9*, 1932. [CrossRef] [PubMed]
22. Boniakowski, A.E.; Kimball, A.S.; Joshi, A.; Schaller, M.; Davis, F.M.; denDekker, A.; Obi, A.T.; Moore, B.B.; Kunkel, S.L.; Gallagher, K.A. Murine macrophage chemokine receptor CCR2 plays a crucial role in macrophage recruitment and regulated inflammation in wound healing. *Eur. J. Immunol.* **2018**, *48*, 1445–1455. [CrossRef] [PubMed]
23. Xuan, Y.; Chen, C.; Wen, Z.; Wang, D.W. The Roles of Cardiac Fibroblasts and Endothelial Cells in Myocarditis. *Front. Cardiovasc. Med.* **2022**, *9*, 882027. [CrossRef] [PubMed]
24. Liu, S.; Liu, C.; Wang, Q.; Liu, S.; Min, J. CC Chemokines in Idiopathic Pulmonary Fibrosis: Pathogenic Role and Therapeutic Potential. *Biomolecules* **2023**, *13*, 333. [CrossRef] [PubMed]
25. Unver, N. Macrophage chemoattractants secreted by cancer cells: Sculptors of the tumor microenvironment and another crucial piece of the cancer secretome as a therapeutic target. *Cytokine Growth Factor. Rev.* **2019**, *50*, 13–18. [CrossRef] [PubMed]
26. Vatrella, A.; Maglio, A.; Pelaia, C.; Ciampo, L.; Pelaia, G.; Vitale, C. Eosinophilic inflammation: An Appealing Target for Pharmacologic Treatments in Severe Asthma. *Biomedicines* **2022**, *10*, 2181. [CrossRef]
27. Kubota, A.; Frangogiannis, N.G. Macrophages in myocardial infarction. *Am. J. Physiol. Cell Physiol.* **2022**, *323*, C1304–C1324. [CrossRef]
28. Lenga, Y.; Koh, A.; Perera, A.S.; McCulloch, C.A.; Sodek, J.; Zohar, R. Osteopontin expression is required for myofibroblast differentiation. *Circ. Res.* **2008**, *102*, 319–327. [CrossRef]
29. Zhang, D.; Liu, B.; Cao, B.; Wei, F.; Yu, X.; Li, G.F.; Chen, H.; Wei, L.Q.; Wang, P.L. Synergistic protection of Schizandrin B and Glycyrrhizic acid against bleomycin-induced pulmonary fibrosis by inhibiting TGF-beta1/Smad2 pathways and overexpression of NOX4. *Int. Immunopharmacol.* **2017**, *48*, 67–75. [CrossRef]
30. Ceccato, T.L.; Starbuck, R.B.; Hall, J.K.; Walker, C.J.; Brown, T.E.; Killgore, J.P.; Anseth, K.S.; Leinwand, L.A. Defining the Cardiac Fibroblast Secretome in a Fibrotic Microenvironment. *J. Am. Heart Assoc.* **2020**, *9*, e017025. [CrossRef]
31. Gibb, A.A.; Lazaropoulos, M.P.; Elrod, J.W. Myofibroblasts and Fibrosis: Mitochondrial and Metabolic Control of Cellular Differentiation. *Circ. Res.* **2020**, *127*, 427–447. [CrossRef] [PubMed]
32. Ruan, H.; Gao, S.; Li, S.; Luan, J.; Jiang, Q.; Li, X.; Yin, H.; Zhou, H.; Yang, C. Deglycosylated Azithromycin Attenuates Bleomycin-Induced Pulmonary Fibrosis via the TGF-beta1 Signaling Pathway. *Molecules* **2021**, *26*, 2820. [CrossRef] [PubMed]
33. Anwar, M.A.; Basith, S.; Choi, S. Negative regulatory approaches to the attenuation of Toll-like receptor signaling. *Exp. Mol. Med.* **2013**, *45*, e11. [CrossRef] [PubMed]
34. Mishima, K.; Kitoh, H.; Ohkawara, B.; Okuno, T.; Ito, M.; Masuda, A.; Ishiguro, N.; Ohno, K. Lansoprazole Upregulates Polyubiquitination of the TNF Receptor-Associated Factor 6 and Facilitates Runx2-mediated Osteoblastogenesis. *EBioMedicine* **2015**, *2*, 2046–2061. [CrossRef] [PubMed]
35. Aguilera, A.; Yanez-Mo, M.; Selgas, R.; Sanchez-Madrid, F.; Lopez-Cabrera, M. Epithelial to mesenchymal transition as a triggering factor of peritoneal membrane fibrosis and angiogenesis in peritoneal dialysis patients. *Curr. Opin. Investig. Drugs* **2005**, *6*, 262–268.
36. Wilson, R.B.; Archid, R.; Reymond, M.A. Reprogramming of Mesothelial-Mesenchymal Transition in Chronic Peritoneal Diseases by Estrogen Receptor Modulation and TGF-beta1 Inhibition. *Int. J. Mol. Sci.* **2020**, *21*, 4158. [CrossRef]
37. Huang, J.; Tong, X.; Zhang, L.; Zhang, Y.; Wang, L.; Wang, D.; Zhang, S.; Fan, H. Hyperoside Attenuates Bleomycin-Induced Pulmonary Fibrosis Development in Mice. *Front. Pharmacol.* **2020**, *11*, 550955. [CrossRef]
38. Xiong, C.; Liu, N.; Shao, X.; Sharif, S.; Zou, H.; Zhuang, S. Delayed administration of suramin attenuates peritoneal fibrosis in rats. *BMC Nephrol.* **2019**, *20*, 411. [CrossRef]

Disclaimer/Publisher's Note: The statements, opinions and data contained in all publications are solely those of the individual author(s) and contributor(s) and not of MDPI and/or the editor(s). MDPI and/or the editor(s) disclaim responsibility for any injury to people or property resulting from any ideas, methods, instructions or products referred to in the content.

Article

The Role of STAT3 Signaling Pathway Activation in Subconjunctival Scar Formation after Glaucoma Filtration Surgery

Yanxia Li, Jing Zhao, Yuan Yin, Chenchen Zhang, Zhaoying Zhang and Yajuan Zheng *

Department of Ophthalmology, The Second Hospital of Jilin University, Jilin University, Changchun 130041, China; yanxia20@mails.jlu.edu.cn (Y.L.); lhbswqw@jlu.edu.cn (J.Z.); yinyuan215@jlu.edu.cn (Y.Y.); cczhang21@mails.jlu.edu.cn (C.Z.); zhaoyingz22@mails.jlu.edu.cn (Z.Z.)
* Correspondence: yjzheng@jlu.edu.cn

Abstract: Scar formation resulting from overly active wound healing is a critical factor in the success rate of glaucoma filtration surgery (GFS). IL-6 and TGF-β have been implicated in the pathogenesis of fibrogenesis. In addition, the signal transducer and activator of transcription 3 (STAT3) can be activated by numerous cytokines and growth factors, including IL-6 and TGF-β1. Thus, STAT3 activation may integrate common profibrotic pathways to promote fibrosis. In this study, an increase in p-STAT3 was observed in activated HTFs. Inhibiting STAT3 in cultured HTFs by pharmacological inactivation reversed the fibrotic responses, such as fibroblast migration, the differentiation of resting fibroblasts into myofibroblasts and the deposition of ECM, mediated by IL-6 and TGF-β1. Moreover, the expression of suppressor of cytokine signaling 3 (SOCS3) was decreased in HTFs cultured with IL-6 and TGF-β1, and SOCS3 overexpression rescued ECM deposition, α-SMA expression and migration in IL-6- and TGF-β1-stimulated HTFs by inactivating STAT3. Finally, S3I-201 treatment inhibited profibrotic gene expression and subconjunctival fibrosis in a rat model of GFS. In conclusion, our data suggests that STAT3 plays a central role in fibrosis induced by different profibrotic pathways and that STAT3 is a potential target for antifibrotic therapies following GFS.

Keywords: signal transducer and activator of transcription 3 (STAT3); glaucoma filtration surgery; fibrosis; suppressor of cytokine signaling 3 (SOCS3)

Citation: Li, Y.; Zhao, J.; Yin, Y.; Zhang, C.; Zhang, Z.; Zheng, Y. The Role of STAT3 Signaling Pathway Activation in Subconjunctival Scar Formation after Glaucoma Filtration Surgery. *Int. J. Mol. Sci.* **2023**, *24*, 12210. https://doi.org/10.3390/ijms241512210

Academic Editors: Margherita Sisto and Sabrina Lisi

Received: 6 June 2023
Revised: 26 July 2023
Accepted: 26 July 2023
Published: 30 July 2023

Copyright: © 2023 by the authors. Licensee MDPI, Basel, Switzerland. This article is an open access article distributed under the terms and conditions of the Creative Commons Attribution (CC BY) license (https://creativecommons.org/licenses/by/4.0/).

1. Introduction

Glaucoma is a chronic progressive optic neuropathy and a major cause of blindness worldwide. It is anticipated that 111.8 million patients will suffer from glaucoma by 2040 [1]. An increase in intraocular pressure (IOP) is a major risk factor for glaucoma and permanent loss of vision [2]. Currently, lowering IOP through medical and surgical methods remains the only evidence-based treatment available [3]. Glaucoma filtration surgery (GFS), which is a surgical procedure that involves the guidance of aqueous humor into the space between the sclera and conjunctiva, is widely used to lower IOP. However, fibrotic responses such as aberrant fibroblast proliferation, myofibroblast differentiation, and excessive deposition of extracellular matrix (ECM) induced by GFS-induced injury in the subconjunctival area could result in surgical failure [4]. Although antimetabolic agents, such as mitomycin C (MMC), have been commonly used to inhibit scar formation at the filtering site, some patients continue to have poor prognoses after GFS [5]. Additionally, current antifibrotic therapies also have large side-effect profiles, including endophthalmitis, cornea toxicity, blebitis and wound leakage [6–8]. Thus, better antifibrotic therapies that can improve the efficacy and safety of GFS are needed to avoid blindness due to glaucoma.

Transforming growth factor β (TGF-β) plays a central role in modulating fibrosis by stimulating fibroblasts and the differentiation of myofibroblasts. Additionally, interleukin-6 (IL-6), which is a multifunctional cytokine involved in inflammatory responses, was shown to promote fibrosis in previous studies [9–11]. IL-6 signal transduction involves the classic signaling pathway and trans-signaling pathway, which are mediated by membrane-bound IL-6R (mIL-6R) and soluble IL-6R (sIL-6R), respectively [12–14]. Although mIL-6R is expressed only on specific cell types, IL-6 can transduce its signal to nearly all cells through sIL-6R. In response to IL-6 stimulation, signal transducer and activator of transcription 3 (STAT3) proteins are recruited and phosphorylated, and then the homodimerized STAT3 translocates to the nucleus, where it triggers the transcription of various target genes, including cellular differentiation, apoptosis, angiogenic factors and cytokines [15,16]. Interestingly, there is substantial evidence that TGF-β can also activate STAT3 [17,18], suggesting that STAT3 activation may integrate common profibrotic pathways to promote fibrosis.

As a result, there is considerable interest in determining whether STAT3 inhibition alleviates subconjunctival fibrosis following GFS. The purpose of this study was to examine the effect of inhibiting STAT3 on the development of subconjunctival fibrosis in two complementary fibrosis models of human tenon fibroblasts (HTFs): one involving inflammation-dependent pathways (induced by IL-6) and the other involving fibroblast activation devoid of inflammation (induced by TGF-β1). S3I-201, a selective small molecule inhibitor of STAT3 [19], was used to test the efficacy of STAT3 inhibition on the prevention of GFS fibrosis in vivo and in vitro. Our findings provide the first evidence that pharmacological inhibition of STAT3 by S3I-201 prevents scar formation following GFS.

2. Results

2.1. Phosphorylated STAT3 Is Increased in Activated Fibroblasts Induced by IL-6 and TGF-β1

HTFs, which are key cells in subconjunctival fibrosis, are typically used to study fibrosis associated with GFS in vitro [20]. Activation of HTFs induced by various cytokines leads these cells to subsequently reenter the cell cycle, migrate and transform into myofibroblasts [21]. Myofibroblasts are characterized by the expression of α-SMA and produce collagen I and fibronectin, which are essential for ECM deposition [22]. Consistent with previous studies, our results demonstrated that the expression of collagen I, fibronectin and α-SMA was increased in response to IL-6/sIL-6R and TGF-β1 (Figure 1a–d). We further evaluated the level of phosphorylated STAT3 in the two complementary fibrosis models of HTFs: one involving inflammation-dependent pathways (induced by IL-6) and the other involving fibroblast activation devoid of inflammation (induced by TGF-β1). The results showed that the level of phosphorylated STAT3 was increased in IL-6- and TGF-β1-stimulated HTFs (Figure 1e–h). Immunofluorescence staining demonstrated nuclear localization of p-STAT3 and further confirmed the activation of STAT3 signaling in IL-6- and TGF-β1-stimulated fibroblasts (Figure 1e,f). Given the consistent activation of STAT3 in cultured HTFs induced by IL-6 and TGF-β1, we hypothesized that STAT3 was a core pathway for fibrogenesis induced by various cytokines.

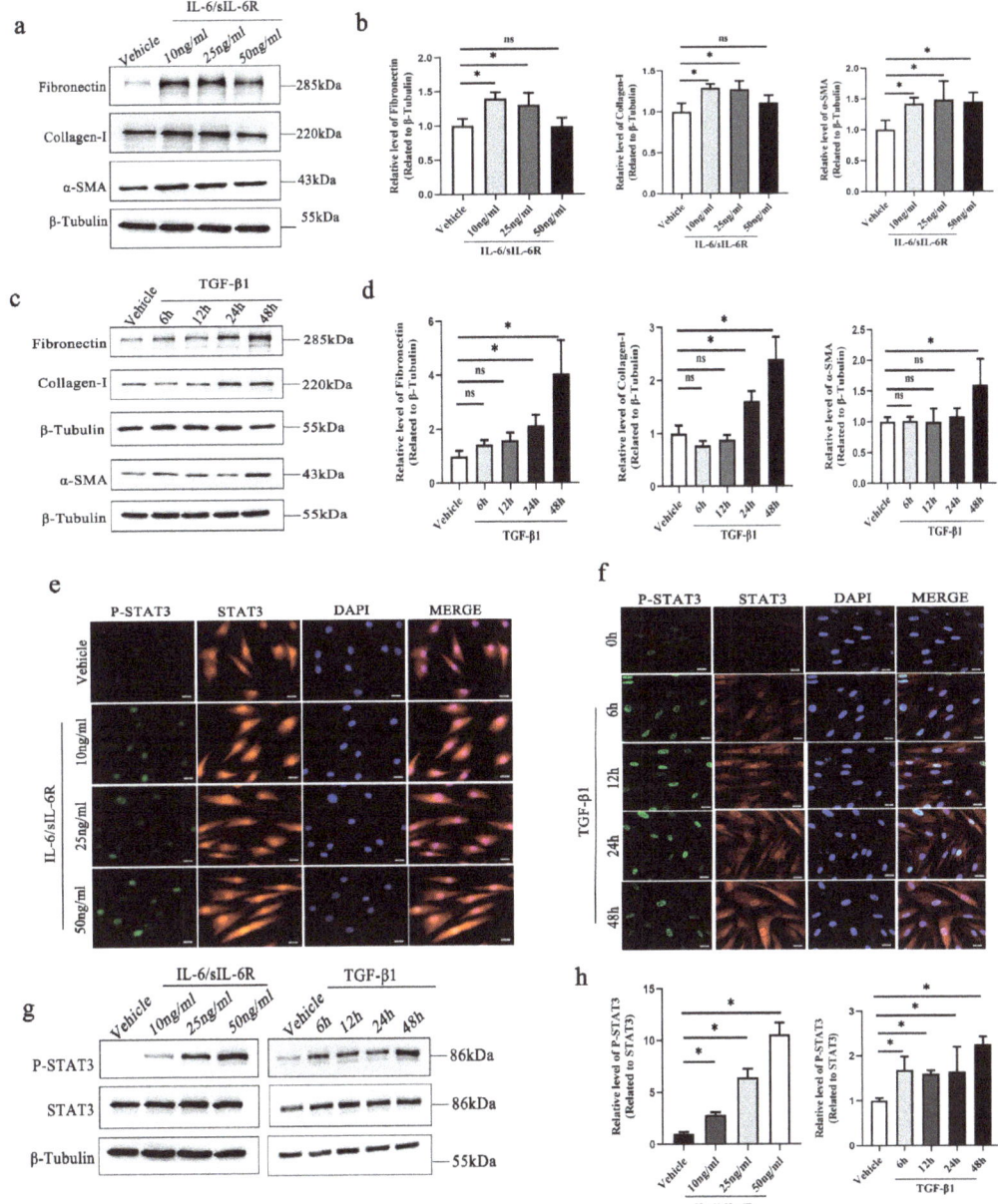

Figure 1. Increased expression of p-STAT3 in activated fibroblasts induced by IL-6 and TGF-β1. Western blot and quantitative analysis of the levels of fibronectin, collagen-I and α-SMA in fibroblasts after IL-6/sIL-6R (**a**,**b**) or TGF-β1 (**c**,**d**) stimulation. Immunofluorescence analysis (scale bar, 20 μm) of p-STAT3 (green) and total STAT3 (red) expression in cultured fibroblasts stimulated by IL-6/sIL-6R (**e**) or TGF-β1 (**f**). (**g**) Western blot and (**h**) quantitative analysis of p-STAT3 and total STAT3 in fibroblasts treated with TGF-β1 or IL-6/sIL-6R. * $p < 0.05$, ns $p > 0.05$. Significance was determined by one-way ANOVA.

2.2. STAT3 Regulates IL-6– and TGF-β1–Mediated Fibroblast Activation

S3I-201, which is a small molecule inhibitor, can block STAT3 phosphorylation and STAT3 DNA binding by binding to the STAT3–SH2 domain. To further clarify the role of STAT3 in the fibrotic process induced by IL-6 and TGF-β1, we used S3I-201 to target STAT3 signaling in cultured HTFs treated with TGF-β1 or the IL-6/sIL-6R complex. Here, we found that collagen I, fibronectin and α-SMA levels in HTFs treated with S3I-201 were markedly reduced in both complementary fibrosis models. (Figure 2a,b). In addition, another well-known mechanism of subconjunctival fibrogenesis involves fibroblast migration from the periphery of the surgery site and proliferation in the resident location. In the present study, the chemotactic migration of fibroblasts was detected using wound healing assays and transwell chamber assays. We found that HTFs treated with S3I-201 exhibited less migration than those treated with IL-6 and TGF-β1 (Figure 2c,d). These data indicate that STAT3 signaling plays a central role in regulating IL-6–mediated and TGF-β1–mediated fibroblast activation, suggesting that STAT3 activation may integrate common profibrotic pathways to promote fibrosis.

2.3. SOCS3 Overexpression Suppresses STAT3 Activation and the Fibrotic Response Mediated by TGF-β1 and IL-6

Previous studies have shown that STAT3 plays a key role in the fibrotic response of HTFs. Suppressors of cytokine signaling 3 (SOCS3) can bind to JAKs to inhibit their kinase activity, preventing subsequent activation of STAT3 and the transcription of STAT3-dependent target genes [23]. We tested the hypothesis that SOCS3 may negatively regulate fibroblast activation by inhibiting STAT3 signaling induced by TGF-β1 or IL-6. The results showed that the expression of SOCS3 was downregulated in cultured HTFs treated with TGF-β1 and the IL-6/sIL-6R complex (Figure 3a,b). To further confirm the role of SOCS3 in the fibrotic process, SOCS3 was overexpressed in HTFs through lentivirus infection. We found that the expression of p-STAT3 was decreased, and the activated phenotype of fibroblasts was completely rescued. More specifically, SOCS3 overexpression significantly reversed the protein expression of collagen, fibronectin and α-SMA in HTFs treated with TGF-β1 or the IL-6/sIL-6R complex (Figure 3e,f). Furthermore, the effect of TGF-β1 or IL-6 on the migration of HTFs cells was prevented by SOCS3 overexpression (Figure 3c,d).

2.4. S3I-201 Is Safe for Subconjunctival Injection in Rat Eyes

MMC, which is a cytotoxic antiproliferative agent, is typically used to reduce scarring following GFS [24]. However, the failure rate is still approximately 50% at 5 years [25]. Moreover, the potential risk of complications such as endophthalmitis and corneoscleral toxic effects limits the long-term use of these agents [26]. We sought to determine the potential safety of S3I-201 in treating subconjunctival scar formation in a rat model of GFS. Rats with GFS were randomly assigned to three groups, and each group of animals was treated with S3I-201, MMC, or DMSO (designated the vehicle group). The treatment scheme is shown in Figure 4a. A TUNEL assay was used to evaluate the toxicity of S3I-201 and MMC to rat eyes after GFS. As expected, many TUNEL-positive cells in the subconjunctival area were observed in the MMC group. In contrast, few conjunctival TUNEL-positive cells were found in the S3I-201 group (Figure 4b). Based on this observation, we concluded that S3I-201 had almost no toxic effects on the normal ocular tissues of rats.

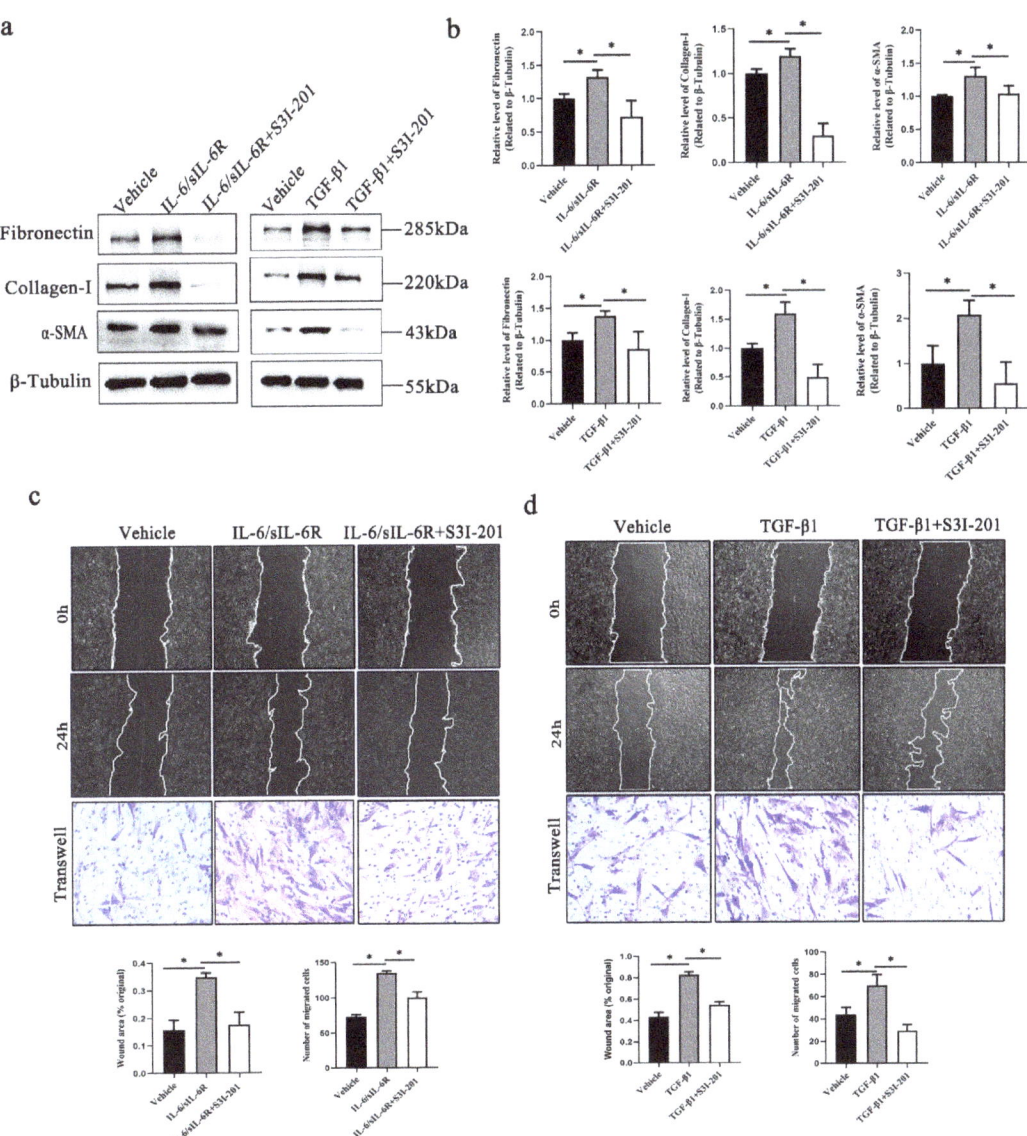

Figure 2. S3I-201 inhibits fibrogenesis in both IL-6/sIL-6R- and TGF-β1-cultured HTFs. (**a**) Western blot and (**b**) quantitative analysis of the expression of fibronectin, collagen-I and α-SMA in IL-6/sIL-6R- or TGF-β1-cultured fibroblasts treated with S3I-201. Wound healing (scale bar, 200 μm) and transwell chamber analysis (scale bar, 50 μm) of migration in (**c**) IL-6/sIL-6R- or (**d**) TGF-β1-cultured fibroblasts treated with S3I-201. * $p < 0.05$. Significance was determined by one-way ANOVA.

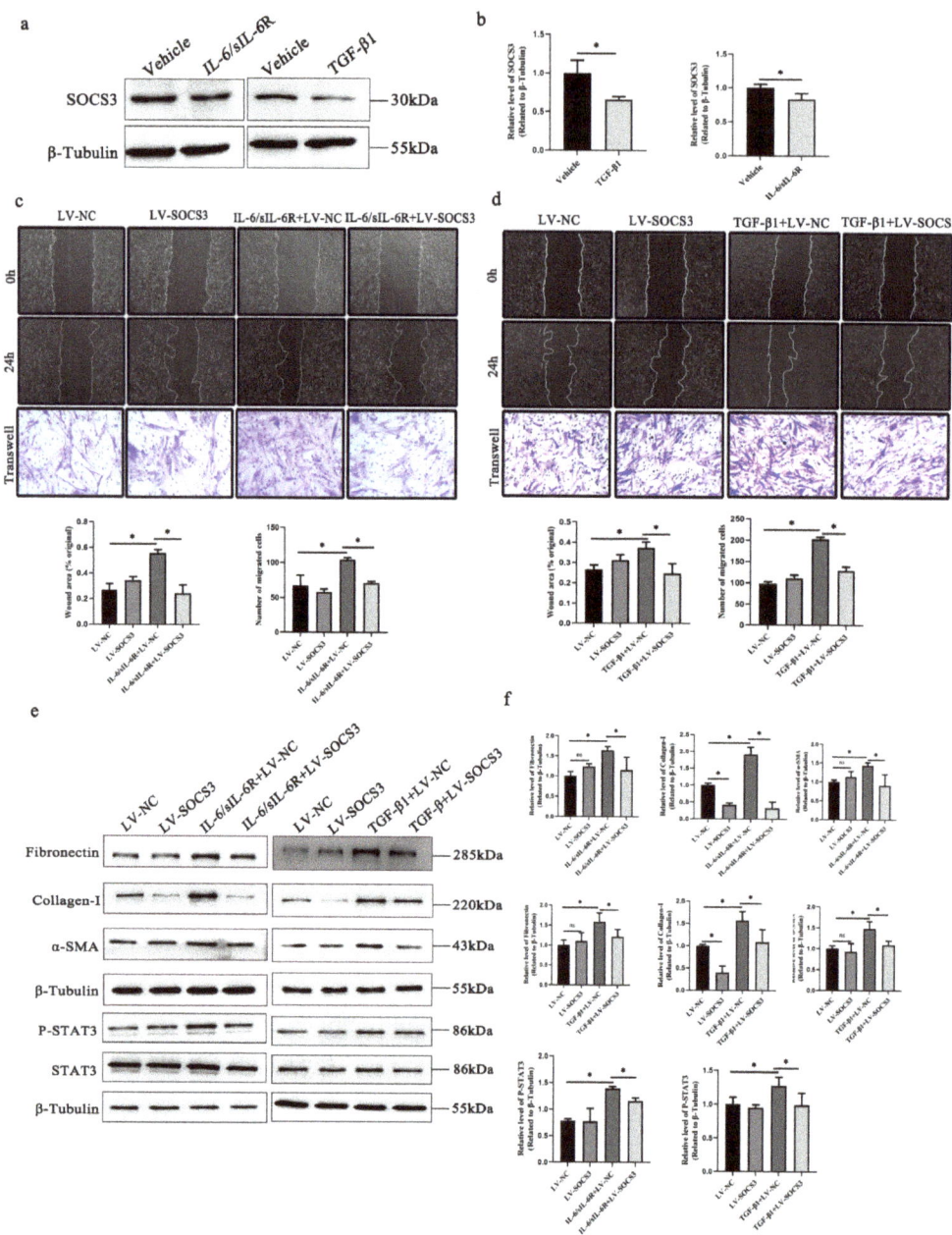

Figure 3. SOCS3 overexpression inhibits STAT3 activation and fibrogenesis in IL-6/sIL-6R- and TGF-β1-cultured HTFs. (**a**) Western blot and (**b**) quantitative analysis of the expression of SOCS3 in IL-6/sIL-6R- or TGF-β1-cultured fibroblasts. Wound healing (scale bar, 200 μm) and transwell chamber analysis (scale bar, 50 μm) of migration in (**c**) IL-6/sIL-6R- or (**d**) TGF-β1-cultured overexpressed SOCS3 fibroblasts. (**e**) Western blot and (**f**) quantitative analysis of the expression of fibronectin, collagen-I, α-SMA, P-STAT3 and total STAT3 in IL-6/sIL-6R- or TGF-β1-cultured overexpressed SOCS3 fibroblasts. * $p < 0.05$, ns $p > 0.05$. Significance was determined by one-way ANOVA.

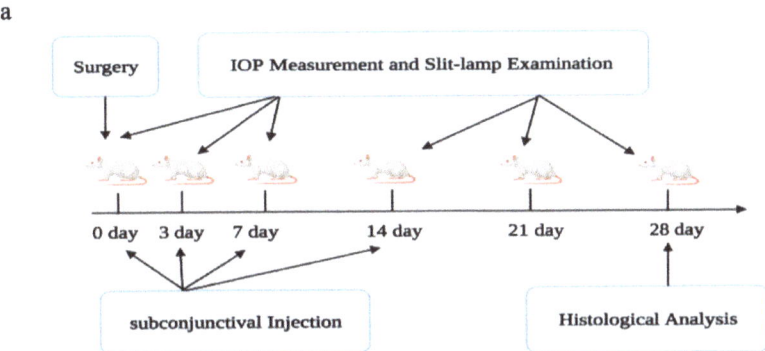

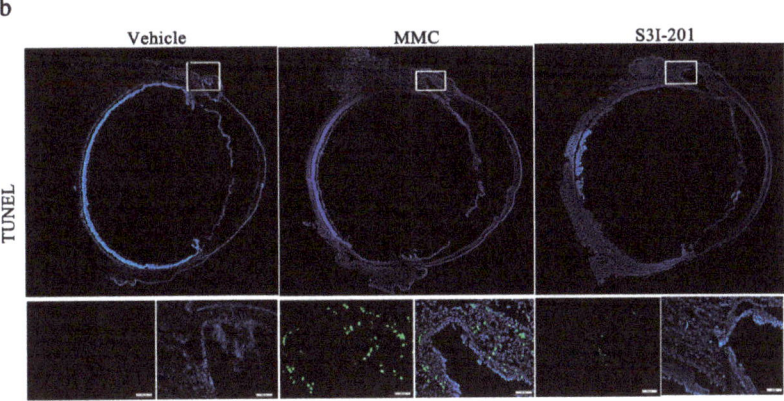

Figure 4. Assessment of the toxicity of S3I-201 in rat eyes. (**a**) Experimental scheme for the rats after GFS. (**b**) Representative images of TUNEL staining (upper panel scale bar, 200 μm; lower panel scale bar, 20 μm) of the subconjunctival tissues of surrounding the filtering areas.

2.5. S3I-201 Treatment Ameliorates Rebound IOP Elevation and Prolongs Filtering Bleb Survival after GFS

The obstruction of aqueous humor outflow resulting from subconjunctival fibrosis following GFS leads to rebound IOP elevation, which is the main factor responsible for the failure of filtering surgery. To ascertain the potential effect of S3I-201 treatment on subconjunctival scar formation, we first evaluated the effect of S3I-201 treatment on the stabilization of the reduction in IOP in a rat model of GFS. After GFS, a drastic decrease in IOP was observed on postoperative day 3 in animals in all three surgical groups (vehicle 13.7 ± 1.1 mmHg, MMC 14.2 ± 1.2 mmHg, S3I-201 15.7 ± 1.1 mmHg). Animals in the vehicle group showed a complete rebound IOP elevation to a level equivalent to the baseline on days 21 to 28. Animals in the S3I-201- and MMC-treated groups also exhibited a rebound IOP elevation. However, the rebound occurred at a slower rate than that in the vehicle group. S3I-201-treated rats showed the slowest rate of rebound IOP elevation during the 28 days after GFS (Figure 5c).

Maintenance of a functional filtering bleb is crucial for IOP control following GFS. Thus, we evaluated the effect of S3I-201 treatment on the survival of filtering blebs following GFS. Bleb survival analysis demonstrated a significant difference in the survival distribution among the vehicle, MMC, and S3I-201 groups ($p < 0.05$). Rats in the vehicle group exhibited rapid loss of filtering blebs after GFS, and the filtering blebs were completely lost in this group on day 28. Compared with the vehicle group, filtering blebs in the MMC and S3I-201 groups survived a significantly longer period of time. Loss of filtering blebs in the MMC group started on day 14, and the survival rate of filtering blebs was 50% on day 28. The loss of blebs in the S3I-201 group occurred on day 21, and the survival rate of filtering blebs was 62.5% on day 28. S3I-201 and MMC treatment significantly improved the survival of blebs compared with the vehicle group (Figure 5a,b). Collectively, these data suggest that S3I-201 treatment benefits the stabilization of IOP reduction and the survival of filtering blebs after GFS.

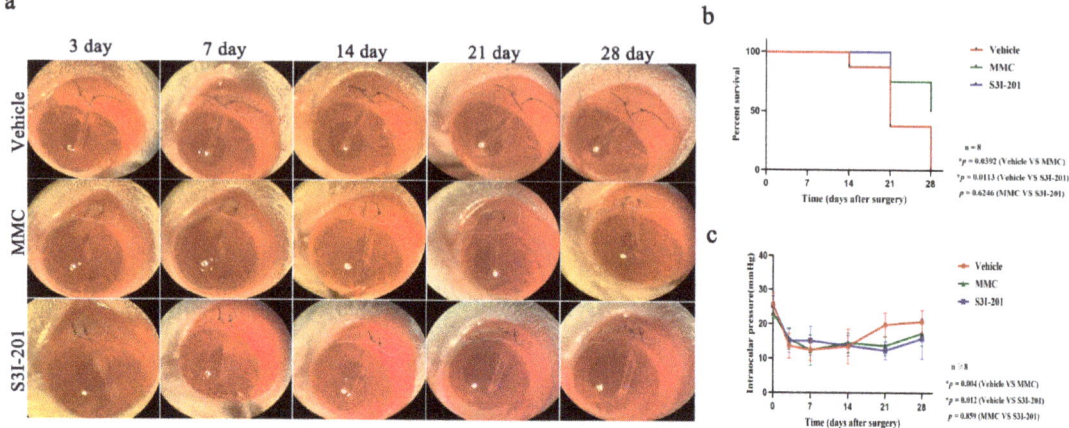

Figure 5. S3I-201 promoted functional bleb formation and maintained the decrease in IOP for a period in rats after GFS. (**a**) Representative morphological images of filtering blebs at different time points in the vehicle, MMC and SI-201 treatment groups; bleb area (white line). (**b**) Kaplan–Meier survival curve of filtering blebs in the vehicle, MMC and SI-201 treatment groups (log rank test; overall $p < 0.05$, vehicle vs. MMC $p < 0.05$, vehicle vs. S3I-201 $p < 0.05$, * $p < 0.05$). (**c**) IOPs of the eyes at different time points in the vehicle, MMC and SI-201 treatment groups (repeated measure ANOVA; vehicle vs. MMC $p < 0.05$, vehicle vs. S3I-201 $p < 0.05$, * $p < 0.05$).

2.6. S3I-201 Treatment Alleviates Subconjunctival Fibrosis following GFS

To observe the pathological changes in the subconjunctival tissues following GFS, HE and Masson staining were performed. HE staining showed that the subconjunctival tissue in the vehicle group was significantly thickened with marked hyperplasia of fibrous connective tissue, inflammatory cells and fibroblasts were of a high density and grew in clumps. Compared with the control group (no surgery), Masson staining showed that collagen deposition was increased in the vehicle group. In the MMC group, the subconjunctival fibrous layer was looser and thinner, forming cavities, with few cells and reduced collagen deposition. In the S3I-201 group, the structure of fibrous connective tissue was loose, with the formation of voids, few cells and reduced collagen deposition, as observed by HE and Masson staining (Figure 6a,b).

To further investigate the subconjunctival fibrotic response to surgical injury induced by GFS, immunofluorescence was used to analyze the expression of collagen-I and fibronectin in the bleb tissues of rat eyes. Immunofluorescence analysis showed that the expression of collagen-I and fibronectin was significantly increased after GFS in the vehicle group, indicating that GFS activated the expression of profibrotic genes. MMC and S3I-201 treatment significantly attenuated surgical injury-induced expression of these genes, suggesting that similar to MMC, S3I-201 inhibits subconjunctival scar formation after GFS. Consistent with the Masson staining results, immunofluorescence staining demonstrated less collagen deposition in the subconjunctival region after GFS in the S3I-201 and MMC groups than in the vehicle group (Figure 6c).

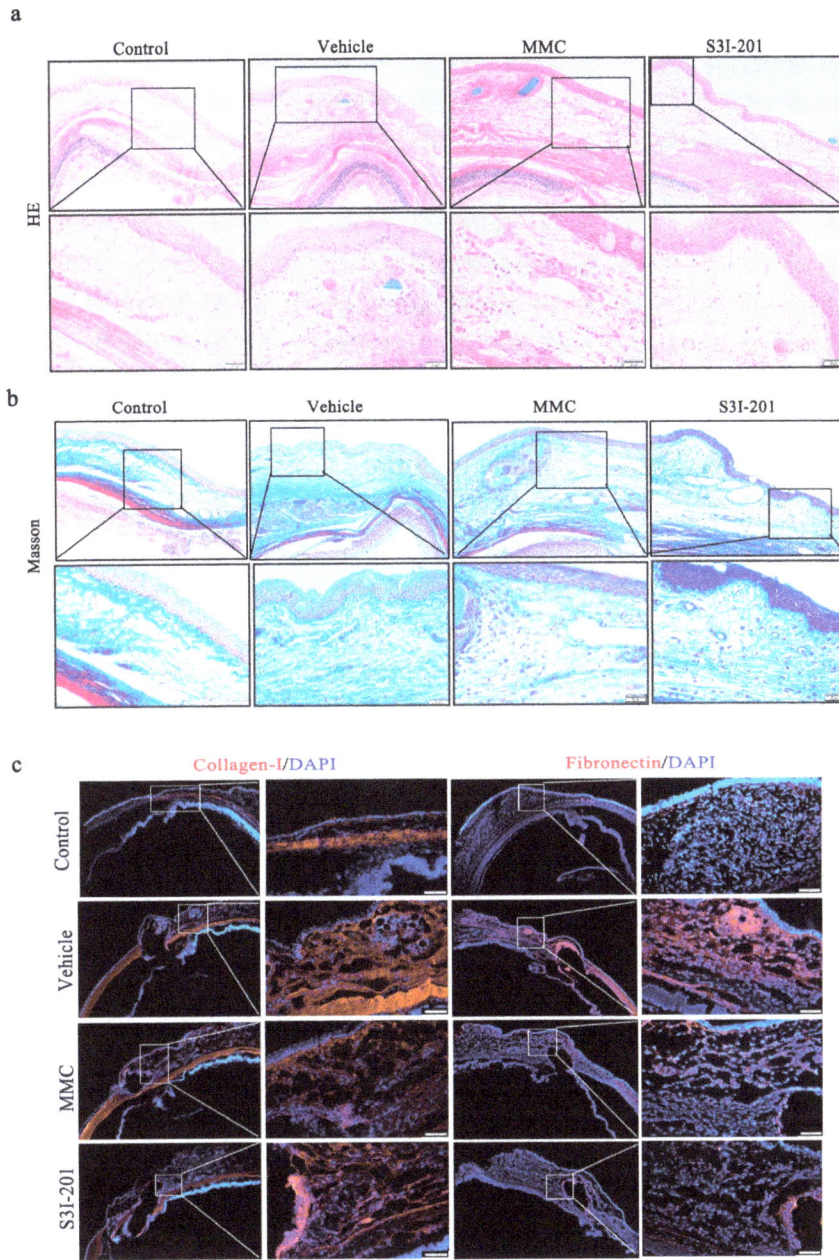

Figure 6. S3I-201 attenuated subconjunctival fibrosis in rats after GFS. (**a**) Representative HE staining images (upper panel scale bar, 100 μm; lower panel scale bar, 20 μm) and (**b**) Masson's trichrome staining images (upper panel scale bar, 100 μm; lower panel scale bar, 20 μm) of the operative area. (**c**) Representative immunofluorescent images (left panel scale bar, 200 μm; right panel scale bar, 20 μm) for specific markers, including collagen-I, fibronectin (red) and DAPI (blue), in the subconjunctival tissues of surrounding filtering areas.

3. Discussion

In the current study, we identified STAT3 as a key intracellular mediator of profibrotic effects. Inhibiting STAT3 in cultured HTFs by pharmacological inactivation or siRNA prevents fibroblast proliferation and migration and the differentiation of resting fibroblasts into myofibroblasts and significantly reduces the deposition of ECM. It was also demonstrated that STAT3 signaling plays a central role in regulating IL-6–and TGF-β–mediated fibroblast activation, suggesting that STAT3 activation may integrate common profibrotic pathways to promote fibrosis. Finally, we provided in vivo evidence that pharmacological inactivation of STAT3 inhibits profibrotic gene expression and subconjunctival fibrosis in a rat model of GFS. Together, these data demonstrate that STAT3 is a core pathway for fibrogenesis.

The current study used in vitro fibrosis models induced by IL-6 and TGF-β1 to advance our understanding of how STAT3 may regulate the development of fibrosis. These models allow for the investigation of different phases or pathways involved in wound healing. The acute wound healing process after trauma or surgery, including GFS, is traditionally divided into four overlapping phases: hemostasis, inflammation, proliferation, and remodeling [27]. In the inflammatory phase, recruited neutrophils and macrophages secrete proinflammatory cytokines, such as IL-6, into wound sites [28,29]. Subsequently, the proliferative phase of wound healing is initiated by an influx of fibroblasts and is characterized by the transdifferentiation of fibroblasts into myofibroblasts, which is driven by TGF-β [30]. In vitro studies with HTFs clearly demonstrate that STAT3 activation can regulate the fibrotic phenotype of HTFs triggered by IL-6 or TGF-β1. These data suggested that inhibiting STAT3 decreases fibrosis in the inflammatory phase and proliferative phase during the wound healing process.

STAT3 signaling is strictly regulated by the suppressor of cytokine signaling 3 (SOCS3) proteins [31]. Hence, we hypothesized that fibroblast activation induced by TGF-β1 or IL-6 may be associated with the suppression of SOCS3. Our results showed that the expression of SOCS3 was downregulated in HTFs treated with IL-6 and TGF-β1. To further confirm that SOCS3 participates in fibrosis mediated by STAT3 activation, SOCS3 was overexpressed in HTFs. Indeed, we found that SOCS3 overexpression significantly rescued the fibrotic responses induced by IL-6 and TGF-β1 due to the inactivation of STAT3. These data suggested that SOCS3 may play a suppressive role in fibrosis promoted by STAT3.

S3I-201 is a selective and potent inhibitor that binds to the SH2 domain of STAT3 to block the constitutive activation of STAT3 and the expression of STAT3 target genes [32]. S3I-201 inhibits transcriptional activity in cells that contain constitutive STAT3 activation by blocking STAT3–STAT3 complex formation and STAT3 DNA binding. However, other commonly used STAT3 inhibitors, such as AG490, act on Janus kinases, which are upstream activators of STAT3, and other STAT isoforms [33,34]. Unlike STAT3, recent studies using knockout mice showed that STAT1 and STAT5 play key roles in promoting liver and lung fibrosis after injury [35,36]. Thus, S3I-201 was used to treat GFS fibrosis in our study because its characteristics distinguished it from other STAT3 inhibitors.

Fibrosis in filtering area prevention is critical for the success of GFS. In this study, S3I-201 significantly reduced the IOP and prolonged bleb survival compared with the vehicle. Histologic examination showed that the numbers of inflammatory cells and myofibroblasts were reduced in the S3I-201 group compared with the vehicle group. Collagen deposition was also reduced in the S3I-201 group. Additionally, only a few TUNEL-positive cells were observed in the conjunctiva of the S3I-201-treated group; conversely, many TUNEL-positive cells were observed in the MMC group due to significant toxicity. These results indicate that S3I-201 was safe as an antiproliferative medication for use in the GFS rat model. The in vivo and in vitro results strongly suggest that STAT3 is a potential target in fibrotic responses after GFS.

However, importantly, this study has its own limitations. First, the optimal dosage, concentration, and application method of S3I-201 are needed to further improve the surgical outcome after GFS. Second, the sample size was small, and the follow-up time was short. More samples and longer follow-ups are needed in future studies. Finally, the inhibitory effect of S3I-201 on GFS was not superior to MMC. However, S3I-201 was observed to have a less toxic effect on eye tissues than MMC. We can expect that the use of S3I-201 can improve the safety profile of surgery compared with MMC. Additionally, S3I-201 combined with MMC may further inhibit the fibrotic response after GFS and lower the dose and exposure time of MMC.

In conclusion, STAT3 integrates profibrotic signals from different mechanisms, suggesting that STAT3 may be a promising target for antifibrotic therapies following GFS. Indeed, the findings of the current study demonstrate that inactivating STAT3 signaling exerts potent antifibrotic effects on GFS-induced fibrosis in vitro and in vivo. Furthermore, STAT3 inactivation inhibits IL-6-induced fibrosis in the early inflammatory phases of wound healing but also exerts potent antifibrotic effects on TGF-β-induced fibrosis in the late noninflammatory stages of wound healing. Thus, targeting STAT3 may be beneficial for different wound healing stages in subconjunctival fibrosis following GFS.

4. Materials and Methods

4.1. The Preparation of Tissues and Cells

The human subconjunctival Tenon's capsules that were used to isolate HTFs were obtained from patients undergoing strabotomy. Under sterile conditions, the tissues were cut into 3 to 4 mm pieces. The samples were then placed into a fibroblast medium (FM) (ScienCell, San Diego, CA, USA) containing 10% fetal bovine serum (FBS), 100 U/mL penicillin, 100 μg/mL streptomycin and fibroblast growth supplement. Once the growth of the HTFs was well established, the cells were passaged in a monolayer after being treated with trypsin/EDTA. Cells that were between the third and fifth passages were used in the studies. The study followed the principles outlined in the Helsinki Declaration and was examined and approved by the Second Hospital of Jilin University's Institutional Review Board.

4.2. Cell Stimulation

HTFs were stimulated with recombinant TGF-β1 (R&D Systems, Minneapolis, MN, USA), recombinant IL-6 (R&D Systems, Minneapolis, MN, USA) and recombinant sIL-6R (R&D Systems, Minneapolis, MN, USA). sIL-6R is required for intracellular IL-6 signal transduction because mIL-6R is not expressed by HTFs [37].

4.3. Overexpression Experiment

GeneCopoeia (Guangzhou, China) provided lentiviruses to overexpress SOCS3 or those containing a control vector. Lentivirus-containing media was added and mixed with the HTFs once the cells reached 30% confluence in 6-well plates. Supernatants in the wells were removed after 16 h of incubation and replaced with fresh FM, and the cells were then cultured for 48 h for further analysis.

4.4. Western Blot

Protein samples were separated by SDS-polyacrylamide gel electrophoresis and electrotransferred onto polyvinylidene fluoride (PVDF) membranes (Merck Millipore, Burlington, MA, USA). After being blocked, the membranes were incubated with rabbit anti-collagen I (clone E8F4L, catalog 72026, Cell Signaling Technology, Danvers, MA, USA), rabbit anti-fibronectin (clone EPR23110-46, catalog ab268020, Abcam, Cambridge, UK), rabbit anti-α-SMA (clone EPR5368, catalog ab124964, Abcam, Cambridge, UK), rabbit anti-SOCS3 (polyclonal, catalog AF8025, Beyotime, Shanghai, China), mouse anti-STAT3 (clone 124H6, catalog 9139, Cell Signaling Technology, Danvers, MA, USA), or rabbit anti-p-STAT3 (Tyr705) (clone D3A7, catalog 9145, Cell Signaling Technology, Danvers, MA, USA) overnight at 4 °C. Equal protein loading was confirmed by incubation with mouse anti-β-tubulin (clone 5G3, catalog 44032, Signalway antibody, Greenbelt, MD, USA). Horseradish peroxidase-conjugated secondary antibodies (Signalway antibody, Greenbelt, MD, USA) were incubated with the membranes at room temperature for 1 h. Finally, the bands were visualized with electrochemiluminescence (Merck Millipore, Burlington, MA, USA) and analyzed by ImageJ software 1.46r.

4.5. Immunofluorescence Staining

Tissue sections or cultured HTFs were stained with rabbit anti-collagen I (clone E8F4L, catalog 72026, Cell Signaling Technology, Greenbelt, MD, USA), rabbit anti-fibronectin (clone EPR23110-46, catalog ab268020, Abcam, Cambridge, UK), mouse anti-STAT3 (clone 124H6, catalog 9139, Cell Signaling Technology, Greenbelt, MD, USA), rabbit anti–p-STAT3 (Tyr705) (clone D3A7, catalog 9145, Cell Signaling Technology, Greenbelt, MD, USA), and DAPI (catalog C0060, Solarbio, Beijing, China). The samples were analyzed using a fluorescence microscope (Olympus Corporation, Tokyo, Japan).

4.6. Wound Healing Assay

First, the supernatant of HTFs was aspirated, and a wound was created with a sterile 200 µL pipette tip in the bottom of the 6-well plates. Subsequently, the cells were washed twice with PBS and immediately photographed with an inverted microscope (Olympus Corporation, Tokyo, Japan). Following 24 h of incubation, images were taken again, and the migration distance was calculated. The formula for calculating the cell migration rates is as follows: cell migration rate (%) = (the distance prior to healing − the distance following healing/the distance prior to healing) × 100%.

4.7. Transwell Chamber Assay

Trypsin was used to digest the HTFs in each group, and 600 µL of medium containing 20% FBS was added to the lower chamber of a transwell chamber (Labselect, Hangzhou, China) in a 24-well plate. Next, the cells were resuspended in a serum-free medium and transferred to the upper chamber (1×10^4 cells/well). After 24 h of incubation at 37 °C, the cells that had not invaded from the upper chamber were removed with a cotton swab. Paraformaldehyde and crystal violet staining solutions were used to fix and stain the invasive cells in the lower chamber at room temperature for 20 min. Five fields were randomly selected and photographed with an inverted microscope. The number of invading cells was counted by ImageJ software 1.46r.

4.8. Animal Experiments

Adult male Sprague–Dawley rats that weighed 180–200 g were purchased from Changchun Yisi Biotechnology. The rats were anesthetized by an intraperitoneal injection of 3 mL/kg 10% chloral hydrate followed by ocular surface anesthesia using 0.5% oxybuprocaine hydrochloride eye drops (Santen, Osaka, Japan). All animals were treated in accordance with the ARVO Statement for the Use of Animals in Ophthalmic and Vision Research.

GFS was performed on bilateral eyes as previously described [38,39]. To create a conjunctival fornix-based conjunctival flap 3–5 mm behind the limbus, a conjunctival incision was made, and then the underlying Tenon's capsule was bluntly dissected. Subsequently, a 30-G needle was inserted into the anterior chamber to create a full-thickness scleral tunnel. During the process, great care was taken to avoid iridal blood vessels. Viscoelastic solution (Alcon, Fort Worth, TX, USA) was injected through the needle to maintain the depth of the anterior chamber. A beveled micro cannula (external diameter, 0.3 mm) was then inserted through the scleral tunnel. Finally, the conjunctiva and Tenon's capsule were closed using a 10-0 monofilament nylon suture (Ethicon, Suzhou, China). All surgeries were performed by the same surgeon. Eyes exhibiting subsequent slippage or dislocation of the cannula were excluded.

After surgery, the eyes were treated with dimethyl sulfoxide (DMSO) as a vehicle, S3I-201 (Sigma–Aldrich, Burlington, MA, USA), and MMC (Hanhui, Shanghai, China) ($n \geq 8$) separately. S3I-201 (10 mg/mL, 5 µL) was injected into the subconjunctival space with a 10-µL Hamilton syringe connected to a 33-G needle (Hamilton Company, Reno, NV, USA) immediately after surgery and on days 3, 7, and 14. MMC was applied at a dose of

0.4 mg/mL with a small piece of cellulose sponge for 5 min. Then, irrigation of the treated area was performed with 2 mL 0.9% sodium chloride using a syringe. Each treatment was performed on at least 8 eyes (Vehicle group n = 10; S3I-201 group n = 11; MMC group n = 8). The IOP, bleb appearance and survival, and complications were observed on days 3, 7 14, 21 and 28. At the end of the experiment, 8 eyes were collected for histochemical and immunofluorescence analyses.

4.9. Clinical Examination and Analysis of Blebs

The rats were anesthetized prior to IOP measurement. Intraocular pressure was measured in each rat with a Tono-Pen tonometer (Reichert, Depew, NY, USA) according to the manufacturer's instructions. All IOP measurements were taken 5 to 10 times in each eye.

Slit-lamp examinations were performed on subconjunctival bleb morphology and determined survival time. A bleb was judged to have failed if the surgical site appeared flat and vascularized by slit-lamp analysis.

4.10. Hematoxylin and Eosin (HE) and Masson's Trichrome Staining

On day 28 after surgery, the rats were euthanized, and both eyes were immediately enucleated. The eyes were fixed in FAS fixative (Wuhan Servicebio, Wuhan, China), which contains formaldehyde, acetic acid, and saline, for 48 h, and then embedded in paraffin. The eyes were cut into sequential 4 μm tissue sections and then dewaxed with xylene. The sections were analyzed by HE and Masson's trichrome staining to evaluate histopathology and collagen expression in the subconjunctival tissues surrounding the filtration passage. Representative images were taken using an inverted microscope.

4.11. TUNEL Assay

Tissue sections of rat eyes were processed with a TUNEL kit (Beyotime, Shanghai, China) according to the manufacturer's instructions, and the images were analyzed using fluorescence microscopy.

4.12. Statistical Analysis

The results are presented as the means ± standard deviations (SD) of at least three independent experiments. One-way analysis of variance (ANOVA) and t-tests were performed with SPSS to analyze the statistical results. Bleb survival analysis was performed with Kaplan–Meier and Mantel–Cox pairwise comparison tests. Values of $p < 0.05$ were considered significant.

Author Contributions: Conceptualization, Y.L. and Y.Z.; methodology, Y.L., J.Z. and Y.Y.; software, C.Z. and Z.Z.; writing—original draft preparation, Y.L.; writing—review and editing, Y.Z.; funding acquisition, J.Z. and Y.Z. All authors have read and agreed to the published version of the manuscript.

Funding: This work was supported by the National Natural Science Foundation of China (82000919).

Institutional Review Board Statement: The human and animal studies were approved by the Institutional Review Board of the Second Hospital of Jilin University (approval code: 2020-129, date: 23 November 2020).

Informed Consent Statement: Informed consent was obtained from all subjects involved in the study.

Data Availability Statement: The novel findings reported in this study can be found within the article. For additional information or inquiries, please contact the corresponding author.

Conflicts of Interest: The authors declare no conflict of interest.

References

1. Tham YC: Li, X.; Wong, T.Y.; Quigley, H.A.; Aung, T.; Cheng, C.Y. Global prevalence of glaucoma and projections of glaucoma burden through 2040: A systematic review and meta-analysis. *Ophthalmology* **2014**, *121*, 2081–2090. [CrossRef]
2. Liu, P.; Wang, F.; Song, Y.; Wang, M.; Zhang, X. Current situation and progress of drugs for reducing intraocular pressure. *Ther. Adv. Chronic Dis.* **2022**, *13*, 20406223221140392. [CrossRef] [PubMed]
3. Komáromy, A.M.; Koehl, K.L.; Park, S.A. Looking into the future: Gene and cell therapies for glaucoma. *Veter Ophthalmol.* **2021**, *24* (Suppl. S1), 16–33. [CrossRef]
4. Zada, M.; Pattamatta, U.; White, A. Modulation of Fibroblasts in Conjunctival Wound Healing. *Ophthalmology* **2018**, *125*, 179–192. [CrossRef]
5. Bell, K.; Bezerra, B.D.P.S.; Mofokeng, M.; Montesano, G.; Nongpiur, M.E.; Marti, M.V.; Lawlor, M. Learning from the past: Mitomycin C use in trabeculectomy and its application in bleb-forming minimally invasive glaucoma surgery. *Surv. Ophthalmol.* **2021**, *66*, 109–123. [CrossRef]
6. DeBry, P.W.; Perkins, T.W.; Heatley, G.; Kaufman, P.; Brumback, L.C. Incidence of Late-Onset Bleb-Related Complications following Trabeculectomy with Mitomycin. *Arch. Ophthalmol.* **2002**, *120*, 297–300. [CrossRef] [PubMed]
7. Beckers, H.J.; Kinders, K.C.; Webers, C.A. Five-year results of trabeculectomy with mitomycin C. *Graefe's Arch. Clin. Exp. Ophthalmol.* **2003**, *241*, 106–110. [CrossRef] [PubMed]
8. Anand, N.; Arora, S.; Clowes, M. Mitomycin C augmented glaucoma surgery: Evolution of filtering bleb avascularity, transconjunctival oozing, and leaks. *Br. J. Ophthalmol.* **2006**, *90*, 175–180. [CrossRef]
9. Fielding, C.A.; Jones, G.W.; McLoughlin, R.M.; McLeod, L.; Hammond, V.J.; Uceda, J.; Williams, A.S.; Lambie, M.; Foster, T.L.; Liao, C.-T.; et al. Interleukin-6 Signaling Drives Fibrosis in Unresolved Inflammation. *Immunity* **2014**, *40*, 40–50. [CrossRef] [PubMed]
10. O'reilly, S.; Ciechomska, M.; Cant, R.; Hügle, T.; van Laar, J.M. Interleukin-6, its role in fibrosing conditions. *Cytokine Growth Factor Rev.* **2012**, *23*, 99–107. [CrossRef]
11. Kumar, S.; Wang, G.; Zheng, N.; Cheng, W.; Ouyang, K.; Lin, H.; Liao, Y.; Liu, J. HIMF (Hypoxia-Induced Mitogenic Factor)-IL (Interleukin)-6 Signaling Mediates Cardiomyocyte-Fibroblast Crosstalk to Promote Cardiac Hypertrophy and Fibrosis. *Hypertension* **2019**, *73*, 1058–1070. [CrossRef] [PubMed]
12. Mihara, M.; Hashizume, M.; Yoshida, H.; Suzuki, M.; Shiina, M. IL-6/IL-6 receptor system and its role in physiological and pathological conditions. *Clin. Sci.* **2012**, *122*, 143–159. [CrossRef] [PubMed]
13. Hunter, C.A.; Jones, S.A. IL-6 as a keystone cytokine in health and disease. *Nat. Immunol.* **2015**, *16*, 448–457. [CrossRef]
14. Rose-John, S.; Neurath, M.F. IL-6 trans-signaling: The heat is on. *Immunity* **2004**, *20*, 2–4. [CrossRef]
15. Heinrich, P.C.; Behrmann, I.; Haan, S.; Hermanns, H.M.; Müller-Newen, G.; Schaper, F. Principles of interleukin (IL)-6-type cytokine signalling and its regulation. *Biochem. J.* **2003**, *374*, 1–20. [CrossRef] [PubMed]
16. Yu, H.; Pardoll, D.; Jove, R. STATs in cancer inflammation and immunity: A leading role for STAT3. *Nat. Rev. Cancer* **2009**, *9*, 798–809. [CrossRef]
17. Pedroza, M.; Le, T.T.; Lewis, K.; To, S.; George, A.T.; Blackburn, M.R.; Tweardy, D.J.; Agarwal, S.K.; Karmouty-Quintana, H. STAT-3 contributes to pulmonary fibrosis through epithelial injury and fibroblast-myofibroblast differentiation. *FASEB J.* **2016**, *30*, 129–140. [CrossRef]
18. Chakraborty, D.; Šumová, B.; Mallano, T.; Chen, C.-W.; Distler, A.; Bergmann, C.; Ludolph, I.; Horch, R.E.; Gelse, K.; Ramming, A.; et al. Activation of STAT3 integrates common profibrotic pathways to promote fibroblast activation and tissue fibrosis. *Nat. Commun.* **2017**, *8*, 1130. [CrossRef]
19. Pang, M.; Ma, L.; Gong, R.; Tolbert, E.; Mao, H.; Ponnusamy, M.; Chin, Y.E.; Yan, H.; Dworkin, L.D.; Zhuang, S. A novel STAT3 inhibitor, S3I-201, attenuates renal interstitial fibroblast activation and interstitial fibrosis in obstructive nephropathy. *Kidney Int.* **2010**, *78*, 257–268. [CrossRef]
20. Przekora, A.; Zarnowski, T.; Ginalska, G. A simple and effective protocol for fast isolation of human Tenon's fibroblasts from a single trabeculectomy biopsy—A comparison of cell behaviour in different culture media. *Cell. Mol. Biol. Lett.* **2017**, *22*, 5. [CrossRef]
21. Stahnke, T.; Löbler, M.; Kastner, C.; Stachs, O.; Wree, A.; Sternberg, K.; Schmitz, K.-P.; Guthoff, R. Different fibroblast subpopulations of the eye: A therapeutic target to prevent postoperative fibrosis in glaucoma therapy. *Exp. Eye Res.* **2012**, *100*, 88–97. [CrossRef] [PubMed]
22. Yang, L.; Zheng, S.; Ge, D.; Xia, M.; Li, H.; Tang, J. LncRNA-COX2 inhibits Fibroblast Activation and Epidural Fibrosis by Targeting EGR1. *Int. J. Biol. Sci.* **2022**, *18*, 1347–1362. [CrossRef] [PubMed]
23. Liu, Z.-K.; Li, C.; Zhang, R.-Y.; Wei, D.; Shang, Y.-L.; Yong, Y.-L.; Kong, L.-M.; Zheng, N.-S.; Liu, K.; Lu, M.; et al. EYA2 suppresses the progression of hepatocellular carcinoma via SOCS3-mediated blockade of JAK/STAT signaling. *Mol. Cancer* **2021**, *20*, 79. [CrossRef] [PubMed]
24. Wolters, J.E.J.; Van Mechelen, R.J.S.; Al Majidi, R.; Pinchuk, L.; Webers, C.A.B.; Beckers, H.J.M.; Gorgels, T.G.M.F. History, presence, and future of mitomycin C in glaucoma filtration surgery. *Curr. Opin. Ophthalmol.* **2020**, *32*, 148–159. [CrossRef] [PubMed]
25. Gedde, S.J.; Schiffman, J.C.; Feuer, W.J.; Herndon, L.W.; Brandt, J.D.; Budenz, D.L.; Tube versus Trabeculectomy Study Group. Treatment Outcomes in the Tube Versus Trabeculectomy (TVT) Study after Five Years of Follow-up. *Am. J. Ophthalmol.* **2012**, *153*, 789–803.e2. [CrossRef]

26. Katz, L.J.; Cantor, L.B.; Spaeth, G.L. Complications of surgery in glaucoma. Early and late bacterial endophthalmitis following glaucoma filtering surgery. *Ophthalmology* **1985**, *92*, 959–963. [CrossRef]
27. Schlunck, G.; Meyer-Ter-Vehn, T.; Klink, T.; Grehn, F. Conjunctival fibrosis following filtering glaucoma surgery. *Exp. Eye Res.* **2016**, *142*, 76–82. [CrossRef]
28. Yamanaka, O.; Kitano-Izutani, A.; Tomoyose, K.; Reinach, P.S. Pathobiology of wound healing after glaucoma filtration surgery. *BMC Ophthalmol.* **2015**, *15*, 19–27. [CrossRef]
29. Mateo, R.B.; Reichner, J.S.; Albina, J.E. Interleukin-6 activity in wounds. *Am. J. Physiol.* **1994**, *266*, R1840–R1844. [CrossRef]
30. Gurtner, G.C.; Werner, S.; Barrandon, Y.; Longaker, M.T. Wound repair and regeneration. *Nature* **2008**, *453*, 314–321. [CrossRef]
31. Starr, R.; Hilton, D.J. SOCS: Suppressors of cytokine signalling. *Int. J. Biochem. Cell Biol.* **1998**, *30*, 1081–1085. [CrossRef]
32. Siddiquee, K.; Zhang, S.; Guida, W.C.; Blaskovich, M.A.; Greedy, B.; Lawrence, H.R.; Yip, M.L.R.; Jove, R.; McLaughlin, M.M.; Lawrence, N.J.; et al. Selective chemical probe inhibitor of Stat3, identified through structure-based virtual screening, induces antitumor activity. *Proc. Natl. Acad. Sci. USA* **2007**, *104*, 7391–7396. [CrossRef] [PubMed]
33. Meydan, N.; Grunberger, T.; Dadi, H.; Shahar, M.; Arpaia, E.; Lapidot, Z.; Leeder, J.S.; Freedman, M.; Cohen, A.; Gazit, A.; et al. Inhibition of acute lymphoblastic leukaemia by a Jak-2 inhibitor. *Nature* **1996**, *379*, 645–648. [CrossRef]
34. Shawky, A.M.; Almalki, F.A.; Abdalla, A.N.; Abdelazeem, A.H.; Gouda, A.M. A Comprehensive Overview of Globally Approved JAK Inhibitors. *Pharmaceutics* **2022**, *14*, 1001. [CrossRef] [PubMed]
35. Hosui, A.; Kimura, A.; Yamaji, D.; Zhu, B.M.; Na, R.; Hennighausen, L. Loss of STAT5 causes liver fibrosis and cancer development through increased TGF-{beta} and STAT3 activation. *J. Exp. Med.* **2009**, *206*, 819–831. [CrossRef]
36. Walters, D.M.; Antao-Menezes, A.; Ingram, J.L.; Rice, A.B.; Nyska, A.; Tani, Y.; Kleeberger, S.R.; Bonner, J.C. Susceptibility of Signal Transducer and Activator of Transcription-1-Deficient Mice to Pulmonary Fibrogenesis. *Am. J. Pathol.* **2005**, *167*, 1221–1229. [CrossRef] [PubMed]
37. Watanabe-Kitamura, F.; Ogawa, A.; Fujimoto, T.; Iraha, S.; Inoue-Mochita, M.; Watanabe, T.; Takahashi, E.; Tanihara, H.; Inoue, T. Potential roles of the IL-6 family in conjunctival fibrosis. *Exp. Eye Res.* **2021**, *210*, 108708. [CrossRef]
38. Pandav, S.; Akella, M.; Thattaruthody, F. A modified model of glaucoma filtering surgery in Sprague-Dawley rats. *Indian J. Ophthalmol.* **2022**, *70*, 662–664. [CrossRef]
39. Sherwood, M.B.; Esson, D.W.; Neelakantan, A.; Samuelson, D.A. A New Model of Glaucoma Filtering Surgery in the Rat. *J. Glaucoma* **2004**, *13*, 407–412. [CrossRef]

Disclaimer/Publisher's Note: The statements, opinions and data contained in all publications are solely those of the individual author(s) and contributor(s) and not of MDPI and/or the editor(s). MDPI and/or the editor(s) disclaim responsibility for any injury to people or property resulting from any ideas, methods, instructions or products referred to in the content.

Article

The Enzyme 15-Hydroxyprostaglandin Dehydrogenase Inhibits a Shift to the Mesenchymal Pattern of Trophoblasts and Decidual Stromal Cells Accompanied by Prostaglandin Transporter in Preeclampsia

Huiyuan Pang, Di Lei, Tingting Chen, Yujie Liu and Cuifang Fan *

Department of Obstetrics and Gynecology, Renmin Hospital of Wuhan University, Wuhan 430060, China
* Correspondence: fancuifang@whu.edu.cn; Tel.: +86-139-7105-0592

Citation: Pang, H.; Lei, D.; Chen, T.; Liu, Y.; Fan, C. The Enzyme 15-Hydroxyprostaglandin Dehydrogenase Inhibits a Shift to the Mesenchymal Pattern of Trophoblasts and Decidual Stromal Cells Accompanied by Prostaglandin Transporter in Preeclampsia. *Int. J. Mol. Sci.* **2023**, *24*, 5111. https://doi.org/10.3390/ijms24065111

Academic Editors: Margherita Sisto and Sabrina Lisi

Received: 6 December 2022
Revised: 20 February 2023
Accepted: 21 February 2023
Published: 7 March 2023

Copyright: © 2023 by the authors. Licensee MDPI, Basel, Switzerland. This article is an open access article distributed under the terms and conditions of the Creative Commons Attribution (CC BY) license (https://creativecommons.org/licenses/by/4.0/).

Abstract: Preeclampsia (PE) is a pregnancy complication beginning after 20 weeks of pregnancy that involves high blood pressure (systolic > 140 mmHg or diastolic > 90 mmHg), with or without proteinuria. Insufficient trophoblast invasion and abnormal decidualization are involved in PE development. However, whether unhealthy placenta and decidua have the same biological activities is unclear. The enzyme 15-hydroxyprostaglandin dehydrogenase (15-PGDH; encoded by *HPGD*) degrades prostaglandin, and prostaglandin transporter (PGT), as a candidate molecule of prostaglandin carriers, helps transport prostaglandin into cells. Whether 15-PGDH and PGT are involved in PE has not been researched. In this study, we investigated the shared pathogenesis of foetal placenta and maternal decidua from the perspective of epithelial–mesenchymal transition (EMT)/mesenchymal–epithelial transition (MET) and explored the combined effects of 15-PGDH and PGT on the EMT/MET of trophoblasts and decidual stromal cells (DSCs). Here, we demonstrated that placental development and decidualization both involved EMT/MET. In PE, both trophoblasts and DSCs show more epithelial patterns. Moreover, 15-PGDH expression was downregulated in the placentas but upregulated in the deciduas of PE patients. Inhibiting 15-PGDH promotes a shift to a mesenchymal pattern of trophoblasts and DSCs depending on the PGT-mediated transport of prostaglandin E2 (PGE2). In conclusion, our results showed that inhibiting 15-PGDH promotes a shift to the mesenchymal pattern of trophoblasts and DSCs and may provide a new and alternative therapy for the treatment of PE.

Keywords: preeclampsia; placenta; EMT; decidua; MET; 15-PGDH; PGT

1. Introduction

Development of the maternal–foetal interface is the basis of healthy pregnancy, which depends heavily on placental development and decidualization. The pathogenesis of PE is caused by inhibition of the coordinated development of the maternal–foetal interface. Previous studies have mostly focused on the decidualization of DSCs [1,2] or the invasion of trophoblasts separately [3]; however, since half of a foetus's genes originate from the mother, a few studies have focused on the underlying link between decidualization and placental development. The processes of trophoblast differentiation and decidualization are both involved in EMT/MET [4–6]. During placental development, trophoblasts undergo continuous differentiation. Cytotrophoblasts (CTBs) differentiate into extravillous trophoblasts (EVTs) which invade the decidua. The process of differentiation from CTBs to EVTs involves EMT [4]. DSCs contained in decidual tissue undergoing decidualization do not depend on the implantation of human embryos in each menstrual cycle. The process of decidualization involves MET [6]. However, as a common biological behaviour of placental development and decidualization, the shared pathology of trophoblast EMT and DSC MET in PE has seldom been researched.

15-PGDH is a key factor in the degradation of PGE2, and inhibiting 15-PGDH impairs the degradation of PGE2 [7]. Embryonic development depends on the appropriate prostaglandin concentration [8]. Concentrations of prostaglandins that are too high or too low will inhibit embryo implantation. The metabolic regulation of PGs is very important. Most previous studies have focused on the roles of COX-2, a key enzyme in the synthase of prostaglandins, but there have been few studies on 15-PGDH, another important member of the enzyme family that regulates PGE2. Recently, 15-PGDH was found to be closely related to cancer invasion and tissue regeneration [7,9,10]. The decomposition of PGE2 relies on PGT-mediated transport [11]. PGT is a transporter with 12 transmembrane domains and a lactate-PG transport mechanism [12]. However, there is no study about how 15-PGDH and PGT cooperate and affect trophoblast and DSC EMT/MET in PE.

In this study, the abnormal placental development and decidualization in PE were both related to the shift between epithelial and mesenchymal patterns, which helps with the exploration of therapeutic targets of PE. We mainly studied the role of 15-PGDH in PE. We found that 15-PGDH protein expression displays opposite patterns in the foetal placenta and maternal decidua in normal pregnancy, and this expression is altered in PE. We then explored the effect of 15-PGDH on the biological behaviour of trophoblasts and DSCs. Furthermore, we explored the cooperative effect of 15-PGDH and PGT in trophoblasts and DSCs.

2. Results

2.1. Partial EMT Occurs during Placental Development

In the first-trimester villi, we performed coimmunofluorescence to demonstrate that the development of CTBs into EVTs involved partial EMT. Human leukocyte antigen (HLA-G) is gradually expressed when CTB develops into EVT, where it mediates maternal–foetal immune tolerance during pregnancy [13]. HLA-G is a biomarker of CTB development. Since HLA-G is specifically expressed on EVTs in the placenta, HLA-G (pink in Figure 1A, green in Figure 1C) specifically represents EVTs, and the adjacent region represents CTBs and syncytiotrophoblasts (STBs), which are formed by the cell–cell fusion of CTBs. We compared the expression of cadherin-1 (CDH1) and cadherin-2 (CDH2) in HLA-G$^+$ EVT with HLA-G$^-$ CTB to determine how EMT actually occurs in CTB development (Figure 1B,D). CDH1 (red) protein expression decreased during HLA-G$^-$ CTB development into HLA-G$^+$ EVT (Figure 1A,B), while CDH2 (red) did not show a significant difference between CTB and EVT (Figure 1C,D). Therefore, when CTBs develop into EVTs, CTBs gradually lose their epithelial pattern but do not gain their mesenchymal pattern significantly, which demonstrates that partial EMT occurs during trophoblast development.

2.2. MET Occurs during Decidualization

Human decidual stromal cells (hESCs) were induced to decidualize (decidualized hESC: dhESC) according to the above methods [14]. Western blots revealed that prolactin (PRL) and insulin-like growth factor binding protein 1 (IGFBP1), the traditional markers of decidualization, increased (Figure 1E), indicating that in vitro hESC decidualization was successful; moreover, CDH1 increased, and vimentin decreased during hESC decidualization in vitro (Figure 1E,G). Phalloidin-specific binding of F-actin further demonstrated that during decidualization, hESCs gradually lost fibroblastic structural characteristics and gained typical polygonal morphology (the shape change of hESCs is labelled with arrows in Figure 1F). During decidualization, hESCs undergo MET-like changes.

2.3. Insufficient EMT Changes in the Placenta and Excess MET-Like Changes in Decidual Tissue in Preeclampsia

We collected placentas and decidua specimens and performed immunohistochemical staining. Immunohistochemical staining showed CDH1 and CDH2 changes in trophoblasts but not in other cells and revealed that CDH1 protein expression was increased in the placentas of the PE group compared to those of the normal group ($n = 6$; $p \leq 0.01$) (Figure 2A,B,E), while CDH2 was decreased in the placenta of the PE group compared to the normal group ($n = 6$; $p \leq 0.01$) (Figure 2C–E).

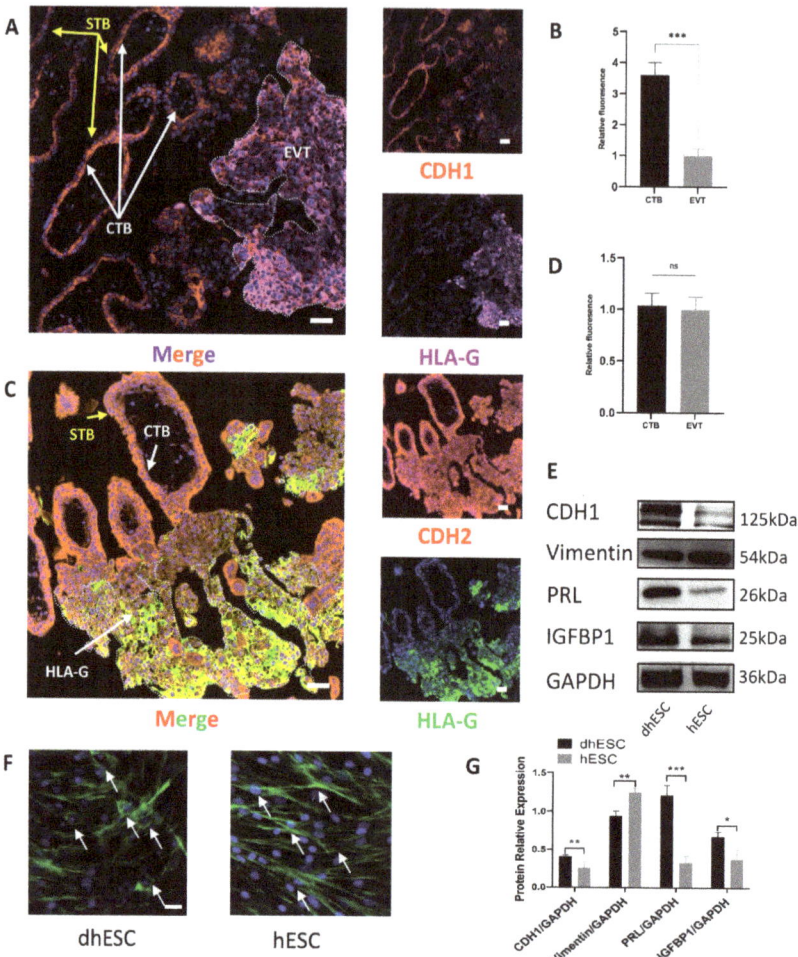

Figure 1. Partial EMT occurs in placental development, while MET occurs during decidualization. (**A**,**B**) CDH1 location and expression levels in CTBs and EVTs were measured by coimmunofluorescence in early pregnancy villi ($n = 3$). Scale bar = 50 nm in (**A**,**B**). (**C**,**D**) CDH2 location and expression levels in CTBs and EVTs were measured by coimmunofluorescence in early pregnancy villi ($n = 3$). Scale bar = 50 nm in (**C**,**D**). *** $p < 0.001$; ns, not significant. (**E**,**G**) Western blot tests of MET and decidualization markers during decidualization in vitro. The experiment was repeated three times independently. dhESC: decidualized hESC. GAPDH served as a loading control. Band intensities were quantified and normalized to the GAPDH values. Values are the mean ± SD. * $p < 0.05$; ** $p < 0.01$; *** $p < 0.001$; ns, not significant. (**F**) F-actin staining of dhESCs and hESCs shows cytoskeletal changes during decidualization. Scale bar = 50 nm. CTB: cytotrophoblast; EVT: extravillous trophoblasts; EMT: epithelial–mesenchymal transition.

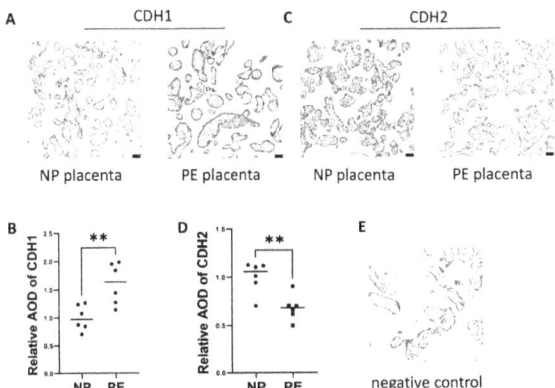

Figure 2. Excess epithelial pattern in trophoblasts in PE patients. (**A,B**) CDH1 protein expression in placental tissues from women with NP and PE was measured by immunohistochemical staining. (**C,D**) CDH2 protein expression in placental tissues from women with NP and PE was measured by immunohistochemical staining. (**E**) represents negative control. Values are the mean ± SD. ** $p < 0.01$; ns, not significant. $n = 6$ each. Scale bar = 50 nm.

We also demonstrated that abnormal decidua in PE is accompanied by excessive MET (Figure 3). In this study, we only focused on DSCs, which are the exact cells that undergo decidualization. Therefore, we also identified DSCs specifically. Since the decidual composition is complex and includes DSCs, trophoblasts and other cells including immune cells, we first identified CK7$^-$vimentin$^+$ DSCs [2] (Cytokeratin 7: CK7), (CK7: green; vimentin: red) by coimmunofluorescence before further study (Figure 3A). Vimentin helped determine that the larger and rounder cells were DSCs by immunohistochemistry (Figure 3B). Therefore, in further research, we will only focus on these DSCs (the area of DSCs is labelled with arrows, and non-DSCs are labelled with triangles in Figure 3C,E). CDH1 protein expression was increased in the DSCs of the PE group compared to those of the normal group ($n = 6$, $p \leq 0.05$). In contrast, CDH2 and vimentin were decreased in the DSCs of the PE group compared to those of the normal group ($n = 6$, $p \leq 0.05$) (Figure 3C–E). We demonstrated that abnormal decidualization in PE (Supplementary Figure S1) is accompanied by excessive MET.

2.4. Expression of 15-PGDH Is Downregulated in the Placenta but Upregulated in the DSCs of PE Patients

Immunohistochemistry revealed that 15-PGDH was downregulated in trophoblasts in PE patients (Figure 4A,E) ($n = 6$; $p \leq 0.05$). Immunohistochemical staining showed that 15-PGDH was upregulated in the DSCs of patients with PE (Figure 4B,F) ($n = 6$; $p \leq 0.05$). Therefore, 15-PGDH is involved in both placental and DSC abnormalities in PE.

2.5. Location of 15-PGDH in the Villi and DSCs

To explore the location of 15-PGDH in trophoblasts, we used first-trimester villi to detect 15-PGDH location. CTB development is accompanied by gradual HLA-G expression and invasion into the maternal decidua. Therefore, in the first-trimester villi, we can clearly identify CTBs and EVTs to explore the 15-PGDH expression pattern in these two kinds of cells. Thus, how 15-PGDH expression changes during CTB development can be explored. Therefore, we first used villous coimmunofluorescence to reveal that 15-PGDH expression increased significantly during the development of CTBs into EVTs (the area of CTBs is labelled with +, and the area of EVTs is labelled with triangles), which also demonstrated that 15-PGDH was specifically located in trophoblasts instead of the stroma and was closely linked to trophoblast EMT (CK7: green; 15-PGDH: red) (Figure 4C). CK7 is a marker of trophoblasts [15].

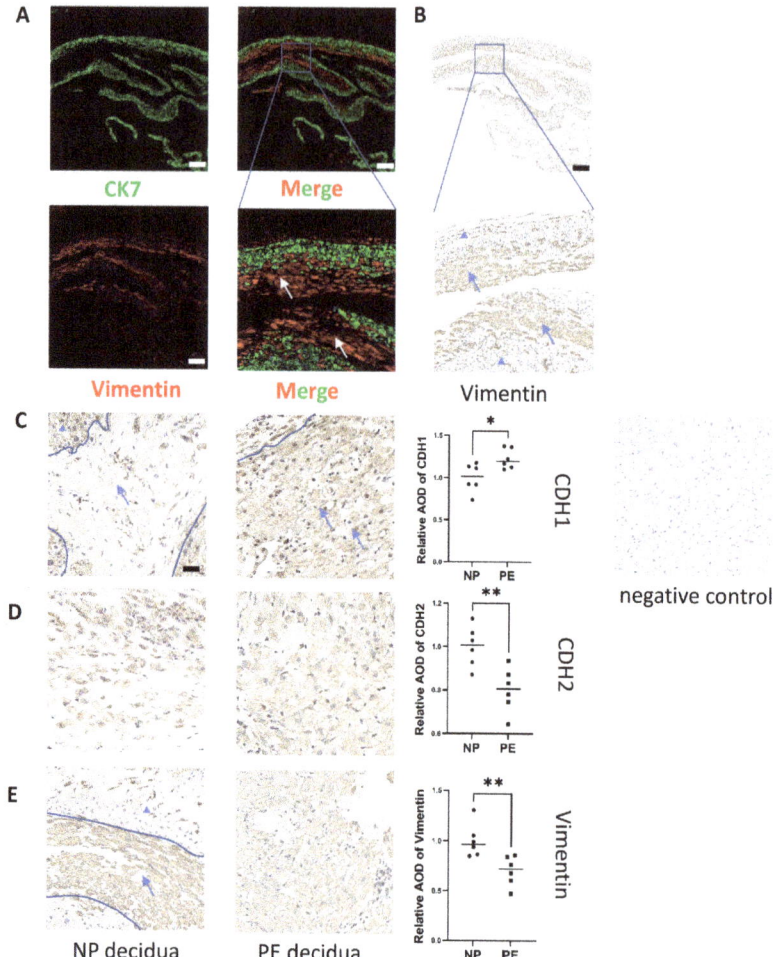

Figure 3. Excess epithelial pattern in DSCs in PE patients. (**A**,**B**) Immunofluorescence and immunohistochemical staining identified DSCs in decidual tissues. Red: vimentin; Green: CK7, scale bar = 500 nm. The experiment was repeated 3 times independently. (**C**,**D**,**E**) Measurement of CDH1, CDH2 and vimentin protein expression in DSCs from women with NP and PE by immunohistochemical staining. The area of DSCs is labelled by arrows and non-DSCs are labelled with triangles, Scale bar = 20 nm. Values are the mean ± SD. * $p < 0.05$; ** $p < 0.01$; ns, not significant. $n = 6$ each. PE: preeclampsia; NP: normal pregnancy; DSC: decidual stromal cell.

To explore the location of 15-PGDH in the decidua, we used decidual tissue obtained during caesarean section since decidualization is not accompanied by DSC location changes. In decidual tissue, coimmunofluorescence revealed that 15-PGDH was located in the cytoplasm of the DSCs (vimentin: green; 15-PGDH: red) (Figure 4D).

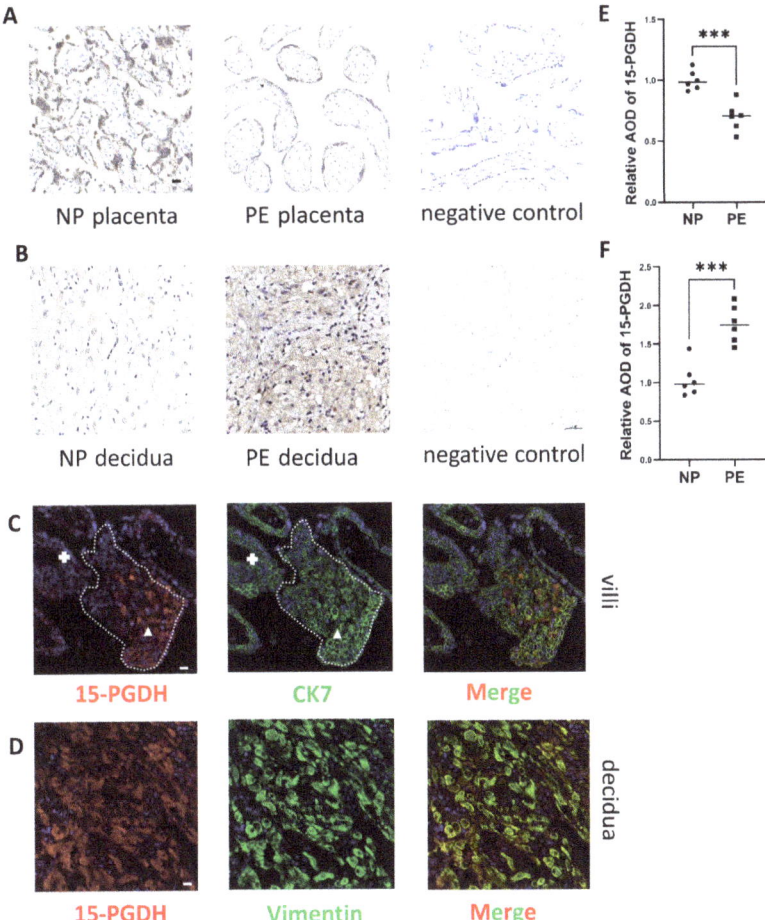

Figure 4. Differential expression of 15-PGDH in PE and the location of 15-PGDH. (**A,B,E,F**) Immunohistochemical staining showed the relative quantification of 15-PGDH in the placenta and decidua of the PE and NP groups. Brown staining represents the target protein. Scale bar = 20 nm, $n = 6$, Values are the mean ± SD. *** $p < 0.001$. $n = 6$ each. (**C**) Differential expression of 15-PGDH in CTBs and EVTs was identified by immunofluorescence, red: 15-PGDH; green: CK7, Bar = 20 nm. The area of CTBs is labelled with +, and the area of EVTs is labelled with triangles. (**D**) Specific location of 15-PGDH in DSCs was identified by immunofluorescence, red: 15-PGDH; green: vimentin, Bar = 20 nm.

2.6. SW033291 Inhibits 15-PGDH and Upregulates PGE2 Expression

SW033291, a 15-PGDH inhibitor, was used to treat Jeg3 cells and hESCs at different levels (0, 50, 100, 200, 350 and 500 nM) to detect the relationship between SW033291 concentration and PGE2 concentration. Then, cell lysates were used to perform ELISA to determine PGE2 concentrations. The PGE2 concentration increased with SW033291 in a dose-dependent manner (Figure 5A). Therefore, we chose these concentrations of SW033291 for further research.

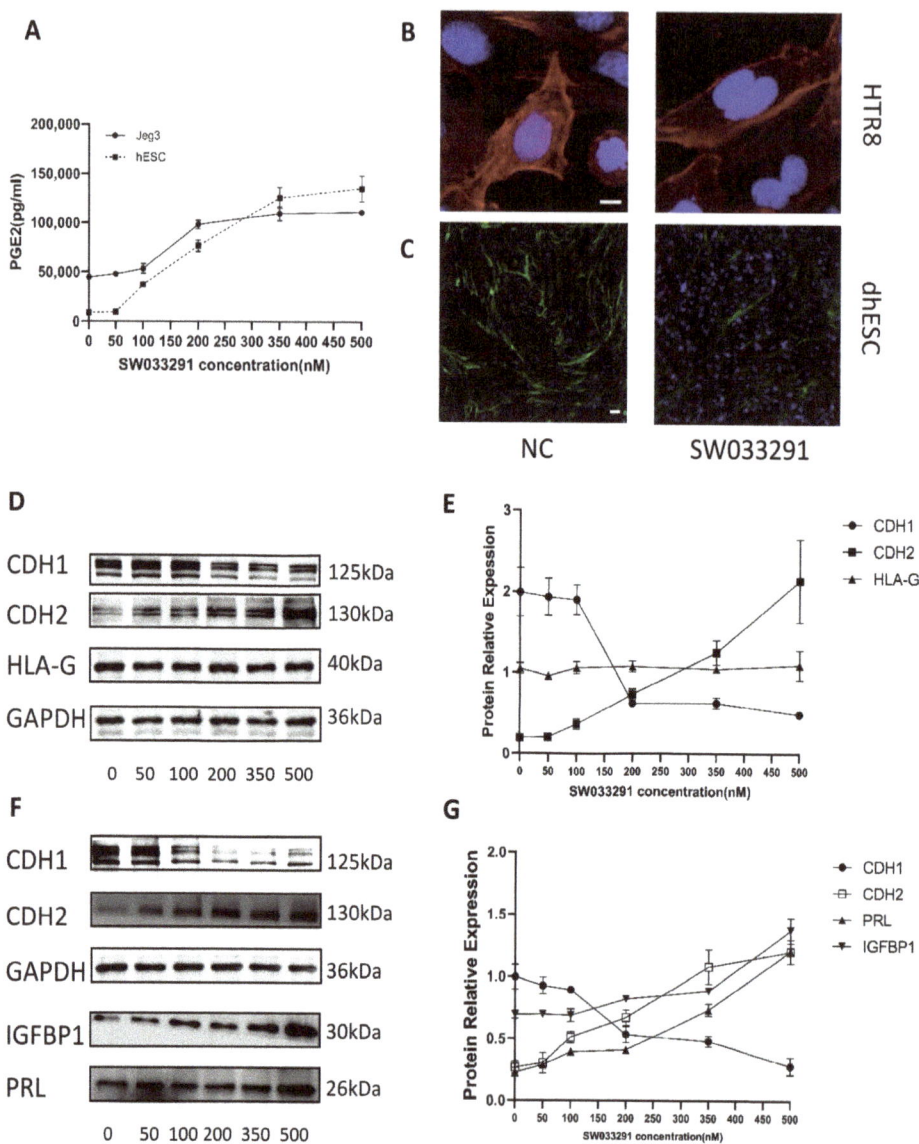

Figure 5. Function of 15-PGDH. (**A**) PGE2 levels in cell lysates were quantified by ELISA. The results are representative of at least three independent experiments. (**B,C**) F-actin staining revealed cytoskeletal changes in dhESCs and HTR8 cells in the NC (normal control) group and SW033291 group. (**D,E**) Western blot assays show the relative levels of EMT markers and HLA-G in Jeg3. (**F,G**) Western blot assays show the relative levels of EMT markers and decidualization markers in hESC. Each experiment was independently performed three times. CTB: cytotrophoblast; EVT: extravillous trophoblasts; EMT: epithelial–mesenchymal transition; DSC: decidual stromal cell. The experiment was repeated 3 times independently. Band intensities were quantified and normalized to the GAPDH values. Values are the mean ± SD.

2.7. Inhibiting 15-PGDH Promotes a Shift to a Mesenchymal Pattern in Trophoblasts and DSCs

To explore the function of 15-PDGH in trophoblasts and DSCs, we added SW033291 (0, 50, 100, 200, 350, and 500 nM) to the culture medium of Jeg3 cells and dhESCs, and Western blotting was used to reveal the expression of EMT/MET markers, decidualization markers (PRL, IGFBP1) and the CTB development marker (HLA-G). After the addition of SW033291, CDH1 decreased while CDH2 increased in dhESCs in a dose-dependent manner, as shown in Figure 5F. We chose the trophoblast Jeg3 cell line to test this effect (Figure 5D,E) since Jeg3 is the only trophoblast cell line that expresses HLA-G. HLA-G did not show a significant change during the EMT of trophoblasts promoted by SW033291 (Figure 5E). In dhESCs, PRL and IGFBP1 also increased in a dose-dependent manner (Figure 5F,G). Among them, PRL expression increased after exposure to SW033291, while IGFBP1 expression increased significantly at 500 nM (Figure 5F,G).

To demonstrate the role of 15-PGDH, we used F-actin staining to evaluate cytoskeletal reorganization, another important characteristic of EMT/MET. Compared with those of the normal control group, HTR8 and dhESCs treated with SW033291 (500 nM) showed fibroblastic-like shapes, which are characterized by strong F-actin stress fibres arranged longitudinally through the major axis (Figure 5B,C).

2.8. PGT Is Upregulated When 15-PGDH Is Inhibited

To explore the effect of 15-PGDH on PGT, we performed Western blotting to determine how PGT expression changed when 15-PGDH was inhibited. Western blot analysis showed that in both Jeg3 cells and hESCs, PGT expression was upregulated when cells were treated with SW033291, as shown in Figure 6A,B, which showed that the inhibition of 15-PGDH promotes the mesenchymal pattern involving PGT.

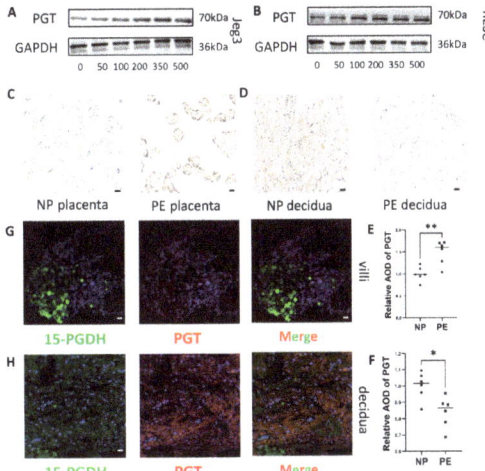

Figure 6. 15-PGDH inhibitor increased PGT expression and PGT differential expression in PE patients. (**A**) Western blot assays showed that PGT expression increased in Jeg3 cells when treated with concentration of SW033291. (**B**) Western blot assays show that PGT expression increased in hESCs treated with concentration of SW033291. (**C,E**) Immunohistochemical staining showed the relative quantification of PGT in the placenta of the PE and NP groups ($n = 6$). Scale bar = 20 nm. (**D,F**) Immunohistochemical staining showed the relative quantification of PGT in the decidua of the PE and NP groups ($n = 6$). Scale bar = 20 nm. Values are the mean ± SD. * $p < 0.05$; ** $p < 0.01$. Each experiment was independently performed three times. (**G,H**) 15-PGDH and PGT, both located in trophoblasts and DSCs, were identified by immunofluorescence; red: PGT; green: 15-PGDH. G:Scale bar = 20 nm; H: Scale bar = 50 nm. The experiment was repeated 3 times independently.

2.9. PGT Is Upregulated in the Placenta but Downregulated in the DSCs of PE Patients

Immunohistochemistry also revealed that PGT was upregulated in the placentas of the PE group (Figure 6C,E) ($n = 6$; $p \leq 0.05$) but downregulated in the decidua of the PE group (Figure 6D,F) ($n = 6$; $p \leq 0.05$). Therefore, PGT expression changed along with 15-PGDH both in vitro and in vivo.

2.10. PGT Coexpression with 15-PGDH in the Villous and DSCs

To further determine the possible synergistic role of 15-PGDH and PGT, we conducted coimmunofluorescence and revealed that 15-PGDH and PGT were both located in trophoblasts (Figure 6G) and DSCs (Figure 6H). Coimmunofluorescence was performed to reveal that PGT was located in trophoblasts (Supplementary Figure S2A) and DSCs (Supplementary Figure S2B). However, how 15-PGDH and PGT affect each other requires further study.

2.11. Inhibition of 15-PGDH Promotes the Mesenchymal Pattern Depending on the Normal Transport Function of PGT

To further demonstrate the relationship between 15-PGDH and PGT, we treated Jeg3, HTR8 and dhESCs with both SW033291 and indocyanine green (ICG), a PGT inhibitor.

We first performed ELISA to detect PGE2 in the cell lysates of Jeg3 cells and dhESCs from three groups: the NC, SW033291 and SW033291 + ICG groups. Both PGE2 in Jeg3 cells and dhESC cells in the SW033291 group were significantly increased; however, when treated with ICG, the concentration of PGE2 then decreased, which demonstrated that the PGE2 increase depended on the lactate-PG transport mechanism of PGT (Figure 7A,B).

We also performed Western blotting to reveal changes in the expression of related molecules. First, in both Jeg3 cells and dhESCs, SW033291 promoted mesenchymal markers, such as CDH2, while inhibiting the epithelial marker CDH1; however, this effect was counterbalanced by ICG (67 μM) (Figure 7E–H). Second, in Jeg3 cells, HLA-G still did not show a significant change when Jeg3 cells transitioned between an epithelial or mesenchymal pattern, indicating that the concentration of PGE2 and the expression of 15-PGDH have no effect on HLA-G (Figure 7G,H). In regard to PRL expression, ICG can counterbalance exposure to SW033291 (Figure 7E,F).

To corroborate these effects, we also used F-actin staining to evaluate cytoskeletal reorganization (Figure 7C,D). HTR8 (Figure 7D) and dhESCs (Figure 7C) treated with SW033291 (500 nM) showed fibroblastic-like shapes, while HTR8 and dhESCs treated with SW033291 (500 nM) and ICG (67 μM) exhibited a change in the localization of actin to the edge of the cell membrane, which represents a typical polygonal morphology (Figure 7C,D). ICG can reverse the effect of SW033291. PGT serves as an exchange for PGs with lactate. When PGT is inhibited by ICG and cells cannot take up PGE2, even inhibiting 15-PGDH is still not useful to promote the mesenchymal pattern.

2.12. Preeclampsia Rat Model Construction and Expression Detection of 15-PGDH and PGT

Given clinical ethics, we could not obtain a complete maternal–foetal interface in patients with PE. However, in the animal model, we obtained the complete maternal–foetal interface as described above. We successfully constructed two kinds of PE rat models according to the above methods. The blood pressure of the reduced uterine perfusion pressure (RUPP) group [16] and NG-nitroarginine methyl ester hydrochloride (L-NAME) group increased significantly (approximately 20 mmHg) (Figure 8A,B). The CK7$^-$vimentin$^+$ zone is consistent with decidual cells, and CK7$^+$ indicates the labyrinth zone, which is consistent with the placentas of humans. HE staining showed that the structure of the maternal–foetal interface in the L-NAME and RUPP groups was significantly changed. First, the placenta was degenerated (Figure 8E,G), the sinus-like structure in the decidua layer became larger, and the number of cells was reduced (Figure 8E,F). Second, the uterine volume was reduced (Supplementary Figure S3A), the growth and development of foetal rats in the RUPP group were restricted and the growth and development in the L-NAME group were slightly restricted (Supplementary Figure S3B). Moreover, HE staining revealed that the structure of the kidney was damaged as well, with widening visible glomerular cysts and weakening of the glomerulus (Supplementary Figure S3C, glomerulus labelled with triangles).

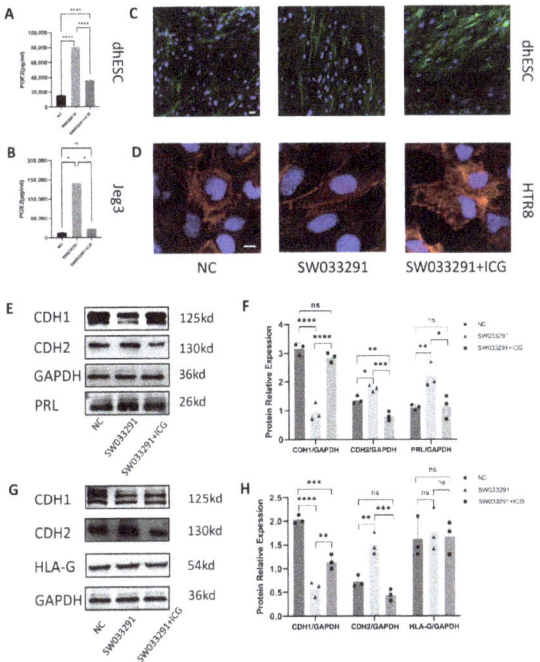

Figure 7. Function of the 15-PGDH inhibitor dependent on PGT. (**A,B**) PGE2 levels in cell lysates were quantified by ELISA. (**A**) represents PGE2 levels in dhESCs; (**B**) represents PGE2 levels in Jeg3 cells. (**C,D**) F-actin staining revealed cytoskeletal changes in dhESCs and HTR8 cells in the NC (normal control) group, SW033291 group, and SW033291+ICG group. Bar= 20 nm. (**E,F**) Western blot assays showed the relative levels of EMT markers and decidualization markers. (**G,H**) Western blot assays showed the relative levels of EMT markers and HLA-G. Each experiment was independently performed three times. GAPDH served as a loading control. Band intensities were quantified and normalized to the GAPDH values. The results are representative of at least three independent experiments. * $p < 0.05$; ** $p < 0.01$; *** $p < 0.001$; **** $p < 0.0001$; ns, not significant (one-way ANOVA). NC: normal control; ICG: indocyanine green; NP: normal pregnancy; PE: preeclampsia.

Immunohistochemistry demonstrated that 15-PGDH was significantly downregulated in the placenta (Figure 9B) ($n = 6; p \leq 0.05$) but upregulated in the decidua of the RUPP rat model and L-NAME PE rat model group (Figure 9A) ($n = 6; p \leq 0.05$). PGT was significantly upregulated in the placenta (Figure 9D) ($n = 6; p \leq 0.05$) but downregulated in the decidua of the RUPP rat model and L-NAME PE rat model (Figure 9C) ($n = 6; p \leq 0.05$). Figure 9E represents negative control of placenta and decidua.

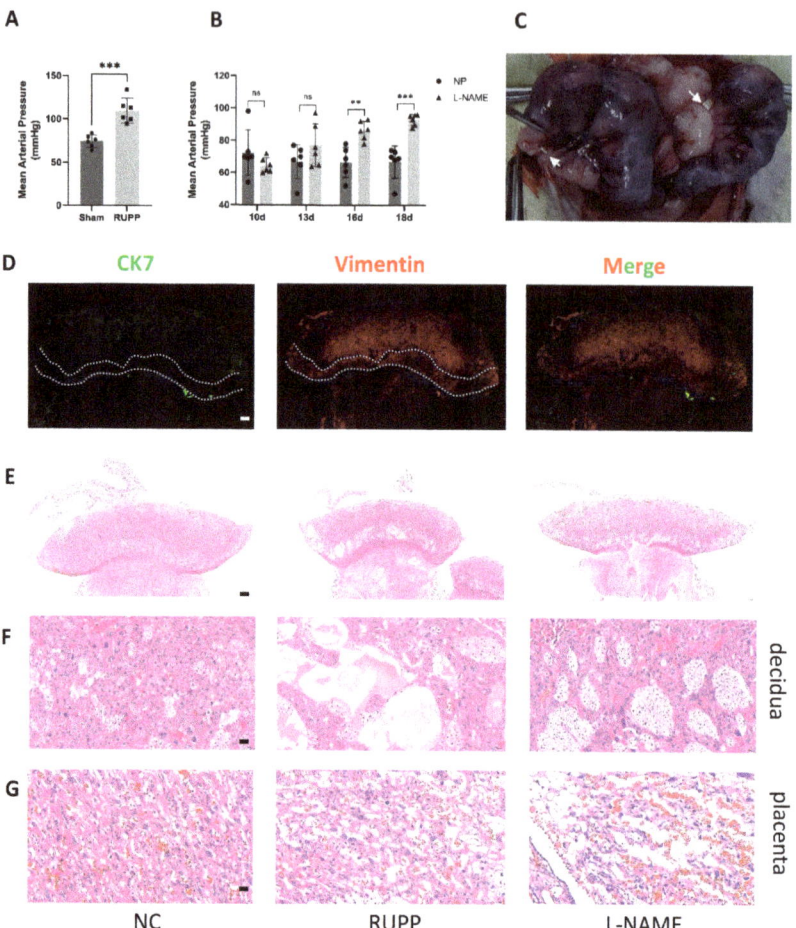

Figure 8. Preeclampsia animal model construction (**A,B**) The success of the animal model was confirmed by statistically analysing blood pressure, $n = 6$ for each group. ** $p < 0.01$; *** $p < 0.001$; ns, not significant. (**C**) Location of two silver clips on vessels between the ovary and uterus. (**D**) Determination of cell components of the maternal–foetal interface of SD rats. Green: CK7; red: vimentin, Bar = 500 nm. (**E**–**G**) HE staining showed the maternal–foetal interface structure change of RUPP and L-NAME rats vs. normal SD rats. $n = 3$ for each group, (**E**) shows the whole picture, scale bar = 500 nm, (**F**) shows the decidua, scale bar = 50 nm, and (**G**) shows the placenta, scale bar = 20 nm.

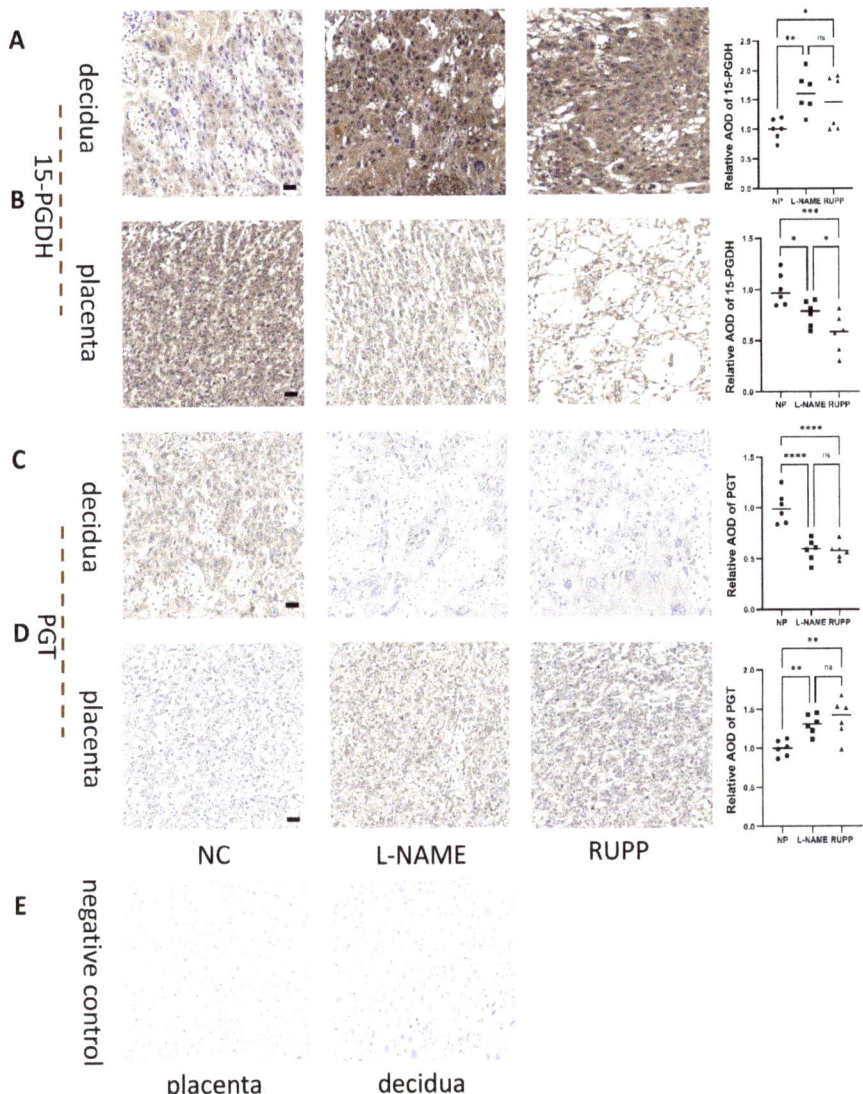

Figure 9. Enzyme 15-PGDH and PGT enzymes in the RUPP and L-NAME models. (**A**) Immunohistochemical staining showed the localization and relative quantification of 15-PGDH in the decidua of the RUPP and L-NAME rat groups. Brown staining represents the target protein, and (**B**) immunohistochemical staining showed the localization and relative quantification of 15-PGDH in the labyrinth zone of the RUPP and L-NAME vs. NP rat groups. $n = 6$ for each group. (**C**) Immunohistochemical staining showed the localization and relative quantification of PGT in the decidua of the RUPP and L-NAME rat groups. Brown staining represents the target protein, and (**D**) immunohistochemical staining showed the localization and relative quantification of PGT in the labyrinth zone of the RUPP and L-NAME vs. NP rat groups. The results are representative of at least three independent experiments. (**E**) represents negative control of placenta and decidua. Scale bar = 50 nm. * $p < 0.05$; ** $p < 0.01$; *** $p < 0.001$; **** $p < 0.0001$; ns, not significant (one-way ANOVA). $n = 6$ for each group. NP: normal pregnancy.

3. Discussion

In our study, we verified EMT/MET during placental development and decidualization. Partial EMT occurs during trophoblast differentiation, and MET occurs during decidualization. Then, we found that both the trophoblasts and DSCs of PE patients tended to have more epithelial patterns, indicating insufficient EMT of trophoblasts and excessive MET during decidualization of DSCs. 15-PGDH was differentially expressed in the placenta and decidua of PE patients. Inhibiting 15-PGDH promoted trophoblast differentiation and DSC decidualization and led to a shift to a mesenchymal pattern in both groups in a dose-dependent manner. Inhibiting 15-PGDH can upregulate PGT, which can lead to the uptake of more PGE2. Inhibiting 15-PGDH promotes a mesenchymal pattern depending on the lactate-PG transport mechanism of PGT.

EMT/MET was once thought to be a transition between a complete epithelial pattern or mesenchymal pattern. However, with more research, EMT/MET should be further elucidated. EMTs/METs are multistep, reversible, dynamic biological processes of cell differentiation and dedifferentiation, with cells transitioning along various stages, including various partial EMT states, which are also characterized by cytoskeleton and molecular marker changes. In regard to trophoblast development, in our study, CDH1 was decreased in the process of CTB EMT, while CDH2 was not significantly increased, which was consistent with some previous dissertations. In 2015, a study found that EMT occurring in placental development lacked traditional characteristics of EMT types 1–3 and defined trophoblast EMT as type 0 [17]. A review published in 2019 in BMJ defined trophoblast EMT as partial EMT [18]. In our research, we used coimmunofluorescence to reveal partial EMT during placental development.

Whether EMT/MET is the basis for cells to undertake different biological functions and cell differentiation, is parallel to cell differentiation, or only prepares for invasion or metastasis is an interesting question. HLA-G is expressed during CTB EMT and is a marker of CTB development, which is a kind of human leukocyte antigen that plays a major role in mediating immune tolerance. This molecule can prevent the embryo from being attacked by immune cells in the decidua. The expression of HLA-G is decreased in patients with PE [19], which is consistent with the decrease in EMT in patients with PE. However, in our research, we found that HLA-G expression did not significantly change when trophoblasts changed along the epithelial and mesenchymal spectra. In addition, the blots of CDH1 show numerous panels, which may be caused by alternative splicing or protein modification. Whether modification of CDH1 involves new mechanism of EMT/MET needs further research.

In the process of decidualization, MET occurs with PRL and IGFBP1 expression. In DSCs in PE patients, excess MET is accompanied by PRL and IGFBP1 deficiency; in an in vitro cell model, insufficient MET was accompanied by PRL and IGFBP1 increases. These phenomena seem contrary, which may be due to the oversimplification of EMT/MET. The relationship between EMT/MET and decidualization is complex. A review of EMT/MET in 2016 may help explain these phenomena [20]. The authors of this review predicted and described hypothetical EMT/MET transitional states, among which there are several special states, including intermediate state (EM) 1 and EM3, indicating that cells are on more thermodynamic peaks with more metastable states than complete epithelial or mesenchymal patterns. Between EM1 and EM3, EM2 has a higher mesenchymal/epithelial score (M/E score) than EM1 and a lower score than EM3 but with lower energy, which means EM2 is more stable than EM1 [20]. Therefore, we predict that the "fitting curve" between EMT/MET and IGFBP1 and PRL is a "sinusoidal curve" instead of a "linear regression curve", which means that IGFBP1 and PRL do not increase with increasing M/E scores. Among the decidualization procedures, there may be different stages with different PRL or IGFBP1 expression or M/E scores. If we understand EMT/MET in this way, many seemingly contradictory conclusions are reasonable. However, due to the limitations of the experimental model, we could not build decidualization with multiple partial EMT states to detect the relationship between EMT/MET. Whether EMT/MET is the basis of HLA-G, IGFBP1, or PRL expression or a parallel occurrence still deserves further study.

The placentas and deciduas of patients with PE showed an excessive epithelial pattern and insufficient mesenchymal pattern, indicating that EMT of the placenta is insufficient, while MET of the decidua is excessive. In the mesenchymal pattern, the loss of stress fibres from the centre of the cell body promotes the movement of cells and makes it easier for trophoblasts and decidual cells to invade each other and for an embryo to attach to the mother, completing vascular remodelling and forming a healthy maternal–foetal interface. This phenomenon may be one of the reasons for the shallow implantation of trophoblasts. Our experiment proved that the epithelial pattern in the deciduas of PE patients was excessive, which was consistent with another study [21]. However, studies have shown that the deciduas of recurrent spontaneous abortion (RSA) show a shift to an excess mesenchymal pattern [22]. Previous studies have shown that RSA and PE share a similar pathogenesis; however, from the perspective of decidua MET, there are essential differences between the decidualization abnormalities of RSA and PE patients. However, this conclusion still needs multicentre and multisource confirmation.

Prostaglandins play an important role in embryo implantation, formation of the maternal–foetal interface and initiation of labour. Prostaglandins regulate embryo implantation at lower concentrations and inhibit embryo implantation at higher concentrations. Therefore, the precise regulation of prostaglandin metabolism is very important for the formation of the maternal–foetal interface. Aspirin, as an inhibitor of prostaglandin synthase COX-2, has been widely used in the clinic to prevent PE [23,24]. PGT and 15-PGDH also play important roles in regulating prostaglandin concentrations. PGT and 15-PGDH may also become targets for the treatment of PE, similar to aspirin.

However, the expression of 15-PGDH is increased in the placentas of PE patients, but it promotes the epithelial–mesenchymal transformation of trophoblasts. This conclusion seems contradictory, and the possible reasons are listed. 1. Lesions exist in early pregnancy in PE patients; however, for ethical reasons, research on PE relies on the placenta at the time of delivery. Therefore, 15-PGDH may be compensated in a long gestational duration, or the expression of 15-PGDH may be affected by aspirin, which is commonly used. 2. The other explanation involves a more controversial viewpoint. Is the placenta an organ that causes PE or an injured organ that is harmed by PE [1,25]? This study tested the hypothesis that decidual defects are an important determinant of the placental phenotype [1]. In our study, we demonstrated that 15-PGDH, which inhibits mesenchymal patterns and decidualization, is upregulated in the deciduas but downregulated in the placentas of PE patients. This conclusion suggests that the decidua and the microenvironment of trophoblasts are potential causes of PE. However, testing only one molecule is not enough, and this issue requires further research.

4. Materials and Methods

4.1. Patients and Sample Collection

First trimester placental villi were obtained from healthy women undergoing elective surgical termination of their pregnancies from 6–8 weeks of gestation. A total of 16 placental tissue samples from patients with PE and matched healthy controls were collected in the Obstetrics and Gynaecology Department, Renmin Hospital of Wuhan University (Wuhan, China) from March 2022 to May 2022 (Supplementary Table S1), and informed consent was obtained from all the patients in advance. Placental tissues and decidual tissues were collected from women in the third trimester during caesarean section. The placenta and decidua specimens were washed with sterile PBS, fixed in 4% paraformaldehyde, or quick-frozen in liquid nitrogen for later use. Human sample collection was authorized by the Ethical Review Board of Renmin Hospital, Wuhan University (WDRY2021-K177) and performed in accordance with the Declaration of Helsinki.

4.2. Cell Culture and Differentiation

Human trophoblast cell lines (Jeg3, HTR8) and human decidual stromal cell line (hESC) were purchased from the Cell Bank of the Chinese Academy of Sciences (Shanghai, China). Jeg3 and HTR8 cells were cultured in a 5% humidified carbon dioxide atmosphere at 37 °C in Dulbecco's modified Eagle's medium (DMEM)/F-12 (Gibco, Life Technologies, Grand Island, NY, USA) with 10% foetal bovine serum (Gibco, Life Technologies, Grand Island, NY, USA), 50 mg/mL streptomycin, and 50 U/mL penicillin. Jeg3 is the only trophoblast cell line that expresses human leukocyte antigen G (HLA-G), which is gradually expressed during development from CTBs to EVTs.

hESCs were cultured in a 5% humidified carbon dioxide atmosphere at 37 °C in phenol red-free Dulbecco's modified Eagle's medium (DMEM)/F-12 (Meilunbio, Dalian, China) with 10% foetal bovine serum, 50 mg/mL streptomycin, and 50 U/mL penicillin. One micromolar medroxyprogesterone-17-acetate (MPA) (HY-B0469, MedChemExpress, NJ, USA) and 0.5 mM N6,20-O-dibutyryladenosine cAMP sodium salt (db-cAMP) (HY-B0764, MedChemExpress, Monmouth Junction, NJ, USA) were added to the culture for 6 days to induce hESC decidualization in vitro [14].

4.3. SW033291 and ICG Treatment

Cells were treated with 500 nM SW033291 (HY-16968, MedChemExpress, Monmouth Junction, NJ, USA) [8] to inhibit 15-PGDH or with 67 µM indocyanine green (ICG) (HY-D0711, MedChemExpress, Monmouth Junction, NJ, USA) to inhibit PGT.

4.4. Western Blot Analysis

Total protein was extracted from cells with RIPA buffer, PMSF protease inhibitors, and a phosphatase inhibitor (all from Servicebio, Wuhan, China) and then ultrasonicated and centrifuged at 12,000 rpm for 10 min at 4 °C. The supernatants, fixed with loading buffer (Elabscience, Wuhan, China), were heated for 5 min at 100 °C and then kept at −20 °C. Protein from each sample was resolved through 10% SDS-PAGE (PG212 Omni-EastTM, EpiZyme, Shanghai, China) and then transferred to polypropylene difluoride membranes (Millipore, USA) for 30 min (PS108P, EpiZyme, Shanghai, China). The membranes were blocked in blocking buffer (PS108P, EpiZyme, Shanghai, China) for 15 min at room temperature and then immunoblotted with primary antibodies against 15-PGDH (11035-1-AP, Proteintech, Wuhan, China), PGT (ab150788, Abcam, Cambridge, UK), E-cadherin, also named CDH1, (20874-1-AP, Proteintech, Wuhan, China), N-cadherin, also named CDH2, (22018-1-AP, Proteintech, Wuhan, China), vimentin (10366-1-AP, Proteintech, Wuhan, China), IGFBP1 (Ab-DF7130, Affinity, Jiangsu, China), prolactin/PRL (Ab-DF6506, Affinity, Jiangsu, China), HLA-G (66447-1-Ig, Proteintech, Wuhan, China), and GAPDH (10494-1-AP, 1:5000, Proteintech, Wuhan, China) overnight at 4 °C, followed by incubation with goat anti-rabbit IgG (H + L) or goat anti-mouse IgG (H + L) (GB23303, 1:1000, Servicebio Technology Co., Wuhan, China) for 1.5 h at 4 °C. Protein expression was detected by a chemiluminescent detection system (Bio-Rad, Hercules, CA, USA) using ECL Plus reagents (Servicebio Technology Co., Wuhan, China). The expression levels of targeted proteins were normalized to GAPDH. Western blot analysis was conducted using ImageJ Pro Plus version 6.0 software.

4.5. Immunohistochemistry

Paraffin-embedded placental tissues and decidua tissues were sectioned at a thickness of 10 nm, dewaxed, rehydrated and blocked with BSA. Sections were incubated overnight with primary antibodies as described for Western blotting. Sections for negative control were incubated without primary antibody. After washing with phosphate-buffered saline (pH 7.4), the sections were incubated for 2 h with HRP-conjugated secondary antibodies: Alexa Fluor 488-conjugated goat anti-mouse IgG (A32723; Thermo Fisher Scientific, Waltham, MA, USA) and Alexa Fluor 568-conjugated goat anti-rabbit IgG (A11011; Thermo Fisher Scientific, Waltham, MA, USA). Nuclei were visualized with DAPI (Beyotime, Shanghai, China). The digital image processing system ImageJ Pro Plus version 6.0 was then employed to evaluate Area and IntDen of IHC. Average optical density (AOC) = IntDen/area, and AOD was calculated.

After deparaffinization and rehydration, standard H&E staining was performed for morphological analysis.

4.6. Immunofluorescence

The steps before incubation with primary antibodies were the same as those used for immunohistochemistry. The samples were incubated with the desired dilutions of primary antibodies overnight at 4 °C. The samples were then incubated with fluorescence-labelled secondary antibody for 1 h (Beyotime, Shanghai, China) at room temperature and counterstained with 4′-6-diamidino-2-phenylindole (DAPI) (Beyotime, Shanghai, China). The primary antibodies used in the study were the same as those used for Western blotting. The secondary antibodies used included anti-rabbit IgG (H + L) Alexa Fluor 555 (Invitrogen, San Diego, CA, USA; A-31572) and anti-goat IgG (H + L) Alexa Fluor Plus 488 (Invitrogen; A32814). A confocal laser scanning microscope (Olympus FV1000, Tokyo, Japan) was used to observe the fluorescence signal. Five visual fields with tissue were selected for analysis. The pixel intensity per unit area was assessed using ImageJ (1.52a, National Institutes of Health, Rockville, MD, USA).

4.7. ELISA

The level of PGE2 in cell lysates was identified by ELISA. Ultrasonicated cells were centrifuged at 12,000 rpm for 10 min at 4 °C. PGE2 concentrations in the cells were detected by a PGE2 competitive ELISA kit (EK8103/2, Multi Science, Hangzhou, China). All the above-mentioned analyses were performed according to the relevant manufacturer's instructions.

4.8. F-Actin Staining

After treatment with SW033291 or ICG, all round coverslip samples were washed with PBS and fixed in 3.7% paraformaldehyde (Servicebio, Wuhan, China) for 15 min, and the coverslips were washed three times with PBS. Then, the coverslips were permeabilized with 0.1% Triton X-100 for 10 min and stained with rhodamine conjugated to phalloidin (Phalloidin-iFluor 647 Reagent, Abcam, Cambridge, UK) for 30 min at 37 °C. After three washes with PBS, the nuclei were visualized with DAPI (100 nM) for 10 min.

4.9. Preeclampsia Rat Model Construction

Twenty-five Sprague–Dawley female rats and fifteen Sprague–Dawley male rats, purchased from Beijing Vital River Laboratory Animal Technology Co., Ltd., China, were used in our studies. After adapting to culture for 7 days in specific pathogen-free (SPF) conditions at Renmin Hospital of Wuhan University, the rats were mated, and the day was recorded as day 0.5 of gestation. Pregnant rats were randomly divided into 4 groups according to their body weight. The L-NAME group was subcutaneously injected with NG-nitroarginine methyl ester hydrochloride (L-NAME) (HY-18729A, MedChemExpress, Monmouth Junction, NJ, USA) from the 10th day of pregnancy to the 18th day of pregnancy (100 mg/kg × day), while the normal pregnancy control group was injected subcutaneously with physiological saline. Blood pressure was measured on gestational day 10, 13, 16 and 18 by noninvasive tail-cuff system (CODA system, Kent Scientific, Torrington, CT, USA). Rats in RUPP group were operated on at 14.5 days of pregnancy. The surgical procedure was performed as described previously [16,26], and the skin was cut along the midline of the abdomen. The omentum was gently pushed with two cotton swabs, and the intestinal tube was pushed with wet gauze to expose the posterior abdominal wall. The abdominal aorta was found, the surrounding fascial tissue was separated, and then, the abdominal aortic silver clip was slid to 0.5 cm above the abdominal aortic bifurcation. Then, the ovarian artery silver clips were placed, as Figure 5C shows (white triangles indicate the position of the silver clips). No silver clip was placed in rats in the sham group. The sham group means negative control group. On gestational day 18, carotid arterial catheters were inserted for blood pressure measurements. After blood pressure measurement, tissues were collected. Collection of the maternal–foetal interface: we cut the uterus along the opposite side of the uterine blood vessels, peeled off the amnion, and cut off the umbilical cord. Without separating the placenta and the uterus, we completely preserved the maternal–foetal interface and maintained its morphology. Kidneys and other organs were also collected. The specimens were washed with sterile PBS and then fixed in 4% paraformaldehyde for later use. All animal studies were approved by the ethics committee for laboratory animal welfare (IACUC) of Renmin Hospital of Wuhan University [No. WDRM animal (f) No. 2022103C].

4.10. Statistical Analysis

Statistical significance was determined by SPSS 20.0 software, and $p = 0.05$ was the threshold. Student's *t* test or one-way ANOVA was used to analyse differences between two or more groups.

5. Conclusions

We demonstrate for the first time that abnormal 15-PGDH and PGT expression could be associated with abnormal EMT/MET in patients with preeclampsia. 15-PGDH inhibition improves the mesenchymal pattern of both trophoblasts and decidual stromal cells, which are the most important components of the maternal–foetal interface. Inhibiting 15-PGDH upregulated PGT expression on the cell membrane. 15-PGDH promotes PE relying on PGT, which functions as an electrogenic anion exchanger for PG with lactate. The present study provides new insights into the potential role of 15-PGDH and PGT in PE treatment.

Supplementary Materials: The following supporting information can be downloaded at: https://www.mdpi.com/article/10.3390/ijms24065111/s1.

Author Contributions: H.P. proposed this study, performed cell culture experiments and designed the study. H.P., T.C. and D.L. performed most of the animal experiments with assistance from Y.L., D.L. and Y.L. acquired and analyzed human samples. C.F. supervised the research and revised the manuscript written by H.P. All authors provided critical feedback and edited the manuscript. All authors have read and agreed to the published version of the manuscript.

Funding: This research received no external funding.

Institutional Review Board Statement: Human sample collection was authorized by the Ethical Review Board of Renmin Hospital, Wuhan University (WDRY2021-K177) and performed in accordance with the Declaration of Helsinki. All animal studies were approved by the ethics committee for laboratory animal welfare (IACUC) of Renmin Hospital of Wuhan University [No. WDRM animal (f) No. 2022103C].

Informed Consent Statement: Informed consent was obtained from all subjects involved in the study.

Data Availability Statement: All data associated with this study are available in the main text or the supplementary materials.

Conflicts of Interest: The authors declare no conflict of interest.

References

1. Garrido-Gomez, T.; Dominguez, F.; Quiñonero, A.; Diaz-Gimeno, P.; Kapidzic, M.; Gormley, M.; Ona, K.; Padilla-Iserte, P.; McMaster, M.; Genbacev, O.; et al. Defective decidualization during and after severe preeclampsia reveals a possible maternal contribution to the etiology. *Proc. Natl. Acad. Sci. USA* **2017**, *114*, E8468–E8477. [CrossRef] [PubMed]
2. Garrido-Gomez, T.; Quiñonero, A.; Dominguez, F.; Rubert, L.; Perales, A.; Hajjar, K.A.; Simon, C. Preeclampsia: A defect in decidualization is associated with deficiency of Annexin A2. *Am. J. Obstet. Gynecol.* **2020**, *222*, 376.e1–376.e17. [CrossRef] [PubMed]
3. Reale, S.C.; Camann, W.R. Preeclampsia. *N. Engl. J. Med.* **2022**, *387*, 286–287. [PubMed]
4. Davies, J.E.; Pollheimer, J.; Yong, H.E.; Kokkinos, M.I.; Kalionis, B.; Knöfler, M.; Murthi, P. Epithelial-mesenchymal transition during extravillous trophoblast differentiation. *Cell Adhes. Migr.* **2016**, *10*, 310–321. [CrossRef] [PubMed]

5. Illsley, N.P.; DaSilva-Arnold, S.C.; Zamudio, S.; Alvarez, M.; Al-Khan, A. Trophoblast invasion: Lessons from abnormally invasive placenta (placenta accreta). *Placenta* **2020**, *102*, 61–66. [CrossRef]
6. Owusu-Akyaw, A.; Krishnamoorthy, K.; Goldsmith, L.T.; Morelli, S.S. The role of mesenchymal-epithelial transition in endometrial function. *Hum. Reprod. Update* **2019**, *25*, 114–133. [CrossRef]
7. Palla, A.R.; Ravichandran, M.; Wang, Y.X.; Alexandrova, L.; Yang, A.V.; Kraft, P.; Holbrook, C.A.; Schürch, C.M.; Ho, A.T.V.; Blau, H.M. Inhibition of prostaglandin-degrading enzyme 15-PGDH rejuvenates aged muscle mass and strength. *Science* **2021**, *371*, eabc8059. [CrossRef]
8. Chan, S.Y. Effects of prostaglandin E2 and F2 alpha on peri-implantation development of mouse embryos in vitro. *Prostaglandins* **1991**, *42*, 321–336. [CrossRef]
9. Zhang, Y.; Desai, A.; Yang, S.Y.; Bae, K.B.; Antczak, M.I.; Fink, S.P.; Tiwari, S.; Willis, J.E.; Williams, N.S.; Dawson, D.M.; et al. TISSUE REGENERATION. Inhibition of the prostaglandin-degrading enzyme 15-PGDH potentiates tissue regeneration. *Science* **2015**, *348*, aaa2340. [CrossRef]
10. Fan, L.; Li, Y.; Zhang, X.; Wu, Y.; Song, Y.; Zhang, F.; Zhang, J.; Sun, H. Time-resolved proteome and transcriptome of paraquat-induced pulmonary fibrosis. *Pulm. Pharmacol. Ther.* **2022**, *75*, 102145. [CrossRef]
11. Nakanishi, T.; Nakamura, Y.; Umeno, J. Recent advances in studies of SLCO2A1 as a key regulator of the delivery of prostaglandins to their sites of action. *Pharmacol. Ther.* **2021**, *223*, 107803. [CrossRef] [PubMed]
12. Chan, B.S.; Endo, S.; Kanai, N.; Schuster, V.L. Identification of lactate as a driving force for prostanoid transport by prostaglandin transporter PGT. *Am. J. Physiol. Ren. Physiol.* **2002**, *282*, F1097–F1102. [CrossRef] [PubMed]
13. Ferreira, L.M.R.; Meissner, T.B.; Tilburgs, T.; Strominger, J.L. HLA-G: At the Interface of Maternal-Fetal Tolerance. *Trends Immunol.* **2017**, *38*, 272–286. [CrossRef] [PubMed]
14. Valatkaitė, E.; Baušytė, R.; Vitkevičienė, A.; Ramašauskaitė, D.; Navakauskienė, R. Decidualization Potency and Epigenetic Changes in Human Endometrial Origin Stem Cells During Propagation. *Front. Cell Dev. Biol.* **2021**, *9*, 765265. [CrossRef]
15. Lin, Y.; Zeng, Y.; Di, J.; Zeng, S. Murine CD200+ CK7+ trophoblasts in a poly (I:C)-induced embryo resorption model. *Reproduction* **2005**, *130*, 529–537. [CrossRef] [PubMed]
16. Pang, H.; Lei, D.; Huang, J.; Guo, Y.; Fan, C. Elevated PGT promotes proliferation and inhibits cell apoptosis in preeclampsia by Erk signaling pathway. *Mol. Cell. Probes* **2023**, *67*, 101896. [CrossRef]
17. DaSilva-Arnold, S.; James, J.L.; Al-Khan, A.; Zamudio, S.; Illsley, N.P. Differentiation of first trimester cytotrophoblast to extravillous trophoblast involves an epithelial-mesenchymal transition. *Placenta* **2015**, *36*, 1412–1418. [CrossRef]
18. Burton, G.J.; Redman, C.W.; Roberts, J.M.; Moffett, A. Pre-eclampsia: Pathophysiology and clinical implications. *BMJ (Clin. Res. Ed.)* **2019**, *366*, l2381. [CrossRef]
19. Wedenoja, S.; Yoshihara, M.; Teder, H.; Sariola, H.; Gissler, M.; Katayama, S.; Wedenoja, J.; Häkkinen, I.M.; Ezer, S.; Linder, N.; et al. Fetal HLA-G mediated immune tolerance and interferon response in preeclampsia. *eBioMedicine* **2020**, *59*, 102872. [CrossRef]
20. Nieto, M.A.; Huang, R.Y.; Jackson, R.A.; Thiery, J.P. EMT: 2016. *Cell* **2016**, *166*, 21–45. [CrossRef]
21. Chen, J.; Ren, W.; Lin, L.; Zeng, S.; Huang, L.; Tang, J.; Bi, S.; Pan, J.; Chen, D.; Du, L. Abnormal cGMP-dependent protein kinase I-mediated decidualization in preeclampsia. *Hypertens. Res. Off. J. Jpn. Soc. Hypertens.* **2021**, *44*, 318–324. [CrossRef] [PubMed]
22. Chen, J.; Liu, J.; Gao, S.; Qiu, Y.; Wang, Y.; Zhang, Y.; Gao, L.; Qi, G.; Wu, Y.; Lash, G.E.; et al. Role of Slit2 upregulation in recurrent miscarriage through regulation of stromal decidualization. *Placenta* **2021**, *103*, 1–9. [CrossRef] [PubMed]
23. Davidson, K.W.; Barry, M.J.; Mangione, C.M.; Cabana, M.; Caughey, A.B.; Davis, E.M.; Donahue, K.E.; Doubeni, C.A.; Kubik, M.; Li, L.; et al. Aspirin Use to Prevent Preeclampsia and Related Morbidity and Mortality: US Preventive Services Task Force Recommendation Statement. *JAMA* **2021**, *326*, 1186–1191.
24. Rolnik, D.L.; Wright, D.; Poon, L.C.; O'Gorman, N.; Syngelaki, A.; de Paco Matallana, C.; Akolekar, R.; Cicero, S.; Janga, D.; Singh, M.; et al. Aspirin versus Placebo in Pregnancies at High Risk for Preterm Preeclampsia. *N. Engl. J. Med.* **2017**, *377*, 613–622. [CrossRef] [PubMed]
25. Melchiorre, K.; Giorgione, V.; Thilaganathan, B. The placenta and preeclampsia: Villain or victim? *Am. J. Obstet. Gynecol.* **2022**, *226*, S954–S962. [CrossRef]
26. Granger, J.P.; LaMarca, B.B.; Cockrell, K.; Sedeek, M.; Balzi, C.; Chandler, D.; Bennett, W. Reduced uterine perfusion pressure (RUPP) model for studying cardiovascular-renal dysfunction in response to placental ischemia. *Methods Mol. Med.* **2006**, *122*, 383–392.

Disclaimer/Publisher's Note: The statements, opinions and data contained in all publications are solely those of the individual author(s) and contributor(s) and not of MDPI and/or the editor(s). MDPI and/or the editor(s) disclaim responsibility for any injury to people or property resulting from any ideas, methods, instructions or products referred to in the content.

Article

Altered Mesenchymal Stem Cells Mechanotransduction from Oxidized Collagen: Morphological and Biophysical Observations

Regina Komsa-Penkova [1,*], Adelina Yordanova [2], Pencho Tonchev [3], Stanimir Kyurkchiev [2], Svetla Todinova [4], Velichka Strijkova [4,5], Mario Iliev [6], Borislav Dimitrov [1] and George Altankov [7,*]

[1] Department of Biochemistry, Medical University Pleven, 5800 Pleven, Bulgaria
[2] Tissue Bank BulGen, 1330 Sofia, Bulgaria
[3] Department of Surgery, Medical University Pleven, 5800 Pleven, Bulgaria
[4] Institute of Biophysics and Biomedical Engineering, Bulgarian Academy of Sciences, 1113 Sofia, Bulgaria
[5] Institute of Optical Materials and Technologies, Bulgarian Academy of Sciences, 1113 Sofia, Bulgaria
[6] Faculty of Physics, Sofia University, St. Clément Ohnishi, 1164 Sofia, Bulgaria
[7] Research Institute, Medical University Pleven, 5800 Pleven, Bulgaria
* Correspondence: rkomsa@gmail.com (R.K.-P.); altankov@abv.bg (G.A.)

Citation: Komsa-Penkova, R.; Yordanova, A.; Tonchev, P.; Kyurkchiev, S.; Todinova, S.; Strijkova, V.; Iliev, M.; Dimitrov, B.; Altankov, G. Altered Mesenchymal Stem Cells Mechanotransduction from Oxidized Collagen: Morphological and Biophysical Observations. *Int. J. Mol. Sci.* **2023**, *24*, 3635. https://doi.org/10.3390/ijms24043635

Academic Editors: Sabrina Lisi and Margherita Sisto

Received: 8 January 2023
Revised: 5 February 2023
Accepted: 8 February 2023
Published: 11 February 2023

Copyright: © 2023 by the authors. Licensee MDPI, Basel, Switzerland. This article is an open access article distributed under the terms and conditions of the Creative Commons Attribution (CC BY) license (https://creativecommons.org/licenses/by/4.0/).

Abstract: Extracellular matrix (ECM) provides various mechanical cues that are able to affect the self-renewal and differentiation of mesenchymal stem cells (MSC). Little is known, however, how these cues work in a pathological environment, such as acute oxidative stress. To better understand the behavior of human adipose tissue-derived MSC (ADMSC) in such conditions, we provide morphological and quantitative evidence for significantly altered early steps of mechanotransduction when adhering to oxidized collagen (Col-Oxi). These affect both focal adhesion (FA) formation and YAP/TAZ signaling events. Representative morphological images show that ADMSCs spread better within 2 h of adhesion on native collagen (Col), while they tended to round up on Col-Oxi. It also correlates with the lesser development of the actin cytoskeleton and FA formation, confirmed quantitatively by morphometric analysis using ImageJ. As shown by immunofluorescence analysis, oxidation also affected the ratio of cytosolic-to-nuclear YAP/TAZ activity, concentrating in the nucleus for Col while remaining in the cytosol for Col-Oxi, suggesting abrogated signal transduction. Comparative Atomic Force Microscopy (AFM) studies show that native collagen forms relatively coarse aggregates, much thinner with Col-Oxi, possibly reflecting its altered ability to aggregate. On the other hand, the corresponding Young's moduli were only slightly changed, so viscoelastic properties cannot explain the observed biological differences. However, the roughness of the protein layer decreased dramatically, from R_{RMS} equal to 27.95 ± 5.1 nm for Col to 5.51 ± 0.8 nm for Col-Oxi ($p < 0.05$), which dictates our conclusion that it is the most altered parameter in oxidation. Thus, it appears to be a predominantly topographic response that affects the mechanotransduction of ADMSCs by oxidized collagen.

Keywords: mesenchymal stem cells; mechanotransduction; collagen; oxidation; YAP/TAZ; focal adhesion

1. Introduction

The extracellular matrix (ECM) initiates several mechanical cues that are able to activate intracellular signaling events through cell–matrix interactions [1–4]. It is generally agreed that the quality and quantity of ECM determine its physical parameters, affecting cellular response [5], which also applies to the behavior of stem cells [3,6–9]. Each organ or tissue provides specific mechanical cues [3] that have to be understood in the context of the entire multicellular organization [5,10]. The situation is different, however, when cells interact with surfaces (in a 2D system), which is often the case with implanted biomaterials. Here the surface roughness and surface energy (unified as nanotopography) play a pivotal role [7–10]. The surface nanotopography strongly influences osteoblastic proliferation, differentiation, and extracellular matrix protein expression [5]. A line of research proves that surface roughness modification of titanium implants improves bone-to-implant contact [11,12]. Other examples are the nanofibers [13] or other linear structures, where the organization of adhesive sites dictates the cellular response [3]. Collectively, a growing body of evidence suggests that both surface stiffness and surface topography affect cell fate, gene expression, and whole cell cycle progression in various cell types [9,14–16]. The cell–matrix interaction is mediated by focal adhesions (FA), the main hub for mechanotransduction, connecting the ECM proteins, integrins, and the cytoskeleton [17]. Focal adhesions, however, develop better on 2D surfaces, driven by the stiffness, topography, and surface energy [18], as well as by the organization of adsorbed adhesive proteins [13,19]. The forces exerted on cell adhesion molecules further regulate the RhoA signaling pathway by controlling the activities of guanine nucleotide exchange factors (GEFs) and GTPase activating proteins (GAPs) [1]. Recently, it has become clear that the intracellular Hippo signaling pathway is the next hub that regulates a number of important biological processes, including cellular proliferation, survival, and differentiation [20–23], and thus determines organ size and tissue homeostasis [20–22]. Originally discovered in Drosophila melanogaster, the Hippo pathway is highly conserved across species, as equivalent genes and their products can be found in mammals [20] as a complex cascade of serine/threonine-protein kinases STK3 and STK4 [2,4]. They form a complex with the adaptor/Salvador protein (SAV1) that can phosphorylate and activate the effector protein, large tumor suppressor 1/2 (LATS1/2). At the same time, it inhibits the transcription cofactors "Yes" associated protein (YAP1) and its transcriptional co-activator with PDZ-binding motif (TAZ) [19,20,24–27].

A growing body of evidence suggests that YAP/TAZ signaling is the next intracellular key for driving cell behavior via the Hippo pathway. When "off," the phosphorylated YAP/TAZ retains in the cytoplasm where it could undergo proteolytic degradation [5,19] but when "on," the unphosphorylated YAP/TAZ moves into the nucleus and binds to transcription factors called TEA DNA-binding proteins (TEAD1–4), regulating various proliferative and pro-survival genes, thus having a critical impact on cell behavior [20–26].

Research from the past decade has tremendously expanded our knowledge about mesenchymal stem cell (MSC) physiology in response to physical signals in the environment [3,6–8]. MSCs are a group of progenitor cells characterized by their ability for self-renewal and directed differentiation [3,28]. Within their local tissue microenvironment or niche, MSCs communicate with the ECM accepting various chemical, physical, and mechanical cues to regulate their fate and behavior [3,7,8,27,29–31]. Nowadays, it is generally agreed that MSCs perceive their microenvironment through soluble (diffusible) signals and mechanical cues, such as ECM stiffness, nanotopography, or confined adhesiveness [3,23–27,29]. For example, MSCs have the ability to differentiate into neuroblast, chondrocyte, osteoblast, adipocyte, and numerous other cell types when they reside within matrices that mimic the stiffness of their native substrate [10,20,24,29]. However, it reflects their physiological environment providing specific viscoelastic properties, while the topographic response is less studied—though it is proposed that this may also determine the local response of stem cells toward tissue repair and regeneration [24]. Collagen is the most abundant protein in the ECM, critical for its mechanical properties, including stiffness, roughness, extracellular forces, and topography, thus affecting various cell functions and communications [32–34]. Though our knowledge of the composition of natural ECM is continuously growing, the impact of its structural organization on the adjacent cellular microenvironment is not well understood, particularly in pathological conditions [32,35].

Oxidative stress is one such condition known to affect the collagen structure and turnover strongly [36,37], including its extracellular processing [32], and remodeling [38,39]. Although these processes are extensively studied, direct investigations utilizing adsorbed collagen layers are rather sparse. From this point, our recent study showed that the oxidation of adsorbed type I collagen alters its remodeling by stem cells [39], opening the door for further research.

The use of adipose tissue-derived MSCs (ADMSCs) as a cellular model also draws considerable attention as they combine relatively easy availability and less donor site morbidity, and possess the characteristic multi-potency making them very suitable for tissue engineering applications [31,39,40]. Here, we provide morphological and quantitative (morphometric) evidence for the altered mechanotransduction of ADMSCs adhering to oxidized collagen involving both focal adhesions (FA) and YAP/TAZ signaling pathways, aiming to understand better the stem cells' behavior in conditions of acute oxidative stress.

2. Results

2.1. Initial Cell Attachment

Glass coverslips were coated with native (Col) or oxidized collagen (Col-Oxi) produced by previously described procedure [41] to follow the initial attachment of ADMSCs after 2 h of incubation in a serum-free medium. This study was focused on the early signaling events (see below) since, at later stages, it was expectable that cells would produce a plethora of matrix proteins that may scramble the collagen effect alone. For the same reason, the serum was omitted from the medium. After incubation, the cells were fixed and permeabilized before being stained according to protocol 1 (see Methods section) for actin (to view the cytoskeleton), vinculin (to visualize the focal adhesions), and the nucleus. Phase contrast pictures of living cells were also captured; representative images are shown in Figure 1. As evident from the low magnification phase contrast pictures (A and D) and the morphometric data in Table 1, ADMSCs attached equally well on both substrates but spread differently: much better on native Col (A) with a spreading area of 216.0 μM^2 versus 179.6 μM^2 for the oxidized samples (D). At the same time, the cells display more rounded morphology on Col-Oxi, as pointed at with yellow arrows on (D), and are confirmed quantitatively by the higher CSI index for the oxidized samples (0.33), lowering to (0.25) for the native ones, as presented in Table 1. Note that the CSI index tends to 1.0 for a circle and 0 for a line, thus quantitatively reflecting the tendency for the delayed spreading of ADMSC (more rounded morphology) when attached to Col-Oxi. This effect corroborates with the less actin cytoskeleton development and focal adhesion formation in the oxidized samples compared to native ones, pointed out with white arrows in Figure 1 (B and C, respectively).

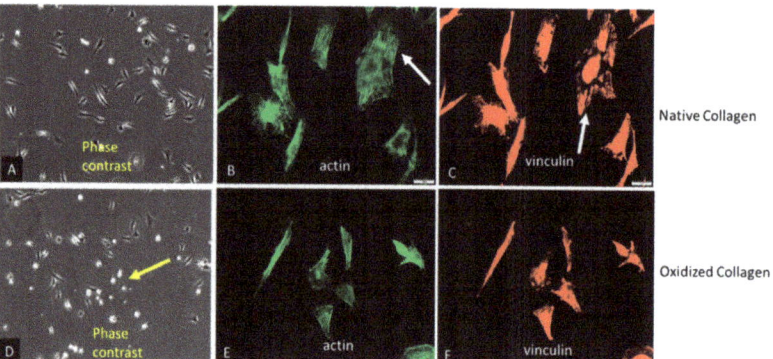

Figure 1. Initial adhesion of ADMSCs to native (**A–C**) and oxidized collagen (**D–F**) for 2 h in a serum-free medium. The samples were viewed at 10× phase contrast (**A,D**) or 40× using the green channel of a fluorescent microscope to view the actin cytoskeleton (**B,E**) or red for focal adhesions (**C,F**). The yellow arrow on (**D**) points to a delayed spreading of ADMSC on oxidized collagen (more rounded cells), while the arrows on (**B,C**) point to the better actin cytoskeleton development and focal adhesions formation in native collagen samples when compared to oxidized counterparts ((**E**) and (**F**), respectively). Bars on (**B,C,E,F**) are 20 μm.

Table 1. Morphometric analysis of ADMSC adhering to native and oxidized collagen, including Cell Spreading Area, Cell Shape Index (CSI), Aspect Ratio (AR), and corresponding p values.

Cellular Parameters	Col	Col-Oxi	p
Cell Spreading Area (μm^2)	216.0	179.6	$p > 0.05$
Cell Shape Index (CSI)	0.25	0.33	$p < 0.05$
Cell Aspect Ratio (CAR)	3.33	2.61	$p > 0.05$

The quantitative data for FA formation are presented in Table 2. These confirm again the significantly higher values for ADMSC adhering to native collagen, namely, the number of FA, total FA area, and the mean area per a single FA, amounting, respectively, to 317, 1085, and 3.66 vs. 143, 406, and 2.84 for the oxidized samples. These data were calculated from the images [42–44].

Table 2. Focal adhesion formation: number of FA, total area of FA, and mean area per FA.

Cell	Col	Col-Oxi	p
Number of FA	317	143	$p < 0.05$
Total FA Area (μm^2)	1085	406	$p < 0.05$
Mean Area per FA (μm^2)	3.66	2.84	$p > 0.05$

We also conducted a study covering the later stages of adhesion. As evident from Figure 2, all these morphological differences did not persist at the 24th hour of incubation, apparent from the phase contrast pictures (A,D), equally developed actin cytoskeleton (B,E), and focal adhesions formation (C,E) on both the Col (upper row) and Col-Oxi samples (bottom row). Quantitative analyses were not performed here. In fact, this result supported our initial desire to focus on the early stages of cellular interaction and signaling events giving the option to evaluate the specific effect of collagen oxidation.

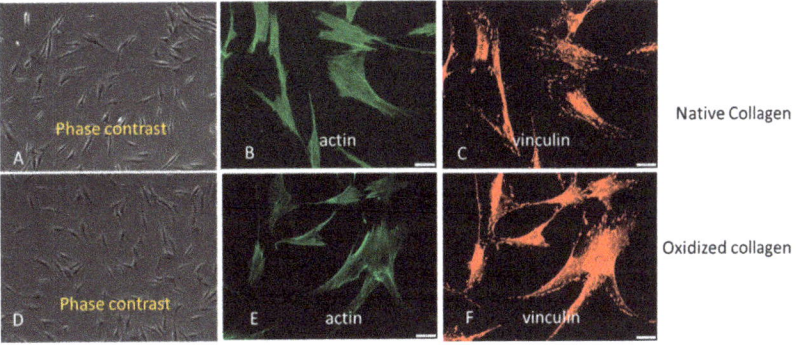

Figure 2. The approximately equal cell spreading (**A,D**) corroborates with the similar actin cytoskeleton development (**B,E**) and focal adhesions formation (**C,F**) at the 24th h of ADMSC adhesion to native (upper row) and oxidized collagen (bottom row). Bar 20 µm.

2.2. YAP/TAZ Signaling Events

To follow the intracellular signaling events upon adhesion of ADMSCs, we fixed and permeabilized the cells at the second hour of incubation before staining simultaneously for actin, nucleus, and TAZ activity (see protocol 2 in the Section 4.5.3).

Figure 3 shows typical images of cells examined sequentially on the red, blue, and green channels.

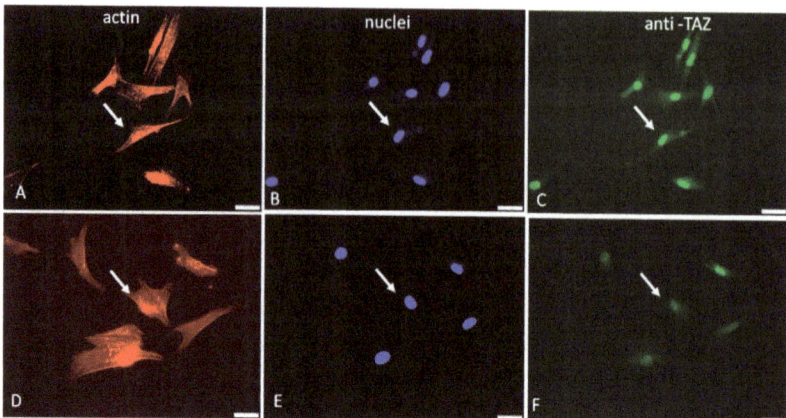

Figure 3. Immunofluorescent visualization of YAP/TAZ activity in ADMSCs adhering to native (**A–C**) and oxidized collagen samples (**D–F**). The same field was viewed at different channels of the microscope: red (actin cytoskeleton), blue (nucleus), and green (for TAZ activity). Arrows on all the images show the location of the nucleus. Bar 20 µm.

As evident from these images (Figure 3) and the supporting quantitative analysis presented in Table 3, the YAP/TAZ activity that coincides with the nucleus is apparently higher for native Col (pointed with arrows in Figure 3B,C) compared to Col-Oxi (pointed with arrows in Figure 3E,F), corresponding to 229.3 versus 178.2 intensity of pixels and ratio 176.2 versus 14.4, respectively, as shown in Table 3. These values suggest a significantly better signal transmission to the nucleus for native collagen samples: the cytosolic TAZ accumulation was only 1.3 pixels versus 12.4 for the Col-Oxi ones. Conversely, the nuclear TAZ accumulation was almost 230 pixels for the native samples vs. 178 pixels for Col-Oxi, resulting in a significantly higher ratio of TAZ nuclei/TAZ cytosol of 176.2 vs. 14.4 for Col-Oxi ($p < 0.05$).

Table 3. Nuclear and Cytosolic TAZ values and their ratio.

Parameters	Col	Col-Oxi	p
Nuclear TAZ	229.3	178.2	$p < 0.05$
Cytosolic TAZ	1.3	12.4	$p < 0.05$
Ratio TAZ Nuclei/TAZ Cytosol	176.2	14.4	$p < 0.05$

Taken together, this uneven distribution of TAZ between the nucleus and cytosol confirms the markedly different mechanical signal transduction between native and oxidized samples, being significantly suppressed in the latter.

Another interesting observation from these images was the difference in the overall nuclear shape: the nuclei in native collagen were more flattened compared to the Col-Oxi ones (see Figure 3B vs. Figure 3E), suggesting a more significant pressure from the cytoskeleton for ADMSC adhering on native collagen. This observation was partly confirmed by the nuclear size and shape analysis presented in Table 4: the mean nuclear area per cell was significantly lower in native collagen samples compared to the oxidized ones, amounting to 17.0 μM^2 versus 21.2 μM^2 ($p < 0.05$). The effect on the nuclear shape was not so pronounced, and we found only a nonsignificant trend for lowering the mean NSI (0.81 versus 0.86) and the change in NAR from 1.68 versus 1.34 for ADMSC adhering to native vs. oxidized collagen, respectively.

Table 4. Morphometric analysis of ADMSC nuclei adhering to native and oxidized collagen: nuclear area per cell, nuclear shape index (NSI), Nuclear Aspect Ratio (NAR), and corresponding p values.

Nuclear Parameters	Col	Col-Oxi	p
Nuclear Area per cell (μM^2)	17.01	21.2	$p < 0.05$
Nuclear Shape Index (NSI)	0.81	0.86	$p > 0.05$
Nuclear Aspect Ratio (NAR)	1.68	1.34	$p > 0.05$

2.3. Comparative Atomic Force Microscopy (AFM) Study

The morphology and the mechanical properties of adsorbed native and oxidized collagen, compared with the denatured one, were further examined in the nanoscale using AFM.

For these experiments, collagen was adsorbed to glass coverslips at identic conditions with cellular studies (at 37 °C for 1 h). The measurements were performed by AFM operating in contact mode at room temperature in the air.

As shown in Figure 4, the native collagen forms relatively large linear structures (Figure 4A) resembling thick interlaced fibers. More detailed 3D analysis, however, showed that these structures are sooner coarse aggregates growing to the z-direction (Figure 4D,G). These linear structures were much thinner on oxidized collagen samples (Figure 4B), showing a tendency for network formation, combined with less growth in the z-direction (Figure 4E,H). In the denatured collagen samples, most of these structures were absent (Figure 4C,F,I). Thermal denaturation curves (Figure 4J–L) obtained by Differential Scanning Calorimetry (DSC) also confirm the relatively minute structural changes in Col-Oxi, thus matching our previous investigation for the appearance of a small pre-peak at 35 °C, apart from the complete absence of any thermal changes in denatured collagen.

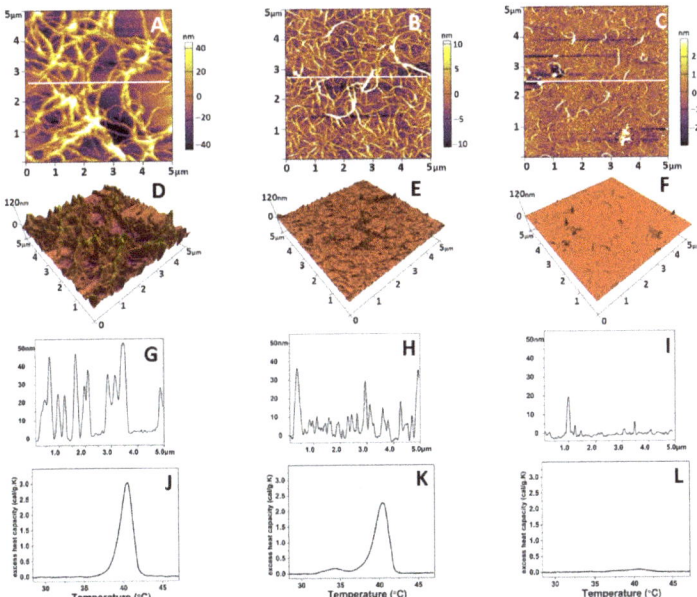

Figure 4. Representative 2D AFM images of native (**A**), oxidized (**B**), and denatured (**C**) collagen; the corresponding 3D topographical images (**D–F**) of the images of (**A–C**) and cross-section plot shapes (**G–I**) corresponding to the white lines in (**A–C**). The images were taken in tapping mode in the air at room temperature. The denaturation DSC profiles of native, oxidized, and denatured collagens were presented in panels (**J–L**), respectively.

The same AFM scans were used to calculate the roughness values (R_{RMS}) and Young's modules (Ea) to follow the mechanical properties of the obtained collagen features. Force–distance curves taken on the collagen samples were selected manually from the force images. Only indentation curves at the top of the fibrils and the overlap region of the collagen aggregates (Figure 4D–F) were considered. Young's modulus of the native collagen was 56.6 ± 8 MPa, whereas that for the oxidized collagen was 66.8 ± 5 MPa (Table 5), respectively. However, no significant difference ($p > 0.05$) was evident between the two collagen forms. Conversely, the data in Table 5 shows that upon oxidation, the roughness of adsorbed collagen features drops dramatically (approx. seven times) from $27.95 + 5.1$ to $5.51 + 0.8$ nm, reflecting a strongly altered ability of the protein to aggregate, apart from the native collagen forming large aggregates with a peak to valley distance of about 28 nm ($p < 0.05$). The denaturation leads to a relatively vague assembly of collagen in aggregates.

Table 5. Roughness values and Young's modulus for native, oxidized, and denatured collagen.

Samples	R_{RMS} (nm)	Ea (MPa)
Collagen Native	27.95 ± 5.1	56.6 ± 8
Col-Oxidized	5.51 ± 0.8	66.8 ± 5
Col Denatured	1.04 ± 0.6	3610 ± 59

3. Discussion

It is widely accepted that ECM anchors cells and directs cell functions not only by biochemical signals but also via specific mechanical cues [2,3]. Stem cells receive such cues from their microenvironment in the niche through mechanosensing and mechanotransduction [7,8,10,13,45], where collagens play a significant role [30,32,46]. Collagen is crucial because it determines most of the mechanical properties of the tissues and organs [5,32,40,47–49].

It is clear today that the physical cues affect proliferation, self-renewal, and the differentiation of MSCs into specific cell fates [3,8,9]; however, little is known about how these cues work in pathological environments, such as the acute oxidative stress that affects numerous homeostatic parameters in the body [36,37]. Recently, we developed a useful in vitro model to study the effect of collagen oxidation on MSC behavior. More specifically, we used adsorbed collagen of either native or preoxidized form [39] as a substratum for ADMSCs adhesion to follow their behavior under conditions that mimic acute oxidative stress [41]. Using this model, we found that oxidation leads to significant suppression of extracellular collagen remodeling by ADMSCs due to minute changes in collagen structure, which opens the door for further applications of this model. Here we show that it may relate to altered mechanical signal transduction in the cell interior. A reasonable question arises: how do MSCs sense such altered collagen structure?

Collagen binding is primarily provided by integrins, mainly α1β1 and α2β1 but also α10β1 and α11β1 [33,34], with an affinity for RGD and GFOGER-like sequences in collagen molecules [33,34]. Integrins are a family of major cell surface receptors generally involved in mediating the cellular response to ECM binding [5]. Composed of alpha and beta subunits, integrins form structural and functional linkages between the ECM fibrils and the intracellular cytoskeletal linker proteins [34]. Binding to immobilized collagen promotes integrin activation and clustering in focal adhesions, which are further associated with intracellular actin filaments through the above-mentioned linker proteins [5]. One such protein is vinculin, a cytoskeletal constituent associated with cell–cell and cell–matrix junctions. It is the most used marker for focal adhesions in anchoring F-actin to the membrane [46,50]. Our results show that ADMSCs hardly develop vinculin-containing focal contacts upon attachment to oxidized collagen, apart from the native collagen, where these structures are well pronounced. It correlates with the substantially diminished cell spreading and cell polarization, which were monitored once morphologically (Figure 1) and confirmed by ImageJ morphometry analysis (Tables 1–4). We show a significantly reduced Cell Spreading Area (from 216 to 179 μM^2) and CSI tending to 0.33 (i.e., to a more circular shape) compared to 0.25 for native Col (Table 1). It has to be noted, however, that this morphological difference was valid only for the initial stages of cell spreading, as at the 24th hour it was no longer observed, and ADSCs attached and spread equally well on both substrates (Figure 2), actually confirming our previous investigation [39]. We are prone to explain it by the constitutive ability of ADMSCs to produce very soon their own matrix, containing many other adhesive proteins capable of obliterating the initial collagen effect. As noted above, this was the reason we focused the present study on the initial stage of cell adhesion and the related signaling events, just to be sure that ADSCs attach to collagen only.

Another interesting finding was the observed tendency for flattening of ADMSCs nuclei in samples with native collagen, while on oxidized ones, the nuclei were larger and visibly rounding—a trend confirmed quantitatively by morphometry analysis (Table 1). It is well documented that focal adhesions and stress fibers generated on stiff substrate transduce mechanical forces to the nucleus, leading to nuclear flattening [5,46,47]. Thus, we got additional evidence for the successful transmission of the mechanical signal to the cell nuclei but this was working better for ADMSC adhering to native collagen than on oxidized collagen. Presumably, and there is proof in this direction, nuclear deformation increases the nuclear import of signaling molecules by decreasing the mechanical restriction in nuclear pores [16,47]. Obviously, it also happens in our system, as judged by the nuclear accumulation of TAZ activity. In fact, using immunofluorescent visualization of the YAP/TAZ signaling cascade (anti-TAZ antibody), we demonstrated its more substantial accumulation in the nuclear region at the second hour of incubation (Figure 3), again valid mainly for the ADMSC adhering to native collagen; moreover, it was confirmed statistically with morphometry analysis (Table 3), showing that on the oxidized samples, the signal was considerably fainter ($p < 0.05$).

However, the question still remains: why does the adsorbed native collagen provide a better signal to ADMSC during both the initial recognition phase and in the subsequent steps of signal transmission to the nucleus? Even if we accept the version that oxidized collagen is partly denatured upon oxidation, there is no direct proof that such collagen is worse recognized by the cells. On the contrary, there is proof that the unwinding of the collagen molecule releases additional RGD sequences, which improve cellular interaction [49].

On the other hand, the analysis of the DSC curves suggests relatively minute changes in the collagen structure on oxidation; the thermogram splits with the appearance of an additional transition, with added melting temperatures of 33.6 °C (Figure 4J) to a native transition characteristic for collagen at 40.1 °C (melting) (Figure 4K), confirming our previous investigation [41].

This structural change in oxidized collagen, however, is far from the curve of denatured collagen, where the complete absence of temperature transitions was observed (Figure 4L). Data in the literature regarding the binding of cells to collagen are quite extensive and sometimes contradictory. As an abundant ECM protein, collagen binds with at least five different groups of cell receptors, including first integrins but also DDR, Glycoprotein VI, Osteoclast-associated receptor (OSCAR), LAIR-1, and uPARAP/Endo180 [48]; therefore, it is difficult to assume that oxidized collagen is not recognized by cells, which directs our thinking rather to the physical parameters of adsorbed collagen. It was most likely to turn our attention to its viscoelastic properties, which are known to affect mechanotransduction significantly [2–5,10]. To our surprise, however, the changes in Young's modules (Ea) for the oxidized collagen were relatively slight, with a deviation in Ea of about 20% in the direction of hardening (Table 2); moreover, these were statistically insignificant ($p > 0.05$).

On deeper analysis, however, we decided that such a fact should not puzzle us, considering that this is an adsorbed protein and the role of the underlying substrate stiffness can hardly be ignored. In contrast, the AFM data demonstrated a significant change in surface roughness measured over the adsorbed collagen molecules: from relatively thick linear structures, characterized as coarse aggregates in 3D images, they visibly switch to much thinner linear features on oxidized samples (Figure 4A–F). Moreover, the calculated roughness values (R_{RMS}) showed that upon oxidation, the roughness of adsorbed collagen features drops dramatically to about 5.5 nm (pick to valley distance), compared to 28 nm for native collagen samples (i.e., approx. seven times less) reflecting a significantly altered ability ($p < 0.05$) of oxidized protein to aggregate under these conditions. It has to be noted here that the adsorption of proteins was performed at 37 °C for 1 h, i.e., in conditions identical to the cellular studies, meaning that it represented the natural roughness that cells experience from the substratum.

Though not directly related to collagen, a line of studies confirms the topographic response of stem cells [7–9,11,16,17]. It necessitates the conclusion that the most altered parameter to which cells are exposed in our conditions is the roughness of adsorbed protein, i.e., per se, it is a kind of response of ADMSCs to substrate topography, which determines the impaired mechanotransduction from oxidized collagen.

4. Materials and Methods

4.1. Collagen Preparation

Collagen type I was produced from rat tail tendon by acetic acid extraction and salting out with NaCl, as described elsewhere [39,41]. After centrifugation at 4000 rpm at 4 °C, the pellets were redissolved in 0.05 M acetic acid. The excess NaCl was removed by dialysis against 0.05 M acetic acid. All procedures were performed at 4 °C. Thus, a nearly monomolecular composition of collagen solution, in which the collagen content approaches 100% of the total dry mass, was prepared. The collagen concentration in the solutions was measured by optical absorbance at 220–230 nm [41].

4.2. Collagen Oxidation Procedure

The collagen solution (2 mg/mL) was incubated in 0.05M acetic acid, pH 4.3, with 50 µM $FeCl_2$ and 5 mM H_2O_2 for 18 h at room temperature, as previously described [41]. The oxidant solutions were freshly prepared and 10 mM EDTA was used to stop the oxidation reaction, followed by intensive dialysis versus 0.05 M acetic acid to remove the excess oxidants. The oxidized collagen, Col-Oxi, was freshly prepared before the experiments.

4.3. Cells

Human ADMSCs of passage 1 were received from Tissue Bank BulGen using healthy volunteers with written consent before liposuction. The cells were maintained in DMEM/F12 medium containing 1% GlutaMAX™, 1% Antibiotic-Antimycotic solution, and 10% Gibco Fetal Bovine Serum (FBS), all purchased from Thermo Fisher Scientific (Waltham, MA, USA). Every two days, the medium was replaced until the cells reached approximately 90% confluency to be used for the experiments up to the 7th passage.

4.4. Morphological Study

For the morphological observations, collagen (100 µg/mL) dissolved in 0.05 M acetic acid was used to coat regular glass coverslips (12 × 12 mm, ISOLAB Laborgeräte GmbH, Eschau, Germany) for 60 min at 37 °C, placed in 6-well TC plates (Sensoplate, Greiner Bio-one, Meckenheim, Germany). Then, the cells were seeded at 5×10^4 cells/well density in the final volume of 3 mL serum-free medium before being incubated for 2 h or 24 h. In a later case, 10% FBS was added at the end of the 2nd hour. The initial cell adhesion and overall cell morphology were studied at the 2nd hour and imaged under phase contrast using an inverted microscope, Leica DM 2900, or processed for immunofluorescent analysis in two protocols, as follows:

4.4.1. First Protocol (Cell Spreading and FA Formation)

After incubations (2 or 24 h) the samples were fixed with 4% paraformaldehyde and permeabilized with 0.5% Triton X-1000 before fluorescence staining. Green fluorescent Alexa fluorTM 444 Phalloidin (Invitrogen, Thermo Fisher Scientific Inc Branchburg, NJ,

USA) was used to visualize the actin cytoskeleton, while the cell nuclei were stained by Hoechst 33342 (dilution 1:2000) (Sigma-Aldrich/Merck KGaA, Darmstadt, Germany).

Darmstadt, Germany). Focal adhesions were viewed with Anti-Vinculin Mouse Monoclonal Antibody (Clone: hVIN-1, Thermo Fisher Scientific, Waltham, MA, USA) IgG (1:150) followed by fluorescent Alexa Fluor 555 conjugated goat anti-mouse IgG (minimal x-reactivity) antibody (both provided by Sigma-Aldrich) used in dilution 1:100;

4.4.2. Second Protocol (YAP/TAZ Signaling Events)

To follow the YAP/TAZ signaling events, separate samples from the same series were stained with a rabbit polyclonal anti-TAZ antibody followed by green fluorescent Alexa FluorTM 444 conjugated goat anti-rabbit antibody (both provided by Sigma-Aldrich, Merck KGaA, Darmstadt, Germany) used in dilution 1:100, further counterstained for cell nuclei with Hoechst 33342 and red fluorescent Rhodamine Phalloidin (Sigma-Aldrich, Merck KGaA, Darmstadt, Germany) to view actin cytoskeleton, using dilutions as above.

Finally, all samples were mounted upside down on glass slides with Mowiol and viewed for 1st protocol using the blue (nuclei), green (actin cytoskeleton), and red (vinculin) channels of an inverted fluorescent microscope (Olympus BX53, Upright Microscope Olympus Corporation, Shinjuku Ku, Tokyo, Japan)) with objectives UPlan FLN (40×/0.50). TAZ samples were viewed separately in the green (TAZ activity), blue (nuclei), and red (actin) channels. A minimum of three representative images were obtained for each sample. The respective image processing software merged the different colors. All experiments were quadruplicated.

4.5. Image Analysis

4.5.1. Quantitative Analysis of Raw Format Images by ImageJ

All image analysis was performed per cell using ImageJ, which provides a wide range of processing and analysis approaches. The fluorescence intensity of the fibrillary arrays was measured based on raw format images of cells captured from at least three separate images under the same conditions. Pixel-based treatments are performed to highlight the regions of interest (ROIs) and allow the removal of artifacts. A default black-and-white threshold was used in the segmentation module. Images of equal size (W:1600 px/H: 1200 px) were examined. All measurements were performed at the respective channel of the two or three colored images.

4.5.2. Quantification of Overall Morphological Parameters

Four metrics were acquired: Spread Area (SA), Cell Shape Index (SCI), Aspect ratio (AR), and Focal Adhesion size. The individual cellular domains were determined by generating binary masks using Otsu's intensity-based thresholding method from fluorescent actin images. Cellular masks were then used to calculate ADMSC SA and CSI. The CSI was calculated using the formula:

$$CSI = 4\pi \times A/P^2$$

where A is the mean cell area and P is the mean cell perimeter.

With this metric, a line and a circle have CSI values of 0 (indicating an elongated polygon) and 1 (indicating a circle), respectively. AR was calculated as the ratio of the largest and smallest side of a bounding rectangle encompassing the cell. The same counter function was used to calculate the nuclear surface area (NSA) and overall nuclear shape index (NSI) as important morphometric characterization for each cell.

4.5.3. YAP/TAZ Signaling

To quantify the YAP/TAZ nuclear-to-cytosolic ratio, binary masks of the nuclei were generated using Otsu's intensity-based thresholding method from fluorescent Hoechst images and were superimposed with corresponding actin masks to generate masks that encompass the cytosol yet exclude the nucleus. Fluorescent TAZ images were then superimposed either with the nuclear- or cytosol-only masks to isolate the TAZ signal in the nucleus or cytosol, respectively. Integral TAZ signal intensity was determined in these domains, and their ratio was normalized to the corresponding areas. The ratio of TAZ activity in the nucleus versus cytoplasm was further calculated and compared for both native and oxidized samples.

4.5.4. Quantification of Focal Adhesions (FA)

Focal adhesions were estimated following the procedure described by Horzum et al. The steps of image processing were carried out using ImageJ. Briefly, the raw fluorescent images were processed in several steps, as follows [42]: choose the sliding paraboloid option with the rolling ball radius set to 50 pixels [43]; enhance the local contrast of the image using the following values, block size = 19, histogram bins = 256, maximum slope = 6, no mask, and fast [44]; apply mathematical exponential (exp) to minimize the background further; adjust brightness and contrast automatically; run log3d (Laplacian of Gaussian or Mexican Hat) filter, here we define the size of log3Step 1b of log3d filter as sigma X = 5 and sigma Y = 5); run log3d (Laplacian of Gaussian or Mexican Hat) filter; execute analysis of particles command using the following parameters, size = 50, infinity and circularity = 0.00–0.99.

4.6. AFM Studies

AFM imaging and force–distance curves of native, oxidized, and denatured collagen were performed using Atomic Force Microscopy (MFP-3D, Asylum Research, Oxford Instruments, Santa Barbara, CA 93117, USA). All measurements were taken in the air and at room temperature. Silicon AFM tips (Nanosensors, type qp-Bio) of 50 kHz resonance frequency and 0.3 N/m nominal spring constant were used.

For all imaging experiments, collagen solutions were deposited on a clean glass coverslip and incubated at 37 °C for one hour to ensure maximal adsorption for each sample. Afterward, the collagen-coated glasses were washed gently with distilled water to avoid buffer crystallization on the surface. Morphometrical (roughness value) and nanomechanical characterization were accomplished using IgorPro 6.37 software. The mechanical properties of the three types of collagen were assumed by Young's modulus defined by the force–distance (f–d) curves. The value of the elastic modulus was obtained by fitting the force–indentation data to the Hertz model with the embedded IgorPro software, considering the Poisson's ratio to be ≈ 0.5:

$$E = 3F(1-\nu^2)/4\sqrt{(r\delta^3)} \tag{1}$$

where F is the applied force on the sample, δ is the indentation depth, r is the tip radius, and E and ν are Young's modulus and Poisson's ratio, respectively.

4.7. DSC Measurements

DSC measurements were performed using DASM4's (Privalov, BioPribor, Moscow, Russia) built-in, high-sensitivity calorimeter with a cell volume of 0.47 mL. The collagen concentration was adjusted to 2 mg/mL in 0.05 M acetic acid. A constant pressure of 2 atm was applied to the cells to prevent any degassing of the solution. The samples were heated with a scanning rate of 1.0 °C/min from 20 °C to 65 °C and preceded by a baseline run with buffer-filled cells. Each collagen solution was reheated after cooling from the first scan to evaluate the reversibility of the thermally induced transitions. The calorimetric curve corresponding to the second (reheating) scan was used as an instrumental baseline and was subtracted from the first scans, as collagen thermal denaturation is irreversible. The calorimetric data were analyzed using the Origin Pro 2018 software package.

4.8. Statistical Analysis

All experiments were conducted with at least 3 independent series with 3–4 cells per group. One-analysis of variance (ANOVA) followed by Tukey-HSD posthoc tests were performed on all data sets. Error is reported in bar graphs as the standard error of the mean unless otherwise noted. Significance was indicated by *, corresponding to $p < 0.05$.

Author Contributions: G.A. and R.K.-P.—Conceptualization of the study; G.A., R.K.-P., S.T., V.S., A.Y., M.I. and B.D.—Methodology and Investigation; G.A., R.K.-P., S.K., P.T. and S.T.—Analysis and interpretation; P.T.—ImageJ analysis; G.A. and R.K.-P.—Writing, Original Draft Preparation; G.A., R.K.-P.—Review and Editing; R.K.-P., S.K.—Project Administration, funding acquisition; G.A. and R.K.-P.—Primary responsibility for the final content. All authors have read and agreed to the published version of the manuscript.

Funding: Financed by European Regional Development Fund, Operative program "Science and education for intelligence growth" Multidisciplinary project No BG05M2OP001-1.002-0010-C03. The study was further financially supported by the Internal project No 16-2020 of Medical University-Pleven, Bulgaria.

Institutional Review Board Statement: This study was conducted according to the guidelines of the Declaration of Helsinki and approved by the Institutional Ethics Committee of Medical University-Pleven, (APPROVAL N 601-KENID 20/05/19).

Informed Consent Statement: Not applicable.

Data Availability Statement: Not applicable.

Acknowledgments: We acknowledge the support and donations of materials used for experiments given by Tissue Bank BulGen, Bulgaria. We acknowledge that research equipment of the Distributed Research Infrastructure INFRAMAT, part of the Bulgarian National Roadmap for Research Infrastructures, supported by the Bulgarian Ministry of Education and Science, was used in this investigation.

Conflicts of Interest: The authors declare no potential conflict of interest.

References

1. Burridge, K.; Monaghan Benson, E.; Graham, D.M. Mechanotransduction: From the cell surface to the nucleus via RhoA. *Phil. Trans. R. Soc.* **2019**, *374*, 20180229. [CrossRef]
2. Yamashiroa, Y.; Quoc Thangb, B.; Ramireza, K.; Jae Shina, S.; Kohatae, T.; Ohataf, S.; Anh Vu Nguyena, T.; Ohtsukie, S.; Nagayamaf, K.; Yanagisawaa, H. Matrix mechanotransduction mediated by thrombospondin-1/integrin/YAP in the vascular remodeling. *Proc. Natl. Acad. Sci. USA* **2020**, *117*, 9896–9905. [CrossRef] [PubMed]
3. Vining, K.H.; Mooney, D.J. Mechanical forces direct stem cell behavior in development and regeneration. *Nat. Rev. Mol. Cell Biol.* **2017**, *18*, 728–742. [CrossRef] [PubMed]
4. Halder, G.; Dupont, S.; Piccolo, S. Transduction of mechanical and cytoskeletal cues by YAP and TAZ. *Nat. Rev. Mol. Cell Biol.* **2012**, *13*, 591–600. [CrossRef] [PubMed]
5. Humphrey, J.D.; Dufresne, E.R.; Schwartz, M.A. Mechanotransduction and extracellular matrix homeostasis. *Nat. Rev. Mol. Cell Biol.* **2014**, *15*, 802–812. [CrossRef]
6. Engler, A.J.; Sen, S.; Sweeney, H.L.; Discher, D.E. Matrix elasticity directs stem cell lineage specification. *Cell* **2006**, *126*, 677–689. [CrossRef]
7. Li, D.; Zhou, J.; Chowdhury, F.; Cheng, J.; Wang, N.; Wang, F. Role of mechanical factors in fate decisions of stem cells. *Regen. Med.* **2011**, *6*, 229–240. [CrossRef]
8. Lee, J.H.; Park, H.K.; Kim, K.S. Intrinsic and extrinsic mechanical properties related to the differentiation of mesenchymal stem cells. *Biochem. Biophys. Res. Commun.* **2016**, *473*, 752–757. [CrossRef]
9. Hou, Y.; Xie, W.; Yu, L.; Cuellar Camacho, L.; Nie, C.; Zhang, M.; Haag, R.; Wei, Q. Surface Roughness Gradients Reveal Topography-Specific Mechanosensitive Responses in Human Mesenchymal Stem Cells. *Small* **2020**, *16*, 1905422. [CrossRef]
10. Wang, Y.; Wang, G.; Luo, X.; Qiu, J.; Tang, C. Substrate stiffness regulates the proliferation, migration, and differentiation of epidermal cells. *Burns* **2012**, *38*, 414–420. [CrossRef]
11. Shalabi, M.M.; Gortemaker, A.; Van't Hof, M.A.; Jansen, J.A.; Creugers, N.H.J. Implant surface roughness and bone healing: A systematic review. *J. Dent. Res.* **2006**, *85*, 496–500. [CrossRef]
12. Matos, G.R.M. Surface Roughness of Dental Implant and Osseointegration. *Maxillofac. Oral Surg.* **2021**, *20*, 1–4. [CrossRef]
13. Llopis-Hernandez, V.; Rico, P.; Moratal, D.; Altankov, G.; Salmeron-Sanchez, M. Role of Material-Driven Fibronectin Fibrillogenesis. *Acta Biomater.* **2013**, *77*, 74–84.
14. Klein, E.A.; Yin, L.; Kothapalli, D.; Castagnino, P.; Byfield, F.J.; Xu, T.; Levental, I.; Hawthorne, E.; Janmey, P.A.; Assoian, R.K. Cell-cycle control by physiological matrix elasticity and in vivo tissue stiffening. *Curr. Biol.* **2009**, *19*, 1511–1518. [CrossRef]
15. Nedjari, S.; Awaja, F.; Altankov, G. Three Dimensional Honeycomb Patterned Fibrinogen Based Nanofibers Induce Substantial Osteogenic Response of Mesenchymal Stem Cells. *Sci. Rep.* **2017**, *7*, 15947. [CrossRef]
16. Kim, D.; Provenzano, P.P.; Smith, C.L.; Levchenko, A. Matrix nanotopography as a regulator of cell function. *J. Cell Biol.* **2012**, *197*, 351. [CrossRef]
17. Sun, Z.; Guo, S.S.; Fässler, R. Integrin-mediated mechanotransduction. *J. Cell Biol.* **2016**, *215*, 445–456. [CrossRef]

18. Majhy, B.; Priyadarshinia, P.; Sen, A.K. Effect of surface energy and roughness on cell adhesion and growth—Facile surface modification for enhanced cell culture. *RSC Adv.* **2021**, *11*, 15467. [CrossRef]
19. Horbett, T.A. The role of adsorbed proteins in animal cell adhesion. *Colloids Surf. B Biointerfaces* **1994**, *2*, 225–240. [CrossRef]
20. Dupont, S.; Morsut, L.; Aragona, M.; Enzo, E.; Giulitti, S.; Cordenonsi, M.; Zanconato, F.; Le Digabel, J.; Forcato, M.; Bicciato, S.; et al. Role of YAP/TAZ in mechanotransduction. *Nature* **2011**, *474*, 179–183. [CrossRef]
21. Meng, Z.; Moroishi, T.; Guan, K. Mechanisms of Hippo pathway regulation. *Genes Dev.* **2016**, *30*, 1–17. [CrossRef] [PubMed]
22. Zinatizadeh, M.R.; Miri, S.R.; Zarandi, P.K.; Chalbatani, G.M.; Rapôso, C.; Mirzaei, H.R.; Akbari, M.E.; Mahmoodzadeh, H. The Hippo Tumor Suppressor Pathway (YAP/TAZ/TEAD/MST/LATS) and EGFR-RAS-RAF-MEK in cancer metastasis. *Genes Dis.* **2019**, *8*, 48–60. [CrossRef] [PubMed]
23. Pobbati, A.; Hong, W. A combat with the YAP/TAZ-TEAD oncoproteins for cancer therapy. *Theranostics* **2020**, *10*, 3622–3635. [CrossRef] [PubMed]
24. Janmey, P.A.; Wells, R.G.; Assoian, R.K.; McCulloch, C.A. From tissue mechanics to transcription factors. *Differentiation* **2013**, *86*, 112–120. [CrossRef]
25. Johnson, R.; Halder, G. The two faces of Hippo: Targeting the Hippo pathway for regenerative medicine and cancer treatment. *Nat. Rev. Drug Discov.* **2014**, *13*, 63–79. [CrossRef]
26. Wang, K.C.; Yeh, Y.T.; Nguyen, P.; Limqueco, E.; Lopez, J.; Thorossian, S.; Guan, K.L.; Li, Y.S.J.; Chien, S. Flow-dependent YAP/TAZ activities regulate endothelial phenotypes and atherosclerosis. *Proc. Natl. Acad. Sci. USA* **2016**, *113*, 11525–11530. [CrossRef]
27. Li, Y.; Wang, J.; Zhong, W. Regulation and mechanism of YAP/TAZ in the mechanical microenvironment of stem cells. *Mol. Med. Rep.* **2021**, *24*, 506. [CrossRef]
28. Samsonraj, R.M.; Raghunath, M.; Nurcombe, V.; Hui, J.H.; van Wijnen, A.J.; Cool, S.M. Concise Review: Multifaceted Characterization of Human Mesenchymal Stem Cells for Use in Regenerative Medicine. *Stem Cells Transl. Med.* **2017**, *6*, 2173–2185. [CrossRef]
29. Mohri, Z.; Del Rio Hernandez, A.; Krams, R. The emerging role of YAP/TAZ in mechanotransduction. *J. Thorac. Dis.* **2017**, *9*, E507–E509. [CrossRef] [PubMed]
30. Dupont, S. Role of YAP/TAZ in cell-matrix adhesion-mediated signalling and mechanotransduction. *Exp. Cell Res.* **2016**, *343*, 42–53. [CrossRef]
31. Bao, M.; Xie, J.; Huck, W.T.S. Recent Advances in Engineering the Stem Cell Microniche in 3D. *Adv. Sci.* **2018**, *5*, 1800448. [CrossRef]
32. Mouw, J.K.; Ou, G.; Weaver, V.M. Extracellular matrix assembly: A multiscale deconstruction. *Nat. Rev. Mol. Cell Biol.* **2014**, *15*, 771–785. [CrossRef]
33. Heino, J. Cellular signaling by collagen-binding integrins. *Adv. Exp. Med. Biol.* **2014**, *819*, 143–155.
34. Zeltz, C.; Gullberg, D.J. The integrin-collagen connection—A glue for tissue repair? *J. Cell Sci.* **2016**, *129*, 1284. [CrossRef]
35. Myllyharju, J. Intracellular Post-Translational Modifications of Collagens. In *Collagen Topics in Current Chemistry*; Brinckmann, J., Notbohm, H., Müller, P.K., Eds.; Springer: Berlin/Heidelberg, Germany, 2005; Volume 247. [CrossRef]
36. Kennett, E.C.; Chuang, C.Y.; Degendorfer, G.; Whitelock, J.M.; Davies, M.J. Mechanisms and consequences of oxidative damage to extracellular matrix. *Biochem. Soc. Trans.* **2011**, *39*, 1279–1287. [CrossRef]
37. Boin, F.; Erre, G.L.; Posadino, A.M.; Cossu, A.; Giordo, R.; Spinetti, G.; Passiu, G.; Emanueli, C.; Pintus, G. Oxidative stress-dependent activation of collagen synthesis is induced in human pulmonary smooth muscle cells by sera from patients with scleroderma-associated pulmonary hypertension. *Orphanet J. Rare Dis.* **2014**, *9*, 123. [CrossRef]
38. Lu, P.; Takai, K.; Weaver, V.M.; Werb, Z. Extracellular matrix degradation and remodeling in development and disease. *Cold Spring Harb. Perspect. Biol.* **2011**, *3*, a005058. [CrossRef]
39. Komsa-Penkova, R.; Stavreva, G.; Belemezova, K.; Kyurkchiev, S.; Todinova, S.; Altankov, G. Mesenchymal Stem-Cell Remodeling of Adsorbed Type-I Collagen-The Effect of Collagen Oxidation. *Int. J. Mol. Sci.* **2022**, *23*, 3058. [CrossRef]
40. Magin, C.M.; Alge, D.L.; Anseth, K.S. Bio-inspired 3D microenvironments: A new dimension in tissue engineering. *Biomed. Mater.* **2016**, *11*, 022001. [CrossRef] [PubMed]
41. Komsa-Penkova, R.; Koynova, R.; Kostov, G.; Tenchov, B. Discrete reduction of type I collagen thermal stability upon oxidation. *Biophys. Chem.* **2000**, *83*, 185–195. [CrossRef]
42. Horzum, U.; Ozdil, B.; Pesen-Okvur, D. Step-by-step quantitative analysis of focal adhesions. *MethodsX* **2014**, *1*, 56–59. [CrossRef] [PubMed]
43. Saalfeld, S. CLAHE (Contrast Limited Adaptive Histogram Equalization). 2009. Available online: http://rsbweb.nih.gov/ij/plugins/clahe/index.html (accessed on 8 January 2023).
44. Sage, D.; Neumann, F.R.; Hediger, F.; Gasser, S.M.; Unser, M. Automatic tracking of individual fluorescence particles: Application to the study of chromosome dynamics. *IEEE Trans. Image Process.* **2005**, *14*, 1372–1383. [CrossRef] [PubMed]
45. Park, J.; Kim, D.H.; Shah, S.R.; Kim, H.N.; Kshitiz; Kim, P.; Quiñones-Hinojosa, A.; Levchenko, A. Switch-like enhancement of epithelial-mesenchymal transition by YAP through feedback regulation of WT1 and Rho-family GTPases. *Nat. Commun.* **2019**, *10*, 2797. [CrossRef] [PubMed]
46. Wang, J.H.; Thampatty, B.P.; Lin, J.S.; Im, H.J. Mechanoregulation of gene expression in fibroblasts. *Gene* **2007**, *391*, 1–15. [CrossRef]

47. Wolfenson, H.; Yang, B.; Sheetz, M.P. Steps in Mechanotransduction Pathways that Control Cell Morphology. *Annu. Rev. Physiol.* **2019**, *81*, 585. [CrossRef]
48. Elango, J.; Hou, C.; Bao, B.; Wang, S.; Maté Sánchez de Val, J.E.; Wenhui, W. The Molecular Interaction of Collagen with Cell Receptors for Biological Function. *Polymers* **2022**, *14*, 876. [CrossRef]
49. Ruggiero, F.; Champliaud, M.F.; Garrone, R.; Aumailley, M. Interactions between Cells and Collagen V Molecules or Single Chains Involve Distinct Mechanisms. *Exp. Cell Res.* **1994**, *210*, 215–223. [CrossRef]
50. Burridge, K.; Feramisco, J.R. Microinjection and localization of a 130K protein in living fibroblasts: A relationship to actin and fibronectin. *Cell* **1980**, *19*, 587–595. [CrossRef]

Disclaimer/Publisher's Note: The statements, opinions and data contained in all publications are solely those of the individual author(s) and contributor(s) and not of MDPI and/or the editor(s). MDPI and/or the editor(s) disclaim responsibility for any injury to people or property resulting from any ideas, methods, instructions or products referred to in the content.

Article

Evaluation of Epithelial–Mesenchymal Transition Markers in Autoimmune Thyroid Diseases

Pablo Sacristán-Gómez [1,2], Ana Serrano-Somavilla [1,2], Lía Castro-Espadas [1], Nuria Sánchez de la Blanca Carrero [1,2], Miguel Sampedro-Núñez [1,2], José Luis Muñoz-De-Nova [3], Francisca Molina-Jiménez [4], Alejandra Rosell [5], Mónica Marazuela [1,2,*] and Rebeca Martínez-Hernández [1,2,6,*]

[1] Department of Endocrinology, Hospital Universitario de la Princesa, Instituto de Investigación Princesa, Universidad Autónoma de Madrid, C/Diego de León 62, 28006 Madrid, Spain
[2] Centro de Investigación Biomédica en Red de Enfermedades Raras (CIBERER GCV14/ER/12), 28029 Madrid, Spain
[3] Department of General and Digestive Surgery, Hospital Universitario de la Princesa, Instituto de Investigación Princesa, Universidad Autónoma de Madrid, C/Diego de León 62, 28006 Madrid, Spain
[4] Gastroenterology Research Unit, Hospital Universitario de la Princesa, Instituto de Investigación Princesa, Universidad Autónoma de Madrid, C/Diego de León 62, 28006 Madrid, Spain
[5] Pathology Unit, Hospital Universitario de la Princesa, Instituto de Investigación Princesa, Universidad Autónoma de Madrid, C/Diego de León 62, 28006 Madrid, Spain
[6] Faculty of Medicine, Universidad San Pablo CEU, Urbanización Montepríncipe, Alcorcón, 28925 Madrid, Spain
* Correspondence: monica.marazuela@gmail.com (M.M.); rbk_mar@yahoo.es (R.M.-H.)

Citation: Sacristán-Gómez, P.; Serrano-Somavilla, A.; Castro-Espadas, L.; Sánchez de la Blanca Carrero, N.; Sampedro-Núñez, M.; Muñoz-De-Nova, J.L.; Molina-Jiménez, F.; Rosell, A.; Marazuela, M.; Martínez-Hernández, R. Evaluation of Epithelial–Mesenchymal Transition Markers in Autoimmune Thyroid Diseases. *Int. J. Mol. Sci.* **2023**, *24*, 3359. https://doi.org/10.3390/ijms24043359

Academic Editors: Sabrina Lisi and Margherita Sisto

Received: 20 January 2023
Revised: 1 February 2023
Accepted: 3 February 2023
Published: 8 February 2023

Copyright: © 2023 by the authors. Licensee MDPI, Basel, Switzerland. This article is an open access article distributed under the terms and conditions of the Creative Commons Attribution (CC BY) license (https://creativecommons.org/licenses/by/4.0/).

Abstract: A state of chronic inflammation is common in organs affected by autoimmune disorders, such as autoimmune thyroid diseases (AITD). Epithelial cells, such as thyroid follicular cells (TFCs), can experience a total or partial transition to a mesenchymal phenotype under these conditions. One of the major cytokines involved in this phenomenon is transforming growth factor beta (TGF-β), which, at the initial stages of autoimmune disorders, plays an immunosuppressive role. However, at chronic stages, TGF-β contributes to fibrosis and/or transition to mesenchymal phenotypes. The importance of primary cilia (PC) has grown in recent decades as they have been shown to play a key role in cell signaling and maintaining cell structure and function as mechanoreceptors. Deficiencies of PC can trigger epithelial–mesenchymal transition (EMT) and exacerbate autoimmune diseases. A set of EMT markers (E-cadherin, vimentin, α-SMA, and fibronectin) were evaluated in thyroid tissues from AITD patients and controls through RT-qPCR, immunohistochemistry (IHC), and western blot (WB). We established an in vitro TGF-β–stimulation assay in a human thyroid cell line to assess EMT and PC disruption. EMT markers were evaluated in this model using RT-qPCR and WB, and PC was evaluated with a time-course immunofluorescence assay. We found an increased expression of the mesenchymal markers α-SMA and fibronectin in TFCs in the thyroid glands of AITD patients. Furthermore, E-cadherin expression was maintained in these patients compared to the controls. The TGF-β-stimulation assay showed an increase in EMT markers, including vimentin, α-SMA, and fibronectin in thyroid cells, as well as a disruption of PC. The TFCs from the AITD patients experienced a partial transition to a mesenchymal phenotype, preserving epithelial characteristics associated with a disruption in PC, which might contribute to AITD pathogenesis.

Keywords: autoimmune thyroid diseases; Graves' disease; Hashimoto's Thyroiditis; TGF-β; EMT; primary cilia

1. Introduction

Autoimmune thyroid diseases (AITD) are the most prevalent autoimmune disorders in the global population, with a 5% prevalence, and are most common among middle-aged women [1,2]. They develop as a consequence of tolerance loss against self-thyroid antigens and can be classified into two main types with opposite clinical phenotypes: Hashimoto's Thyroiditis (HT) and Graves' disease (GD). In HT, the thyroid gland is seriously damaged by a heavy infiltration of immune cells, which results in thyroid cell apoptosis and hypothyroidism [1]. On the other hand, GD is characterized by follicular hyperplasia and excessive production of thyroid hormones leading to hyperthyroidism. Hyperthyroidism is mainly due to the overactivation of the thyroid-stimulating hormone receptor (TSH-R) by stimulating antibodies (TSH-R-Ab) [1]. An extrathyroidal manifestation commonly observed in GD patients is Graves' ophthalmopathy (GO). GO is characterized by both an excessive production of glycosaminoglycans or extracellular matrix (ECM) components by orbital fibroblasts (OF) and an immune cell infiltration [3,4]. In these phenotypes, other autoimmune features are present, such as an increase in anti-thyroperoxidase (TPO) and anti-thyroglobulin (TG) antibodies [1,3,5].

After years of research and despite their high prevalence, the molecular mechanisms underlying these diseases are still not completely understood [5]. Several studies have shed light on the genetic predisposition to AITD, reporting relationships of these diseases with several polymorphisms in different genes [6], as well as epigenetic variations [7]. In addition, age, sex, and environmental factors have been related to their development [2,8]. AITD are characterized by a disruption in immune homeostasis, where immune cells such as regulatory T cells (Tregs) show a reduced immune suppression activity, whereas the function of cells with inflammatory phenotypes, such as T helper (Th) 1, Th2, or Th17 cells, is enhanced [5,9–11]. These changes, together with the influence of several cytokines, such as interferon gamma (IFN-γ) [12], interleukin 1 beta (IL-1β), tumor necrosis factor alpha (TNF-α), and transforming growth factor beta (TGF-β), create a chronic inflammation environment that can lead to thyroid cell destruction [13–17]. These cytokines can also lead to the acquisition of a mesenchymal phenotype by epithelial cells [18–20].

Epithelial–mesenchymal transition (EMT) is a process that occurs under both physiological and pathological conditions, and it is characterized by the loss of epithelial characteristics and the acquisition of mesenchymal features by epithelial cells [21]. Accordingly, this process results in changes in cell behavior, morphology, polarity, cytoskeletal organization, or molecular components in epithelial cells, which lead to cells with increased motility, migration, plasticity, and secretion of ECM components [22]. Briefly, cells lose epithelial markers such as E-cadherin, cytokeratin, or Zonula occludens-1 (ZO-1) and acquire mesenchymal markers such as fibronectin, alpha-smooth muscle actin (α-SMA), fibroblast specific protein-1 (FSP-1/S100A4), or vimentin, among others. Also, mesenchymal cells synthesize ECM components, such as collagen I, which may trigger fibrosis caused by excessive fibrous connective tissue deposition [22,23]. The accumulation of fibrotic components can impair the organ affected. Indeed, EMT is a major feature in several autoimmune diseases, such as rheumatoid arthritis (RA), inflammatory bowel disease (IBD), and primary Sjögren syndrome (pSS) [24–29]. In this context, although fibrosis is one of the main pathological characteristics of HT and the transition of orbital fibroblasts to myofibroblasts contributes to the pathogenesis of GO [3], studies on EMT in AITD are scarce [19]. EMT is triggered by several factors, such as genetic/epigenetic alterations, chronic inflammation, and cytokines, such as TGF-β, among others [30–35].

By integrating miRNA and mRNA data from AITD thyroid tissue samples, we recently reported dysregulation of primary cilia (PC) as a novel susceptibility pathway that controls AITD pathogenesis. Indeed, the number of PC was dramatically reduced in AITD, and, in some cases, they almost disappeared [36]. PC are defined as individual organelles in a protrusion in the apical surface of the cell. PC trigger several intracellular signal transduction cascades that are indispensable for cell development, proliferation, differentiation, survival, and migration [37]. In thyroid follicular cells (TFCs), PC can also play a role in modulating hormone secretion [38,39]. In addition, it was recently described that the deficiency of primary cilia triggers EMT under resting condition and exacerbates it under the influence of fibrotic signals such as TGF-β [40,41].

Keeping all this in mind and regarding our previous results on primary cilia defects in AITD [36] and their possible involvement in EMT, we evaluated the expression of epithelial and mesenchymal markers in thyroid tissue samples from AITD and correlated their expression with patients' clinical outcomes. Our data indicate that there is an increase in the acquisition of mesenchymal markers by TFCs in AITD that could contribute to the pathogenesis of these diseases. Furthermore, EMT induction by TGF-β in thyroid cells suggests the potential usefulness of this pathway as a novel therapeutic avenue to treat AITD.

2. Results

2.1. EMT Markers in AITD

Concomitant expression of epithelial and mesenchymal markers is often used to identify cells that are undergoing EMT. Thus, we first analyzed RNA levels of epithelial and mesenchymal markers within thyroid tissue from AITD patients and controls. E-cadherin (*CDH1*), a marker of epithelial cells, was significantly downregulated in HT tissue in comparison to the control and GD thyroid tissue (mean relative expression 0.05 in HT vs. 0.13 in controls and 0.14 in GD; $p = 0.001$ and 0.0001, respectively) (Figure 1A). Interestingly, when we correlated the expression levels of the different markers with clinical parameters, we observed that *CDH1* expression had a strong inverse correlation with thyrotropin (TSH) ($r = -0.7680$; $p < 0.0001$) and a significant positive correlation with levels of free-T4 (FT4) ($r = 0.7833$; $p = 0.0172$) and TSH-R-Ab ($r = 0.73$ and $p = 0.045$) (Figure 1B).

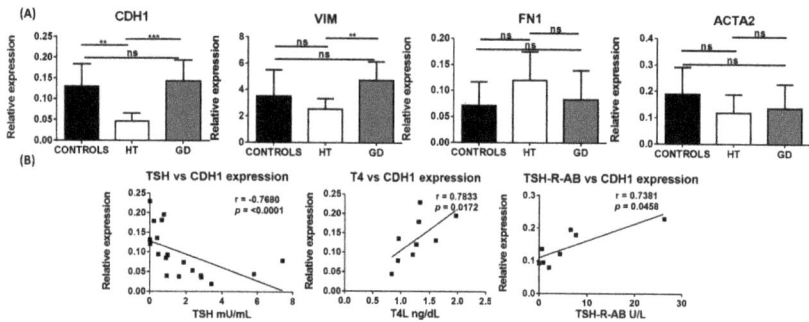

Figure 1. RNA expression of EMT-associated markers in thyroid tissue from controls, HT, and GD patients. (**A**) RT-qPCR expression of E-cadherin (*CDH1*), vimentin (*VIM*), fibronectin (*FN1*), and α-SMA (*ACTA2*) in bulk thyroid tissue samples. Data correspond to the arithmetic mean ± SD. (**B**) Significant correlation analysis of the genes analyzed by RT-qPCR with different clinical laboratory parameters. Abbreviations—ns: not significant, TSH: thyroid-stimulating hormone; FT4: free-T4 hormone; TSH-R-Ab: TSH receptor antibody. ** $p < 0.01$; *** $p < 0.005$.

Regarding markers of mesenchymal cells, vimentin (*VIM*) was significantly upregulated in GD compared to HT tissue (mean relative expression 4.75 vs. 2.53, respectively; $p = 0.0062$); however, no significant differences were found when compared to the controls. Although RNA levels of α-SMA (*ACTA2*) and fibronectin (*FN1*) did not exhibit a significant variation between AITD and control thyroid tissue, *FN1* had a tendency to be upregulated in HT compared to control samples (Figure 1A).

In order to confirm these results and identify cells undergoing EMT, we studied the protein expression of these markers with immunohistochemistry in 50 thyroid samples. Regarding the epithelial marker E-cadherin, we did not observe significant differences between thyroid follicular cells from HT, GD, and control tissues (Figure 2A). Regarding mesenchymal markers, we observed a significant increase in AITD for fibronectin (mean immunohistochemistry [IHC] score of 1.15 in HT and 1.25 in GD vs. 0.1563 in control thyroid tissues; $p = 0.0001$ in both cases) and α-SMA (mean IHC score of 1.68 in HT and

1.80 in GD vs. 0.89 in control thyroid tissues; $p = 0.0194$ and $p = 0.0070$, respectively) (Figure 2B–D). Although we did not detect an increase in vimentin expression, we observed a significant differential distribution pattern in the GD tissue with a location change from a cytoplasmic and perinuclear staining to a peripheral cell distribution toward the basal membrane (basal vimentin mean expression 1.06 in GD vs. 0.22 in control thyroid tissues; $p < 0.0001$) (Figure 2E). Regarding correlations between EMT markers and clinical parameters, positive significant correlations were observed between fibronectin and basal vimentin, fibronectin and α-SMA, and α-SMA and TSH-R-Ab (Figure 2F).

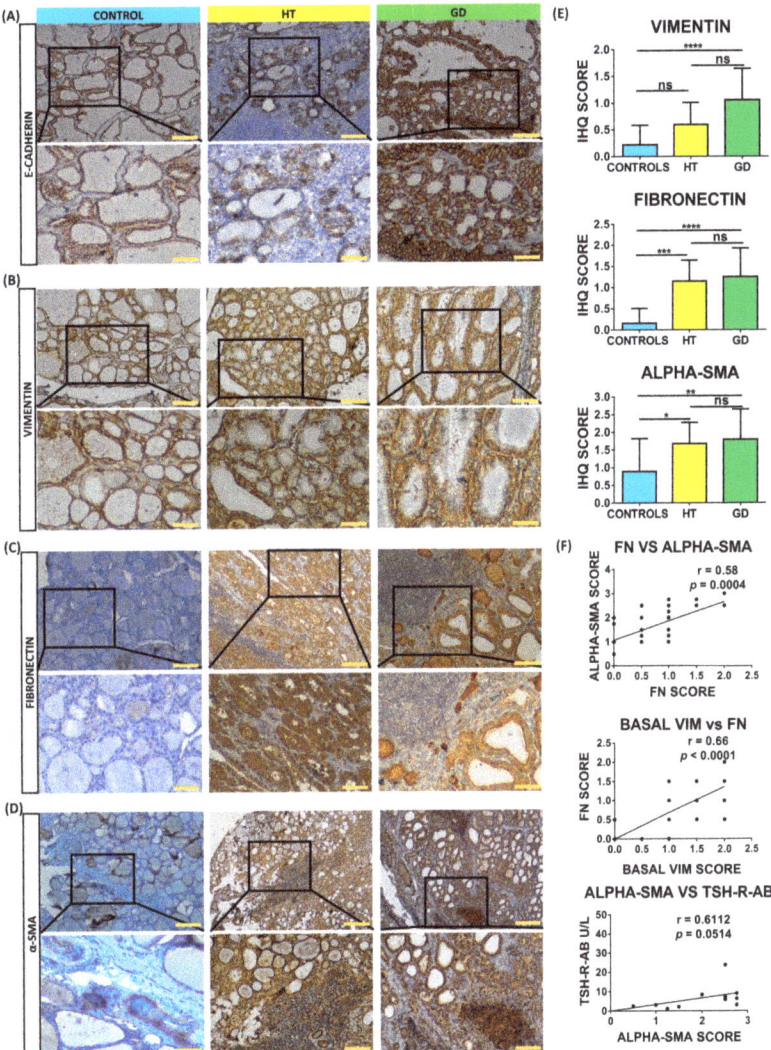

Figure 2. Immunostaining analysis of EMT markers in thyroid tissue from controls, HT, and GD patients. (**A–D**) Immunohistochemistry analysis of E-cadherin, vimentin (VIM), fibronectin (FN), and α-SMA. Scale bar A, B, and C: 200 μm, zoom 100 μm. Scale Bar D: 500 μm, zoom 100 μm. (**E**) Immunohistochemistry (IHC) score quantitation of basal vimentin, fibronectin, and α-SMA. (**F**) Correlation of marker IHC score with clinical laboratory parameters. Data correspond to the arithmetic mean ± SD. Abbreviations—ns: not significant. * $p < 0.05$; ** $p < 0.01$; *** $p < 0.005$; **** $p < 0.0001$.

Although an increase in fibronectin and α-SMA was found by immunohistochemistry, this increase was not corroborated by western blot (WB). However, we detected an increase in total vimentin (mean expression 1.51 in HT, 1.10 in GD, and 1.21 in AITD vs. 0.50 in control tissues; $p = 0.0091$, $p = 0.0279$ and $p = 0.0026$, respectively) and cleaved vimentin (mean expression 0.90 in GD and 0.68 in AITD vs. 0.24 in controls; $p = 0.0135$, $p = 0.0047$, respectively). We also evaluated the expression of ADP-ribosylation factors, such as GTPase 13B (Arl13b), which localize to the cilia. We observed a significantly decreased expression of Arl13b in the AITD tissues compared to the controls (0.23 in HT and 0.25 in AITD tissue vs. 1.27 in controls; $p = 0.0176$ and $p = 0.0227$, respectively) (Figure 3).

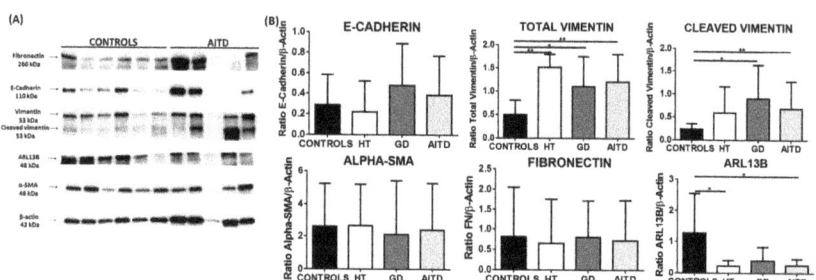

Figure 3. Western blot analysis of EMT markers in thyroid tissue from controls, HT, and GD patients. (**A**) Western blot analysis of E-cadherin, vimentin, fibronectin, α-SMA, and Arl13b. (**B**) Protein quantitation of the EMT markers indicated above. Data correspond to the arithmetic mean ± SD. * $p < 0.05$; ** $p < 0.01$.

2.2. TGF-β Stimulation of Cultured Thyroid Cells

TGF-β is one of the main agents involved in the acquisition of mesenchymal markers by epithelial cells and in their loss of epithelial characteristics [30,31]. In fact, adding TGF-β to epithelial cells in vitro is a suitable method to induce EMT in different cell models [33,42]. To study the possible role of EMT in AITD using in vitro models, we induced EMT in the thyroid cell line NThy-ORi 3.1 via stimulation with TGF-β for 48 and 72 h, as previously described [24].

In the TGF-β stimulated cells, we observed an upregulation of RNA levels of *FN1* (0.54 in controls vs. 3.65 in TGF-β stimulated cells; $p = 0.0022$) and *ACTA2* genes (0.0038 vs. 0.0099; $p = 0.0087$). Although we did not observe significant changes in *CDH1* and *VIM*, *VIM* expression had a tendency to increase in TGF-β stimulated cells compared to the controls ($p = 0.0649$) (Figure 4A).

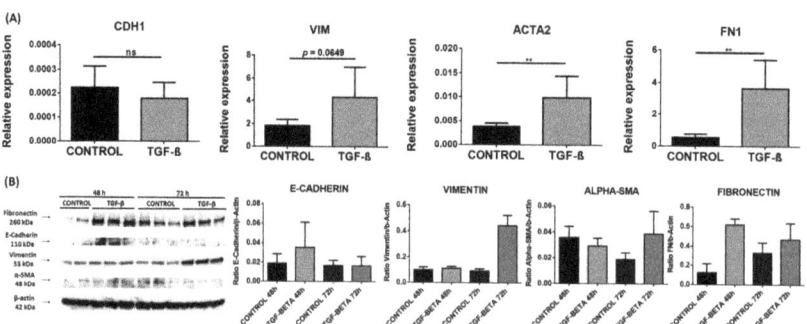

Figure 4. EMT markers in the NThy-ORi 3.1 cell line stimulated with TGF-β. (**A**) RT-qPCR expression of E-cadherin (*CDH1*), vimentin (*VIM*), fibronectin (*FN*), and α-SMA (*ACTA2*). Data correspond to the arithmetic mean ± SD. (**B**) Western blot analysis of E-cadherin, vimentin, fibronectin, α-SMA, and Arl13b. Quantitation is indicated below. Abbreviations—ns: not significant. ** $p < 0.01$.

Next, we analyzed the protein levels of these markers in cell homogenates using WB. The significant differences observed in RNA were not found in protein levels, which showed a tendency to an increased expression of FN and VIM at 72 h (Figure 4B).

2.3. Primary Cilia Disruption after TGF-β Stimulation

Considering the possible role of PC in the pathogenesis of AITD [36] and the association of these structures with EMT [40,41], we performed a morphometric analysis of PC in serum-starved NThy-ORi 3.1 cultured cells at 24, 48, and 72 h after EMT induction by TGF-β. Arl13b antibody was used to identify PC (Figure 5A). As expected, in non-stimulated cells, we observed an increase in cilia length and number in a time-dependent manner. Interestingly, during the TGF-β stimulation, we observed a significantly reduced cilia length (from 2.9 μm to 2.57 μm, $p = 0.0019$) and frequency of cilia (from 35% to 24%, $p = 0.0266$) at 24 h compared to non-stimulated cells. As stimulation time progressed, this reduction was maintained at 48 h (mean length 2.67 μm vs. 2.33 μm, $p = 0.0002$; frequency 41% vs. 20%, $p = 0.0003$) and at 72 h (mean length 3.07 μm vs. 2.5 μm, $p < 0.0001$, and frequency 55% to 24%, $p < 0.0001$) (Figure 5B).

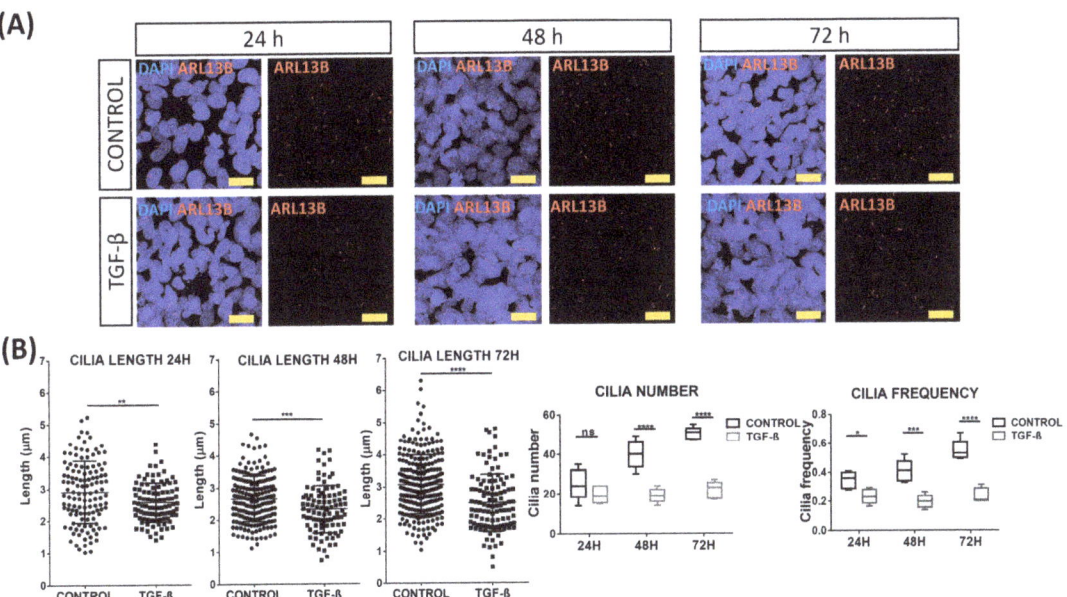

Figure 5. Disruption of primary cilia by TGF-β stimulation. (**A**) Immunofluorescence microscopy analysis of Arl13b (red) expression in absence or presence of TGF-β (10 ng/mL). Cell nuclei are stained with DAPI (blue). Scale bar: 25 μm. (**B**) Quantitation of length, number, and frequency of cilia. Data correspond to the arithmetic mean ± SD. Abbreviations: ns: not significant. * $p < 0.05$; ** $p < 0.01$; *** $p < 0.005$; **** $p < 0.0001$.

3. Discussion

TGF-β plays a pivotal role in normal human immune response and is involved in the pathophysiological spectrum of thyroid autoimmunity [5,13,17,43]. Based on the clear interplay between TGF-β, EMT, and primary ciliogenesis [30,31,33,40,41], we studied EMT markers in AITD and found that thyroid follicular cells can acquire mesenchymal markers and still preserve their epithelial phenotype. Furthermore, the stimulation of human thyroid cell lines with TGF-β upregulated the mesenchymal markers and disrupted primary cilia, suggesting a possible role of this mechanism in the pathogenesis of AITD.

EMT can be classified into three types: developmental (Type I), fibrosis and wound healing (Type II), and pathological (Type III). Type III is usually associated with cancer progression and inflammation [22,23]. In autoimmune disorders, the pro-inflammatory environment affects cells within tissues. If this environment persists, alteration in the wound-healing process can lead to the accumulation of mesenchymal cells, resulting in fibrosis and atrophy of the organ. This scenario leads to organ failure and the spreading of fibrotic cells to nearby healthy areas [44].

E-cadherin is a calcium-dependent tight-junction protein expressed on the cell membrane. Its main function is to maintain cell–cell adhesion, and its loss is one of the main hallmarks of EMT [45,46]. However, in some kinds of tumors, such as pancreatic cancers, tumor cells do not experience a downregulation of E-cadherin expression and yet preserve their epithelial phenotype, also exhibiting motility and the ability to migrate to other tissues [47]. Studies on rheumatoid arthritis (RA), an autoimmune disorder characterized by synovial tissue hyperplasia, have described a widely spread E-cadherin pattern in patients' synovial tissue that was related to cell hyperplasia. This expression pattern confirmed that synoviocytes had both epithelial and mesenchymal features due to the influence of arthritic synovial fluid [24]. In the context of AITD, in a study performed in HT samples with RET gene rearrangements, the authors observed a decreased expression of *CDH1* in RET$^+$ HT patients, suggesting an association between RET activation and the loss of cell adhesion [48]. In our analysis, *CDH1* expression levels were downregulated in HT tissue compared to the controls and GD tissue. Furthermore, *CDH1* expression levels had a direct correlation with FT4 and TSH-R-Ab levels and an inverse correlation with TSH levels. In GD, hypertrophy and hyperfunction of thyroid follicular cells secondary to the presence of TSH-R-Abs lead to increased FT4 levels. Thus, the correlation of CDH1 levels with FT4 and TSH-R-Ab is probably related to the increase in the number and functions of TFCs. On the contrary, the reduction of this marker in HT can be related to the partial loss of TFCs with epithelial phenotype and the increase in fibrosis associated with HT.

Vimentin is a major component of intermediate filaments, and it is widely expressed in mesenchymal cells. It is overexpressed in several epithelial cancers and is recognized as one of the EMT markers. The upregulation of this protein is associated with an increase in focal adhesions, cell motility, and cytoskeletal reorganization [49,50]. Vimentin cleavage by caspases produces a form of the protein that is associated with a disruption in cytoskeletal organization and, at a final stage, with apoptosis [51]. Indeed, a more diffuse cytoplasmic distribution pattern of vimentin with a stronger staining near the basal membrane has been previously described in AITD. This pattern was attributed to a more proliferative state or hyperplasia [52]. Although in later studies these changes were not considered to be indicative of a specific thyroid pathologic condition [53], we found an increase in vimentin expression levels in AITD patients. Furthermore, the increase in cleaved protein could be explained by the apoptosis of TFCs associated with the progression of HT and by the basal distribution pattern observed in GD tissue samples.

α-SMA is a protein commonly expressed in vascular smooth-muscle cells or myofibroblasts with a main role in fibrogenesis [54,55]. α-SMA expression in fibroblasts correlates with their activation state; i.e., α-SMA levels positively correlate with the number of extracellular matrix proteins produced by fibroblasts [56]. In our study, although α-SMA expression assessed by RNA and WB did not change in AITD samples compared to controls, we could observe a clear increase of α-SMA expression in TFCs from AITD through immunohistochemical analysis. These results suggest that some TFCs experience a partial phenotype transition to mesenchymal cells. Indeed, an increased α-SMA expression was reported in thyroid tissue fibroblasts of patients with GD and HT, showing that these cells had many similar features to orbit fibroblasts that differentiate from myofibroblasts [57].

Fibronectin (FN) is an extracellular protein that acts as a scaffold between cells and components of the extracellular matrix [58]. FN is commonly upregulated among other markers in in vitro EMT-induction models [59]. However, the use of this protein as an EMT marker is partially limited as it is produced by many cell types, including not only fibroblasts but also epithelial cells or mononuclear cells [60,61]. In RA, the increased expression of FN was associated with the induction of pro-inflammatory responses and disease progression [62]. In our study, we found a significantly increased staining for FN in AITD TFCs when compared to control tissues. However, as observed with α-SMA, the analysis of bulk tissue using RNA and WB did not show a significant change.

Regarding the markers evaluated by immunohistochemistry, staining was heterogeneous, and some mesenchymal markers were not expressed in the whole tissue, with some areas presenting an increased staining, especially those closer to immune infiltrates or connective tissue. This could be explained by the fact that, in the context of AITD, immune cells secrete cytokines such as IL1-β, TGF-β, or TNF-α that could contribute to the induction of genes related to a mesenchymal phenotype. For example, the synergy between IL-1β and TGF-β could lead to an increase in TGF-β-induced EMT [18]. Regarding AITD, a study performed in a mouse model of granulomatous experimental autoimmune thyroiditis (g-EAT) showed that, at the early stages of the disease, TGF-β induces an immunosuppressive environment. However, at the final stages, this cytokine plays a profibrotic role. Thus, g-EAT thyroid samples with fibrosis showed a higher presence of TGF-β and TNF-α than the control samples [19].

Although we observed the acquisition of these markers through immunohistochemistry, our analysis using WB did not corroborate these results, showing only an increase in cleaved vimentin and a decrease in Arl13b in protein lysates. One of the advantages of immunohistochemistry is the detection of the exact location (namely, specific cells) of a target protein within a tissue sample. On the other hand, WB gathers quantitative information on global protein levels in the tissue. Thus, the differences observed with the different methods could also be related to the effect on WB determinations of a more vascularized tissue or of a higher content of fibrous or connective tissue in AITD, which also express these markers.

Another key point is the dynamic characteristics of EMT, in which cells progressively acquire mesenchymal markers without a concomitant complete loss of epithelial markers. The expression of both mesenchymal and epithelial markers reflects the plasticity of cells depending on their environment [63,64]. Therefore, EMT does not define the final fate of a cell since this process is reversible and there is also a mesenchymal–epithelial transition (MET) where mesenchymal cells can reacquire an epithelial phenotype. In the context of chronic epithelial degradation, only a few cells experiment with a transition to a mesenchymal state [65], immunostaining being the gold standard technique to analyze them, as bulk tissue analysis would not detect these alterations.

Regarding the possible role of PC in the pathogenesis of AITD [36], we also analyzed the effect of TGF-β on these structures. We showed a decrease in the number and length of PC in TGF-β stimulated cells. PC are involved in TGF-β signaling, as the receptors (TGFBRI and TGFBRII) for this cytokine are expressed at the cilia base [66]. In chondrocytic cells, TGF-β was reported to reduce the levels of intraflagellar transport 88 (IFT88), which is expressed in the cilia, leading to a reduction in cilia length and frequency [67]. Furthermore, PC were disrupted in a kidney epithelial cell line undergoing a TGF-β induced EMT, as they changed their morphology to become longer and increased the expression levels of α-SMA and collagen III genes [40]. In light of these results, we can consider that TGF-β may also be involved in PC alteration.

This study has some limitations. First, we analyzed the main commonly used EMT markers to study EMT. Other markers, such as N-cadherin; different types of collagen, i.e., collagen III, fibroblast secreted protein 1 (FSP-1); transcription factors, such as Snail or Twist family; signaling pathways, such as the Sonic hedgehog (Shh) pathway or the Wnt-β-catenin signaling pathway, can be included in future studies. Second, in culture models, we tried to establish an in vitro model with primary TFCs derived from patients; however, since the number of samples was a limiting factor, an alternative based on a human thyroid cell line was chosen.

To conclude, we have reported the acquisition of mesenchymal features by TFCs in patchy areas of the thyroid, which can be attributed to a transition to myofibroblasts expressing mesenchymal markers. TGF-β, a cytokine involved in thyroid autoimmunity, can be related to the acquisition of this phenotype. Finally, primary cilia disruption may represent a potential research area within the study of AITD and the acquisition of mesenchymal phenotypes by TFCs.

4. Materials and Methods

4.1. Patient Samples

Thyroid tissue samples were collected from surgeries from AITD patients at the Hospital Universitario de la Princesa and from non-thyroid pathology laryngectomy samples or healthy organ donors at the Institut d'Investigació en Ciències de la Salut Germans Trias i Pujol (IGTP-HUGTIP) Biobank. Clinical diagnoses were all reviewed by a single experienced endocrinologist based on standard clinical, laboratory, and histological criteria. Serum free thyroxine (FT4), thyroid-stimulating hormone (TSH), and antibodies against thyroglobulin (TG), thyroperoxidase (TPO), and TSH receptor (TSH-R) were determined in all patients at the time of the surgery. Clinical data are shown in Table 1.

Table 1. Clinical parameters of AITD patients included in RT-qPCR analyses.

Parameters	HT	GD
N	10	10
Gender (F/M)	10/0	9/1
Age, years	62 (57–70)	47 (40–57)
Ophthalmopathy	0	7
TSH, mU/mL	2.59 (1.68–3.26)	0.44 (0.01–0.8)
T4, ng/dL	-	1.29 (1.02–1.54)
TG-Ab, UI/mL	626 (143.5–728.5)	20 (20–2279)
TPO-Ab, UI/mL	713.5 (434.75–1421.25)	169 (20–578.5)
TSH-R-Ab, U/L	-	5.63 (0.91–7.67)

Values are categorical values and median (interquartile intervals 25–75) for continuous variables. Abbreviations: F, female; M, male T4, thyroxine (normal range = 0.93–1.7); TG-Ab, anti-thyroglobulin antibody (negative < 344); TPO-Ab, anti-thyroid peroxidase antibody (negative < 100); TSH, thyrotropin (normal range = 0.27–4.20); TSH-R-Ab, anti-thyrotropin receptor antibody (negative < 0.7).

This study was approved by the Internal Ethical Review Committee of Hospital Universitario de la Princesa (Committee Register Number: 2796, approval date: 26 May 2016), and written informed consent was obtained from all patients in accordance with the Declaration of Helsinki.

4.2. RNA Isolation and RT-qPCR

RNA from fresh-frozen thyroid tissues (10 HT, 10 GD, and 10 control samples) were isolated with the miRNeasy Mini Kit (Qiagen) according to the manufacturer's instructions, and the quality and quantity of RNA were evaluated by NanoDrop ND-1000 analysis. First-strand cDNA was generated with a high-capacity cDNA reverse transcription kit with a ribonuclease inhibitor (Applied Biosystems. Waltham, MA, USA), and quantitative reverse transcription–polymerase chain reaction (RT-qPCR) was performed in triplicate using SYBR Green qPCR Master Mix (Thermo Fisher Scientific. Waltham, MA, USA). A list of primers is included in Table 2, and the reaction was performed with the CFX384 Touch Real-Time PCR Detection System (Bio-Rad. Hercules, CA, USA). Ct values were normalized by the Ct of housekeeping genes such as β-actin and GAPDH.

Table 2. List of primers used in RT-qPCR analyses.

Primer	Orientation	Sequence
CDH1	FORWARD	GCCGAGAGCTACACGTTCAC
	REVERSE	ACTTTGAATCGGGTGTCGAG
VIM	FORWARD	CTCCCTCTGGTTGATACCCAC
	REVERSE	GGTCATCGTGATGCTGAGAAG
FN1	FORWARD	CCTCAATTGTTGTTCGCTGGAGCA
	REVERSE	GGTGACGGAGTTTGCAGTTTC
ACTA2	FORWARD	TGGCTATCCAGGCGGTGCTGTCT
	REVERSE	ATGGCATGGGGCAAGGCATAGC
GAPDH	FORWARD	GCCCAATACGACCAAATCC
	REVERSE	AGCCACATCGCTCAGACAC
β-ACTIN	FORWARD	GCCGACAGGATGCAGAAGGA
	REVERSE	CGGAGTACTTGCGCTCAGGA

4.3. Tissue Microarrays

A total of 49 formalin-fixed, paraffin-embedded (FFPE) tissues were evaluated using tissue microarrays (TMAs). Of these, 30 were AITD thyroid samples with pathological diagnosis of HT and GD (17 and 16, respectively), and 16 corresponded to control thyroid samples from surgeries at the Hospital Universitario de la Princesa. All samples had a duplicate in the same TMA and were taken and managed in accordance with local regulations with the approval of the local institutional review board. Clinical data are shown in Table 3.

Table 3. Clinical parameters of patients included in TMA.

Parameters	HT	GD	Controls
N	17	16	16
Gender (F/M)	12/5	15/1	10/6
Age, years	62 (42–66)	48 (43–59)	57 (43–61)
Ophthalmopathy	0	8	0
Smoking	1	6	2
TSH, mU/mL	3.48 (2.15–4.70)	0.15 (0.01–5.04)	1.63 (1.32–2.95)
T4, ng/dL	1.34 (1.11–1.45)	1.05 (0.95–1.51)	1.34 (1.06–1.81)
TG-Ab, UI/mL	80 (20.75–216.25)	95 (23–117)	12 (12–12)
TPO-Ab, UI/mL	86 (22.5–297)	175 (16–223)	4 (4–10)
TSH-R-Ab, U/L	-	5.22 (2.98–8.11)	-

Values are categorical values and median (interquartile intervals 25–75) for continuous variables. Abbreviations: F, female; M, male; T4, thyroxine (normal range = 0.93–1.7); TG-Ab, anti-thyroglobulin antibody (negative < 344); TPO-Ab, anti-thyroid peroxidase antibody (negative < 100); TSH, thyrotropin (normal range = 0.27–4.20); TSH-R-Ab, anti-thyrotropin receptor antibody (negative < 0.7).

4.4. Immunohistochemistry

FFPE samples from healthy controls and patients with AITD were collected and processed in order to obtain tissue sections 3 µm in thickness. In the case of TMAs, a preincubation of the slides with Clear Rite at 65 °C for 15 min was performed with the aim of removing the excess of paraffin. Antigen retrieval was performed in an Agilent Dako PTlink (Agilent. Santa Clara, CA, USA) in a basic or acid buffer, depending on the antibody requirements. Endogenous peroxidase was inhibited with a peroxidase-blocking solution (Dako. Santa Clara, CA, USA). Then, tissue sections were incubated overnight at 4 °C with primary antibodies against E-cadherin (Thermo Fisher Scientific. Waltham, MA, USA. Cat# 33-4000, RRID:AB_2533118), Vimentin (Thermo Fisher Scientific. Waltham, MA, USA. Cat# PA5-27231, RRID:AB_2544707), α-SMA (Thermo Fisher Scientific. Waltham, MA, USA. Cat# PA5-18292, RRID: AB_10980764), and Fibronectin (Thermo Fisher Scientific. Waltham, MA, USA. Cat# PA5-29578, RRID:AB_2547054). The following day, sections were incubated with the proper secondary antibodies conjugated to horseradish peroxidase. Finally, tissue sections were incubated with 3,3′-Diaminobenzidine (DAB), counterstained with hematoxylin (Sigma-Aldrich. San Luis, MO, USA), dehydrated in alcohol, cleared with xylene, and mounted.

4.5. Immunohistochemistry Score

Immunohistochemistry quantification was determined by analyzing the intensity of staining in the case of α-SMA and fibronectin and assessing the intensity and basal or cytoplasmatic expression for vimentin. The IHC score was graded as follows: for α-SMA, 0 is for negative staining, 1 is for light staining, 2 is for moderate staining, and 3 is for intense staining; for fibronectin, 0 is for negative staining, 1 is for light staining, 2 is for intense staining; and for vimentin, 0 indicates cytoplasmatic expression, 1 is for low basal expression, and 2 is for wide basal expression.

4.6. Thyroid Cell Cultures

Cell cultures were performed with the human thyroid cell line NThy-ORi 3-1 (ECACC 90011609, kindly provided by Dr. Pilar Santisteban, Instituto de Investigaciones Biomédicas "Alberto Sols", Madrid, Spain). The NThy-ORi 3-1 cell line was cultured in RPMI 1640 medium supplemented with Gluta-MAX, 10% fetal bovine serum or FBS (Hyclone. Logan, UT, USA), and 1% of penicillin/streptomycin (Gibco. Carlsbad, CA, USA).

4.7. TGF-B Stimulation Assays

Cells were cultured until reaching confluence. Thereafter, cells were stimulated or not with TGF-β 10 ng/mL (Miltenyi Biotec. Bergisch Gladbach, Germany) in serum-free DMEM. TGF-β was left for 24 h, 48 h, and 72 h. Then, cells were washed and collected in TRiZol for RNA extraction, then scrapped and resuspended in RIPA + Protease and phosphatase inhibitor cocktail Halt^{TM}. (ThermoFisher Scientific. Waltham, MA, USA) for WB and in round coverslips for immunofluorescence analysis.

4.8. Immunofluorescence Microscopy Analysis

Cells were cultured on round coverslips in 6-well plates, as previously described [7]. Briefly, cells were washed with PBS and fixed with 4% paraformaldehyde. Later, cells were permeabilized with PBS 0.1% Triton X-100 at room temperature and blocked with 5% bovine serum albumin and 10% BSA-PBS.

Cells were incubated with an anti-Arl13b antibody (Proteintech. Rosemont, IL, USA). Cat# 17711-1-AP, RRID:AB_2060867) overnight at 4 °C. Then, slides were incubated for 1 h with an Alexa Fluor 568 labeled goat anti-mouse IgG antibody (Thermo Fisher Scientific. Waltham, MA, USA. Cat# A-11031, RRID: AB_144696). Finally, cell nuclei were counterstained with 4′,6-diamidino-2-phenylindole (DAPI) and analyzed in a Leica Sp5 confocal microscope (Leica Biosystems. Wetzlar, Germany).

The frequency of cilia was estimated manually by analyzing Z-stacked images captured in a confocal microscope. The frequency of ciliated cells was estimated by analyzing the relative number of PC vs. the number of nuclei. A total of 1307 nuclei in non-stimulated cells and 1372 nuclei in stimulated cells were analyzed. The PC length was measured using the ROI measurement tool of ImageJ 1.52i software (National Institutes of Health. Bethesda, MD, USA) for a total of 576 cilia in non-stimulated and 302 in stimulated cells.

4.9. Western Blot Analysis

Thyroid tissue samples were mechanically disaggregated in liquid nitrogen and resuspended in RIPA buffer (Sigma-Aldrich. San Luis, MO, USA) containing a protease inhibitor cocktail Halt^{TM} (Thermo Fisher Scientific. Waltham, MA, USA). After 30 min on ice, samples were sonicated, and cell lysates were centrifuged at 4 °C. The supernatant was recovered and stored at −80 °C until use.

Protein samples obtained from the NThy-ORi 3-1 cell line were lysed in RIPA buffer with protease inhibitors at 4 °C in shaking conditions. Later, cells were scrapped and sonicated, followed by a centrifugation step. The resultant supernatant was transferred to another tube and stored at −80 °C until use.

WB was performed in an 8–15% mini-protean TGX precast gel (Bio-Rad. Hercules, CA, USA) and transferred to nitrocellulose membranes. Membranes were blocked and incubated overnight at 4 °C with primary antibodies against E-cadherin (Thermo Fisher Scientific. Waltham, MA, USA. Cat# 33-4000, RRID:AB_2533118), Vimentin (Thermo Fisher Scientific. Waltham, MA, USA. Cat# PA5-27231, RRID:AB_2544707), α-SMA (Thermo Fisher Scientific. Waltham, MA, USA. Cat# PA5-18292, RRID: AB_10980764), fibronectin (Thermo Fisher Scientific. Waltham, MA, USA. Cat# PA5-29578, RRID:AB_2547054), and Arl13b (Proteintech. Rosemont, IL, USA. Cat# 17711-1-AP, RRID:AB_2060867). The next day, membranes were washed with TBS-Tween, incubated with secondary antibodies conjugated to horseradish peroxidase, and visualized using the Pierce™ ECL Western Blotting Substrate chemiluminescent detection reagent kit (Thermo Fisher Scientific. Waltham, MA, USA). After that, membranes were stripped with Restore™ Plus Stripping Buffer (Thermo Fisher Scientific. Waltham, MA, USA) at 37 °C in a shaking incubator, followed by washing steps with TBS-Tween. Finally, membranes were blocked and incubated with anti-β–actin-HRP (Santa Cruz Biotechnology. Dallas, TX, USA. Cat# sc-47778, RRID:AB_626632) polyclonal antibody. ImageJ 1.52i software (National Institutes of Health. Bethesda, MD, USA) was used to quantify the amount of protein in each band.

4.10. Statistics

Results were expressed as the arithmetic mean and standard deviation (SD), and differences between groups were compared by the Mann–Whitney or unpaired *t*-test for two-population experiments and one-way ANOVA or Kruskal–Wallis analyses for experiments with more than two populations. Spearman's rho analyses were performed to detect correlations between the different markers examined by immunohistochemistry and clinical parameters. In addition, p values < 0.05 were considered statistically significant. All statistical analyses were performed with the GraphPad Prism 6.0 software (GraphPad Software. Boston, MA, USA).

5. Conclusions

In conclusion, an interconnection between EMT, TGF-β, and AITDs is described in this manuscript. TFCs experience a transition or partial transition to mesenchymal cells as they acquire mesenchymal markers, such as fibronectin and α-SMA. This effect is probably caused by the pro-inflammatory microenvironment in AITD and mainly by the influence of TGF-β. TFCs do not lose their epithelial phenotype completely. Finally, PC disruption by TGF-β can contribute to the acquisition of mesenchymal markers by TFCs.

Author Contributions: Conceptualization: P.S.-G., R.M.-H. and M.M.; methodology: P.S.-G., A.S.-S., L.C.-E., F.M.-J., R.M.-H. and M.M.; validation: P.S.-G., A.S.-S., L.C.-E. and F.M.-J.; formal analysis: P.S.-G., N.S.d.l.B.C. and M.S.-N.; investigation: P.S.-G., R.M.-H. and M.M.; Resources: M.S.-N., J.L.M.-D.-N., A.R., R.M.-H. and M.M.; data curation: P.S.-G. and R.M.-H.; writing—original draft preparation: P.S.-G. and R.M.-H.; writing—review and editing: P.S.-G., R.M.-H. and M.M.; visualization: P.S.-G., A.S.-S., R.M.-H. and M.M.; supervision: R.M.-H. and M.M.; project administration: R.M.-H. and M.M.; funding acquisition: M.S.-N., R.M.-H. and M.M. All authors have read and agreed to the published version of the manuscript.

Funding: This work was supported by the following grants: Proyectos de Investigación en Salud (PI) PI19-00584 and PI22/01404, and Proyectos de investigación de Medicina Personalizada de Precisión (PMP) PMP22/00021 (funded by Instituto de Salud Carlos III); iTIRONET P2022/BMD7379 (funded by Comunidad de Madrid); FEDER funds to M.M and R.M.-H. (cofinanced); predoctoral fellowship funded by Instituto de Salud Carlos III and FSE+ funds (FI20/00035) to P.S-G; and predoctoral fellowships funded by Comunidad de Madrid (PEJ-2020-AI_BMD-18292) to N.S.d.l.B.C. The funders had no role in the study design, data collection, data analysis, interpretation, or writing of the report.

Institutional Review Board Statement: The study was conducted in accordance with the Declaration of Helsinki, and approved by the the Internal Ethical Review Committee of Hospital Universitario de la Princesa (Committee Register Number: 2796, approval date: 26 May 2016).

Informed Consent Statement: Informed consent was obtained from all subjects involved in the study.

Data Availability Statement: Some or all datasets generated and/or analyzed during the current study are not publicly available but are available from the corresponding author upon reasonable request.

Acknowledgments: We warmly thank Manuel Gómez for English corrections.

Conflicts of Interest: The authors declare no conflict of interest.

References

1. Weetman, A.; DeGroot, L.J. *Autoimmunity to the Thyroid Gland*; MDText.com, Inc.: South Dartmouth, MA, USA, 2000.
2. Mammen, J.S.R.; Cappola, A.R. Autoimmune Thyroid Disease in Women. *JAMA* **2021**, *325*, 2392–2393. [CrossRef]
3. Bahn, R.S. Graves' Ophthalmopathy. *N. Engl. J. Med.* **2010**, *362*, 726–738. [CrossRef]
4. Burch, H.B.; Perros, P.; Bednarczuk, T.; Cooper, D.S.; Dolman, P.J.; Leung, A.M.; Mombaerts, I.; Salvi, M.; Stan, M.N. Management of Thyroid Eye Disease: A Consensus Statement by the American Thyroid Association and the European Thyroid Association. *Eur. Thyroid J.* **2022**, *11*, e220189. [CrossRef] [PubMed]
5. Ramos-Leví, A.M.; Marazuela, M. Pathogenesis of Thyroid Autoimmune Disease: The Role of Cellular Mechanisms. *Endocrinol. Nutr.* **2016**, *63*, 421–429. [CrossRef]
6. Xiaoheng, C.; Yizhou, M.; Bei, H.; Huilong, L.; Xin, W.; Rui, H.; Lu, L.; Zhiguo, D. General and Specific Genetic Polymorphism of Cytokines-Related Gene in AITD. *Mediat. Inflamm.* **2017**, *2017*, 3916395. [CrossRef]
7. Sacristán-Gómez, P.; Serrano-Somavilla, A.; González-Amaro, R.; Martínez-Hernández, R.; Marazuela, M. Analysis of Expression of Different Histone Deacetylases in Autoimmune Thyroid Disease. *J. Clin. Endocrinol. Metab.* **2021**, *106*, 3213–3227. [CrossRef]
8. Prummel, M.F.; Wiersinga, W.M. Smoking and Risk of Graves' Disease. *JAMA* **1993**, *269*, 479–482. [CrossRef]
9. Rodríguez-Muñoz, A.; Vitales-Noyola, M.; Ramos-Levi, A.; Serrano-Somavilla, A.; González-Amaro, R.; Marazuela, M. Levels of Regulatory T Cells CD69(+)NKG2D(+)IL-10(+) Are Increased in Patients with Autoimmune Thyroid Disorders. *Endocrine* **2016**, *51*, 478–489. [CrossRef]
10. García-López, M.A.; Marazuela, M.; Sánchez-Madrid, F.; de la Fuente, H.; Monsiváis-Urenda, A.; Alvarado-Sánchez, B.; Figueroa-Vega, N.; González-Amaro, R. Regulatory T Cells in Human Autoimmune Thyroid Disease. *J. Clin. Endocrinol. Metab.* **2006**, *91*, 3639–3646. [CrossRef]
11. Nanba, T.; Watanabe, M.; Inoue, N.; Iwatani, Y. Increases of the Th1/Th2 Cell Ratio in Severe Hashimoto's Disease and in the Proportion of Th17 Cells in Intractable Graves' Disease. *Thyroid Off. J. Am. Thyroid Assoc.* **2009**, *19*, 495–501. [CrossRef]
12. ITO, C.; WATANABE, M.; OKUDA, N.; WATANABE, C.; IWATANI, Y. Association between the Severity of Hashimoto's Disease and the Functional +874A/T Polymorphism in the Interferon-γ Gene. *Endocr. J.* **2006**, *53*, 473–478. [CrossRef]
13. Widder, J.; Dorfinger, K.; Wilfing, A.; Trieb, K.; Pirich, K.; Loebenstein, R.; Niederle, B.; Gessl, A.; Spitzauer, S.; Grubeck-Loebenstein, B. The Immunoregulatory Influence of Transforming Growth Factor Beta in Thyroid Autoimmunity: TGF β Inhibits Autoreactivity in Graves' Disease. *J. Autoimmun.* **1991**, *4*, 689–701. [CrossRef]
14. Gianoukakis, A.G.; Khadavi, N.; Smith, T.J. Cytokines, Graves' Disease, and Thyroid-Associated Ophthalmopathy. *Thyroid* **2008**, *18*, 953–958. [CrossRef]
15. Ganesh, B.B.; Bhattacharya, P.; Gopisetty, A.; Prabhakar, B.S. Role of Cytokines in the Pathogenesis and Suppression of Thyroid Autoimmunity. *J. Interferon Cytokine Res.* **2011**, *31*, 721–731. [CrossRef]
16. Vural, P.; Değirmencioğlu, S.; Doğru-Abbasoğlu, S.; Baki, M.; Özderya, A.; Karadağ, B.; Uysal, M. Arg25Pro (c.915G>C) Polymorphism of Transforming Growth Factor B1 Gene Suggests an Association with Increased Risk for Hashimoto's Thyroiditis. *Int. Immunopharmacol.* **2015**, *28*, 521–524. [CrossRef]
17. Kutluturk, F.; Yarman, S.; Sarvan, F.O.; Kekik, C. Association of Cytokine Gene Polymorphisms (IL6, IL10, TNF-α, TGF-β and IFN-γ) and Graves' Disease in Turkish Population. *Endocr. Metab. Immune Disord. Drug Targets* **2013**, *13*, 163–167. [CrossRef]
18. Zhang, S.; Fan, Y.; Qin, L.; Fang, X.; Zhang, C.; Yue, J.; Bai, W.; Wang, G.; Chen, Z.; Renz, H.; et al. IL-1β Augments TGF-β Inducing Epithelial-Mesenchymal Transition of Epithelial Cells and Associates with Poor Pulmonary Function Improvement in Neutrophilic Asthmatics. *Respir. Res.* **2021**, *22*, 216. [CrossRef]
19. Chen, K.; Wei, Y.; Sharp, G.C.; Braley-Mullen, H. Mechanisms of Spontaneous Resolution versus Fibrosis in Granulomatous Experimental Autoimmune Thyroiditis. *J. Immunol.* **2003**, *171*, 6236–6243. [CrossRef] [PubMed]
20. Yu, L.; Mu, Y.; Sa, N.; Wang, H.; Xu, W. Tumor Necrosis Factor α Induces Epithelial-Mesenchymal Transition and Promotes Metastasis via NF-ΚB Signaling Pathway-Mediated TWIST Expression in Hypopharyngeal Cancer. *Oncol. Rep.* **2014**, *31*, 321–327. [CrossRef]
21. Thiery, J.P.; Acloque, H.; Huang, R.Y.J.; Nieto, M.A. Epithelial-Mesenchymal Transitions in Development and Disease. *Cell* **2009**, *139*, 871–890. [CrossRef] [PubMed]
22. Kalluri, R.; Weinberg, R.A. The Basics of Epithelial-Mesenchymal Transition. *J. Clin. Investig.* **2009**, *119*, 1420–1428. [CrossRef] [PubMed]

23. Yang, J.; Antin, P.; Berx, G.; Blanpain, C.; Brabletz, T.; Bronner, M.; Campbell, K.; Cano, A.; Casanova, J.; Christofori, G.; et al. Guidelines and Definitions for Research on Epithelial–Mesenchymal Transition. *Nat. Rev. Mol. Cell Biol.* **2020**, *21*, 341–352. [CrossRef] [PubMed]
24. Steenvoorden, M.M.; Tolboom, T.C.; van der Pluijm, G.; Löwik, C.; Visser, C.P.; DeGroot, J.; Gittenberger-DeGroot, A.C.; DeRuiter, M.C.; Wisse, B.J.; Huizinga, T.W.; et al. Transition of Healthy to Diseased Synovial Tissue in Rheumatoid Arthritis Is Associated with Gain of Mesenchymalfibrotic Characteristics. *Arthritis Res. Ther.* **2006**, *8*, R165. [CrossRef] [PubMed]
25. Lefèvre, S.; Knedla, A.; Tennie, C.; Kampmann, A.; Wunrau, C.; Dinser, R.; Korb, A.; Schnäker, E.-M.; Tarner, I.H.; Robbins, P.D.; et al. Synovial Fibroblasts Spread Rheumatoid Arthritis to Unaffected Joints. *Nat. Med.* **2009**, *15*, 1414–1420. [CrossRef]
26. Bataille, F.; Rohrmeier, C.; Bates, R.; Weber, A.; Rieder, F.; Brenmoehl, J.; Strauch, U.; Farkas, S.; Fürst, A.; Hofstädter, F.; et al. Evidence for a Role of Epithelial Mesenchymal Transition during Pathogenesis of Fistulae in Crohn's Disease. *Inflamm. Bowel Dis.* **2008**, *14*, 1514–1527. [CrossRef] [PubMed]
27. Scharl, M.; Weber, A.; Fürst, A.; Farkas, S.; Jehle, E.; Pesch, T.; Kellermeier, S.; Fried, M.; Rogler, G. Potential Role for SNAIL Family Transcription Factors in the Etiology of Crohn's Disease-Associated Fistulae. *Inflamm. Bowel Dis.* **2011**, *17*, 1907–1916. [CrossRef] [PubMed]
28. Sisto, M.; Lorusso, L.; Ingravallo, G.; Ribatti, D.; Lisi, S. TGFβ1-Smad Canonical and -Erk Noncanonical Pathways Participate in Interleukin-17-Induced Epithelial-Mesenchymal Transition in Sjögren's Syndrome. *Lab. Investig. J. Tech. Methods Pathol.* **2020**, *100*, 824–836. [CrossRef]
29. Sisto, M.; Lorusso, L.; Ingravallo, G.; Tamma, R.; Ribatti, D.; Lisi, S. The TGF-B1 Signaling Pathway as an Attractive Target in the Fibrosis Pathogenesis of Sjögren's Syndrome. *Mediat. Inflamm.* **2018**, *2018*, 1965935. [CrossRef]
30. Hao, Y.; Baker, D.; Ten Dijke, P. TGF-β-Mediated Epithelial-Mesenchymal Transition and Cancer Metastasis. *Int. J. Mol. Sci.* **2019**, *20*, 2767. [CrossRef]
31. Xu, J.; Lamouille, S.; Derynck, R. TGF-β-Induced Epithelial to Mesenchymal Transition. *Cell Res.* **2009**, *19*, 156–172. [CrossRef]
32. Kim, K.K.; Kugler, M.C.; Wolters, P.J.; Robillard, L.; Galvez, M.G.; Brumwell, A.N.; Sheppard, D.; Chapman, H.A. Alveolar Epithelial Cell Mesenchymal Transition Develops in Vivo during Pulmonary Fibrosis and Is Regulated by the Extracellular Matrix. *Proc. Natl. Acad. Sci. USA* **2006**, *103*, 13180–13185. [CrossRef] [PubMed]
33. Miettinen, P.J.; Ebner, R.; Lopez, A.R.; Derynck, R. TGF-Beta Induced Transdifferentiation of Mammary Epithelial Cells to Mesenchymal Cells: Involvement of Type I Receptors. *J. Cell Biol.* **1994**, *127*, 2021–2036. [CrossRef] [PubMed]
34. Willis, B.C.; Liebler, J.M.; Luby-Phelps, K.; Nicholson, A.G.; Crandall, E.D.; du Bois, R.M.; Borok, Z. Induction of Epithelial-Mesenchymal Transition in Alveolar Epithelial Cells by Transforming Growth Factor-Beta1: Potential Role in Idiopathic Pulmonary Fibrosis. *Am. J. Pathol.* **2005**, *166*, 1321–1332. [CrossRef] [PubMed]
35. Fan, J.-M.; Ng, Y.-Y.; Hill, P.A.; Nikolic-Paterson, D.J.; Mu, W.; Atkins, R.C.; Lan, H.Y. Transforming Growth Factor-β Regulates Tubular Epithelial-Myofibroblast Transdifferentiation in Vitro. *Kidney Int.* **1999**, *56*, 1455–1467. [CrossRef]
36. Martínez-Hernández, R.; Serrano-Somavilla, A.; Ramos-Leví, A.; Sampedro-Nuñez, M.; Lens-Pardo, A.; Muñoz De Nova, J.L.; Triviño, J.C.; González, M.U.; Torné, L.; Casares-Arias, J.; et al. Integrated MiRNA and MRNA Expression Profiling Identifies Novel Targets and Pathological Mechanisms in Autoimmune Thyroid Diseases. *EBioMedicine* **2019**, *50*, 329–342. [CrossRef]
37. Anvarian, Z.; Mykytyn, K.; Mukhopadhyay, S.; Pedersen, L.B.; Christensen, S.T. Cellular Signalling by Primary Cilia in Development, Organ Function and Disease. *Nat. Rev. Nephrol.* **2019**, *15*, 199–219. [CrossRef]
38. Goetz, S.C.; Anderson, K.V. The Primary Cilium: A Signalling Centre during Vertebrate Development. *Nat. Rev. Genet.* **2010**, *11*, 331–344. [CrossRef]
39. Martin, A.; Hedinger, C.; Häberlin-Jakob, M.; Walt, H. Structure and Motility of Primary Cilia in the Follicular Epithelium of the Human Thyroid. *Virchows Arch. B Cell Pathol.* **1988**, *55*, 159–166. [CrossRef]
40. Han, S.J.; Jung, J.K.; Im, S.-S.; Lee, S.-R.; Jang, B.-C.; Park, K.M.; Kim, J.I. Deficiency of Primary Cilia in Kidney Epithelial Cells Induces Epithelial to Mesenchymal Transition. *Biochem. Biophys. Res. Commun.* **2018**, *496*, 450–454. [CrossRef]
41. Ehnert, S.; Sreekumar, V.; Aspera-Werz, R.H.; Sajadian, S.O.; Wintermeyer, E.; Sandmann, G.H.; Bahrs, C.; Hengstler, J.G.; Godoy, P.; Nussler, A.K. TGF-β(1) Impairs Mechanosensation of Human Osteoblasts via HDAC6-Mediated Shortening and Distortion of Primary Cilia. *J. Mol. Med. Berl. Ger.* **2017**, *95*, 653–663. [CrossRef]
42. Valcourt, U.; Kowanetz, M.; Niimi, H.; Heldin, C.-H.; Moustakas, A. TGF-beta and the Smad Signaling Pathway Support Transcriptomic Reprogramming during Epithelial-Mesenchymal Cell Transition. *Mol. Biol. Cell* **2005**, *16*, 4. [CrossRef] [PubMed]
43. Kardalas, E.; Maraka, S.; Papagianni, M.; Paltoglou, G.; Siristatidis, C.; Mastorakos, G. TGF-β Physiology as a Novel Therapeutic Target Regarding Autoimmune Thyroid Diseases: Where Do We Stand and What to Expect. *Med. Kaunas Lith.* **2021**, *57*, 621. [CrossRef] [PubMed]
44. Tolboom, T.C.A.; van der Helm-Van Mil, A.H.M.; Nelissen, R.G.H.H.; Breedveld, F.C.; Toes, R.E.M.; Huizinga, T.W.J. Invasiveness of Fibroblast-like Synoviocytes Is an Individual Patient Characteristic Associated with the Rate of Joint Destruction in Patients with Rheumatoid Arthritis. *Arthritis Rheum.* **2005**, *52*, 1999–2002. [CrossRef]
45. Gheldof, A.; Berx, G. Cadherins and Epithelial-to-Mesenchymal Transition. In *Progress in Molecular Biology and Translational Science*; Elsevier: Amsterdam, The Netherlands, 2013; Volume 116, pp. 317–336. ISBN 978-0-12-394311-8.
46. Loh, C.-Y.; Chai, J.; Tang, T.; Wong, W.; Sethi, G.; Shanmugam, M.; Chong, P.; Looi, C. The E-Cadherin and N-Cadherin Switch in Epithelial-to-Mesenchymal Transition: Signaling, Therapeutic Implications, and Challenges. *Cells* **2019**, *8*, 1118. [CrossRef]

47. Liu, X.; Huang, H.; Remmers, N.; Hollingsworth, M.A. Loss of E-Cadherin and Epithelial to Mesenchymal Transition Is Not Required for Cell Motility in Tissues or for Metastasis. *Tissue Barriers* **2014**, *2*, e969112. [CrossRef] [PubMed]
48. Smyth, P.; Sheils, O.; Finn, S.; Martin, C.; O'Leary, J.; Sweeney, E.C. Real-Time Quantitative Analysis of E-Cadherin Expression in Ret/PTC-1-Activated Thyroid Neoplasms. *Int. J. Surg. Pathol.* **2001**, *9*, 265–272. [CrossRef] [PubMed]
49. Liu, C.-Y.; Lin, H.-H.; Tang, M.-J.; Wang, Y.-K. Vimentin Contributes to Epithelial-Mesenchymal Transition Cancer Cell Mechanics by Mediating Cytoskeletal Organization and Focal Adhesion Maturation. *Oncotarget* **2015**, *6*, 15966–15983. [CrossRef]
50. Wu, Y.; Zhang, X.; Salmon, M.; Lin, X.; Zehner, Z.E. TGFβ1 Regulation of Vimentin Gene Expression during Differentiation of the C2C12 Skeletal Myogenic Cell Line Requires Smads, AP-1 and Sp1 Family Members. *Biochim. Biophys. Acta BBA—Mol. Cell Res.* **2007**, *1773*, 427–439. [CrossRef]
51. Byun, Y.; Chen, F.; Chang, R.; Trivedi, M.; Green, K.J.; Cryns, V.L. Caspase Cleavage of Vimentin Disrupts Intermediate Filaments and Promotes Apoptosis. *Cell Death Differ.* **2001**, *8*, 443–450. [CrossRef]
52. Henzen-Logmans, S.C.; Mullink, H.; Ramaekers, F.C.; Tadema, T.; Meijer, C.J. Expression of Cytokeratins and Vimentin in Epithelial Cells of Normal and Pathologic Thyroid Tissue. *Virchows Arch. A Pathol. Anat. Histopathol.* **1987**, *410*, 347–354. [CrossRef]
53. Viale, G.; Dell'Orto, P.; Coggi, G.; Gambacorta, M. Coexpression of Cytokeratins and Vimentin in Normal and Diseased Thyroid Glands. Lack of Diagnostic Utility of Vimentin Immunostaining. *Am. J. Surg. Pathol.* **1989**, *13*, 1034–1040. [CrossRef] [PubMed]
54. Lepreux, S.; Desmoulière, A. Human Liver Myofibroblasts during Development and Diseases with a Focus on Portal (Myo)Fibroblasts. *Front. Physiol.* **2015**, *6*, 173. [CrossRef] [PubMed]
55. van den Borne, S.W.M.; Diez, J.; Blankesteijn, W.M.; Verjans, J.; Hofstra, L.; Narula, J. Myocardial Remodeling after Infarction: The Role of Myofibroblasts. *Nat. Rev. Cardiol.* **2010**, *7*, 30–37. [CrossRef] [PubMed]
56. Shinde, A.V.; Humeres, C.; Frangogiannis, N.G. The Role of α-Smooth Muscle Actin in Fibroblast-Mediated Matrix Contraction and Remodeling. *Biochim. Biophys. Acta Mol. Basis Dis.* **2017**, *1863*, 298–309. [CrossRef]
57. Smith, T.J.; Padovani-Claudio, D.A.; Lu, Y.; Raychaudhuri, N.; Fernando, R.; Atkins, S.; Gillespie, E.F.; Gianoukakis, A.G.; Miller, B.S.; Gauger, P.G.; et al. Fibroblasts Expressing the Thyrotropin Receptor Overarch Thyroid and Orbit in Graves' Disease. *J. Clin. Endocrinol. Metab.* **2011**, *96*, 3827–3837. [CrossRef] [PubMed]
58. Hynes, R.O.; Yamada, K.M. Fibronectins: Multifunctional Modular Glycoproteins. *J. Cell Biol.* **1982**, *95*, 369–377. [CrossRef]
59. Yang, Z.; Zhang, X.; Gang, H.; Li, X.; Li, Z.; Wang, T.; Han, J.; Luo, T.; Wen, F.; Wu, X. Up-Regulation of Gastric Cancer Cell Invasion by Twist Is Accompanied by N-Cadherin and Fibronectin Expression. *Biochim. Biophys. Res. Commun.* **2007**, *358*, 925–930. [CrossRef]
60. Dvorak, H.F. Tumors: Wounds That Do Not Heal. Similarities between Tumor Stroma Generation and Wound Healing. *N. Engl. J. Med.* **1986**, *315*, 1650–1659. [CrossRef] [PubMed]
61. Zeisberg, M.; Strutz, F.; Müller, G.A. Renal Fibrosis: An Update. *Curr. Opin. Nephrol. Hypertens.* **2001**, *10*, 315–320. [CrossRef]
62. Zhang, X.; Chen, C.T.; Bhargava, M.; Torzilli, P.A. A Comparative Study of Fibronectin Cleavage by MMP-1, -3, -13, and -14. *Cartilage* **2012**, *3*, 267–277. [CrossRef]
63. Pastushenko, I.; Brisebarre, A.; Sifrim, A.; Fioramonti, M.; Revenco, T.; Boumahdi, S.; Van Keymeulen, A.; Brown, D.; Moers, V.; Lemaire, S.; et al. Identification of the Tumour Transition States Occurring during EMT. *Nature* **2018**, *556*, 463–468. [CrossRef] [PubMed]
64. Williams, E.D.; Gao, D.; Redfern, A.; Thompson, E.W. Controversies around Epithelial–Mesenchymal Plasticity in Cancer Metastasis. *Nat. Rev. Cancer* **2019**, *19*, 716–732. [CrossRef]
65. Acloque, H.; Adams, M.S.; Fishwick, K.; Bronner-Fraser, M.; Nieto, M.A. Epithelial-Mesenchymal Transitions: The Importance of Changing Cell State in Development and Disease. *J. Clin. Investig.* **2009**, *119*, 1438–1449. [CrossRef] [PubMed]
66. Clement, C.A.; Ajbro, K.D.; Koefoed, K.; Vestergaard, M.L.; Veland, I.R.; Henriques de Jesus, M.P.R.; Pedersen, L.B.; Benmerah, A.; Andersen, C.Y.; Larsen, L.A.; et al. TGF-β Signaling Is Associated with Endocytosis at the Pocket Region of the Primary Cilium. *Cell Rep.* **2013**, *3*, 1806–1814. [CrossRef]
67. Kawasaki, M.; Ezura, Y.; Hayata, T.; Notomi, T.; Izu, Y.; Noda, M. TGF-β Suppresses Ift88 Expression in Chondrocytic ATDC5 Cells: TGF-β REGULATES IFT88. *J. Cell. Physiol.* **2015**, *230*, 2788–2795. [CrossRef] [PubMed]

Disclaimer/Publisher's Note: The statements, opinions and data contained in all publications are solely those of the individual author(s) and contributor(s) and not of MDPI and/or the editor(s). MDPI and/or the editor(s) disclaim responsibility for any injury to people or property resulting from any ideas, methods, instructions or products referred to in the content.

MDPI
St. Alban-Anlage 66
4052 Basel
Switzerland
www.mdpi.com

International Journal of Molecular Sciences Editorial Office
E-mail: ijms@mdpi.com
www.mdpi.com/journal/ijms

Disclaimer/Publisher's Note: The statements, opinions and data contained in all publications are solely those of the individual author(s) and contributor(s) and not of MDPI and/or the editor(s). MDPI and/or the editor(s) disclaim responsibility for any injury to people or property resulting from any ideas, methods, instructions or products referred to in the content.